AF360710

Design and Development of Nanostructured Thin Films

Design and Development of Nanostructured Thin Films

Special Issue Editors

Antonella Macagnano
Fabrizio De Cesare
Sara Cavaliere

MDPI • Basel • Beijing • Wuhan • Barcelona • Belgrade • Manchester • Tokyo • Cluj • Tianjin

Special Issue Editors

Antonella Macagnano
CNR-Institute of Atmospheric
Pollution Research
Italy

Fabrizio De Cesare
Department for Innovation in
Biological, Agro-food and Forest
Systems—University of Tuscia
Italy

Sara Cavaliere
University of
Montpellier—Institut Charles
Gerhardt Montpellier
France

Editorial Office
MDPI
St. Alban-Anlage 66
4052 Basel, Switzerland

This is a reprint of articles from the Special Issue published online in the open access journal *Nanomaterials* (ISSN 2079-4991) (available at: https://www.mdpi.com/journal/nanomaterials/special_issues/nano_thin_film).

For citation purposes, cite each article independently as indicated on the article page online and as indicated below:

LastName, A.A.; LastName, B.B.; LastName, C.C. Article Title. *Journal Name* **Year**, *Article Number*, Page Range.

ISBN 978-3-03928-738-3 (Pbk)
ISBN 978-3-03928-739-0 (PDF)

Cover image courtesy of Antonella Macagnano and Fabrizio De Cesare.

Contents

About the Special Issue Editors

Antonella Macagnano received a degree in Biology from University of Lecce (Italy). Since 2001, she has been a research scientist at the National Research Council (CNR) and leader of the Macro-Activity in her laboratories devoted to high-performance sensors and sensing systems for monitoring air quality and environment at the Institute of Atmospheric Pollutant Research (IIA) of CNR. Her research activities concern the study and the design of chemicals (conductive polymers, hybrid, and composite materials to be housed in advanced sensors to selective interactions with both gases and volatile organic compounds). She is author of numerous papers focused on sensors and smart nanostructured materials for sensing application and book chapters.

Fabrizio De Cesare received a degree in Biology from Sapienza University of Rome (Italy) and a Ph.D. in Agricultural Chemistry from Tuscia University (Italy). He is currently a research assistant professor and aggregate professor at the Deptartment for Innovation in Biological, Agro-Food and Forest Systems of the University of Tuscia. His recent research concerns soil science, agriculture, sensing systems, and Nanomaterials, and the integration of these various disciplines. In detail, his research fields include soil (bio)chemistry; ad–absorption phenomena; soil pollution and (bio)remediation; gas and VOCs sensing systems development for applications in soil science, environment, and agriculture; development of nanomaterials for the creation of advanced (bio)fertilizers; biostimulants; and biopesticides.

Sara Cavaliere's work aims to develop advanced nanostructured materials that can improve performance and durability of electrochemical energy conversion and storage devices, such as fuel cells and water electrolyzers. After a M.S. in Chemistry at the University of Milan (Italy), a Ph.D. at the Lavoisier Institute of Versailles (France), and postdoctoral fellowships at the Universities of Freiburg (Germany) and Lyon (France), Dr. Cavaliere has held a position as a lecturer at the University of Montpellier since 2009. Supported in 2013 by an ERC Starting Grant from the European Research Council, in 2017, she was awarded the CNRS bronze medal and appointed junior member of the Institut Universitaire de France.

Preface to "Design and Development of Nanostructured Thin Films"

Nanostructured thin films have attracted a considerable interest due to their unique size-dependent physicochemical properties, which are appealing for a multitude of applications. A thin film is traditionally defined as a material layer ranging from few nanometers to several micrometers in thickness. Due to their width, thin films are often used to coat surfaces or layers previously deposited. However, recent technological breakthroughs have enabled the development of freestanding thin layers, extending the application perspectives. Depending on both the target applications and the deposition technique used, nanostructured films can be designed and developed by tuning their atomic-molecular 2D and/or 3D aggregation, thickness, crystallinity, and porosity, as well as their optical, mechanical, catalytic, and conductive properties. The resulting structures can be exploited in electronic devices, electronic semiconductor devices (superconductive, semiconductive, magnetic, thermal, and insulating layers), energy storage and conversion (battery, solar, and fuel cells), optical and protective coating, and functional tools (catalysts, sensors, and biosensors).

This book contains 27 scientific articles covering most of the recent advances in designing and preparing nanostructured films for target functions and applications, focusing on the improvements resulting from adjustments of the fabrication processes and the materials employed. The emphasized advances in thin film growth technologies here reported are noteworthy because they addressing environmental issues such as the use of eco-friendly materials and procedures, more efficient adsorption/detection systems, alternative energy applications, and removal of pollutants or toxic substances.

Paul et al. grew a controlled thin layer of cobaltate, Ca_xCoO_2, using a modified two-stage step deposition (co-sputtering and thermal annealing). This structured film is promising for harvesingt energy, thus contributing to environmentally friendly energy systems. The impacts of La_2O_3 passivation layer thickness on the interfacial and electrical properties of HfO_2 insulators and Ge substrates were investigated by Zhao et al., who provide an effective procedure for preparing high-quality dielectric/Ge interfaces for metal-insulator-semiconductor (MIS) devices that are more thermodynamically stable.

The sputtering method has many advantages, including better film growth control, uniformity, reproducibility, low-temperature deposition, and large-scale stability. The properties of sputtered ZnO thin films not only depend on the deposition parameters, but also the post-deposition processes such as thermal treatment. Gandhi et al. described the effects of thermal annealing onto the grain size and the following phonon confinement, relevant for the future development of high-performance nanoscale optoelectronic devices. Jin et al. notably improved the visible transmittance of VO_2 films and thermochromic performance via a simple method of Ar/O plasma post-treatment, which led to effective stoichiometry refinement and microstructure reconstruction in the interfacial area. In general, Ar/O plasma irradiation serves as an additional energy source that promotes oxidation and surface migration of VO_2 molecular groups prompting micro-structure reconstruction at the VO_2 interface. The long-term goal of the authors was energy savings from smart-window optimization, which is particularly suitable for building glass or plastic substrates that cannot sustain the high temperatures present in the annealing process.

Despite its high cost and scarcity, platinum is one of the most active electrocatalysts for the reactions occurring in energy converters, such as the oxygen reduction reaction in proton-exchange membrane fuel cells. Strategies devoted to reducing the amount of this noble metal but maintaining its high electroactivity have been based on extended surfaces formation. An ultra-thin and conformal film of platinum was electrochemically prepared by Haidar et al. through surface-limited redox replacement based on the monolayer-limited nature of underpotential deposition and using Te as the sacrificial metal. Farina et al. modulated the morphology of a Pt film by changing the pulse duration in the high overpotential electrodeposition. Alpuche-Aviles et al. reported that Pt islands, formed spontaneously on highly oriented pyrolytic graphite during immersion in the electrodeposition solution, drove the growth of the 2D metal nanostructures.

Different preparation procedures were applied by Lewandowski et al. to clarify the features of the achieved structures of FeO thin films on Ag(111), revealing interesting tunable thickness-dependent properties and their relationships with electronic, catalytic, and magnetic features. Michalak et al. investigated the strong effects of the substrate crystal surfaces. The study of the growth of FeO and Fe_3O_4 on two close-packed metal single crystal surfaces (Pt(111) and Ru(0001)) revealed the influence of the mutual orientation of adjacent substrate terraces on the morphology of iron oxide films epitaxially grown on top of them.

Sometimes for specific applications, nanowire films can exhibit better performance than flat surfaces. In photovoltaics, solar cells based on CdTe wires exhibited excellent features due to the advantageous carrier transport properties within the wires. Commonly, the growth processes necessitate metal catalyst seed particles to facilitate wire growth, but these compounds can reduce the device quality material. Baines et al. successfully fabricated self-catalyzed CdTe wires in a single step technique and were free from contaminants using a simple industrially scalable PVD technique.

He et al. proposed a simple approach to prepare a conductive and optically active ink suitable for optoelectronic devices such as touch screen, solar cells, and organic light-emitting diodes. The authors prepared an aqueous suspension with SWNTs finely dispersed with rGO and then used a scalable technique as roll-to-roll printing to get the thin layer.

Li et al. grew hexagonal nanostructured back reflectors based on Ag/ZnO and proposed them as candidates for low cost and robust devices in silicon thin film solar cells. Kuo et al. proved how different growth conditions (nanospiral structural arrangement, shape, geometric confinement, and chemical composition) affect the relationship between the exhibiting circular polarization state and its corresponding wavelength of light reflection in thin film indium aluminum nitride.

Mendoza-Galván et al. produced a birefringent and ordered transparent film of cellulose by dragging forces acting during dip-coating deposition.

Surfaces and interfaces relationships and interactions with the surroundings were clarified by a detailed mathematical study by Jin and Li, applied to multilayered ReB_2/TaN structures, which is used as a protective coating.

Constantinou et al. used a combination of plasma enhanced chemical vapor deposition and physical vapor deposition technologies to form nanocomposite structures with tunable coating performance by potentially controlling the carbon bonding, hydrogen content, and the type, size, and percent of metallic nanoparticles, providing new strategies for a broad range of applications with tunable mechanical, physical, biological, and/or combinatorial properties.

Aerosol-based deposition technology is commonly used to generate a layer of nano-microparticles. However, the same turbulence that generates aerosol and triboelectrically charges the microparticles also results in inhomogeneity in the microparticle deposition. Shih et al. stabilized the turbulence with lengthened aerosol nozzles and masks optimizing flow rates and pressures improving the deposition homogeneity. An ordered layer composed of an aggregation of nanoparticles was also achieved by Macagnano et al. via drop-casting onto a thin microfibrous quartz slice and then incorporated into a standard axial sampler to be used as a potential passive sampling device for air mercury adsorption. This adsorbent layer, composed of nanoparticle of titania finely decorated with gold nanoparticles, distributed and anchored onto the quartz fibrous substrate, was found to be a stable, robust, reusable, and promising candidate for novel air mercury samplers. Sato et al. found a simple method to fabricate silver-coated monodisperse polystyrene nanoparticle photonic structures properly embedded into a polydimethylsiloxane matrix, exhibiting a strong green iridescence. Due to the colorimetric shift when exposed to hazardous solvents, these layers are expected have a major impact on global safety measures such as innovative photonic technology for quickly visualizing and identifying the presence of contaminants in the environment.

Thin films are also desirable in packaging. The excellent resistance of a polymer packaging materials functionalized with SiOx barrier layers by plasma deposition in a vacuum against the permeation of oxygen and humidity were prepared and studied by Prikryl et al. for protective garments, resistant against toxic liquids and gases.

The structural and functional properties of polymer nanocomposites are attractive in sensor investigation. Macagnano et al. designed a multi-component nanofibrous thin film using a straightforward procedure based on electrospinning technology, blending two insulating and thermoplastic polymers (polystyrene and polyhydroxybutyrate), a known percentage of conductive mesoporous graphitized carbon and a free-base tetraphenylporphyrin, for designing conductive-thermal-driven gas/VOCs sensors. These polymers were selected, among other features, for their low environmental impact, being recyclable and biodegradable materials. Notably, the possibility of tuning the sensitivity and selectivity of the chemical sensors by adjusting the working temperature was emphasized. Electrospinning technology was also used by De Clerk et al. to fabricate and investigate nanostructured hydrogels by blending polycaprolactone and gelatin; the several combinations of the materials offer promise for biomedical applications. Using different technology, Yang et al. proposed a further and simple environmental-friendly strategy to obtain a polymer network gel of sericin and agarose. When lysozyme was loaded, the gel worked as a promising tool for wound dressing. Electrospinning was also investigated to fabricate innovative filtration systems. The demand is huge for the filtration application of nanofiber layers due to their specific surface, small pore size, and high porosity. Yalcinkaya and Hruza explain the effect of laminating pressure on polyacrylonitrile and polyvinylidenefluoridemultilayer nanofibrous membranes for liquid microfiltration.

Chen et al. suggest an innovative and straightforward strategy to develop mesoporous Al_2O_3 and MgO films (separation and heterogeneous catalysis) on silicon wafer substrates using spin-coated poly(dimethylacrylamide) hydrogels as porogenic matrices.

Metal–organic frameworks (MOFs) are promising materials for a wide range of applications, including gas storage, separation, catalysis, and sensing. Careful selection of the inorganic nodes and organic linkers that form the chemical structure of the material can lead to a variety of highly tunable pore sizes and chemical functionality. Bowser et al. explored how the morphology and crystal structure of a specific MOF (HKUST-1, copper(II) ions and benzene 1,3,5 tricarboxylate) known for its ability to capture and store a wide variety of gases (carbon monoxide, carbon dioxide, nitric oxide, nitrogen, hydrogen, and sulfur dioxide) responded to ammonia gas exposure in a surface anchored-MOF, drop-cast, and bulk powdered film, respectively. They proved that the presence of solvent traces in the structure (or lack thereof) deeply affected the morphology and then the functionality.

A facile hydrothermal method to grow NiAl-layered double hydroxide in situ on Mg alloy for corrosion protection was exploited for the first time by Ye et al. The excellent properties were ascribed to the peculiar orientation of the nanosheets in the coating.

To conclude, this book reports an overview on the design, synthesis, and applications of thin films of different natures and nanostructures, highlighting the latest advances as well as the challenges to be tackled, especially from an environmental standpoint. We hope that it can be useful and inspiring for a vast audience from academia and industry.

Antonella Macagnano, Fabrizio De Cesare, Sara Cavaliere
Special Issue Editors

 *nanomaterials*

Article

Growth of Ca$_x$CoO$_2$ Thin Films by A Two-Stage Phase Transformation from CaO–CoO Thin Films Deposited by Rf-Magnetron Reactive Cosputtering

Biplab Paul *, Jun Lu and Per Eklund

Thin Film Physics Division, Department of Physics Chemistry and Biology (IFM), Linköping University, SE-58183 Linköping, Sweden; jun.lu@liu.se (J.L.); per.eklund@liu.se (P.E.)
* Correspondence: biplab.paul@liu.se; Tel.: +46-(0)-708390137

Received: 14 February 2019; Accepted: 12 March 2019; Published: 15 March 2019

Abstract: The layered cobaltates A$_x$CoO$_2$ (A: alkali metals and alkaline earth metals) are of interest in the area of energy harvesting and electronic applications, due to their good electronic and thermoelectric properties. However, their future widespread applicability depends on the simplicity and cost of the growth technique. Here, we have investigated the sputtering/annealing technique for the growth of Ca$_x$CoO$_2$ (x = 0.33) thin films. In this approach, CaO–CoO film is first deposited by rf-magnetron reactive cosputtering from metallic targets of Ca and Co. Second, the as-deposited film is reactively annealed under O$_2$ gas flow to form the final phase of Ca$_x$CoO$_2$. The advantage of the present technique is that, unlike conventional sputtering from oxide targets, the sputtering is done from the metallic targets of Ca and Co; thus, the deposition rate is high. Furthermore, the composition of the film is controllable by controlling the power at the targets.

Keywords: thin film; nanostructure; Ca$_x$CoO$_2$; sputtering; phase transformation

---

1. Introduction

Thermoelectricity, through its ability to harvest energy from waste heat, can contribute to environmentally friendly energy systems. The thermoelectric efficiency of a material system is determined by a dimensionless parameter, the thermoelectric figure of merit, $ZT = S^2/\rho\kappa$, where S is the Seebeck coefficient, ρ is the electrical resistivity, and κ is the thermal conductivity. Therefore, good thermoelectric materials have a high Seebeck coefficient (for high output voltage from a thermoelectric device for a fixed temperature gradient), low electrical resistivity (to reduce energy loss due to Joule heating), and low thermal conductivity (to sustain a high temperature gradient across thermoelectric device for high output voltage) [1,2]. However, the design of such materials is challenging, because electrically conductive materials are generally also thermally conductive.

For the reduction of thermal conductivity without disturbing electronic transport, different nanostructuring approaches have been investigated [3–9]. In these approaches, different types of nanoscale features have been incorporated. The nanoscale dimension of these features is comparable to the length scale of phonon mean free path, but is higher than electronic mean free path. Therefore, they can selectively scatter phonons without adversely affecting the electronic transport. Thus, ZT can be enhanced. Inherently nanolaminated materials [10,11] and artificially layered materials [12–14] have been investigated to exploit these concepts. The interfacial phonon scattering in these materials systems is reported to drastically reduce their thermal conductivity, leading to the multifold enhancement of ZT [12,13]. The disadvantage of such artificially grown superlattice materials is that they are not thermodynamically stable structures, reducing their stability at high temperature [15]. A related approach is the use of inherently layered materials, e.g., A$_x$CoO$_2$ (A = Na, Ca, Sr, Ba, La Pr, Nd) [16–22].

Such layered materials have a layered structure similar to a superlattice structure and can sustain high temperatures.

The layered cobaltates A_xCoO_2 consist of alternate stacks of A_x sheets and CdI_2-type CoO_2 sheets. Due to the inherently layered structure, A_xCoO_2 exhibits anisotropic electronic and phononic properties. Among the layered cobaltates, Na_xCoO_2 (x~0.7) is reported to have the highest power factor, as high as the standard thermoelectric material Bi_2Te_3, due to its low electrical resistivity (ρ = 0.2 m$\Omega\cdot$cm at 300 K) and high Seebeck coefficient (100 μV K^{-1} at 300 K) [23]. Even with this high power factor, Na_xCoO_2 cannot offer reliable performance, due to its poor chemical stability, because the mobile Na^+ ions tend to be ejected from the material at high temperatures. To achieve stable performance from this material system, the monovalent Na^+ ions should be replaced with divalent Ca^{2+} ions, for example by ion exchange method, producing Ca_xCoO_2 ($0.26 \leq x \leq 0.5$) thin films [24,25]. The same technique has been reported to be useful to grow a series of layered materials A_xCoO_2 (A = Na, Ca, Sr, Ba, La Pr, Nd) [16–22]. Apart from this ion exchange method, physical [26–29] methods have also been investigated to grow Ca_xCoO_2 thin films.

Here, we have investigated the sputtering/annealing method for the growth of Ca_xCoO_2 thin films. In this method, first CaO–CoO thin film is deposited by rf-magnetron reactive cosputtering and then is annealed to form the final phase of Ca_xCoO_2. In our previous work, we have demonstrated the growth of $Ca_3Co_4O_9$ by the sputtering/annealing approach [30–32]. The thermally induced phase transformation leading to the final phase of $Ca_3Co_4O_9$ was found to consist of the following steps:

$$CoO + O_2 \Rightarrow Co_3O_4 \tag{1}$$

$$Co_3O_4 + CaO \Rightarrow Ca_xCoO_2 \tag{2}$$

$$Ca_xCoO_2 + CaO \Rightarrow Ca_3Co_4O_9 \tag{3}$$

Here, we determine how to stop the phase transformation at second step so that final film of Ca_xCoO_2 is obtained. Ca_xCoO_2 thin films can be promising for near room temperature thermoelectric applications, due to its higher power factor (S^2/ρ) as compared to $Ca_3Co_4O_9$ [33].

2. Materials and Methods

Three sets of samples with different Ca:Co ratio (0.25, 0.35, 0.45) were prepared. Prior to deposition, Al_2O_3 (001) substrates were heated to 650 °C for 1 h under vacuum inside the deposition system, and the same substrate temperature was maintained during deposition. Then, CaO–CoO films were reactively cosputtered from metallic targets of Ca (99.95 % pure) and Co (99.99 % pure) onto the Al_2O_3 (001) substrates by rf-magnetron sputtering at 0.27 Pa (2 mTorr) in an oxygen (1.5%)–argon (98.5%) mixture. The deposition system is described elsewhere [34,35]. The power of the cobalt target (50 W) was kept constant, while the power of Ca target was varied to vary the Ca:Co ratio in the films. As-deposited CaO–CoO films were annealed at 650 °C under O_2 gas flow for 3 h to form the final phase of Ca_xCoO_2. The crystal structure and morphology of the films were characterized by θ–2θ XRD analyses using monochromatic Cu Kα radiation (λ = 1.5406 Å), transmission electron microscopy (TEM) by using a FEI Tecnai G2 TF20 UT instrument, from Eindhoven, Netherlands, with a field emission gun operated at 200 kV and with a point resolution of 1.9 Å, and scanning electron microscopy (SEM, LEO 1550 Gemini). The θ–2θ XRD scans were performed with a Philips PW 1820 diffractometer. Ex situ annealing and XRD experiments were performed on a single CaO–CoO thin film with Ca:Co = 0.35. The Ca:Co = 0.35 film was subjected to several sequential annealing steps and θ–2θ XRD scans were performed after each annealing step. The annealing furnace was stabilized at a given set-point temperature prior to inserting the sample for a specified time period. The sample temperature was monitored as a function of time via a thermocouple in contact with the film substrate. The sample was removed from the furnace after the specified annealing time and cooled in ambient air at room temperature. The Ca:Co ratio in the films is confirmed by energy dispersive spectroscopy (EDS) attached to a scanning electron microscope.

3. Results and Discussion

3.1. As-Deposited CaO–CoO and Annealed Ca–Co–O Films

The as-deposited films are yellowish in color and turn dark after annealing in oxygen atmosphere (Figure 1). Figure 2a shows a θ–2θ XRD scan for the as-deposited Ca:Co = 0.35 film. XRD scans of all the as-deposited films are similar in appearance (shown in Figure S1 of Supplementary Information), albeit with varying CaO:CoO peak intensity ratios. It is evident from the XRD analyses that as deposited film consists of CaO and CoO phases. Figure 2b–d show the XRD patterns of all the annealed films. These XRD patterns confirm the formation of Ca_xCoO_2 phase in all the films and d-spacing is calculated to be 5.434 Å irrespective of the Ca to Co ratio in the films, which nearly matches the value reported elsewhere [24]. Apart from the Ca_xCoO_2 phase, the XRD peaks of Co_3O_4 are also visible in all the films irrespective of the Ca to Co ratio. The presence of Co_3O_4 phase in the film Ca:Co = 0.25 is attributed to the excess Co in the film. The presence of Co_3O_4 phase in the post-annealed films Ca:Co = 0.35 and 0.45 is attributed to the inhomogeneous distribution of CaO and CoO, which is also confirmed by TEM analyses of as-deposited films (see below). The small peaks at 2θ angles 8.249 and 26.527° in Figure 2d are attributed to the $Ca_3Co_4O_9$ phase, which might be due to the local enrichment of the film with the CaO phase.

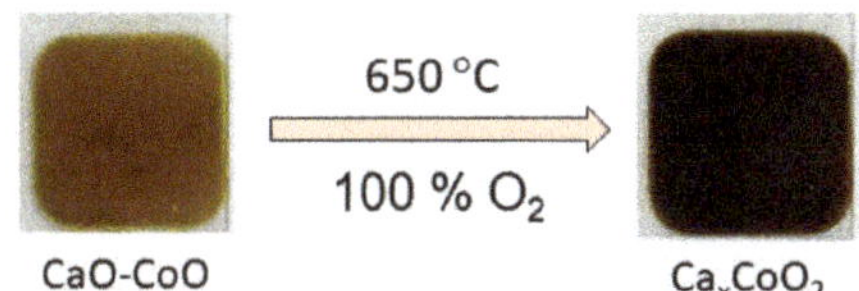

Figure 1. Optical image of as-deposited CaO–CoO film and annealed Ca_xCoO_2 film.

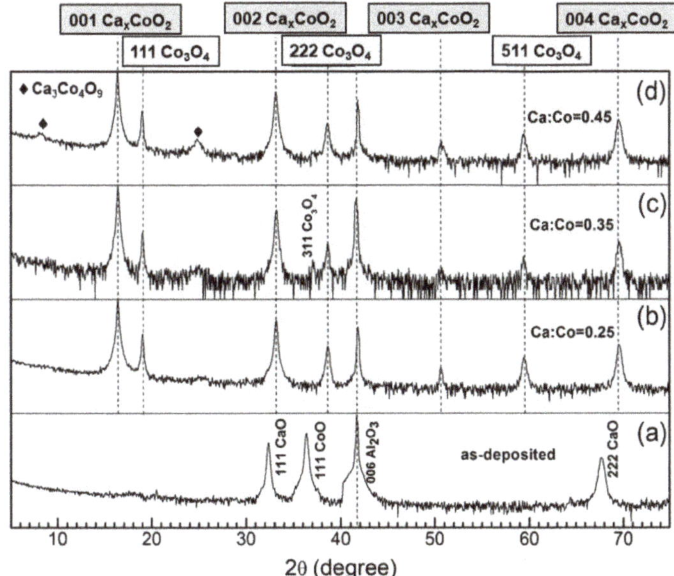

Figure 2. θ–2θ XRD patterns of (**a**) as-deposited CaO–CoO film and annealed films with (**b**) Ca:Co = 0.25, (**c**) Ca:Co = 0.35, (**d**) Ca:Co = 0.45.

Figure 3 shows SEM images of postannealed film Ca:Co = 0.35. The apparent flat surface of the film is attributed to the *c*-axis orientation of the film, which is also consistent with the observation by XRD.

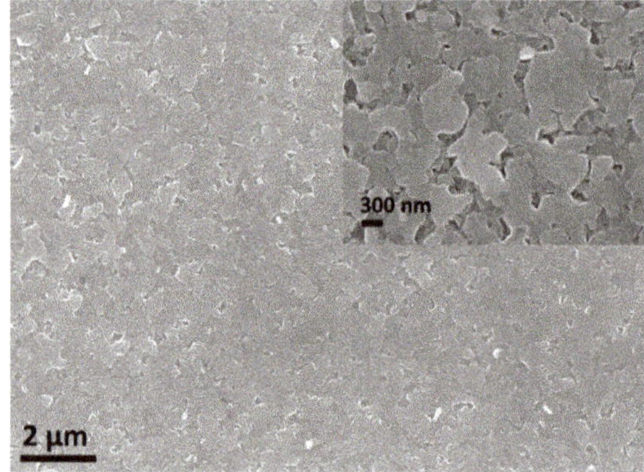

Figure 3. A typical SEM image of the annealed film Ca:Co = 0.35.

Figure 4a shows a typical cross-sectional TEM image of as-deposited film Ca:Co = 0.35. Figure 4b and c show the EDS mapping of the film. The EDS mapping shows the separation into Co-rich and Ca-rich phases, i.e., CoO and CaO. This observation is in consistent with the observation by XRD, i.e., the presence of CaO and CoO phases in as-deposited films. Figure 4b indicates the Ca-deficiency near the interfacial region. This is caused by the segregation of Ca near the surface of the as-deposited films, due to the substrate heating during sputter deposition, consistent with our pervious observation on the growth of $Ca_3Co_4O_9$ thin films [30]. The inhomogeneous distribution of CaO and CoO phases along the in-plane direction is also confirmed by the line scan as shown in Figure 4d.

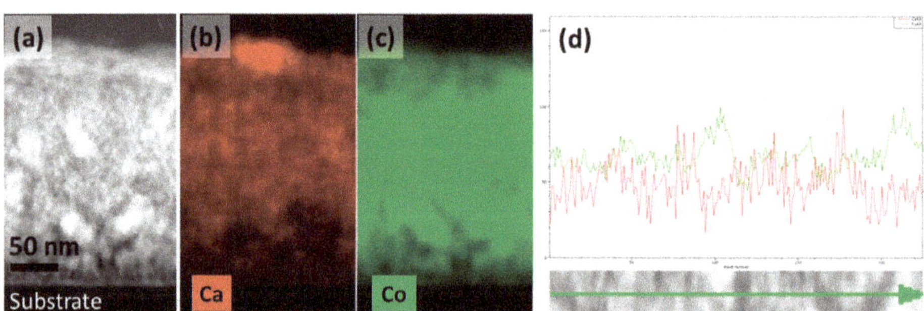

Figure 4. (**a**) TEM image of as-deposited CaO–CoO film, (**b**,**c**) EDS mapping of the corresponding film, (**d**) line scan along in-plane direction of the film; green line represents Co and red line Ca.

Figure 5a shows a TEM image of annealed film Ca:Co = 0.35. The formation of a layered structure is evident from the figure, and there are also grains with nonbasal orientation. The Ca to Co ratio in the layered zone is determined to be ~0.33 by EDS analyses (in TEM), which indicates that the postannealed film Ca:Co = 0.35 is of $Ca_{0.33}CoO_2$ phase. From EDS analyses, all the postannealed films are found to consist of the same $Ca_{0.33}CoO_2$ phase irrespective of the Ca:Co ratio in the as-deposited films, which is consistent with the observation by XRD. Figure 5b shows a high resolution TEM (HRTEM) image of the film and d-spacing of the film is confirmed to be 5.43 Å, which is consistent with the value calculated from XRD. Figure 5c schematically shows the atomic arrangements of alternate layers of $Ca_{0.33}CoO_2$.

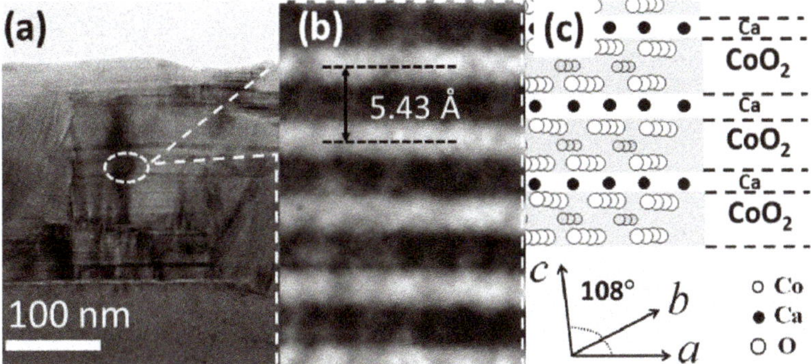

Figure 5. (**a**) TEM image of Ca$_{0.33}$CoO$_2$ film, (**b**) Lattice-resolved TEM image and (**c**) schematic of the atomic arrangement of the layers.

From the above results, it is concluded that Ca$_{0.33}$CoO$_2$ is the favorable composition of Ca$_x$CoO$_2$ at the present conditions. The phase-purity is likely to improve the Ca$_{0.33}$CoO$_2$ film by controlling the composition and ensuring the homogeneous distribution of CaO and CoO phases in the as-deposited CaO–CoO films. It is anticipated that homogeneous distribution of CaO and CoO in as-deposited films can be possible by lowering the deposition temperature.

3.2. Ex-Situ XRD Annealing Experiments

Figure 6a shows the θ–2θ XRD scan for the as-deposited film Ca:Co = 0.35. Figure 6b,c show θ–2θ scans after the film was subjected to annealing temperatures 500 °C and 650 °C, respectively. After annealing at 500 °C (Figure 6b), the appearance of Co$_3$O$_4$ peaks with a concomitant decrease in the peak intensity of CoO indicates the reaction of CoO with oxygen to form Co$_3$O$_4$. The Co$_3$O$_4$ has also reacted with CaO to form the Ca$_{0.33}$CoO$_2$-phase, as confirmed by the presence of the Ca$_x$CoO$_2$ 001, 002, 003 and 004 peaks. These results indicate a competition between the formation and consumption of Co$_3$O$_4$. After annealing at 650 °C (Figure 6c) the Ca$_{0.33}$CoO$_2$ phase becomes more intense and CoO phase has disappeared. However, the low-intensity peaks of Co$_3$O$_4$ still remain. To verify the completion of the phase transformation, the annealing experiment was performed at 650 °C for a longer period (10 h), with no change in XRD peak intensity.

From the above results, it is concluded that two different phase transformation processes (i.e., CoO + O$_2$ $\Rightarrow$ Co$_3$O$_4$; and Co$_3$O$_4$ + CaO $\Rightarrow$ Ca$_{0.33}$CoO$_2$) simultaneously occur at temperatures below 650 °C. At 650 °C, the CaO phase is completely consumed by the reaction to form the final phase of Ca$_{0.33}$CoO$_2$, and thus the phase transformation processes end. The partial presence of Co$_3$O$_4$ is attributed to the local enrichment of Co in the film, due to the inhomogeneous distribution of CaO and CoO phases in the film.

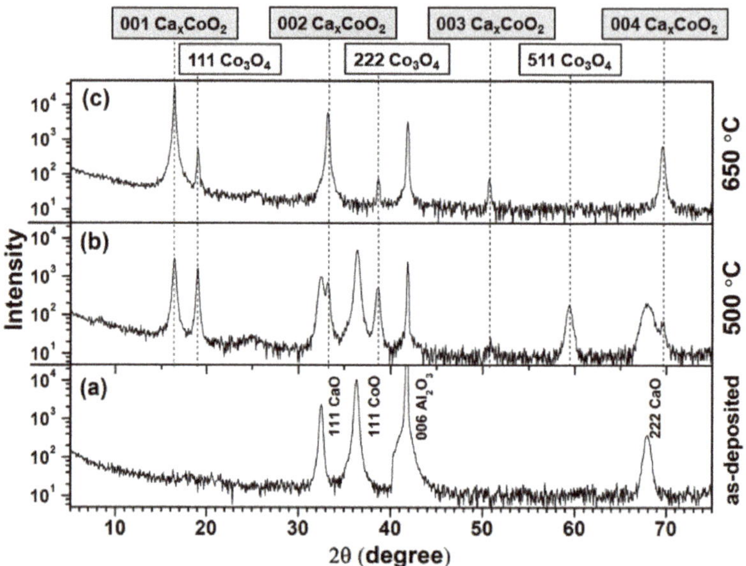

Figure 6. θ–2θ XRD patterns of the Ca:Co = 0.35 thin film as a function of annealing temperature. (**a**) As-deposited film, (**b**) annealed at 500 °C, (**c**) annealed at 650 °C.

4. Conclusions

A two-step sputtering/annealing approach has been demonstrated for the growth of Ca$_x$CoO$_2$ (x = 0.33) thin film. Thermally induced phase transformation from reactively cosputtered CaO–CoO film leads to the formation of the final phase of Ca$_x$CoO$_2$. The phase transformation consists of the following steps:

$$CoO + O_2 \Rightarrow Co_3O_4 \tag{4}$$

$$Co_3O_4 + CaO \Rightarrow Ca_xCoO_2 \tag{5}$$

The composition of Ca$_x$CoO$_2$ is Ca$_{0.33}$CoO$_2$ phase irrespective of the Ca:Co ratio in the as-deposited film, i.e., the Ca$_{0.33}$CoO$_2$ is the most favorable phase in this route of thermally induced solid state phase transformation. The final film of Ca$_{0.33}$CoO$_2$ is not phase pure, due to the partial presence of Co$_3$O$_4$ phase in the film. The Co$_3$O$_4$ content in the postannealed films decreases with the increase in Ca-content in the as-deposited films. The partial presence of Co$_3$O$_4$ phase in the Ca-rich films Ca:Co = 0.35 and 0.45 is attributed to the local enrichment of CoO phase in the as-deposited films. It is expected that the phase pure Ca$_{0.33}$CoO$_2$ can be grown by ensuring a homogeneous distribution of CoO and CaO phases in the initial sputtered deposited films.

Supplementary Materials: The following are available online at http://www.mdpi.com/2079-4991/9/3/443/s1, Figure S1: θ –2θ XRD patterns of as-deposited CaO–CoO film (a) Ca:Co = 0.25, (b) Ca:Co = 0.35, (c) Ca:Co = 0.45.

Author Contributions: B.P. designed the experiments, performed data analysis and interpretations, and wrote the manuscript. J.L. performed the TEM analyses. P.E. contributed to experiment design, planning, discussion and interpretations, and revised the manuscript.

Funding: APC was funded by the Library of Linköping University.

Acknowledgments: The research leading to these results has received funding from the European Research Council (ERC) under the European Community's Seventh Framework Programme (FP/2007-2013)/ERC Grant 335383, the Swedish Research Council (VR) under Project 2016-03365, the Swedish Government Strategic Research Area in Materials Science on Functional Materials at Linköping University (Faculty Grant SFO-Mat-LiU 2009 00971), the Knut and Alice Wallenberg foundation through the Academy Fellow program, the Eurostars project

E!8892 T-to-Power, the Swedish Foundation for Strategic Research (SSF) through the Future Research Leaders 5 program, and funding from the Åforsk foundation.

Conflicts of Interest: The authors declare no conflict of interest.

References

1. Snyder, G.J.; Toberer, E.S. Complex thermoelectric materials. *Nat. Mater.* **2008**, *7*, 105–114. [CrossRef] [PubMed]
2. Snyder, G.J.; Christensen, M.; Nishibori, E.; Caillat, T.; Iversen, B.B. Disordered zinc in Zn_4Sb_3 with phonon-glass and electron-crystal thermoelectric properties. *Nat. Mater.* **2004**, *3*, 458–463. [CrossRef] [PubMed]
3. Ren, G.-K.; Lan, J.-L.; Ventura, K.J.; Tan, X.; Lin, Y.-H.; Nan, C.-W. Contribution of point defects and nano-grains to thermal transport behaviours of oxide-based thermoelectrics. *NPJ Comput. Mater.* **2016**, *2*, 16023. [CrossRef]
4. Yin, Y.; Tudu, B.; Tiwari, A. Recent advances in oxide thermoelectric materials and modules. *Vacuum* **2017**, *146*, 356–374. [CrossRef]
5. He, J.; Tritt, T.M. Advances in thermoelectric materials research: Looking back and moving forward. *Science* **2017**, *357*, 1369. [CrossRef] [PubMed]
6. Tan, G.; Zhao, L.-D.; Kanatzidis, M.G. Rationally designing high-performance bulk thermoelectric materials. *Chem. Rev.* **2016**, *116*, 12123–12149. [CrossRef] [PubMed]
7. Biswas, K.; He, J.; Blum, I.D.; Wu, C.-I.; Hogan, T.P.; Seidman, D.N.; Dravid, V.P.; Kanatzidis, M.G. High-performance bulk thermoelectrics with all-scale hierarchical architectures. *Nature* **2012**, *489*, 414–418. [CrossRef] [PubMed]
8. Paul, B.; Kumar, A.V.; Banerji, P. Embedded Ag-rich nanodots in PbTe: Enhancement of thermoelectric properties through energy filtering of the carriers. *J. Appl. Phys.* **2010**, *108*, 064322. [CrossRef]
9. Berettaa, D.; Neophytouc, N.; Hodgesd, J.M.; Kanatzidisd, M.G.; Narduccie, D.; Martin-Gonzalezf, M.; Beekmang, M.; Balkeh, B.; Cerrettii, G.; Tremelj, W.; et al. Thermoelectrics: From history, a window to the future. *Mater. Sci. Eng. R Rep.* **2018**. [CrossRef]
10. Jood, P.; Ohta, M. Hierarchical architecturing for layered thermoelectric sulfides and chalcogenides. *Materials* **2015**, *8*, 1124–1149. [CrossRef] [PubMed]
11. Ravichandran, J. Thermoelectric and thermal transport properties of complex oxide thin films, heterostructures and superlattices. *J. Mater. Res.* **2017**, *32*, 183–203. [CrossRef]
12. Tan, M.; Deng, Y.; Hao, Y. Enhanced thermoelectric properties and superlattice structure of a bi_2te_3/zrb_2 film prepared by ion-beam-assisted deposition. *J. Phys. Chem. C* **2013**, *117*, 20415–20420. [CrossRef]
13. Beyer, H.; Nurnus, J.; Böttner, H.; Lambrecht, A.; Wagner, E.; Bauer, G. High thermoelectric figure of merit ZT in PbTe and Bi_2Te_3-based superlattices by a reduction of the thermal conductivity. *Phys. E Low Dimens. Syst. Nanostruct.* **2002**, *13*, 965–968. [CrossRef]
14. Saha, B.; Shakouri, A.; Sands, T.D. Rocksalt nitride metal/semiconductor superlattices: A new class of artificially structured materials featured. *Appl. Phys. Rev.* **2018**, *5*, 021101. [CrossRef]
15. Hansen, A.-L.; Dankwort, T.; Winkler, M.; Ditto, J.; Johnson, D.C.; Koenig, J.D.; Bartholomé, K.; Kienle, L.; Bensch, W. Synthesis and thermal instability of high-quality Bi_2Te_3/Sb_2Te_3 superlattice thin film thermoelectrics. *Chem. Mater.* **2014**, *26*, 6518–6522. [CrossRef]
16. Liu, J.F.; Huang, X.Y.; Xu, G.S.; Chen, L.D. Thermoelectric properties of layered $Sr_{0.29}CoO_2$ crystals. *J. Alloys Comp.* **2013**, *576*, 247. [CrossRef]
17. Ishikawa, R.; Ono, Y.; Miyazaki, Y.; Kajitani, T. Low-temperature synthesis and electric properties of new layered cobaltite, Sr_xCoO_2. *Jpn. J. Appl. Phys.* **2002**, *41*, L337–L339. [CrossRef]
18. Liu, J.; Huang, X.; Yang, D.; Wanb, S.; Xu, G. High-temperature thermoelectric properties of layered Ba_xCoO_2. *Scr. Mater.* **2015**, *100*, 63–65. [CrossRef]
19. Liu, J.F.; Huang, X.Y.; Yang, D.F.; Xu, G.S.; Chen, L.D. Synthesis and physical properties of layered Ba_xCoO_2. *Dalton Trans.* **2014**, *43*, 15414–15418. [CrossRef] [PubMed]
20. Brázda, P.; Palatinus, L.; Klementová, M.; Buršík, J.; Knížek, K. Mapping of reciprocal space of $La_{0.30}CoO_2$ in 3D: Analysis of superstructure diffractions and intergrowths with Co_3O_4. *J. Solid State Chem.* **2015**, *227*, 30–34.

21. Kriz̆ek, K.; Hejtma'nek, J.; Marys̆ko, M.; S̆antava', E.; Jira'k, Z.; Burs̆i'k, J.; Kirakci, K.; Beran, P. Structure and properties of a novel cobaltate $La_{0.30}CoO_2$. *J. Solid State Chem.* **2011**, *184*, 2231–2237. [CrossRef]

22. Knížek, K.; Jirák, Z.; Hejtmánek, J.; Maryško, M.; Buršík, J. Structure and properties of novel cobaltates $Ln_{0.3}CoO_2$ (Ln = La, Pr, and Nd). *J. Appl. Phys.* **2012**, *111*, 07D707. [CrossRef]

23. Terasaki, I.; Sasago, Y.; Uchinokura, K. Large thermoelectric power in $NaCo_2O_4$ single crystals. *Phys. Rev. B Condens. Matter* **1997**, *56*, 12685. [CrossRef]

24. Cushing, B.L.; Wiley, J.B. Topotactic routes to layered calcium cobalt oxides author links open overlay panel. *J. Solid State Chem.* **1998**, *141*, 385–391. [CrossRef]

25. Huang, R.; Mizoguchi, T.; Sugiura, K.; Nakagawa, S.; Ohta, H.; Saito, T.; Koumoto, K.; Hirayama, T.; Ikuhara, Y. Microstructure evolution of $Ca_{0.33}CoO_2$ thin films investigated by high-angle annular dark-field scanning transmission electron microscopy. *J. Mater. Res.* **2009**, *24*, 279–287. [CrossRef]

26. Takahashi, K.; Kanno, T.; Sakai, A.; Adachi, H.; Yamada, Y. Gigantic transverse voltage induced via offdiagonal thermoelectric effect in thin films. *Appl. Phys. Lett.* **2010**, *97*, 021906. [CrossRef]

27. Du, Y.; Xu, J.; Paul, B.; Eklund, P. Flexible thermoelectric materials and devices. *Appl. Mater. Today* **2018**, *12*, 366–388. [CrossRef]

28. Takahashi, K.; Kanno, T.; Sakai, A.; Adachi, H.; Yamada, Y. Large crystallographic orientation tilting induced by postoxidation annealing in layered cobaltite Ca_xCoO_2 thin films. *Cryst. Growth Des.* **2012**, *12*, 1708–1712. [CrossRef]

29. Kanno, T.; Yotsuhashi, S.; Adachi, H. Anisotropic thermoelectric properties in layered cobaltite A_xCoO_2 (A = Sr and Ca) thin films. *Appl. Phys. Lett.* **2004**, *85*, 739. [CrossRef]

30. Paul, B.; Schroeder, J.L.; Kerdsongpanya, S.; Nong, N.V.; Schell, N.; Ostach, D.; Lu, J.; Birch, J.; Eklund, P. Mechanism of formation of the thermoelectric layered cobaltate $Ca_3Co_4O_9$ by annealing of CaO-CoO thin films. *Adv. Electron. Mater.* **2015**, *1*, 1400022. [CrossRef]

31. Paul, B.; Lu, J.; Eklund, P. Nanostructural tailoring to induce flexibility in thermoelectric $Ca_3Co_4O_9$ thin films. *ACS Appl. Mater. Interfaces* **2017**, *9*, 25308–25316. [CrossRef] [PubMed]

32. Paul, B.; Björk, E.M.; Kumar, A.; Lu, J.; Eklund, P. Nanoporous $Ca_3Co_4O_9$ thin films for transferable thermoelectrics. *ACS Appl. Energy Mater.* **2018**, *1*, 2261–2268. [CrossRef] [PubMed]

33. Liu, J.; Huang, X.; Li, F.; Liu, R.; Chen, L. Low-temperature magnetic and thermoelectric properties of layered $Ca_{0.33}CoO_2$ crystals. *J. Phys. Soc. Jpn.* **2011**, *80*, 074802. [CrossRef]

34. Trinh, D.H.; Ottosson, M.; Collin, M.; Reineck, I.; Hultman, L.; Högberg, H. Nanocomposite Al_2O_3-ZrO_2 thin films grown by reactive dual radio-frequency magnetron sputtering. *Thin Solid Films* **2008**, *516*, 4977–4982. [CrossRef]

35. Frodelius, J.; Eklund, P.; Beckers, M.; Persson, P.O.Å.; Högberg, H.; Hultman, L. Sputter deposition from a Ti_2AlC target: Process characterization and conditions for growth of Ti_2AlC. *Thin Solid Films* **2010**, *518*, 1621. [CrossRef]

 *nanomaterials* 

Letter

Electrical Properties and Interfacial Issues of HfO₂/Ge MIS Capacitors Characterized by the Thickness of La₂O₃ Interlayer

Lu Zhao, Hongxia Liu *, Xing Wang, Yongte Wang and Shulong Wang

Key Laboratory for Wide Band Gap Semiconductor Materials and Devices of Education, School of Microelectronics, Xidian University, Xi'an 710071, China; lzhaoxd@163.com (L.Z.); xwangsme@xidian.edu.cn (X.W.); mikewyt@163.com (Y.W.); slwang@xidian.edu.cn (S.W.)
* Correspondence: hxliu@mail.xidian.edu.cn; Tel.: +86-29-88204085

Received: 24 February 2019; Accepted: 29 April 2019; Published: 4 May 2019

Abstract: Effects of the La₂O₃ passivation layer thickness on the interfacial properties of high-k/Ge interface are investigated systematically. In a very thin range (0~15 cycles), the increase of La₂O₃ passivation layer deposition cycles improves the surface smoothness of HfO₂/Ge structures. The capacitance-voltage (*C-V*) characteristics show that the thickness of La₂O₃ passivation layer can affect the shift of flat band voltage (V_{FB}), hysteretic behaviors, and the shapes of the dual-swept *C-V* curves. Moreover, significant improvements in the gate leakage current and breakdown characteristics are also achieved with the increase of La₂O₃ interlayer thickness.

Keywords: Ge surface engineering; La₂O₃ passivation layer; atomic layer deposition; electrical properties

---

1. Introduction

Along with the continuing scaling of complementary metal oxide semiconductor (CMOS) technology according to Moore's Law, insulation oxides with higher permittivity have been introduced to replace SiO_2 for acceptable gate leakage current density and low power consumption [1]. However, other than reducing feature size, technological progress to gain high-speed operation is also needed with the development of integrated circuit (IC) technology. Considering this, high mobility semiconductors have been considered as alternative channel materials for obtaining high drive current. Among the high mobility semiconductors, germanium (Ge) is compatible with standard Si CMOS process, and both the electron and hole bulk mobility are higher than those of Si substrate [2,3]. Therefore, Ge has been regarded as one of the most promising alternative channel materials. Unfortunately, different from the perfect interfacial properties of SiO_2/Si interface, the poor thermal stability of GeO_2/Ge interface results in the deterioration of interfacial properties to a great extent [4]. When most gate dielectric materials deposited on the unpassivated Ge substrate, the generation of unstable Ge oxides is unavoidable during the high temperature post-deposition annealing (PDA) process, and the decomposition or desorption from GeO_2 to volatile GeO would bring in a great deal of defects. Consequently, to get acceptable electrical properties of high-k/Ge structures, a suitable surface passivation treatment is a necessary technical issue prior to the deposition of high-k gate dielectrics on Ge substrates [5].

Owing to the much better thermodynamic stability than that of GeO_2/Ge, fewer interface traps exist at the interface between La-based oxides and Ge substrates [6]. Consequently, La-based oxides have been considered as one kind of the alternative gate dielectric materials deposited on Ge to realize good electrical performance of Ge-based metal-insulator-semiconductor (MIS) devices [7,8]. However, the hygroscopicity of La₂O₃ layer limits its application as gate insulators [9]. In a previous study, we have proved that a ~2 nm La_2O_3 interlayer could effectively improve the electrical performance of Ge MIS devices, resulting in more than one order of magnitude decrease in the gate leakage current density [10]. In this work, a further investigation was performed on the impacts of La_2O_3 passivation layer thickness on the interfacial properties of HfO_2 insulators and Ge substrates.

2. Materials and Methods

La_2O_3 and HfO_2 were deposited on Sb-doped n-type Ge (100) wafers with a doping concentration of 1×10^{16} cm^{-3} by atomic layer deposition (ALD) method (R-150, Picosun, Espoo, Finland). Prior to the deposition, all the wafers were treated using acetone and alcohol, followed by dipping into HF (2%) and deionized water by cyclic cleaning for 5 times to remove the defective native Ge oxide. Tris(isopropyl-cyclopentadienyl) lanthanum ($La(^iPrCp)_3$) and tetrakis (ethylmethylamino) hafnium (TEMAH) were used as La and Hf precursor, while H_2O was used as oxidant with reference to the reported literatures [11,12]. Under the chamber temperature of 300 °C, La_2O_3 layers were deposited by alternately introducing $La(^iPrCp)_3$ and H_2O precursors to the reactor chamber using high purity N_2 (>99.999%) as the carrier gas with a typical ALD growth cycle of 0.3 s $La(^iPrCp)_3$ pulse/4 s N_2 purge/0.3 s H_2O pulse/9 s N_2 purge. Using these process parameters, the steady-state growth rate of La_2O_3 was approximately 0.85 Å/cycle. For an ALD growth cycle of HfO_2, TEMAH was pulsed into the chamber by carrier gas for 0.3 s with a 8 s N_2 purge, and then H_2O was pulsed for 0.1 s followed by a 8 s N_2 purge. As a result, a stable growth rate of 0.75 Å/cycle was obtained for HfO_2. The thickness of the La_2O_3 passivation layer was tuned by varying the number of ALD cycles, and the cycle numbers of 5, 10, and 15 cycles were chosen. Then ~6 nm HfO_2 was deposited as the gate insulator materials after the deposition of La_2O_3 passivation layer. For comparison, the sample with only ~6 nm HfO_2 as the gate oxide was also fabricated as a control sample. After the deposition of gate insulators, rapid thermal annealing (RTA) was carried out at 400 °C for 90 s in N_2 ambient for all the wafers. For simplicity, the control sample is assigned as S1, and samples with 5, 10, and 15 ALD cycles of La_2O_3 passivation layers are assigned as S2, S3, and S4, respectively.

The physical thickness of the deposited films was optically measured using the Woollam M2000U spectroscopic ellipsometry (SE, Woollam Co. Inc., Lincoln, NE, USA), and the ellipsometry data were fitted using a Gen-Osc mode consisting of Gaussian and Tauc-Lorentz oscillators. The chemical bonding states of the interfaces between the deposited films and Ge substrates were examined by X-ray photoelectron spectroscopy (XPS, Axis Ultra DLD, Kratos Analytical, UK) measurements. Cross-sectional high-resolution transmission electron microscopy (HRTEM, TECNAI F20 system, ThermoFisher Scientific, Waltham, MA, USA) and energy dispersive X-ray spectroscopy (EDX) line scan measurements were performed to observe the microstructures and atomic compositions of the deposited films. The surface morphology of the films was monitored using atomic force microscopy (AFM, Dimension 3100, Veeco Digital Instruments by Bruker, Billerica, MA, USA) in tapping mode. The electrical properties of the HfO_2/La_2O_3 gate stacks were evaluated using MIS capacitor structures. MIS capacitors were fabricated by ion-beam etching (IBE, IBE-A-150, Beijing Chuangshi Weina Technology Co., Ltd., Beijing, China) of e-beam evaporated 150 nm Al through a shadow mask with a diameter of 300 μm as the top gate electrode and 100 nm Al as the back ohmic contact. Then, the electrical properties of the fabricated MIS capacitors were measured using an Agilent B1500A parameter analyzer (Santa Clara, CA, USA).

3. Results and Discussion

3.1. Chemical Bonding States of the Interfaces between the Deposited Films and Ge Substrates

The chemical bonding states near the interface of HfO_2 and HfO_2/La_2O_3 gate stacks deposited on Ge substrate were investigated by XPS measurements. Prior to the measurements, the wafers S1~S4 were etched by Ar^+ ion beam bombardment for 25 s, 30 s, 30 s, and 35 s, respectively, under the etch rate of ~0.20 nm/s to remove the influence of the surface impurities and to analyze the chemical bonding states in the interfacial region. The XPS data were calibrated by setting the C 1s peak originated from the carbon impurities in the spectra at 284.6 eV for all the wafers. Figure 1 shows the variations of the O 1s XPS spectra for the HfO_2 film and HfO_2/La_2O_3 gate stacks deposited on Ge substrates. For the control sample (S1), the O 1s spectra consist of three Gaussian–Lorentzian line shape peaks. These peaks are located at 530.2, 531.0, and 531.9 eV, corresponding to the chemical bonding states of Hf–O–Hf, Hf–O–Ge, and Ge–O–Ge, respectively [13,14]. The existence of Hf–O–Ge and Ge–O–Ge peaks indicate that $HfGeO_x$ and GeO_x were generated at the HfO_2/Ge interface during the deposition of HfO_2 and post-deposition annealing (PDA) process. As for the HfO_2/La_2O_3/Ge structures, the O 1s core level spectra are deconvoluted into four peaks, three of which are at 529.0 (I), 530.2 (II), and 531.9 (IV) eV. These peaks correspond to the chemical bonds of La–O–La, Hf–O–Hf, and Ge–O–Ge, respectively [15]. The other peak (III) at around 530.9 eV originates from the Hf–O–Ge and/or La–O–Ge bonding states, indicating the generation of germanate ($LaGeO_x$ and $HfGeO_x$) compound at the interface of HfO_2/La_2O_3 gate stacks and Ge substrates. It is worth noting that the position of peak (III) shifts to lower binding energies from sample S1 to S4. This phenomenon is attributed to the transformation from Hf-rich to La-rich interfacial layer (IL) with the increase of La_2O_3 passivation layer thickness, owing to the stronger reaction between La_2O_3 and outdiffused Ge atoms than that of HfO_2 with outdiffused Ge atoms [16,17].

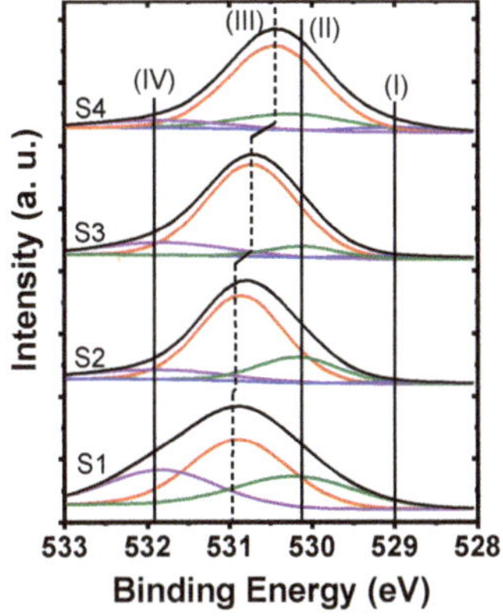

Figure 1. O 1s XPS spectra of HfO_2 film and HfO_2/La_2O_3 gate stacks deposited on Ge substrates.

To further discuss the variation tendency of Ge oxides at high k dielectrics/Ge interfaces with the increase of La_2O_3 passivation layer thickness, investigations on the Ge 3d XPS spectra are shown in Figure 2. A substrate peak (Ge $3d^0$) located at ~28.6 eV with additional peaks at higher binding energies in relation to the IL (consisting of Ge oxides and germanate) could be observed in all spectra. The Ge $3d^0$ substrate peak is fitted with a doublet of Ge $3d_{5/2}$ and Ge $3d_{3/2}$ with spin-orbit splitting of 0.6 eV and intensity ratio of 3:2, respectively [18]. The Ge oxides (GeO_x) consist of four peaks (Ge^{1+}, Ge^{2+}, Ge^{3+}, Ge^{4+}), which are at higher binding energy respect to the Ge $3d^0$ with energy shifts of 0.8, 1.8, 2.6, and 3.4 eV, respectively [19]. Compared with the control sample shown in Figure 2a, the intensity of the Ge^{2+}(GeO) peak decreases with the increase of La_2O_3 passivation layer thickness, while the intensity of the $LaGeO_x$ peak shows an increasing trend. Consequently, with the increase of La_2O_3 passivation layer thickness, the variation tendency of the interfacial components extracted from the Ge 3d XPS spectra is consistent with the results extracted from the O 1s XPS spectra.

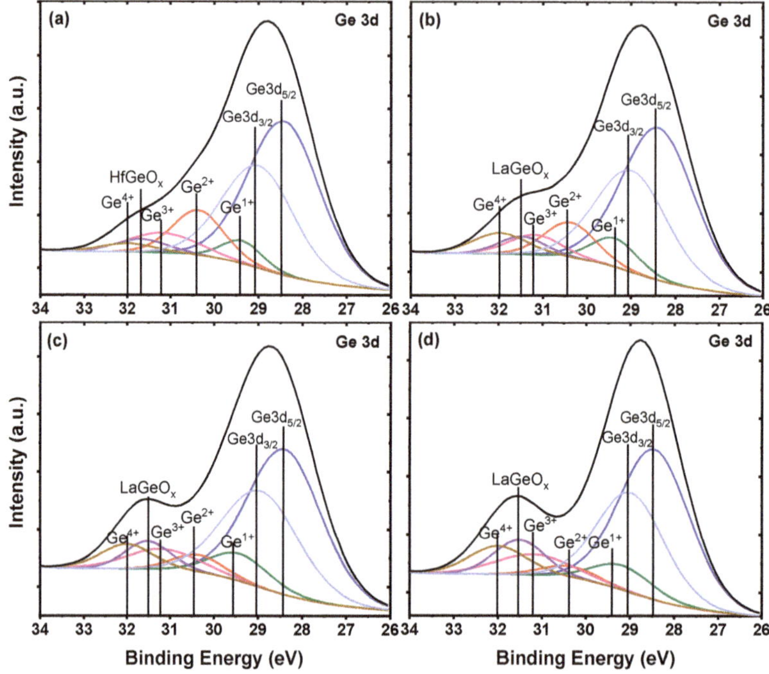

Figure 2. Ge 3d XPS spectra of the deposited gate stacks with different ALD cycles of La_2O_3 passivation layer on Ge substrates. (**a**) The control sample; (**b**) 5 ALD cycles of La_2O_3 passivation layer; (**c**) 10 ALD cycles of La_2O_3 passivation layer; and (**d**) 15 ALD cycles of La_2O_3 passivation layer.

3.2. Microstructures of the Deposited Films on Ge Substrates

Additional microstructure information of the deposited films is provided by cross-sectional HRTEM analysis as shown in Figure 3. For the control sample without La_2O_3 passivation layer, the HfO_2 layer exhibits an amorphous structure as no nanometer-sized crystal or long-range ordered crystal region is observed. However, in sample S4 with 15 ALD cycles of La_2O_3 passivation layer (as shown in Figure 3b), nanometer-sized crystals could be observed in the HfO_2 layer, indicating the incomplete crystallization state of the HfO_2 film. It is noteworthy that upon the same RTA condition at 400 °C for 90 s in N_2 ambient, the HfO_2 films in sample S1 and S4 show different crystalline characteristics. This difference could be attributed to the RTA-induced Ge diffusion from the substrate into the HfO_2 layer as shown in the EDX profiles near the interfaces between the deposited films and Ge substrates (Figure 4). The EDX data were dealt with normalization method after eliminating the influence of impurity elements, and the depth values were calibrated by HRTEM results. During the RTA process, in sample S1 as shown in Figure 4a, the substrate Ge atoms would diffuse easily to the upper HfO_2 layer and react with the oxygen nearby the HfO_2/Ge interface, which would prevent the formation of crystalline HfO_2 precipitates [20] and cause the formation of IL (~0.8 nm) consisting of $HfGeO_x$ and GeO_x, as analyzed in the above XPS results. As for the HfO_2/La_2O_3/Ge structure in sample S4 (Figure 4b), due to the high affinity of La_2O_3 for Ge atoms [21], the out-diffused Ge atoms would react with La_2O_3 passivation layer, leading to the spontaneous formation of stable $LaGeO_x$ IL. As a result, few Ge atoms would diffuse into the HfO_2 layer, and the crystallization phenomenon occurs after the RTA treatment.

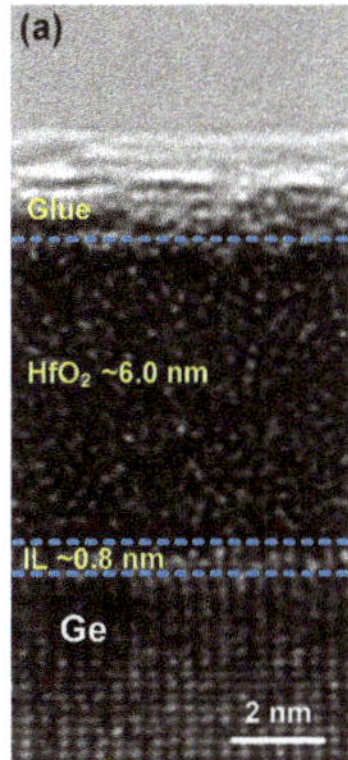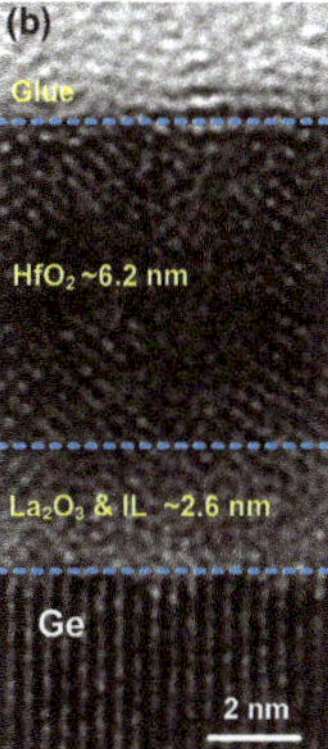

Figure 3. Cross-sectional HRTEM images showing the interfaces with Ge substrates for (**a**) the control sample; (**b**) sample S4 with 15 ALD cycles of La_2O_3 passivation layer.

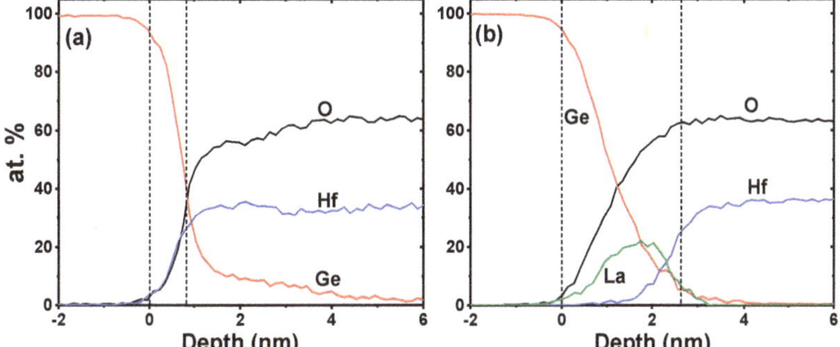

Figure 4. EDX profiles near the interfaces between the deposited films and Ge substrates. (**a**) the control sample; (**b**) sample S4 with 15 ALD cycles of La_2O_3 passivation layer.

3.3. Surface Morphology of HfO₂ and HfO₂/La₂O₃ Gate Stacks on Ge Substrates

The AFM images of HfO_2/La_2O_3 gate stacks on Ge substrates with 0, 5, 10, and 15 ALD cycles of La_2O_3 interfacial passivation layer are shown in Figure 5. The scan size of each AFM image was set as 10×10 μm^2. The AFM images were captured from the AFM data by the software named Gwyddion. The root mean square (RMS) roughness of samples S1~S4 is extracted to be 0.63, 0.51, 0.28, and 0.25 nm, respectively. The small RMS roughness values of all the samples illustrate the smooth and crack-free surfaces of the deposited HfO_2/La_2O_3 gate stacks, indicating that ALD is a good deposition method for high-k dielectric films and stacks. On the basis of the smooth surfaces, a decrease in the RMS roughness values is observed after inserting a La_2O_3 interfacial passivation layer, and the RMS roughness values decrease with the increase of La_2O_3 layer thickness. It has been reported that the outdiffused Ge atoms from substrates and the desorption of volatile species GeO during the high temperature annealing process could bring in large roughness [22]. After inserting La_2O_3 interfacial passivation layer, the stable $LaGeO_x$ layer could suppress the formation of unstable GeO_x layer, which contributes to the decrease of the RMS roughness value. With the increase of La_2O_3 thickness, more La_2O_3 reacts with the outdiffused Ge atoms and unstable Ge oxides, which effectively suppresses desorption from GeO_2 to volatile GeO, resulting in a smoother surface. It is found that when the thickness of La_2O_3 layer increase above 15 cycles, the variation in the RMS roughness values becomes negligible. The variation tendency of RMS roughness illustrates that 15 ALD cycles of La_2O_3 layer are thick enough to improve the surface smoothness for HfO_2/Ge structures.

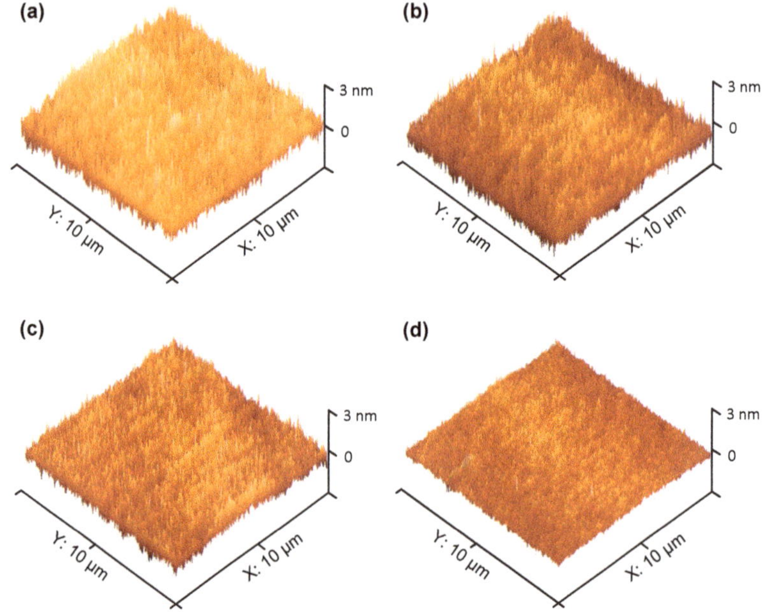

Figure 5. Three-dimensional AFM images of the deposited gate stacks with different ALD cycles of La$_2$O$_3$ passivation layer on Ge substrates. (**a**) The control sample; (**b**) 5 ALD cycles of La$_2$O$_3$ passivation layer; (**c**) 10 ALD cycles of La$_2$O$_3$ passivation layer; and (**d**) 15 ALD cycles of La$_2$O$_3$ passivation layer.

3.4. Electrical Performance of Al/HfO$_2$/Ge and Al/HfO$_2$/La$_2$O$_3$/Ge MIS Capacitors

The capacitance-voltage (*C-V*) and conductance-voltage (*G-V*) characteristics of the fabricated Al/HfO$_2$/n-Ge and Al/HfO$_2$/La$_2$O$_3$/n-Ge MIS capacitors are shown in Figure 6. Dual-swept *C-V* curves were obtained by sweeping from inversion to accumulation (forward) and sweeping back (backward) at the frequency of 100 kHz. *G-V* curves were obtained simultaneously with the *C-V* curves measured forward. To quantitatively characterize the effects of La$_2$O$_3$ passivation layer thickness on electrical performance of HfO$_2$/Ge MOS structures, some parameters are extracted from the dual-swept *C-V* curves. The accumulation capacitance (C_{ac}) values for gate stacks S1~S4 are obtained to be 1.35, 1.48, 1.69, and 1.87 μF/cm^2, respectively. Then, the oxide capacitance (C_{ox}) and dielectric constant values of gate stacks could be calculated following the equations [23,24]:

$$C_{ox} = C_{ac}\left[1 + \left(\frac{G_{ac}}{\omega C_{ac}}\right)^2\right] \tag{1}$$

$$CET = \frac{\varepsilon_0 \varepsilon_{SiO_2} A}{C_{ox}} \tag{2}$$

$$k = \frac{\varepsilon_{SiO_2} t_{ox}}{CET} \tag{3}$$

where C_{ac} is the capacitance value at accumulation region, G_{ac} is the conductance corresponding to the accumulation region of the *C-V* curves, ω is the angular frequency, C_{ox} is the oxide capacitance of gate stacks, A is the area of the top electrode, *CET* is the capacitance equivalent thickness of deposited gate stacks (including IL), t_{ox} is the measured thickness of deposited gate stacks (including IL), ε_0 and ε_{SiO2} are the permittivity values of vacuum and SiO$_2$, respectively. The physical thickness of gate stacks (including IL) in S1, S2, S3, and S4 is optically measured to be 6.71, 7.89, 8.26, and 8.95 nm, respectively. The corresponding 95% confidence interval for the average thickness of the deposited films was shown in Table S1 in the Supplementary Materials to evaluate the discrete degree of thickness testing data. Therefore, according to Equations (1)–(3), the effective *k* values of S1~S4 are achieved to be 10.69, 13.43, 15.97, and 18.46, respectively. It is found that the dielectric constant value of HfO$_2$/La$_2$O$_3$ gate stacks increases with the increase of La$_2$O$_3$ layer thickness. The improvement on the value of dielectric constant for the gate stacks is attributed to the effective suppression of Ge atoms outdiffusion benefiting from the generation of stable LaGeO$_x$, since the *k* value of La$_2$O$_3$ is much higher than that of GeO$_x$ [25]. In addition, as analyzed in the EDX profiles, the HfO$_2$ quality is improved by La$_2$O$_3$ passivation, which should be the other cause that increases the dielectric constant of the deposited films.

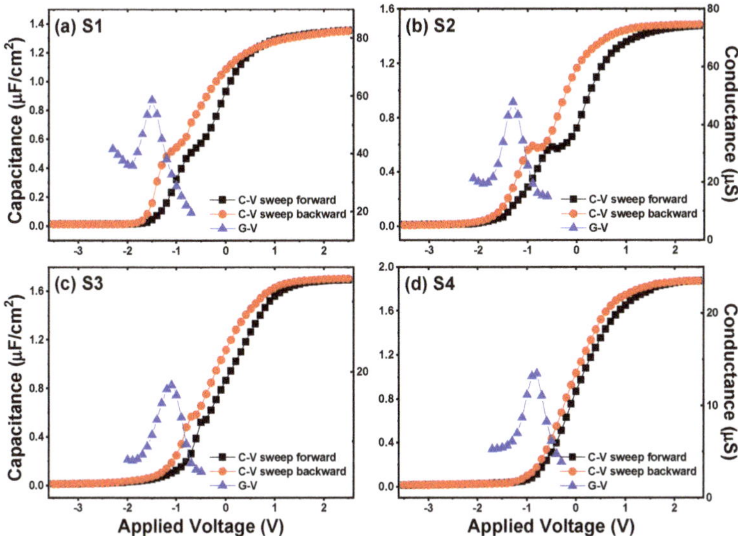

Figure 6. *C-V* curves and *G-V* characteristics of the fabricated Al/HfO$_2$/La$_2$O$_3$/Ge MIS capacitors with different ALD cycles of La$_2$O$_3$ passivation layer. (**a**) The control sample; (**b**) 5 ALD cycles of La$_2$O$_3$ passivation layer; (**c**) 10 ALD cycles of La$_2$O$_3$ passivation layer; and (**d**) 15 ALD cycles of La$_2$O$_3$ passivation layer. The curves were measured at the frequency of 100 kHz.

Moreover, as shown in Figure 6, varying degrees of anomalous humps in the weak inversion region of *C-V* curves could be observed. Compared with the control sample (Figure 6a), the anomalous hump phenomenon turns more serious after inserting 5 ALD cycles of La$_2$O$_3$ passivation layer. While with the further increase of La$_2$O$_3$ passivation layer thickness, the humps of *C-V* curves are obviously reduced. Besides, as the thickness of La$_2$O$_3$ passivation layer increases, different degrees of frequency dispersion phenomenon at the weak inversion regions could be observed from the multi-frequency *C-V* curves in Figure 7, and the degrees of frequency dispersion phenomenon show a similar variation tendency with the anomalous humps in the *C-V* curves measured at 100 kHz. It has been reported that the existence of slow interface traps (Q_{it}) should be responsible for the anomalous phenomenon at the weak inversion regions of *C-V* curves [26,27]. Considering this, the interface state density (D_{it}) values of S1~S4 are discussed using the following relation of single-frequency approximation method through the forward swept *C-V* and *G-V* curves [28]:

$$D_{it} = \frac{2}{q A} \frac{\frac{G_{max}}{\omega}}{\left[\left(\frac{G_{max}}{\omega C_{ox}}\right)^2 + \left(1 - \frac{C_{max}}{C_{ox}}\right)^2\right]} \tag{4}$$

where q is the elementary charge (1.602×10^{-19} C), A is the area of the top electrode, C_{ox} is the gate oxide capacitance as defined in Equation (1), G_{max} is the peak value of G-V curve, and C_{max} is the capacitance corresponding to G_{max} at the same gate applied voltage. Following Equation (4), the D_{it} values for the fabricated MIS capacitors using S1~S4 as insulators are calculated to be 9.18×10^{12}, 9.72×10^{12}, 3.95×10^{12}, and 2.71×10^{12} eV^{-1}cm^{-2}, respectively. The D_{it} value increases slightly after inserting 5 ALD cycles of La$_2$O$_3$ passivation layer, which is in consistent with the variation tendency of the humps in C-V curves. The increment of D_{it} is reported to be caused by an incomplete reaction of the precursor molecules during the first few ALD cycles [29], which brings in extra interface states at the interface. As the ALD cycles of La$_2$O$_3$ passivation layer increase, the improvement of the La$_2$O$_3$ passivation layer on the interfacial properties of gate insulators/Ge interfaces plays a more and more important role, contributing to the decrease of interface traps for the samples S3 and S4. For accuracy, the D_{it} values were also extracted using conductance method [30] measured from 1 kHz to 1 MHz, as shown in Figure 8. It could be observed that for each energy state, the D_{it} value extracted by conductance method increases slightly after inserting 5 ALD cycles of La$_2$O$_3$ passivation layer, while with the further increase of La$_2$O$_3$ passivation layer thickness, the D_{it} value shows decrease trend. This result is in good consistency with the single-frequency approximation method result. The contribution of the La$_2$O$_3$ interfacial passivation layer to the amount of interface traps at the La$_2$O$_3$/Ge interface could be explained as follows: during the ALD process of La(^{i}-PrCp)$_3$ and H$_2$O precursors and the high temperature PDA process, La$_2$O$_3$ is more likely to react with outdiffused Ge atoms and Ge oxides nearby the interface to form stable LaGeO$_x$ compound. The stable LaGeO$_x$ compound could effectively suppress Ge outdiffusion, and further inhibit the decomposition or desorption from GeO$_2$ to volatile GeO (following the reaction equation of GeO$_2$ + Ge $\rightarrow$ 2GeO), contributing to the reduction of structural defects and dangling bonds [31,32]. The D_{it} value (2.71×10^{12} eV^{-1}cm^{-2}) of sample S4 achieved in our work is comparable with the results obtained in the reported literatures at La$_2$O$_3$/Ge interfaces of 3×10^{12} [8] and 3.5×10^{12} eV^{-1}cm^{-2} [25], indicating that a La$_2$O$_3$ passivation layer with 15 ALD cycles (~1.3 nm) can effectively suppress the generation of interface states at high k dielectrics/Ge interfaces.

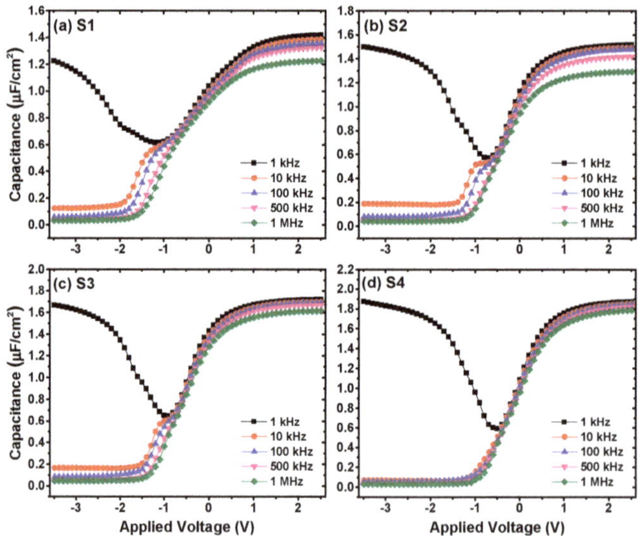

Figure 7. Multi-frequency C-V characteristics measured of the fabricated Al/HfO$_2$/La$_2$O$_3$/Ge MIS capacitors with different ALD cycles of La$_2$O$_3$ passivation layer. (**a**) The control sample; (**b**) 5 ALD cycles of La$_2$O$_3$ passivation layer; (**c**) 10 ALD cycles of La$_2$O$_3$ passivation layer; and (**d**) 15 ALD cycles of La$_2$O$_3$ passivation layer.

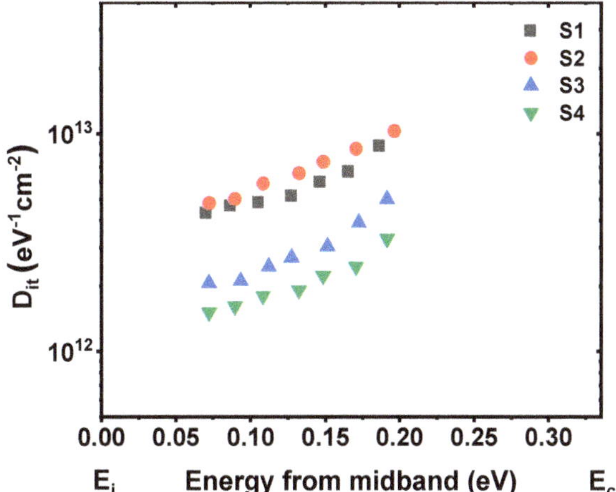

Figure 8. Energy distribution of the interface states at the interface between the deposited films and Ge substrates.

It is also observed from the *C-V* curves that the flat band voltages (V_{FB}) translate toward positive voltages with the increase of La_2O_3 passivation layer deposition cycles. To further investigate this phenomenon in detail, the V_{FB} values of the capacitors were extracted using the Hauser NCSU CVC simulation software taking into account of quantum-mechanical effects [33]. Considering the work function difference between the top electrode aluminum and the n-type Ge substrate with a doping concentration of 1×10^{16} cm^{-3}, the ideal V_{FB} should be 0.03 V. However, the actual V_{FB} swept forward for the control sample is −0.206 V. The negative V_{FB} shift indicates the existence of effective positive oxide charges in the HfO_2 film. The effective positive oxide charges may be attributed to the existence of positive fixed oxide charges and oxide trapped charges mainly consisting of oxygen vacancies and structural defects in the gate insulators and/or nearby the insulators/Ge interface. Compared with the control sample, the forward swept V_{FB} values of MIS capacitors S2~S4 were extracted to be −0.083, −0.071, and −0.048 V, separately. The positive shifts of V_{FB} with the increase of La_2O_3 passivation layer thickness reveal the reduction of oxide charges after the Ge surface passivation treatment using La_2O_3 as the passivation layer. Considering this, the trapped charge density (N_{ot}) is estimated using the midgap charge separation method through the *C-V* hysteresis characteristics following the Equation (5) [34]:

$$N_{ot} = \frac{\Delta V_{FB} C_{ox}}{qA} \tag{5}$$

where ΔV_{FB} is the hysteresis width of V_{FB}, C_{ox} is the oxide capacitance, q is the elementary charge (1.602×10^{-19} C), and A is the area of the electrode. For the control sample, using the exacted ΔV_{FB} value of 685 mV, a trapped oxide charge density of 6.03×10^{12} cm^{-2} is obtained according to Equation (5). After the insertion of La_2O_3 passivation layer, the trapped oxide charge densities of samples S2, S3 and S4 are calculated to be 4.74×10^{12}, 2.87×10^{12}, and 1.79×10^{12} cm^{-2}, respectively. These parameter results concerning the electrical properties are summarized in Table 1. As suspected, the trapped oxide charge density reduces with the increase of La_2O_3 thickness. We ascribe the decrease of trapped charges to the reduction of oxygen vacancies and structural defects in the gate stack region nearby the interface. At the HfO_2/Ge interface, the outdiffused Ge atoms will react with GeO_2 to form unstable GeO, which could bring in a large number of oxygen vacancies and structural defects. The $LaGeO_x$ compound formed by La_2O_3 and Ge could effectively suppress the outdiffusion of Ge atoms and the desorption of volatile GeO, contributing to the reduction of N_{ot} values.

Table 1. The electrical parameters extracted from the fabricated MIS capacitors without and with La_2O_3 interfacial passivation layers.

Sample	C_{ox} ($\mu F/cm^2$)	k	$\triangle V_{FB}$ (mV)	D_{it} ($eV^{-1}cm^{-2}$)	N_{ot} (cm^{-2})
S1	1.411	10.69	685	9.18×10^{12}	6.03×10^{12}
S2	1.510	13.43	504	9.72×10^{12}	4.74×10^{12}
S3	1.712	15.97	269	3.95×10^{12}	2.87×10^{12}
S4	1.883	18.46	152	2.71×10^{12}	1.79×10^{12}

Figure 9 shows the gate leakage current density as a function of the applied effective field for samples S1~S4. It is observed that the gate leakage current density of gate insulator/Ge structures obviously decreases with the increase of La_2O_3 layer thickness. At the applied effective field of 3 MV/cm, compared with the control sample, in the sample S4 the gate leakage current density decreases of almost an order of magnitude. The obvious decrease in the gate leakage current density could be ascribed to the reduction in the structural defects in the HfO_2/La_2O_3 gate stack benefiting from the generation of stable $LaGeO_x$ component. Moreover, the gate leakage current density-voltage (J-V) characteristics show different breakdown behaviors, which suggests that the incorporation of La_2O_3 interfacial passivation layer improves the breakdown field strength of HfO_2/Ge MIS capacitors. The improvement of the breakdown field strength is reported to be associated with a reduction in defects in gate dielectrics [35]. As mentioned above, the existence of La_2O_3 interfacial passivation layer could effectively suppress the generation of structural defects, oxygen vacancies and dangling bonds. The reduction of structural defects, oxygen vacancies and dangling bonds decreases the possibility to create a conduction path by forming a continuous chain connecting the gate to the semiconductor, resulting in higher breakdown field.

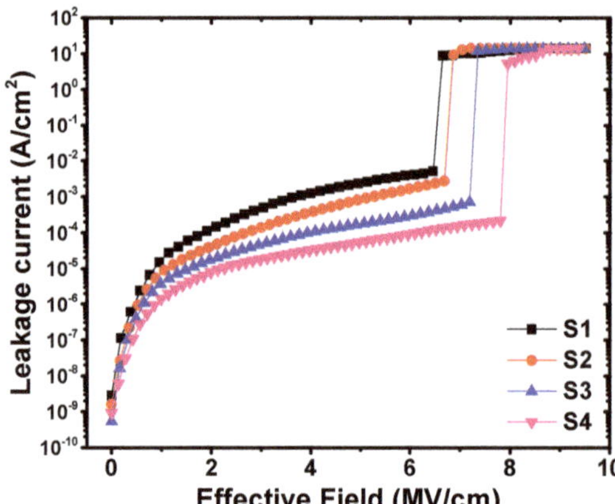

Figure 9. J–V characteristics of $Al/HfO_2/Ge$ and $Al/HfO_2/La_2O_3/Ge$ MIS capacitors.

4. Conclusions

In summary, we investigate the effect of La_2O_3 passivation layer thickness on the interfacial and electrical properties of HfO_2/Ge MIS structures. It is found that the gate leakage current density of the MIS capacitor after inserting 15 ALD cycles (~1.3 nm) of La_2O_3 passivation layer gains almost one order of magnitude decrease than that of the $Al/HfO_2/Ge$ MIS sample. Besides, a relatively low D_{it} value of 2.71×10^{12} $eV^{-1}cm^{-2}$ is achieved. The thickness of the La_2O_3 passivation layer also affects the dielectric constants, V_{FB} shifts, hysteresis behaviors and the shapes of dual-swept C-V curves. The improvement on the interfacial and electrical properties with the increase of La_2O_3 passivation layer deposition cycles is related to the formation of a stable $LaGeO_x$ layer on the Ge surface, which restrains the decomposition or desorption from GeO_2 to volatile GeO, contributing to the decrease of structural defects. These results provide us with quite an effective method for realizing high-quality dielectric/Ge interfaces for Ge-based MIS devices.

Supplementary Materials: The following are available online at http://www.mdpi.com/2079-4991/9/5/697/s1, Table S1: Thickness of the deposited films (including interfacial layer) on Ge substrates.

Author Contributions: L.Z. generated the research idea, analyzed the data, and wrote the paper. L.Z. and Y.W. carried out the experiments and the measurements. S.W. participated in the discussions. H.L. and X.W. have given final approval of the version to be published. All authors read and approved the final manuscript.

Funding: This research was funded by the National Natural Science Foundation of China (Grant No. U1866212), the Foundation for Fundamental Research of China (Grant No. JSZL2016110B003), the China Postdoctoral Science Foundation (Grant No. 2018M633460), and the 111 project (Grant No. B12026).

Acknowledgments: The authors gratefully acknowledge the support provided by the Fundamental Research Funds for the Central Universities and the Innovation Fund of Xidian University.

Conflicts of Interest: The authors declare no conflict of interest.

References

1. Lee, B.H.; Oh, J.; Tseng, H.H.; Jammy, R.; Huff, H. Gate Stack Technology for Nanoscale Devices. *Mater. Today* **2006**, *9*, 32–40. [CrossRef]
2. Kamata, Y. High-K/Ge MOSFETs for Future Nanoelectronics. *Mater. Today* **2008**, *11*, 30–38. [CrossRef]
3. Yi, S.H.; Chang-Liao, K.S.; Wu, T.Y.; Hsu, C.W.; Huang, J.Y. High Performance Ge pMOSFETs with HfO_2/Hf-Cap/GeO_x Gate Stack and Suitable Post Metal Annealing Treatments. *IEEE Electron Device Lett.* **2017**, *38*, 544–547. [CrossRef]
4. Simoen, E.; Mitard, J.; Hellings, G.; Eneman, G.; DeJaeger, B.; Witters, L.; Vincent, B.; Loo, R.; Delabie, A.; Sioncke, S.; et al. Challenges and Opportunities in Advanced Ge pMOSFETs. *Mater. Sci. Semicond. Process.* **2012**, *15*, 588–600. [CrossRef]
5. Houssa, M.; Chagarov, E.; Kummel, A. Surface Defects and Passivation of Ge and III–V Interfaces. *MRS Bull.* **2009**, *34*, 504–513. [CrossRef]
6. Kouda, M.; Suzuki, T.; Kakushima, K.; Ahmet, P.; Iwai, H.; Yasuda, T. Electrical Properties of CeO_2/La_2O_3 Stacked Gate Dielectrics Fabricated by Chemical Vapor Deposition and Atomic Layer Deposition. *Jpn. J. Appl. Phys.* **2012**, *51*, 121101.
7. Liu, Q.Y.; Fang, Z.B.; Liu, S.Y.; Tan, Y.S.; Chen, J.J. Band Offsets of La_2O_3 Films on Ge Substrates Grown by Radio Frequency Magnetron Sputtering. *Mater. Lett.* **2014**, *116*, 43–45. [CrossRef]
8. Abermann, S.; Bethge, O.; Henkel, C.; Bertagnolli, E. Atomic Layer Deposition of ZrO_2/La_2O_3 High-K Dielectrics on Germanium Reaching 0.5 Nm Equivalent Oxide Thickness. *Appl. Phys. Lett.* **2009**, *94*, 262904. [CrossRef]
9. Calmels, L.; Coulon, P.E.; Schamm-Chardon, S. Calculated and Experimental Electron Energy-Loss Spectra of La_2O_3, $La(OH)_3$, and LaOF Nanophases in High Permittivity Lanthanum-Based Oxide Layers. *Appl. Phys. Lett.* **2011**, *98*, 243116. [CrossRef]
10. Zhao, L.; Liu, H.X.; Wang, X.; Wang, Y.T.; Wang, S.L. Improvements on the Interfacial Properties of High-K/Ge MIS Structures by Inserting a La_2O_3 Passivation Layer. *Materials* **2018**, *11*, 2333. [CrossRef]
11. Eom, D.; Hwang, C.S.; Kim, H.J. Thermal Stability of Stack Structures of Aluminum Nitride and Lanthanum Oxide Thin Films. *ECS Trans.* **2006**, *3*, 121–127.
12. Hausmann, D.M.; Kim, E.; Becker, J.; Gordon, R.G. Atomic Layer Deposition of Hafnium and Zirconium Oxides Using Metal Amide Precursors. *Chem. Mater.* **2002**, *14*, 4350–4358. [CrossRef]
13. Cao, D.; Cheng, X.H.; Yu, Y.H.; Li, X.L.; Liu, C.Z.; Shen, D.S.; Mändl, S. Competitive Si and La Effect in HfO_2 Phase Stabilization in Multi-Layer $(La_2O_3)_{0.08}(HfO_2)$ Films. *Appl. Phys. Lett.* **2013**, *103*, 081607. [CrossRef]
14. Mitrovic, I.Z.; Althobaiti, M.; Weerakkody, A.D.; Sedghi, N.; Hall, S.; Dhanak, V.R.; Chalker, P.R.; Henkel, C.; Dentoni, L.E.; Hellström, P.E.; et al. Interface Engineering of Ge Using Thulium Oxide: Band Line-Up Study. *Microelectron. Eng.* **2013**, *109*, 204–207. [CrossRef]
15. Song, J.; Kakushima, K.; Ahmet, P.; Tsutsui, K.; Sugii, N.; Hattori, T.; Iwai, H. Improvement of Interfacial Properties with Interfacial Layer in La_2O_3/Ge Structure. *Microelectron. Eng.* **2007**, *84*, 2336–2339. [CrossRef]

16. Li, X.F.; Liu, X.J.; Cao, Y.Q.; Li, A.D.; Li, H.; Wu, D. Improved Interfacial and Electrical Properties of Atomic Layer Deposition HfO$_2$ Films on Ge with La$_2$O$_3$ Passivation. *Appl. Surf. Sci.* **2013**, *264*, 783–786. [CrossRef]

17. Kim, H.C.; Woo, S.H.; Lee, J.S.; Kim, H.G.; Kim, Y.C.; Lee, H.R.; Jeon, H.T. The Effects of Annealing Ambient on the Characteristics of La$_2$O$_3$ Films Deposited by RPALD. *J. Electrochem. Soc.* **2010**, *157*, H479–H482. [CrossRef]

18. Mitrovic, I.Z.; Althobaiti, M.; Weerakkody, A.D.; Dhanak, V.R.; Linhart, W.M.; Veal, T.D.; Sedghi, N.; Hall, S.; Chalker, P.R.; Tsoutsou, D.; et al. Ge Interface Engineering Using Ultra-Thin La$_2$O$_3$ and Y$_2$O$_3$ Films: A Study into the Effect of Deposition Temperature. *J. Appl. Phys.* **2014**, *115*, 114102. [CrossRef]

19. Schmeisser, D.; Schnell, R.D.; Bogen, A.; Himpsel, F.J.; Rieger, D.; Landgren, G.; Morar, J.F. Surface Oxidation States of Germanium. *Surf. Sci.* **1986**, *172*, 455–465. [CrossRef]

20. Wilka, G.D.; Wallace, R.M. Electrical Properties of Hafnium Silicate Gate Dielectrics Deposited Directly on Silicon. *Appl Phys Lett.* **1999**, *74*, 2854–2856. [CrossRef]

21. Lin, M.H.; Lan, C.K.; Chen, C.C.; Wu, J.Y. Electrical Properties of HfO$_2$/La$_2$O$_3$ Gate Dielectrics on Ge with Ultrathin Nitride Interfacial Layer Formed by in Situ N$_2$/H$_2$/Ar Radical Pretreatment. *Appl. Phys. Lett.* **2011**, *99*, 182105. [CrossRef]

22. Wang, S.K.; Kita, K.; Lee, C.H.; Tabata, T.; Nishimura, T.; Nagashio, K.; Toriumi, A. Desorption Kinetics of GeO from GeO$_2$/Ge Structure. *J. Appl. Phys.* **2010**, *105*, 054104.

23. Bhaisare, M.; Misra, A.; Kottantharayil, A. Aluminum Oxide Deposited by Pulsed-DC Reactive Sputtering for Crystalline Silicon Surface Passivation. *IEEE J. Photovolt.* **2013**, *3*, 930–935. [CrossRef]

24. Cao, D.; Cheng, X.H.; Jia, T.T.; Xu, D.W.; Wang, Z.J.; Xia, C.; Yu, Y.H. Characterization of HfO$_2$/La$_2$O$_3$ Layered Stacking Deposited on Si Substrate. *J. Vac. Sci. Technol. B. Nanotechnol. Microelectron.* **2013**, *31*, 01A113. [CrossRef]

25. Lamagna, L.; Wiemer, C.; Perego, M.; Volkos, S.N.; Baldovino, S.; Tsoutsou, D.; Schamm-Chardon, S.; Coulon, P.E.; Fanciulli, M. O$_3$-Based Atomic Layer Deposition of Hexagonal La$_2$O$_3$ films on Si (100) and Ge (100) Substrates. *J. Appl. Phys.* **2010**, *108*, 084108. [CrossRef]

26. Martens, K.; Chui, C.O.; Brammertz, G.; Jaeger, B.D.; Kuzum, D.; Meuris, M.; Heyns, M.M.; Krishnamohan, T.; Saraswat, K.; Maes, H.E.; et al. On the Correct Extraction of Interface Trap Density of MOS Devices with High-Mobility Semiconductor Substrates. *IEEE Trans. Electron Device* **2008**, *55*, 547–556. [CrossRef]

27. Suzuki, T.; Kouda, M.; Ahmet, P.; Iwai, H.; Kakushima, K.; Yasuda, T. La$_2$O$_3$ Gate Insulators Prepared by Atomic Layer Deposition: Optimal Growth Conditions and MgO/La$_2$O$_3$ Stacks for Improved Metal-Oxide-Semiconductor Characteristics. *J. Vac. Sci. Technol. A* **2012**, *30*, 051507. [CrossRef]

28. Hill, W.A.; Coleman, C.C. A Single-Frequency Approximation for Interface-State Density Determination. *Solid State Electron.* **1980**, *23*, 987–993. [CrossRef]

29. Stesmans, A.; Afanas'ev, V.V. Si Dangling-Bond-Type Defects at the Interface of (100) Si with Ultrathin Layers of SiO$_x$, Al$_2$O$_3$, and ZrO$_2$. *Appl. Phys. Lett.* **2002**, *80*, 1957. [CrossRef]

30. Engel-Herbert, R.; Hwang, Y.; Stemmer, S. Comparison of Methods to Quantify Interface Trap Densities at Dielectric/III–V Semiconductor Interfaces. *J. Appl. Phys.* **2010**, *108*, 124101. [CrossRef]

31. Bethge, O.; Zimmermann, C.; Lutzer, B.; Simsek, S.; Abermann, S.; Bertagnolli, E. ALD Grown Rare-Earth High-K Oxides on Ge: Lowering of the Interface Trap Density and EOT Scalability. *ECS Trans.* **2014**, *64*, 69–76. [CrossRef]

32. Kita, K.; Suzuki, S.; Nomura, H.; Takahashi, T.; Nishimura, T.; Toriumi, A. Direct Evidence of GeO Volatilization from GeO$_2$/Ge and Impact of Its Suppression on GeO$_2$/Ge Metal-Insulator-Semiconductor Characteristics. *Jpn. J. Appl. Phys.* **2008**, *47*, 2349–2353. [CrossRef]

33. Hauser, J.R.; Ahmed, K. Characterization of Ultra-Thin Oxides Using Electrical C-V and I-V Measurements. *AIP Conf. Proc.* **1998**, *445*, 235.

34. SZE, S.M.; NG, K.K. *Physics of Semiconductor Devices*, 3rd ed.; John Wiley & Sons Inc.: Hoboken, NJ, USA, 2006; pp. 223–236.

35. Spahr, H.; Bülow, T.; Nowak, C.; Hirschberg, F.; Reinker, J.; Hamwi, S.; Johannes, H.H.; Kowalsky, W. Impact of Morphological Defects on the Electrical Breakdown of Ultra-Thin Atomic Layer Deposition Processed Al$_2$O$_3$ Layers. *Thin Solid Films* **2013**, *534*, 172–176. [CrossRef]

 nanomaterials

Article

New Insights into the Role of Weak Electron–Phonon Coupling in Nanostructured ZnO Thin Films

Ashish C. Gandhi, Wei-Shan Yeoh, Ming-An Wu, Ching-Hao Liao, Dai-Yao Chiu, Wei-Li Yeh and Yue-Lin Huang *

Department of Physics, National Dong Hwa University, Hualien 97401, Taiwan; acg.gandhi@gmail.com (A.C.G.); letherlightness@gmail.com (W.-S.Y.); swarem610823@gmail.com (M.-A.W.); 410214209@gms.ndhu.edu.tw (C.-H.L.); ella9125710@gmail.com (D.-Y.C.); willy730611@gmail.com (W.-L.Y.)
* Correspondence: huang_yuelin@gms.ndhu.edu.tw; Tel.: +886-3-890-3728

Received: 11 July 2018; Accepted: 15 August 2018; Published: 20 August 2018

Abstract: High-quality crystalline nanostructured ZnO thin films were grown on sapphire substrates by reactive sputtering. As-grown and post-annealed films (in air) with various grain sizes (2 to 29 nm) were investigated by scanning electron microscopy, X-ray diffraction, and Raman scattering. The electron–phonon coupling (EPC) strength, deduced from the ratio of the second- to the first-order Raman scattering intensity, diminished by reducing the ZnO grain size, which mainly relates to the Fröhlich interactions. Our finding suggests that in the spatially quantum-confined system the low polar nature leads to weak EPC. The outcome of this study is important for the development of nanoscale high-performance optoelectronic devices.

Keywords: Raman scattering; quantum confinement; electron–phonon coupling; polar semiconductors; zinc oxide

---

1. Introduction

Bulk ZnO is an n-type semiconductor material with a direct wide band gap energy (3.37 eV) and large exciton binding energy (60 meV) at room temperature. These properties make ZnO one of the most promising oxide semiconductors for high-performance optoelectronic devices, such as touch screens, liquid crystal displays, light-emitting diodes, chemical/biological sensors, dye-sensitized solar cells, and piezoelectric devices [1–3]. The optical properties of nanoscale devices can be controlled by tailoring the average size, shape, and surface modifications of nanometer-sized ZnO crystals [4–6]. However, at the nanoscale, the optical and the electrical properties, such as the energy relaxation rate of excited carriers and phonon reproduction of excitons in the photoluminescence (PL), as well as Raman scattering, are greatly influenced by the electron–phonon coupling (EPC) in semiconductor materials [7]. For instance, the electric field within a material correlates to Coulomb interactions with the exciton, and the strength of EPC will be enhanced if the wavelength of the phonon vibration is comparable to the spatial extent of the exciton [8–13]. Such a quantum-confined spatial system of semiconductors is possible to prepare. However, the optical and the electrical properties of such system differ significantly from their bulk counterparts.

In the past, both theoretical and experimental studies have been done to estimate the EPC strength and its dependence on the crystallite size of semiconductor nanocrystals [7–16]. Theoretically, it has been suggested that the electron–longitudinal optical (LO) phonon coupling should vanish with decreasing nanocrystal size by using a simple charge neutrality model [9]. However, using the non-parabolicity of the bands, it has been shown that EPC strength increases when decreasing the material size [10,11]. Experimentally, using Raman spectroscopy, it has been observed that Fröhlich interactions play an important role in defining the strength of EPC in various ZnO nanocrystals and its value decreases when reducing the crystallite size. However, the dependence of EPC strength on the nanocrystal shape and size of wide bandgap ZnO is still not completely understood [7,17–23]. Therefore, deliberate control of size and shape of ZnO nanostructures is of particular interest for the understanding of the fundamental physics of ZnO nanocrystals and for their application in functional devices.

The sputtering method has many advantages including better film growth control, uniformity, repeatability, low-temperature deposition, and large-scale stability. The properties of sputtered ZnO thin films not only depend on the deposition parameters (RF power, pressure, substrate temperature, ambient atmosphere), but also on the post-deposition processes such as thermal treatment [24–26]. However, the effect of post-annealing on the ZnO film properties has not often been reported in the literature. In this study, we report on the effect of post-annealing on the structural and the optical properties of the magnetron sputtering deposited ZnO thin films. The main focus of this study is to examine the influence of the grain size on the evolution of phonon confinement and the strength of EPC using Raman spectroscopy. Raman spectroscopy is a versatile and fast nondestructive characterization technique sensitive to distortions in the crystal lattice, crystal defects, and phase transformation [27]. Based on inelastic light scattering, it can provide information about the phonon vibrational and rotational mode properties of post-annealed ZnO to get insight into the effect of size reduction on the EPC in ZnO nanocrystals. The results of crystalline structure, grain size, surface morphology, defects, and optical properties of both as-deposited and post-annealed ZnO thin films have been investigated.

2. Materials and Methods

The vapor-phase deposition was performed with reactive sputtering on sapphire substrates with Al_2O_3 (0001) normal orientation using a Zn target of purity 99.995% located about 8 cm away from the substrates and a mixture gas with argon:oxygen = 4:1. During the deposition, a total pressure was maintained at 50 mTorr and monitored by an absolute capacitance manometer. A direct-current power supply was used to drive the magnetron sputtering plasma at 80 W. The substrate was rotated at a speed of three turns per minute to eliminate the effects of slanted deposition. A calibrated quartz crystal microbalance (Model TM-350, Matek Inc., Cypress, CA, USA) operating at 6 MHz was taken to follow the deposited mass in real time, realizing a comparable thickness of 300 nm for each film. The sputtered films (ZnO/ALO) were annealed in air for one hour at temperatures (T_A) from 100 to 800 °C in a step of 100 °C followed by characterizations of structure and properties at room temperature. Structural and surface morphological analysis of films was carried out using X-ray diffraction (XRD, Rigaku D/Max-2500V X-ray Diffractometer, The Woodlands, TX, USA) with CuK α radiation (wavelength: 1.54178 Å) and scanning electron microscopy (SEM) taking 15 keV incidence beam for secondary-electron imaging on JEOL JSM-7000F (Peabody, MA, USA) equipped with a field emission electron source. The mean diameter distribution was estimated from relative SEM images using Nano Measurer software (Nano measurer 1.2.5, Jie Xu, Fudan University, Shanghai, China). Raman scattering was utilized to study the defects and the effect of reduced grain size on phonon confinement and the strength of EPC. Scattering spectra were recorded with a Jobin Yvon T64000 spectrometer (Horiba Scientific, Paris, France) using a continuous-wave laser of 325 nm. The excitation laser power on the sample was ~4 mW with 1 μm diameter of the laser beam spot. The full profile fitting of Raman spectra was carried out using Fityk 0.9.3 software [28].

3. Results and Discussion

3.1. Morphology and Structural Characterization

Figure 1 shows SEM images of as-deposited room temperature (RT) and 100 to 800 °C annealed ZnO/ALO films (in a step of 100 °C), revealing granular nature within nanometric range and an increase of grain size with the increase of T_A. Uniformly distributed nanoplate-like grains of ~10 to 15 nm thickness and ~50 to 80 nm length have been seen from RT and 100 to 300 °C annealed films. Annealing at temperatures from 400 to 500 °C resulted in the transformation of nanoplates into pseudospherical nanoparticles, forming agglomerated sphere-like nanoparticles. However, the annealing at and above 600 °C resulted in the formation of films with interconnected granular surface morphology. The mean diameter <d> of asymmetrically distributed nanoparticles was obtained by fitting the log-normal distribution function $f(d) = \frac{1}{\sqrt{2\pi}d\sigma} \exp\left[-\frac{(\ln d - \ln\langle d\rangle)}{2\sigma^2}\right]$ to the histogram obtained from SEM images, where σ is the standard deviation of the fitted function shown in Figure S1. The fitted values of ($\langle d\rangle$, σ) vary from ~(13 nm, 0.16) to (65 nm, 0.42) with the increase of T_A from RT to 800 °C (see Table S1). The value of $\langle d\rangle$ ~13 nm, which is the width of the nanoplates, remains independent of T_A up to 300 °C; above that a drastic increase in the grain size can be seen, reaching a saturation value of ~62 nm at T_A 600 °C. From the fitted values of σ, it appears that the distribution of small-sized particles is confined within a narrow range at low T_A and evolves with an increase of T_A.

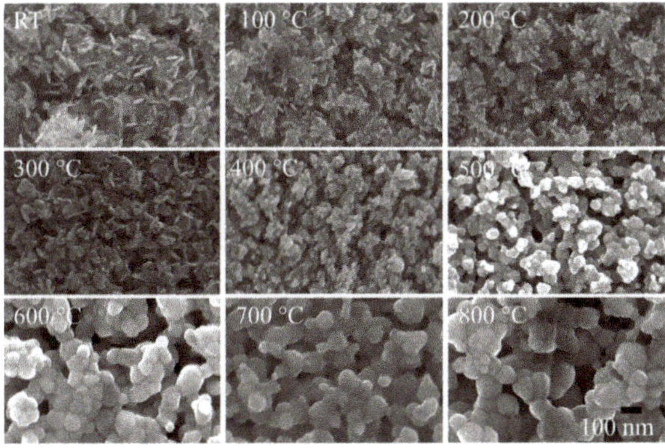

Figure 1. Scanning electron microscopy (SEM) images of RT and 100 °C to 800 °C annealed ZnO/ALO films. (All the images are taken at the same magnification ($\times 10^5$) with a scale bar of 100 nm shown in the image of 800 °C).

Figure 2a shows XRD spectra for RT and 100 to 800 °C annealed ZnO/ALO films, taken under a fixed glancing incident angle of 3°. All Bragg reflections from RT film were identified to emerge from wurtzite ZnO (JCPDS 036-1451), and simple hexagonal Zn (JCPDS PDF#00-004-0831) [29], indicating the high purity of the vapor-phase deposited film. The annealing at temperatures above 100 °C resulted in complete oxidation of Zn and the formation of pure ZnO films. This was also supported by energy-dispersive X-ray spectroscopy, showing that only Zn, O, Al, and C are detectable in the films (data not shown), where Al and C came from the sapphire substrates and surface adsorbents respectively. The crystallinity of film improves with an increase of T_A, resulting in an enhanced Bragg maximum and reduced spectral width. The appearance of weak diffraction peaks corresponding to the (100), (101), (102), (110), (103), (112) and (201) planes of ZnO suggested the presence of some randomly oriented grains. However, there was a preferential growth along the (002)$_{ZnO}$ plane's normal, resulting in the very strong (002) reflection in Figure 2a. The intensity profile of the most intense (002) and (101) reflections can be fitted with a Lorentzian distribution function to determine its full-width at half maximum (FWHM) β. Using Scherrer's formula: $d = k\lambda/\beta \cos\theta$ (where $k = 0.94$ is a Scherrer's constant and θ is the Bragg angle), the calculated grain size ($d_{(002)}$, $d_{(101)}$) of RT, and 100 to 800 °C ZnO films is (2 nm, 19 nm), (5 nm, 23 nm), (7 nm, 25 nm), (8 nm, 22 nm), (10 nm, 20 nm), (16 nm, 22 nm), (26 nm, 28 nm), (24 nm, 27 nm) and (29 nm, 29 nm), respectively (see Table S1). Figure 2b compares the calculated values of $d_{(002)}$, $d_{(101)}$ and estimated <d> from SEM images with respect to T_A. The value of $d_{(002)}$ shows a minimum of 20 nm at $T_A = 400$ °C, whereas $d_{(101)}$ increases with T_A

and at and above 600 °C, $d_{(101)} \approx d_{(002)}$. Above finding indicates a change in grain dimensions with aspect ratio $d_{(002)}/d_{(101)}$ decreasing from ~9 to ~1 due to increasing T_A up to 800 °C. Furthermore, particle size $<d>$ ~13 nm obtained from SEM images (RT to 300 °C) is smaller than that of $d_{(002)}$ ~20 nm, indicating that nanoplate-like grains exhibit fast grown rate along (002) direction, which is consistent with XRD spectra. This trend is consistent with SEM observations, supporting the idea that ZnO grains have developed from small plate-like grains (which could be oriented along the *c*-axis) to large sphere-like ones above 400 °C. Our findings revealed that much finer nanocrystals can be formed during the reactive deposition of ZnO films as compared with ZnO grains ($d_{(101)}$ ~13 to 24 nm) formed through annealing a pure Zn film on Al_2O_3(001) substrates (Zn/ALO) with a comparable film thickness for one hour in air at similar temperature T_A varying from RT to 800 °C. As compared with the ZnO/ALO film, different structural evolution of ZnO from post-annealing of the Zn/ALO film is reflected in the distinct XRD 2-θ scan patterns (Figure S2) as well as the calculated lattice constants and the internal parameter u (see the comparison in Table S1 and Figure S3). This difference is thought to be related with the highly anisotropic early-stage growth of ZnO crystallites emerged during reactive sputtering of the ZnO/ALO film leading to an aspect ratio of $d_{(002)}/d_{(101)}$ as high as ~9. Whereas, ZnO crystallites evolved isotropically from nearly-sphere-shaped crystallites with aspect ratios between 0.9 and 1.1 formed during post-annealing of the Zn/ALO film in the air. In fact, ZnO grain sizes in as-deposited films have been observed to depend on sputtering parameters [30,31]. Our observation of average grain sizes down to ~5 nm in low-temperature annealed ZnO/ALO films are consistent with a previous report that fine crystallites are formed during reactive sputtering of ZnO films on Si(100) substrates using oxygen partial pressures below 30% (20% in our case) [30]. The fact that the calculated grain size using XRD is much smaller than the particle size estimated from SEM images is attributed to the formation of multi-grain ZnO particles due to annealing, particularly at temperatures above 400 °C.

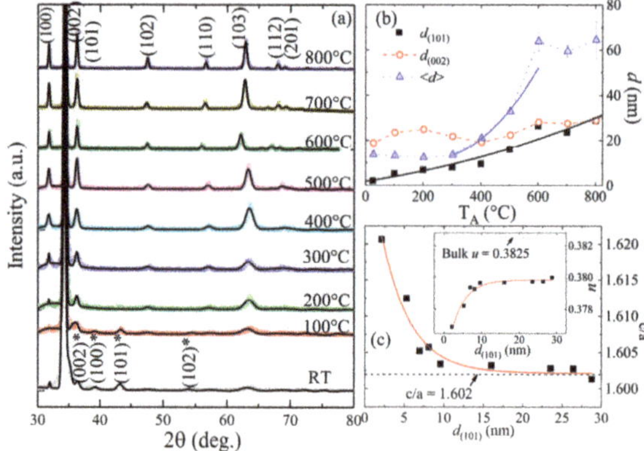

Figure 2. (**a**) "F(calc) weighted" LaBail extraction analysis [32] (solid line) for the X-ray diffraction (XRD) pattern (crosses) of the RT, and 100 to 800 °C annealed ZnO/ALO films (bottom to top) with Bragg reflections of ZnO and Zn (star-marked) phases indicated; (**b**) T_A-dependency of $d_{(002)}$, $d_{(101)}$ calculated from XRD patterns and the fitted values of $<d>$ obtained from SEM images. Black colored solid lines represent the growth law fitting to $d_{(101)}$; (**c**) Grain-sized $d_{(101)}$ dependence of the ratio of lattice constants c/a, where the horizontal lines represent the value of bulk ZnO, and the solid line denotes the exponential fit. The inset in (**c**) shows the corresponding evolution of the internal parameter u.

For detailed structural investigation of the ZnO/ALO films, "F(calc) weighted" LaBail extraction analysis of all samples was carried out by refining XRD spectra using the GSAS software package, as shown in Figure 2a [32]. We confirmed that the Zn and ZnO phase (space group P6$_3$/mmc and

P6$_3$mc) existed in RT and 100 °C films, whereas there is a single ZnO phase in 200 to 800 °C films. The fitted values of lattice constants ($a = b$, c) of the ZnO phase obtained from RT, and 100 to 800 °C films, are summarized in Table 1, showing a lattice contraction along the $a = b$ axis from RT and 100 °C films, whereas the value obtained from 200 to 800 °C films shows a value very near the bulk value of 3.250 Å. On the contrary, the lattice constant along the c-axis shows exponentially decreasing behavior with the increase of T_A and approaches the bulk value of 5.207 Å around ~500 °C, above which it is almost T_A-independent (see Figure S3a,c). The axial ratio c/a and the internal parameter u (u represents the relative position of two hexagonal close-packed sublattices of the wurtzite) can be used as indicators of the polarity as the deviation occurs in the lattice of the wurtzite ZnO structure. Figure 2c shows the $d_{(101)}$ dependence of fitted values of c/a, revealing the development of short- to long-range polar interaction with the increase of grain size. The solid curve in Figure 2c denotes an exponential decay function, namely $c/a = (c/a)_{Bulk} + 0.034(5) \exp\left(-d_{(101)}/\xi_o\right)$, where $(c/a)_{Bulk} = 1.602$ represents the ratio of long-range polar interaction and $\xi_o = 3.6(6)$ nm is the domain size. The obtained domain size is very near that of the quantum-confined dots (~3.5 nm), from which a very weak EPC strength of ~0.39 has been reported [18]. Ignoring the slight length difference between the Zn–O bond along c-axis and the others, the internal parameter can be estimated by $u = \frac{1}{3}\left(\frac{a^2}{c^2}\right) + \frac{1}{4}$. Taking the fitted values of lattice parameters a and c, changes in the estimated u with respect to $d_{(101)}$ (shown in the inset of Figure 2c) are found to evolve from 0.3769 (RT) and show an exponentially increasing behavior with size, resulting in a saturation value of ~0.3800 at ~10 nm ($T_A = 400$ °C), which is smaller than the reported bulk value of 0.3825. In contrast, as discussed before, the annealing of pure Zn films (Zn/ALO) at 100–400 °C for 1 h in air has led to the formation of much larger grains of ZnO ($d_{(100)} \approx$ 11–14 nm) with aspect ratios of ~0.9–1.1 and significantly lower values of the internal parameter u ~0.3797–0.3798 (see Table S1 and Figure S3d). Therefore, both the surface dipole-induced electrostatic effects of the polar wurtzite structure and the vapor-phase ambient conditions during the magnetron sputtering were thought to dominate the anisotropic growth of ZnO nanocrystals during deposition. The observed increase of c/a and the very low value of u in fine-sized granular films (<10 nm) could be due to the effect of compressive strain. The strain (ε) and stress (σ) in the films, along with the c-axis, can be estimated using $\varepsilon = (c_{film} - c_{bulk})/c_{bulk}$ and $\sigma = -2.33 \times 10^{11} \text{GPa} \cdot (c_{film} - c_{bulk})/c_{bulk}$, where c_{film} and $c_{bulk} = 5.207$ Å are the lattice parameters of the film and unstrained ZnO, respectively. The calculated values of ε, σ summarized in Table 1 show decreasing behavior with the increase of grain size. Overall trends seen from SEM images, grain size, and lattice constant can be understood by considering that increasing T_A at and above 400 °C resulted in morphology transition accompanied by lattice relaxation towards equilibrium bulk state as indicated with the bulk value of c/a or u. The above structural finding indicates that the effect of quantum confinement in RT and low-temperature annealed films could also influence the band gap shift and EPC strength.

Table 1. Lattice constants $a = b$ and c for Zn and ZnO phases and the internal parameter u, strain ε, stress σ, and parameters wRp and Rp for fitting the XRD patterns of the ZnO/ALO films based on the "F(calc) weighted" LaBail extraction analysis [32].

T_A (°C)	Zn		ZnO					wRp	Rp	χ^2
	$a = b$ (Å)	c (Å)	$a = b$ (Å)	c (Å)	u	ε (%)	σ (GPa)			
25 (RT)	2.6531(11)	5.1315(16)	3.2358(11)	5.2442(3)	0.3769	0.71	−1.66	0.2638	0.1538	6.132
100	2.6569(13)	5.0304(24)	3.2462(8)	5.2343(7)	0.3782	0.52	−1.22	0.3403	0.2317	5.741
200			3.2514(8)	5.2191(2)	0.3794	0.23	−0.54	0.2639	0.1794	4.989
300			3.2495(7)	5.2179(2)	0.3793	0.21	−0.49	0.2492	0.1597	5.448
400			3.2525(6)	5.2151(5)	0.3797	0.16	−0.36	0.3169	0.2233	4.235
500			3.2449(7)	5.2023(12)	0.3797	−0.09	0.21	0.3092	0.204	3.878
600			3.2485(5)	5.2066(9)	0.3798	−0.01	0.02	0.3743	0.2436	3.426
700			3.2509(2)	5.2106(2)	0.3797	0.07	−0.16	0.3084	0.1907	4.043
800			3.2488(4)	5.2024(6)	0.3800	−0.09	0.21	0.3515	0.2318	3.3

3.2. Electron–Phonon Coupling in ZnO Films

An intense multiphonon scattering of the RT and 100 to 800 °C annealed ZnO/ALO films with various grain sizes was observed with the Raman spectra in Figure 3a, where the three major bands were attributed to longitudinal optical A_1(LO) polar symmetry modes and its overtones A_1(2LO, 3LO). It is interesting to note that as T_A increased, the intensities of the A_1(LO) mode and their overtones decreased, with the spectral width narrowing down. The anomalous behavior was analyzed quantitatively using a profile fitting method. The solid line in Figure 3a represents the fit using the Voigt function covering the whole spectrum, and the fitting parameters are summarized in Table S2. Based on the reported zone-center optical phonon frequencies in ZnO, Raman spectra obtained from RT ZnO film exhibited prominent maxima at 573.9(1) cm^{-1}, 1144.7(5) cm^{-1} and 1724(2) cm^{-1} identified with A_1(1LO), A_1(2LO), and A_1(3LO) respectively, which are consistent with that observed on sol-gel derived ZnO quantum dots of size down to ~3.5 nm [18]. Figure 3b depicts the $d_{(101)}$ dependency of the peak center of A_1(1LO), which can be described very well using $X_C(1LO) = X_{C0}(1LO)\left[1 - 0.016\ \exp\left(-\left(d_{(101)}/d_C\right)\right)\right]$, where $X_{C0}(1LO) = 580$ cm^{-1} is the bulk value of ZnO and $d_C = 7(2)$ nm. Compared with that of the ZnO bulk, a redshift ranging from ~0 to 6 cm^{-1} was obtained as the grain size decreased from 29 to 2 nm. Inset of Figure 3b shows the obtained increasing behavior of full-width at half-maximum ($\Delta\omega$) of A_1(1LO) mode with the decrease of grain size. A similar redshift and the broadening due to decreasing grain size was also observed from the overtones of A_1(1LO) mode. Figure 3c shows the $d_{(101)}$ dependency of the integrated intensity of A_1(LO) mode (I_{1LO}), where the solid line is a guide for the eye. An exponential decay in the intensity can be seen with the increase of grain size (i.e., increase of T_A). Such a pronounced spectral redshift, broadening, and the asymmetry of A_1 mode and its overtones from small sized grains could result either from phonon localization due to intrinsic defects or the spatial confinement within the grain boundaries. However, from the PL spectra of the ZnO/ALO films, we have observed (i) a blueshift in UV emission from 3.25 eV to 3.30 eV and (ii) a redshift in discernible broad and intense green band emission (1.8–2.7 eV) with the decreasing grain size from 10 to 2 nm (data not shown). Since, exciton Bohr radius of bulk ZnO is 2.34 nm (i.e., diameter 4.68 nm), the carrier confinement in the ZnO films with grain size below 9(1) nm is in the moderate to strong confinement regimes. Therefore, the observed blueshift of UV emission with decreasing grain size can be ascribed to quantum confinement effect, whereas redshift of green band emission to the high density of oxygen vacancies at the surface of the grains. Interestingly, both UV and green band emission do not show any size dependency above 10 nm (further details about PL of granular ZnO films will be published in the near future). The above finding suggests that the n-phonon process is contributed to the Raman scattering cross section of low annealed films, which occurs when the energy of incoming or scattered photon matches the real electronic states in the quantum-confined material.

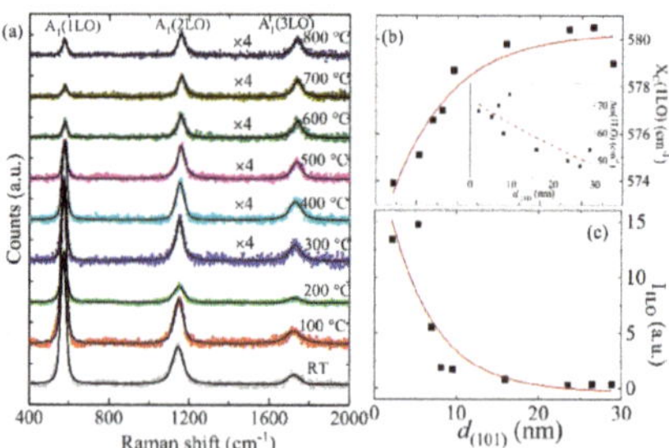

Figure 3. (**a**) Raman spectra of RT and 100 to 800 °C annealed ZnO/ALO films, where the solid line represents the fit using Voigt function; (**b**) The $d_{(101)}$ dependency of peak center of A_1(1LO) phonon mode where the solid line represents the fit. Inset of the figure shows $d_{(101)}$ dependency of the spectral width $\Delta\omega$(1LO), where the dashed line is a guide for the eye; (**c**) The $d_{(101)}$ dependency of the Raman scattering intensity I_{1LO}, where the solid line is a guide for the eye.

The ratio of the relative cross-sections of the first- and second-order Raman scattering $A_1(2LO)$ and $A_1(1LO)$ modes, $R = I_{2LO}/I_{1LO}$, can be used to estimate the EPC strength, which is related to the Huang–Rhys parameter S by taking the scattered intensity $I_{nLO} \propto S^n e^{-S}/n!$ within the Franck–Condon approximation [33]. Along with the grain size dependency of EPC strength observed in this work, Figure 4 also depicts the size dependency of reported values of EPC strength of nanoparticles (9 to 20 nm) [17], quantum dots [18,22] (3.5 to 20 nm), and granular thin film (33 nm) [34]. It is interesting to note that quantum dots with size below 12 nm show a lowest, almost a constant value of EPC strength ~0.4 and above which it increases exponentially, whereas a highest value of EPC from the granular thin film approaches 3.1 [18]. Therefore, the EPC strength obtained from granular ZnO films can be split into two regions, namely, Re-I (~2 to 9 nm) and Re-II (~9 to 29 nm). Contrary to size dependency of quantum dots, the grain size dependency of EPC strength in Re-I shows shallow deep around ~5 nm, which can be described as $R = 0.76 - 0.09d_{(101)} + 0.01d^2_{(101)}$ (represented by the solid curve in Figure 4). However, in Re-II, above ~9 nm, a drastic increase in the EPC strength with value approaching towards a bulk value of 2.85 (represented by a horizontal dotted line) can be seen, where dashed line is guided for the eye. The deformation potential and the Fröhlich potential are responsible for the size dependency of the EPC strength. The TO phonon mode includes a contribution from the deformation potential, involving short-range interaction between electrons and the lattice displacements, whereas the LO phonon mode includes contributions from both the potentials that involve the long-range interaction generated by the macroscopic electric field associated with the LO phonons. However, in the present set of ZnO films, under resonant conditions, the intensity of the LO phonon greatly increases with the decrease of grain size, and that of the TO phonon is almost insensitive. Therefore, the low value of EPC obtained from the spatially quantum-confined grains in Re-I can be attributed to less polar nature as observed from XRD analysis, whereas the linear increasing behavior of EPC in Re-II signaling strong size dependency is mainly related to the long-range Fröhlich interactions.

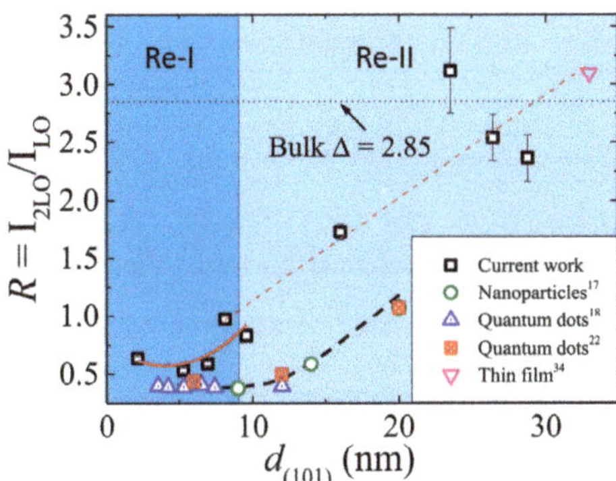

Figure 4. A plot of $d_{(101)}$ dependency of EPC strength in the current work compared with that reported for nanoparticles [17], quantum dots [18,22], and thin granular film [34], where the horizontal dotted line indicates the bulk ZnO value of 2.85. The solid red color line represents a fit as discussed in the text, whereas the straight red and black dashed lines are guides for the eye.

Structural reconstruction has been proposed to take place while annealing reactively sputtered ZnO films with T_A exceeding the melting point of zinc (420 °C for the bulk) due to forming films in a melted state, whereas higher T_A is needed for recrystallization [35]. In contrast to a recrystallization process, we propose that a structural reconstruction process takes place accompanied with out-diffusion of native defects thermally activated at high T_A. This view was corroborated with correlated observations with T_A ~600 °C: (i) annealing at and above 600 °C resulted in the formation of films with well interconnected granular (~62 nm) surface morphology (ii) retrogressive increase of oxygen vacancy (V_O) density in the surface region revealed by XPS (data not shown), and (iii) drastic enhancement of the EPC strength revealed by Raman scattering. Since no surface vibration modes were detected in our study, Raman scattering events were thought to have taken place mainly inside the bulk. The proposed thermally activated out-diffusion is consistent with recently reported density functional theory calculations supporting an energetically favorable migration of V_O in the direction from bulk to surface in bare ZnO nanowires [36]. Our observations were self-consistent in that bulk EPC property was approached at high T_A when lattice relaxation revealed with XRD also proceeded in parallel towards bulk state as discussed in Figure 2c. It is noted, however, that a direct comparison of EPC strengths derived for nanostructures with that of bulk is inappropriate and could lead to conceptual confusion [7,37].

4. Conclusions

Granular ZnO film with nanoplate-like morphology and strong preferential growth along the (002) plane's normal was grown by reactive sputtering technique. Annealing from 100 to 800 °C in an ambient atmosphere leads to an increase of crystalline size $d_{(101)}$ from 2 to 29 nm and a morphology transformation of nanoplate-like shapes to well inter-connected sphere-like grains. The structural analysis carried out by "F(calc) weighted" LaBail extraction analysis of XRD spectra reveal low polar nature of small-sized granular films. The EPC strength deduced from the ratio of the second- to the first-order Raman scattering intensity shows a linear decreasing behavior with the decrease of grain size down to 10 nm, in principle as a result of the Fröhlich interaction. However, below a critical size of 9 (1) nm, a shallow valley with a minimum of EPC strength ~0.53 was obtained around 5 nm. Our finding suggests that in the spatially quantum-confined system the low polar nature leads to weak EPC strength. The outcome of this study is important for the future development of nanoscale high-performance optoelectronic devices.

Supplementary Materials: The following are available online at http://www.mdpi.com/2226-4310/4/2/18/s1, Figure S1: Size-distribution histograms with log-normal fitting for the RT and 100 °C to 800 °C (T_A) annealed ZnO/ALO films; Figure S2: Rietveld analyses (solid line) of XRD pattern (crosses) for the 100 °C to 800 °C (T_A) annealed Zn/ALO films; Figure S3: Lattice parameters (a) $a = b$, (b) c, their ratio (c) c/a, and the internal parameter (d) u of the ZnO/ALO and Zn/ALO films in dependence of T_A (a, b) and $d_{(101)}$ (c, d) respectively; Table S1: $\langle d \rangle$, σ obtained from the log-normal fitting in Figure S1 for the ZnO/ALO films, $d_{(002)}$ and $d_{(101)}$ calculated using the Scherrer's formula for the ZnO/ALO and Zn/ALO films; Table S2: Fitted values of Raman mode frequencies X, spectral widths $\Delta\omega$, and the intensity ratio $R = I_{2LO}/I_{1LO}$ for the ZnO/ALO films.

Author Contributions: Y.-L.H. conceived and designed the research project; W.-S.Y., M.-A.W., C.-H.L., D.-Y.C., and W.-L.Y. performed the experiments; Y.-L.H. and A.C.G. performed the data analyses; A.C.G. accomplished the SEM size analyses, XRD refinement calculations, and Raman profile fitting; Y.-L.H. wrote the original manuscript; A.C.G. and Y.-L.H. reviewed and revised the manuscript.

Funding: This research was funded by the National Science Council of Taiwan, Republic of China, under grant numbers NSC 99-2112-M-259-007-MY3 and NSC 102-2112-M-259-004.

Acknowledgments: The authors are grateful to Sheng Yun Wu of National Dong Hwa University for valuable discussion and help with revising the manuscript.

Conflicts of Interest: The authors declare no conflict of interest.

References

1. Muchuweni, E.; Sathiaraj, T.S.; Nyakotyo, H. Effect of gallium doping on the structural, optical and electrical properties of zinc oxide thin films prepared by spray pyrolysis. *Ceram. Int.* **2016**, *42*, 10066–10070. [CrossRef]
2. Muchuweni, E.; Sathiaraj, T.S.; Nyakotyo, H. Low temperature synthesis of radio frequency magnetron sputtered gallium and aluminium co-doped zinc oxide thin films for transparent electrode fabrication. *Appl. Surf. Sci.* **2016**, *390*, 570–577. [CrossRef]
3. Lee, J.-H.; Ko, K.-H.; Park, B.-O. Electrical and optical properties of ZnO transparent conducting films by the sol–gel method. *J. Cryst. Growth* **2003**, *247*, 119–125. [CrossRef]

4. Yang, Y.; Yan, H.; Fu, Z.; Yang, B.; Zuo, J. Correlation between 577 cm^{-1} Raman scattering and green emission in ZnO ordered nanostructures. *Appl. Phys. Lett.* **2006**, *88*, 191909. [CrossRef]

5. Chassaing, P.-M.; Demangeot, F.; Paillard, V.; Zwick, A.; Combe, N.; Pagès, C.; Kahn, M.L.; Maisonnat, A.; Chaudret, B. Raman study of *E2* and surface phonon in zinc oxide nanoparticles surrounded by organic molecules. *Appl. Phys. Lett.* **2007**, *91*, 053108. [CrossRef]

6. Lu, J.G.; Ye, Z.Z.; Zhang, Y.Z.; Liang, Q.L.; Fujita, S.; Wang, Z.L. Self-assembled ZnO quantum dots with tunable optical properties. *Appl. Phys. Lett.* **2006**, *89*, 023122. [CrossRef]

7. Wang, R.P.; Xu, G.; Jin, P. Size dependence of electron-phonon coupling in ZnO nanowires. *Phys. Rev. B* **2004**, *69*, 113303. [CrossRef]

8. Alivisatos, A.P.; Harris, T.D.; Carroll, P.J.; Steigerwald, M.L.; Brus, L.E. Electron–vibration coupling in semiconductor clusters studied by resonance Raman spectroscopy. *J. Chem. Phys.* **1989**, *90*, 3463–3468. [CrossRef]

9. Schmitt-Rink, S.; Miller, D.A.B.; Chemla, D.S. Theory of the linear and nonlinear optical properties of semiconductor microcrystallites. *Phys. Rev. B* **1987**, *35*, 8113–8125. [CrossRef]

10. Marini, J.C.; Stebe, B.; Kartheuser, E. Exciton-phonon interaction in CdSe and CuCl polar semiconductor nanospheres. *Phys. Rev. B* **1994**, *50*, 14302–14308. [CrossRef]

11. Nomura, S.; Kobayashi, T. Exciton—LO-phonon couplings in spherical semiconductor microcrystallites. *Phys. Rev. B* **1992**, *45*, 1305–1316. [CrossRef]

12. Shiang, J.J.; Risbud, S.H.; Alivisatos, A.P. Resonance Raman studies of the ground and lowest electronic excited state in CdS nanocrystals. *J. Chem. Phys.* **1993**, *98*, 8432–8442. [CrossRef]

13. Shiang, J.J.; Wolters, R.H.; Heath, J.R. Theory of size-dependent resonance Raman intensities in InP nanocrystals. *J. Chem. Phys.* **1997**, *106*, 8981–8994. [CrossRef]

14. Haiming, F.; Bingsuo, Z.; Yulong, L.; Sishen, X. Size effect on the electron–phonon coupling in CuO nanocrystals. *Nanotechnology* **2006**, *17*, 1099.

15. Kelley, A.M. Electron-Phonon Coupling in CdSe Nanocrystals from an Atomistic Phonon Model. *ACS Nano* **2011**, *5*, 5254–5262. [CrossRef] [PubMed]

16. Murphy-Armando, F.; Fagas, G.; Greer, J.C. Deformation Potentials and Electron-Phonon Coupling in Silicon Nanowires. *Nano Lett.* **2010**, *10*, 869–873. [CrossRef] [PubMed]

17. Cheng, H.-M.; Lin, K.-F.; Hsu, H.-C.; Lin, C.-J.; Lin, L.-J.; Hsieh, W.-F. Enhanced Resonant Raman Scattering and Electron−Phonon Coupling from Self-Assembled Secondary ZnO Nanoparticles. *J. Phys. Chem. B* **2005**, *109*, 18385–18390. [CrossRef] [PubMed]

18. Cheng, H.-M.; Lin, K.-F.; Hsu, H.-C.; Hsieh, W.-F. Size dependence of photoluminescence and resonant Raman scattering from ZnO quantum dots. *Appl. Phys. Lett.* **2006**, *88*, 261909. [CrossRef]

19. Hsieh, W.F.; Cheng, H.M.; Lin, K.F.; Hsu, H.-C. Size dependence of band gap variation and electron-phonon coupling in ZnO Quantum Dots. In Proceedings of the Pacific Rim Conference on Lasers and Electro-Optics, Tokyo, Japan, July 2005; pp. 466–468.

20. Lin, K.-F.; Cheng, H.-M.; Hsu, H.-C.; Hsieh, W.-F. Band gap engineering and spatial confinement of optical phonon in ZnO quantum dots. *Appl. Phys. Lett.* **2006**, *88*, 263117. [CrossRef]

21. Ojha, A.K.; Srivastava, M.; Kumar, S.; Hassanein, R.; Singh, J.; Singh, M.K.; Materny, A. Influence of crystal size on the electron-phonon coupling in ZnO nanocrystals investigated by Raman spectroscopy. *Vib. Spectrosc.* **2014**, *72*, 90–96. [CrossRef]

22. Ray, S.C.; Low, Y.; Tsai, H.M.; Pao, C.W.; Chiou, J.W.; Yang, S.C.; Chien, F.Z.; Pong, W.F.; Tsai, M.-H.; Lin, K.F.; et al. Size dependence of the electronic structures and electron-phonon coupling in ZnO quantum dots. *Appl. Phys. Lett.* **2007**, *91*, 262101. [CrossRef]

23. Gaikwad, S.S.; Gandhi, A.C.; Pandit, S.D.; Pant, J.; Chan, T.-S.; Cheng, C.-L.; Ma, Y.-R.; Wu, S.Y. Oxygen induced strained ZnO nanoparticles: An investigation of Raman scattering and visible photoluminescence. *J. Mater. Chem. C* **2014**, *2*, 7264–7274. [CrossRef]

24. Ismail, A.; Abdullah, M.J. The structural and optical properties of ZnO thin films prepared at different RF sputtering power. *J. King Saud Univ. Sci.* **2013**, *25*, 209–215. [CrossRef]

25. Husna, J.; Aliyu, M.M.; Islam, M.A.; Chelvanathan, P.; Hamzah, N.R.; Hossain, M.S.; Karim, M.R.; Amin, N. Influence of Annealing Temperature on the Properties of ZnO Thin Films Grown by Sputtering. *Energy Procedia* **2012**, *25*, 55–61. [CrossRef]

26. Gardeniers, J.G.E.; Rittersma, Z.M.; Burger, G.J. Preferred orientation and piezoelectricity in sputtered ZnO films. *J. Appl. Phys.* **1998**, *83*, 7844–7854. [CrossRef]

27. Gandhi, A.C.; Pant, J.; Pandit, S.D.; Dalimbkar, S.K.; Chan, T.-S.; Cheng, C.-L.; Ma, Y.-R.; Wu, S.Y. Short-Range Magnon Excitation in NiO Nanoparticles. *J. Phys. Chem. C* **2013**, *117*, 18666–18674. [CrossRef]

28. Wojdyr, M. Fityk: A General-purpose Peak Fitting Program. *J. Appl. Cryst.* **2010**, *43*, 1126–1128. [CrossRef]

29. Lupan, O.; Chow, L.; Chai, G.; Heinrich, H. Fabrication and characterization of Zn–ZnO core–shell microspheres from nanorods. *Chem. Phys. Lett.* **2008**, *465*, 249–253. [CrossRef]

30. Ellmer, K. Magnetron sputtering of transparent conductive zinc oxide: Relation between the sputtering parameters and the electronic properties. *J. Phys. D Appl. Phys.* **2000**, *33*, R17–R32. [CrossRef]

31. Hong, R.; Qi, H.; Huang, J.; He, H.; Fan, Z.; Shao, J. Influence of Oxygen Partial Pressure on the Structure and Photoluminescence of Direct Current Reactive Magnetron Sputtering ZnO Thin Films. *Thin Solid Films* **2005**, *473*, 58–62. [CrossRef]

32. Von Dreele, R.B.; Larson, A.C. *General Structure Analysis System (GSAS), Los Alamos National Laboratory Report LAUR 86–748*; Los Alamos National Laboratory: Los Alamos, NM, USA, 2000; p. 221.

33. Huang, K.; Rhys, A. Theory of light absorption and non-radiative transitions in F-centres. *Proc. R. Soc. Lond. Ser. A Math. Phys. Sci.* **1950**, *204*, 406–423. [CrossRef]

34. Zhang, X.T.; Liu, Y.C.; Zhi, Z.Z.; Zhang, J.Y.; Lu, Y.M.; Shen, D.Z.; Xu, W.; Zhong, G.Z.; Fan, X.W.; Kong, X.G. Resonant Raman scattering and photoluminescence from high-quality nanocrystalline ZnO thin films prepared by thermal oxidation of ZnS thin films. *J. Phys. D Appl. Phys.* **2001**, *34*, 3430. [CrossRef]

35. Hsieh, P.T.; Chen, Y.C.; Wang, C.M.; Tsai, Y.Z.; Hu, C.C. Structural and photoluminescence characteristics of ZnO films by room temperature sputtering and rapid thermal annealing process. *Appl. Phys. A Mater. Sci. Process.* **2006**, *84*, 345–349. [CrossRef]

36. Deng, B.; da Rosa, A.L.; Frauenheim, T.; Xiao, J.P.; Shi, X.Q.; Zhang, R.Q.; Van Hove, M.A. Oxygen vacancy diffusion in bare ZnO nanowires. *Nanoscale* **2014**, *6*, 11882–11886. [CrossRef] [PubMed]

37. Kelley, A.M. Electron-Phonon Coupling in CdSe Nanocrystals. *J. Phys. Chem. Lett.* **2010**, *1*, 1296–1300. [CrossRef]

 *nanomaterials*

Article

Notable Enhancement of Phase Transition Performance and Luminous Transmittance in VO_2 Films via Simple Method of Ar/O Plasma Post-Treatment

Jingcheng Jin [1,2,*], Dongping Zhang [1,*], Xiaonan Qin [1], Yu Yang [1], Ying Huang [1], Huan Guan [1], Qicong He [1], Ping Fan [1] and Weizhong Lv [3]

1   Shenzhen Key Laboratory of Advanced Thin Films and Applications, College of Physics and Energy, Shenzhen University, Shenzhen 518060, China; 2150100110@email.szu.edu.cn (X.Q.); 2160100104@email.szu.edu.cn (Y.Y.); 2160100101@email.szu.edu.cn (Y.H.); 2170218820@email.szu.edu.cn (H.G.); hcq13923407098@hotmail.com (Q.H.); fanping@szu.edu.cn (P.F.)
2   Key Laboratory of Optoelectronic Devices and Systems of Ministry of Education and Guangdong Province, College of Optoelectronic Engineering, Shenzhen University, Shenzhen 518060, China
3   College of Chemistry and Environmental Engineering, Shenzhen University, Shenzhen 518060, China; lvwzh@szu.edu.cn
*   Correspondence: jingchengjin@szu.edu.cn (J.J.); zdpsiom@szu.edu.cn (D.Z.)

Received: 16 December 2018; Accepted: 10 January 2019; Published: 16 January 2019

Abstract: Ar/O plasma irradiation is proposed for post-treatment of vanadium dioxide (VO_2) films. Oxidation and surface migration were observed in the VO_2 films following irradiation. This combined effect leads to an effective stoichiometry refinement and microstructure reconstruction in the interfacial area. A notable improvement in luminous transmittance and an enhancement in phase transition performance of the treated VO_2 films were achieved. Compared with that of as-deposited VO_2 films, the electrical phase transition amplitude of treated films increased more than two-fold. The relative improvement in luminous transmittance (380–780 nm) is 47.4% (from 25.1% to 37%) and the increase in solar transmittance is 66.9% (from 29.9% to 49.9%), which is comparable to or better than the previous work using anti-reflection (AR) coatings or doping methods. The interfacial boundary state proved to be crucial and Ar/O plasma irradiation offers an effective approach for further refinement of thermochromic VO_2 films.

Keywords: vanadium dioxide; post-treatment; plasma irradiation; luminous transmittance; phase transition performance

---

1. Introduction

Vanadium dioxide (VO_2) undergoes a reversible metal–insulator transition (MIT) near temperatures of 68 °C [1]. During this phase transition, many of its physical properties change significantly resulting in abrupt changes in near-infrared (IR) transmittance. At high temperatures ($>T_c$), the film blocks IR radiation whereas at low temperatures ($<T_c$), it is allowed to pass. This unique characteristic of the phase transition is employed to regulate the solar heat flux in a manner dependent on the ambient temperature, thereby making VO_2 a very promising material for energy-efficient, smart-window applications [2–4].

Much effort has been devoted to improving the thermochromic properties of VO_2 films and limit global energy consumption in a warming climate. For the successful application of VO_2-based energy-efficient smart windows, the requirements for high luminous transmittance and phase-transition performance must be met simultaneously. High luminous transmittance and notable IR adjustment capability are two opposing aspects in VO_2 films. To fabricate VO_2 films with satisfactory luminous transmittance and phase transition performance is a challenging task [5,6]. Previously, to improve its visible transmittance, much of the work was focused on coating an anti-reflection (AR) layer on a VO_2 surface or doping another low-absorption element with VO_2 [7–10]. These strategies were either complicated or failed to achieve the desired phase-transition performance. In VO_2-based smart-windows research, annealing became a conventional post-treatment method that permitted remarkable modifications of the film stoichiometry as well as surface morphology [11,12]. However, in some instances, the high annealing temperatures were disastrous in fabricating toughened glass and flexible substrates [13].

In this paper, we present a simple method involving room-temperature Ar/O plasma irradiation post-treatment, which not only notable improves the visible transmittance of VO_2 films, but also enhances its thermochromic performance. High-energy plasma irradiation is a commonly used technique to modify material surfaces. Nonetheless, Ar/O plasma irradiation for post-treatment has not been applied to thermochromic VO_2 films. The phase-transition characteristics of VO_2 are sensitive to the film's stoichiometry, crystal quality, defect content, and boundary state [14–17]. The top-surface layer of the VO_2 films bridges light transmission between film and media, and differences in boundary states of the composition and microstructure creates an anisotropy in interfacial admittance and electron mobility. Insufficient attention has been given to this interfacial state modification because controlling the interfacial configuration is difficult even during the optimized physical vapor deposition (PVD) process.

The present work aims to assess the effect of Ar/O plasma irradiation around neighboring areas of treated VO_2 film surfaces and the corresponding impact on its MIT properties. The solo Ar and oxygen plasma have been systematically used in earlier experiments, and the better potential of property improvement was found when using the Ar/O plasma. The ratios of Ar and oxygen gas in the plasma were also systematically optimized based on earlier results. The VO_2 films are deposited by reactive direct current (DC) magnetron sputtering and irradiated with various levels of Ar/O plasma power. We provide the results demonstrating the notable enhancement in luminous transmittance and phase-transition performance from plasma irradiation. Favorable results are achieved after properly controlling the process. Moreover, we show that electrical phase-transition amplitude increases more than two-fold while nearly preserving the thermal hysteresis width and phase-transition temperature.

2. Experimental Details

VO_2 thin films with the thickness of 70 nm were deposited by DC reactive magnetron sputtering on K9 glass (D30 × 3 mm). The vacuum chamber was pumped down to a base pressure of 6×10^{-4} Pa, and the substrate temperature was maintained at 480 °C. The gas flow of Ar and O_2 were 40 sccm and 2 sccm, respectively, and the working pressure was maintained around 0.5 Pa during the deposition.

The Ar/O plasma irradiation was performed in a vacuum chamber system equipped with a microwave ion source and the working frequency was 2.45 GHz (Alpha Plasma Q150). Base pressure of the vacuum system was 1.5 Pa, and the typical working temperature was 30 °C. During the plasma irradiation, the flow of Ar and O_2 was set as 50 sccm and 100 sccm, respectively, the irradiation power range varied from 350 W to 550 W, and this lasted 30 min for each round.

The crystalline phases were identified by X-ray diffractometer (UltimaIV, Rigaku, Tokyo, Japan) with θ-2θ coupled scanning mode (CuKα radiation with a wavelength of 0.15418 nm). The surface roughness was evaluated by Atomic force microscopy (AFM) (EasyScan2, Nanosurf, Liestal, Switzerland) with sampling range 2 μm × 2 μm. A field emission scanning electron microscope (Supra 55Sapphire, SIGMA Essential, Jena, Germany) was used to examine the surface morphology of the samples. Horiba Scientific XploRA PLUS Raman spectrum equipped with a 532 nm laser at an output power of 2 mW was used for structural properties analysis. The transmittance spectra were obtained by a spectrophotometer (Lambda 950, PE, Boston, USA) at 25 °C and 70 °C with a wavelength range of 250-2500 nm, respectively. Temperature dependence of sheet resistances were measured by a four-probe method and the temperature range was adjusted as 25–80 °C for both heating and cooling loops.

3. Results and Discussion

3.1. Structure Characteristic

Figure 1a shows the X-ray diffraction (XRD) patterns of an as-deposited VO_2 film and treated VO_2 films. An intense peak located at $2\theta = 27.86°$ and two weak peaks located at $2\theta = 55.53°$, $57.56°$ are observed in all samples. They correspond to the (011), (220), and (022) planes of VO_2 (M) (PDF#82-0661). This indicates that the as-deposited VO_2 layers exhibit a highly crystallized monoclinic phase with a preferential growth in the (011) lattice orientation. However, an obvious diffraction peak at $2\theta = 20.13°$ (encircled by shaded ellipse) appears and becomes stronger as the plasma power increases. Its presence reveals that (001) V_2O_5 appears when the irradiation power is sufficiently high (>450 W in this instance).

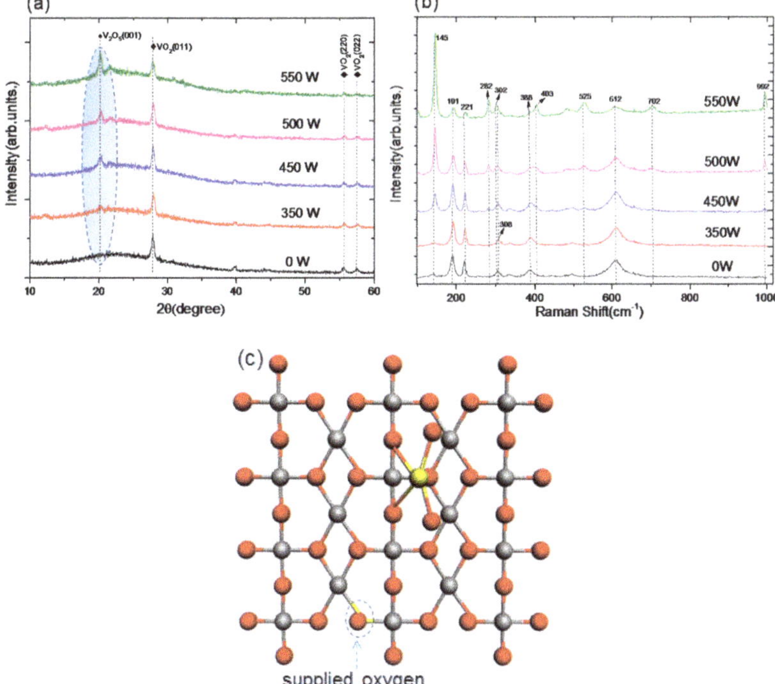

Figure 1. (a) X-ray diffraction (XRD) patterns of VO_2 films irradiated with Ar/O plasma at different powers; (b) Raman spectra of Ar/O plasma irradiated VO_2 films with different powers; (c) top-view of the VO_2 cluster model showing the binding sites, the silver and red spheres represent V(+4) and oxygen atoms, and the solitary yellow sphere a V(+5) atom.

The Raman spectra of VO_2 films bombarded with Ar/O plasma (Figure 1b) show obvious Raman bands at 191, 221, 308, 388, and 612 cm⁻¹, which are associated with the monoclinic phase of VO_2 (M), all five bands corresponding to the A_g symmetry mode. The peaks at 308, 388, and 612 cm⁻¹ correspond to the V–O vibration modes whereas the pronounced peaks of 191 and 221 cm⁻¹ are assigned to the V–V vibration modes and become stronger as the power increases in the range 350–450 W. The implication is that the content of VO_2 (M) increases from oxygen deficiency that occurs during Ar/O plasma irradiation. The new Raman peaks at 145, 282, 525, and 992 cm^{-1} belong to the V = O vibration mode in V_2O_5 that appears noticeably after plasma irradiation, with the increase in intensity as the bombarding power is increased [18]. The results are also in good agreement with the XRD analysis. A top-view of the model (Figure 1c) shows the binding of the clusters occurring in the VO_2 interface area. Oxygen atoms (red spheres) may be supplied and inserted in the oxygen-vacant sites (marked by an arrow) in the VO_2 clusters during Ar/O plasma irradiation, thereby raising the VO_2 concentration. This may be accompanied by some additional content of V (+5) atoms (light yellow sphere) binding onto a nearby area on the surface when the Ar/O plasma irradiation power is high enough.

3.2. Surface Morphology

The surface morphologies of VO_2 films after Ar/O plasma irradiation exhibit significant differences compared with as-deposited films. From the scanning electron microscope (SEM) images (Figure 2) flake-like nanoparticles are seen distributed evenly on the untreated VO_2 surfaces. Small rice-like nanoparticles at first appear on the surface and grow (350 W and 450 W in Figure 2); they then gather into rod-like nanoparticles as the plasma power continues to increase up to 500 W. The rod-like shape and number of crystalline grains of VO_2 films have significantly augmented after Ar/O plasma irradiation at 550 W.

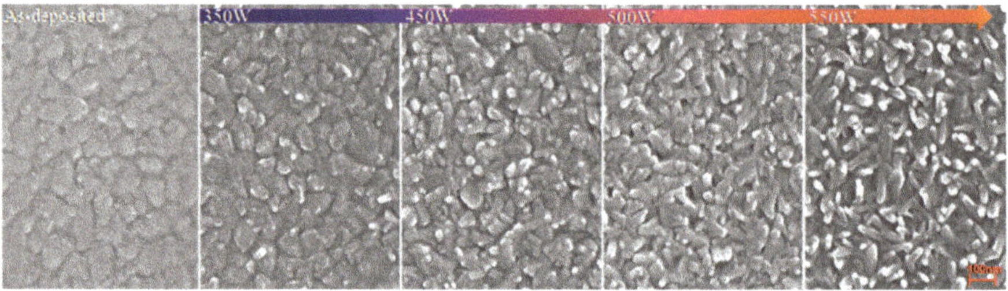

Figure 2. Scanning electron microscope (SEM) images of the surface morphologies inVO_2 films irradiated with Ar/O plasma at different powers.

A comparison between as-deposited VO_2 films and films irradiated with Ar/O plasma power of 500 W of the typical morphology (Figure 3) highlights the corresponding accumulative changes on the VO_2 surface during the irradiation process. The surface roughness is evaluated by determining the average root-mean-square (RMS) roughness and the peak–valley (PV) value obtained from atomic force microscopy (AFM) images. The underlying microstructure has undergone a conversion by the bombardment with the AFM images clearly revealing a nanoporous morphology (Figure 3b) changing into rod-like clustering (Figure 3f). Moreover, a rougher and more bumpy surface forms (Figure 3g) producing a RMS roughness and PV value of 7.6 nm and 58.2 nm, respectively. Both quantities increase sharply as the power rises because of the newly formed rod-like clusters, as evident in the SEM images (Figure 2). A SEM micrograph of a vertical section (Figure 3e) reveals that a change in boundary state results from recrystallization and surface migration in the underlying area between the top surface and the bulk of the VO_2 film, with a penetrating depth of about 20 nm.

The irradiation of the surface causes oxidation because the energy of the co-existing Ar/O ions is high. The Ar$^+$ ions colliding with the VO_2 molecules help break the chemical bonds, and at the same time, the oxygen atoms/ions (such as O$^-$, O$_2^-$) react with the V atoms to fill oxygen vacancy sites or cause further oxidation. A schematic of the interaction in a VO_2 film during Ar/O plasma irradiation is shown in Figure 3d; a schematic enlargement (panel I) shows the crystalline structure of monoclinic VO_2 (M) in an untreated VO_2 film during irradiation. Either the oxygen vacancies are filled or further oxidization takes place in parts of the structure to change VO_2 (M) into orthorhombicV_2O_5 (dotted circle). In addition, the Ar$^+$ ion serves as an extra source of energy that may be transferred to the atomic groups (blue spheres) and induce further surface migration or crystallization near the surface area (panel II). The clusters of interfacial groups gather and reshape the surface to create new functionality. Insufficient attention has been given to this interfacial state modification because controlling the interfacial configuration is difficult during the physical vapor deposition process. Combining these two aspects, the microstructure as well as the composition in the interfacial area is reconstructed and modulated simultaneously.

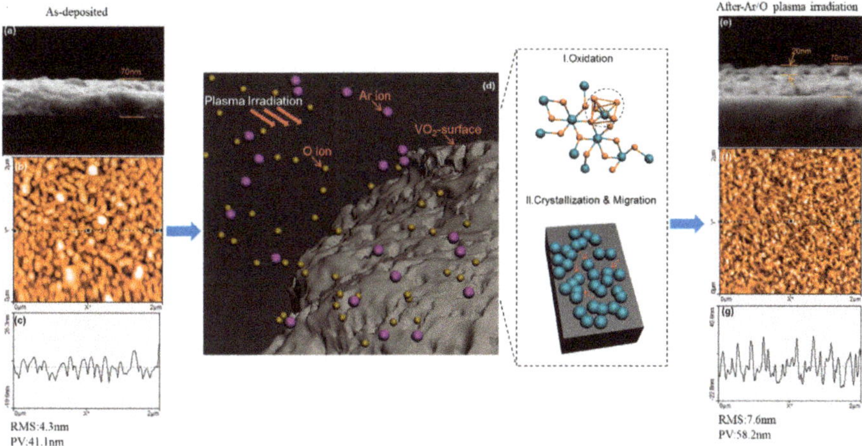

Figure 3. Schematic of Ar/O plasma irradiation and corresponding effect on the VO$_2$ film surface (**d**); Cross-section SEM images of as-deposited VO$_2$ films (**a**) and films irradiated with Ar/O plasma (500 W) (**e**); atomic force microscopy (AFM) images and line profiles of untreated VO$_2$ films (**b**,**c**) and films irradiated with Ar/O plasma (500 W) (**f**,**g**).

3.3. Thermochromic Property

Using the four-point probe technique, temperature-dependent sheet resistances were measured and plotted (Figure 4). The VO$_2$ films exhibit semiconductor characteristics with a sheet resistance of about 10^4–10^5 Ω/square at room temperature, and a sharp decrease in sheet resistance is observed when increasing temperature. The ratio of the film sheet resistance before and after the phase transition determines the film's electrical phase-transition amplitude ΔA. The average transition temperature in the heating and cooling process is defined as the film's phase transition temperature T_{MI}. The power dependence of the thermal hysteresis width ΔH and the phase transition temperature T_{MI} (Figure 5b) were obtained from standard Gaussian fitting methods by taking the derivative of the logarithmic sheet resistance with respect to the temperature of the VO$_2$ films during heating and cooling cycles. Both ΔH and T_{MI} exhibit variations within the temperature ranges 9–10 °C and 53.5–54.5 °C, respectively; both were stable within the limits of error uncertainties.

From Figure 5a, ΔA of the samples obviously increases after irradiation by more than two-fold that of as-deposited VO$_2$ films (from 2.19 × 10^5 to (5–6) × 10^5 in the range 450–550 W). In the SEM images, more boundary discontinuities between the top surface and bulk films were generated after plasma irradiation. With a reduced electron mobility, the sheet resistance of the dielectric states increases. However, the variation in composition of the VO$_2$ film dominates the resistance of the metal state, altering oxygen deficiency sites or further oxidizing the VO$_2$ film during plasma irradiation, as confirmed in the XRD and Raman spectra. Thus, the enhancement in the electrical phase transition amplitudes is mostly attributable to details of the stoichiometry as well as the morphology changes in the VO$_2$ surface area [19].

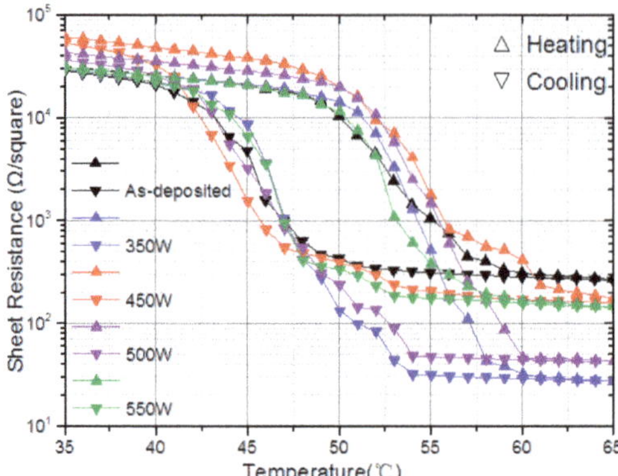

Figure 4. Temperature dependence of sheet-resistance in VO$_2$ films treated with different Ar/O plasma powers.

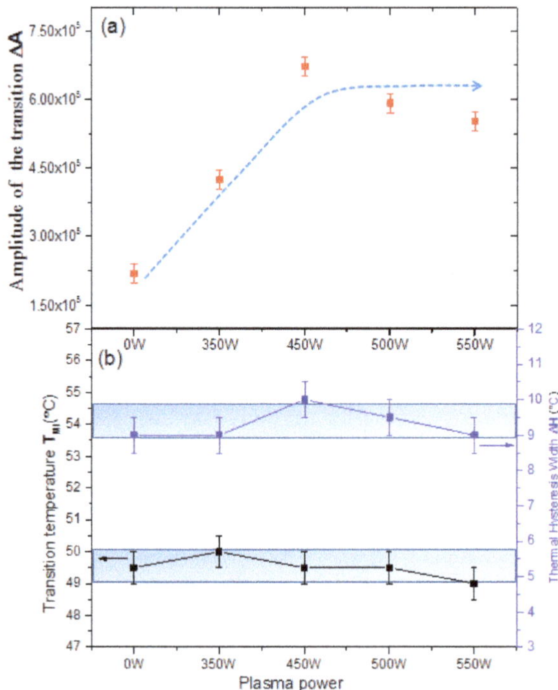

Figure 5. (**a**) Amplitude of transition ΔA and (**b**) phase transition temperature T$_{MI}$ and thermal hysteresis width ΔH of VO$_2$ films irradiated with Ar/O plasma of different powers.

3.4. Optical Performance

The optical transmittance of as-deposited VO$_2$ films and treated film were measured at 25 °C and 70 °C using a spectrophotometer for both insulating and metallic states (Figure 6a). The transmittance spectra for the plasma-treated VO$_2$ films in both the insulating and metallic states show sharp increases (marked by an arrow). In the insulating state, the peak transmittance in the visible range (380 nm–780 nm) increased continuously from 40% to 65%. In the infrared range, the transmittance also improved remarkably, and is only achieved with the extreme-wide-range AR coatings [20].

The transmittance of infrared solar energy and the IR-switching characteristic are calculated using the following formulae,

$$T_{sol} = \int \varphi_{sol}(\lambda)T(\lambda)d(\lambda) / \int \varphi_{sol}(\lambda)d\lambda \tag{1}$$

$$\Delta T_{sol} = T_{sol}\ (25^\circ C) - T_{sol}\ (70^\circ C) \tag{2}$$

where $T(\lambda)$ is the transmittance of wavelength λ, $\varphi_{IR\text{-}sol}(\lambda)$ is the solar irradiation spectrum for AM1.5 in the infrared range (AM1.5 is the global standard spectrum corresponding to the Sun positioned at 37° above the horizon). The solar transmission T_{vis} is also equivalent to the standard luminous transmission T_{lum} for photopic vision in human eyes.

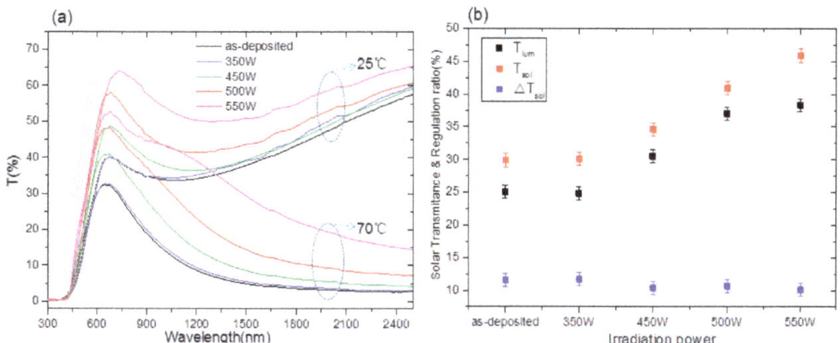

Figure 6. (**a**) The optical transmittance spectra of VO_2 films bombarded with Ar/O plasma were measured at 25 °C and 70 °C using a spectrophotometer for both insulating and metallic states; (**b**) solar transmittance of T_{lum}, T_{sol}, and solar regulation ratio ΔT_{sol} in VO_2 films after Ar/O plasma irradiation.

From the integral transmittance results (Figure 6b), T_{lum} (black square) and T_{sol} (red square) of plasma-treated VO_2 films increase continuously from 25.1% to 38.3% and 29.9% to 45.9%, respectively, as irradiated power increases up to 550 W. The variation in the modulation ratio ΔT_{sol} (blue square) decreases slightly by about one percent at higher power. However, for VO_2 (M) film, a calculation shows that it is difficult to achieve a sufficiently high transmittance while simultaneously preserving ΔT_{sol} [21]. Nevertheless, favorable results are found in the 500W group, with a relative improvement in T_{lum} (380–780 nm) to 47.4% (from 25.1% to 37%) and an increase in T_{sol} of 66.9% (from 29.9% to 49.9%) with a negligible decrease in ΔT_{sol} from 11.6% to 10.7% compared with untreated films. The high absorptance of oxygen vacancy defects can be reversed after Ar/O plasma irradiation and the rod-like clusters are reconstructed, thereby forming a new lower packing density and moth-eye like interface that may act similarly as an 'anti-reflection coating' [22]. Significant improvement has been realized by our simple method of Ar/O plasma post-treatment, which proved to be an efficient alternative solution compared with the former works that used AR coatings or doping methods [5–10,23–25].

4. Conclusions

In general, Ar/O plasma irradiation serves as an extra energy source that promotes oxidation and surface migration of VO_2 molecular groups prompting micro-structure reconstruction at the VO_2 interface. Clear evidence of enhancement in both phase transition performance and luminous transmittance has been realized by appropriately controlling the Ar/O plasma irradiation process. The long-term goal of this work is energy savings from smart-window optimization particularly suitable for building glass or plastic substrates that cannot sustain the high temperatures present in the annealing process. Opportunely, this universal method may also be extended to other situations for which low-temperature, non-destructive, effective solutions are required for further refinements of VO_2 films.

Author Contributions: X.Q., Y.Y., Y.H., H.G. and Q.H. performed the experimental activities and contributed the characterization sections of the paper. J.J., D.Z., P.F. and W.L. reviewed the experimental data, made the analysis and writing revisions to the paper.

Funding: This research was funded by NSFC grant number [61505202], Shenzhen Basic Research Project grant number [JCYJ20150529164656098, ZDSY20170228105421966], Shenzhen Key Lab Fund grant number [ZDSYS20170228105421966] and the Science and Technology plan project Fund of Guangdong Province grant number [2016B090930011].

Conflicts of Interest: The authors declare no conflict of interest.

References

1. Morin, F.J. Oxides which show a metal-to-insulator transition at the neel temperature. *Phys. Rev. Lett.* **1959**, *3*, 34–36. [CrossRef]

2. Gea, L.A.; Boatner, L.A. Optical switching of coherent VO_2 precipitates formed in sapphire by ion implantation and annealing. *Appl. Phys. Lett.* **1996**, *68*, 3081–3083. [CrossRef]

3. Gao, Y.; Luo, H.; Zhang, Z.; Kang, L.; Chen, Z.; Du, J.; Kanehira, M.; Cao, C. Nanoceramic VO_2 thermochromic smart glass: A review on progress in solution processing. *Nano Energy* **2012**, *1*, 221–246. [CrossRef]

4. Huang, Z.; Chen, S.; Lv, C.; Huang, Y.; Lai, J. Infrared characteristics of VO_2 thin films for smart window and laser protection applications. *Appl. Phys. Lett.* **2012**, *101*, 191905. [CrossRef]

5. Sun, G.; Cao, X.; Gao, X.; Long, S.; Liang, M.; Jin, P. Structure and enhanced thermochromic performance of low-temperature fabricated VO_2/V_2O_3 thin film. *Appl. Phys. Lett.* **2016**, *109*, 34. [CrossRef]

6. Zheng, J.; Bao, S.; Jin, P. $TiO_2(R)/VO_2(M)/TiO_2(A)$ multilayer film as smart window: Combination of energy-saving, antifogging and self-cleaning functions. *Nano Energy* **2015**, *11*, 136–145. [CrossRef]

7. Long, S.; Zhou, H.; Bao, S.; Xin, Y.; Cao, X.; Jin, P. Thermochromic multilayer films of $WO_3/VO_2/WO_3$ sandwich structure with enhanced luminous transmittance and durability. *RSC Adv.* **2016**, *6*, 106435–106442. [CrossRef]

8. Tian, R.; Sun, J.; Qi, Y.; Zhang, B.; Guo, S.; Zhao, M. Influence of VO_2 Nanoparticle Morphology on the Colorimetric Assay of H_2O_2 and Glucose. *Nanomaterials* **2017**, *7*, 347. [CrossRef] [PubMed]

9. Hao, Q.; Li, W.; Xu, H.; Wang, J.; Yin, Y.; Wang, H.; Ma, L.; Ma, F.; Jiang, X.; Schmidt, O.G.; Chu, P.K. VO_2/tin plasmonic thermochromic smart coatings for room-temperature applications. *Adv. Mater.* **2018**, *30*, 1705421. [CrossRef]

10. Liu, H.; Wan, D.; Ishaq, A.; Chen, L.; Guo, B.; Shi, S.; Luo, H.; Gao, Y. Sputtering deposition of sandwich-structured V_2O_5/metal (v, w)/V_2O_5 multilayers for the preparation of high-performance thermally sensitive VO_2 thin films with selectivity of VO_2 (b) and VO_2 (m) polymorph. *Appl. Mater. Interfaces* **2016**, *8*, 7884–7890. [CrossRef]

11. Öksüzoğlu, R.M.; Bilgiç, P.; Yildirim, M.; Deniz, O. Influence of post-annealing on electrical, structural and optical properties of vanadium oxide thin films. *Opt. Laser. Technol.* **2013**, *48*, 102–109. [CrossRef]

12. Zimmers, A.; Aigouy, L.; Mortier, M.; Sharoni, A.; Wang, S.; West, K.G.; Ramirez, J.G.; Schuller, I.K. Role of thermal heating on the voltage induced insulator-metal transition in VO_2. *Phys. Rev. Lett.* **2013**, *110*, 056601. [CrossRef] [PubMed]

13. Jung, D.H.; So, H.S.; Ahn, J.S.; Lee, H.; Nguyen, T.T.T.; Yoon, S.; Kim, S.Y.; Jung, H.Y. Low temperature growth of amorphous VO_2 films on flexible polyimide substrates with a TiO_2 buffer layer. *J. Vac. Sci. Technol. A* **2018**, *36*, 3E102. [CrossRef]

14. Brassard, D.; Fourmaux, S.; Jean-Jacques, M.; Kieffer, J.C.; El Khakani, M.A. Grain size effect on the semiconductor-metal phase transition characteristics of magnetron-sputtered VO_2 thin films. *Appl. Phys. Lett.* **2005**, *87*, 34. [CrossRef]

15. Vernardou, D.; Pemble, M.E.; Sheel, D.W. In-situ Fourier transform infrared spectroscopy gas phase studies of vanadium (IV) oxide coating by atmospheric pressure chemical vapour deposition using vanadyl (IV) acetylacetonate. *Thin Solid Films* **2008**, *516*, 4502–4507. [CrossRef]

16. Cui, Y.J.; Ramanathan, S. Substrate effects on metal-insulator transition characteristics of rf-sputtered epitaxial VO_2 thin films. *J. Vac. Sci. Technol. A* **2011**, *29*, 041502. [CrossRef]

17. Louloudakis, D.; Vernardou, D.; Spanakis, E.; Katsarakis, N.; Koudoumas, E. Thermochromic Vanadium Oxide Coatings Grown by APCVD at Low Temperatures. *Phys. Procedia* **2013**, *46*, 137–141. [CrossRef]

18. Huang, Y.; Zhang, D.; Liu, Y.; Jin, J.; Yang, Y.; Chen, T.; Guan, H.; Fan, P.; Lv, W. Phase transition analysis of thermochromic VO_2 thin films by temperature-dependent Raman scattering and ellipsometry. *Appl. Surf. Sci.* **2018**, *456*, 545–551. [CrossRef]

19. Srivastava, A.; Herng, T.S.; Saha, S.; Nina, B.; Annadi, A.; Naomi, N.; Liu, Z.Q.; Dhar, S.; Ariando Ding, J.; Venkatesan, T. Coherently coupled ZnO and VO_2 interface studied by photoluminescence and electrical transport across a phase transition. *Appl. Phys. Lett.* **2012**, *100*, 241907. [CrossRef]

20. Perl, E.E.; Mcmahon, W.E.; Farrell, R.M.; Denbaars, S.P.; Speck, J.S.; Bowers, J.E. Surface structured optical coatings with near-perfect broadband and wide-angle antireflective properties. *Nano Lett.* **2014**, *14*, 5960–5964. [CrossRef]

21. Li, S.Y.; Niklasson, G.A.; Granqvist, C.G. Nanothermochromics: Calculations for VO_2 nanoparticles in dielectric hosts show much improved luminous transmittance and solar energy transmittance modulation. *J. Appl. Phys.* **2010**, *108*, 063525. [CrossRef]

22. Min, W.L.; Jiang, B.; Peng, J. Bioinspired self-cleaning antireflection coatings. *Adv. Mater.* **2010**, *20*, 3914–3918. [CrossRef]

23. Gagaoudakis, E.; Kortidis, I.; Michail, G.; Tsagaraki, K.; Binas, V.; Kiriakidis, G.; Aperathitis, E. Study of low temperature rf-sputtered mg-doped vanadium dioxide thermochromic films deposited on low-emissivity substrates. *Thin Solid Films* **2016**, *601*, 99–105. [CrossRef]

24. Chang, T.; Cao, X.; Jin, P. Atomic layer deposition of VO_2 thin films, largely enhanced luminous transmittance, solar modulation. *Appl. Mater. Interfaces* **2018**, *10*, 26814–26817. [CrossRef]

25. Melnik, V.; Khatsevych, I.; Kladko, V.; Kuchuk, A.; Nikirin, V.; Romanyuk, B. Low-temperature method for thermochromic high ordered VO_2 phase formation. *Mater. Lett.* **2012**, *68*, 215–217. [CrossRef]

 *nanomaterials*

Article

Ultra-Thin Platinum Deposits by Surface-Limited Redox Replacement of Tellurium

Fatima Haidar [1,2], Mathieu Maas [1,2], Andrea Piarristeguy [1], Annie Pradel [1], Sara Cavaliere [2] and Marie-Christine Record [1,3,*]

[1] Institute Charles Gerhardt of Montpellier, UMR CNRS 5253, Chalcogenide Materials and Glasses, University of Montpellier, F-34095 Montpellier CEDEX 5, France; fatima.haidar@umontpellier.fr (F.H.); mathieu.maas@umontpellier.fr (M.M.); andrea.piarristeguy@umontpellier.fr (A.P.); annie.pradel@umontpellier.fr (A.P.)

[2] Institute Charles Gerhardt of Montpellier, UMR CNRS 5253, Aggregates Interfaces and Materials for Energy, University of Montpellier, F-34095 Montpellier CEDEX 5, France; sara.cavaliere@umontpellier.fr

[3] Aix-Marseille University, University of Toulon, CNRS, IM2NP, av. Normandie-Niemen, F-13013 Marseille, France

* Correspondence: m-c.record@univ-amu.fr; Tel.: +33-491-288-313

Received: 15 September 2018; Accepted: 12 October 2018; Published: 15 October 2018

Abstract: Platinum is the most employed electrocatalyst for the reactions taking place in energy converters, such as the oxygen reduction reaction in proton exchange membrane fuel cells, despite being a very low abundant element in the earth's crust and thus extremely expensive. The search for more active electrocatalysts with ultra-low Pt loading is thus a very active field of investigation. Here, surface-limited redox replacement (SLRR) that utilizes the monolayer-limited nature of underpotential deposition (UPD) was used to prepare ultrathin deposits of Pt, using Te as sacrificial metal. Cyclic voltammetry and anodic potentiodynamic scanning experiments have been performed to determine the optimal deposition conditions. Physicochemical and electrochemical characterization of the deposited Pt was carried out. The deposit comprises a series of contiguous Pt islands that form along the grain interfaces of the Au substrate. The electrochemical surface area (ECSA) of the Pt deposit obtained after 5 replacements, estimated to be 18 m^2/g, is in agreement with the ECSA of extended surface catalysts on flat surfaces.

Keywords: Pt thin deposits; galvanic displacement; UPD; SLRR; electrocatalysis

---

1. Introduction

Platinum is the most employed electrocatalyst for the reactions taking place in energy converters, such as proton exchange membrane fuel cells, in particular for the oxygen reduction reaction (ORR) taking place at the cathode side [1–3]. To enhance its surface area and reactivity, it is often used in the form of nanoparticles. Nevertheless, a more effective utilization of platinum is crucial, due to its low abundance in the earth's crust and its high price [4]. Despite their advantages and proven electroactivity, the search for more active electrocatalysts with ultra-low Pt loading highlights several limitations of Pt nanoparticles [5]. Indeed, their ORR activity is related to particle size and the type of exposed crystal planes [6]. Moreover, the limited interface between discrete nanoparticles and the support means there is a tendency towards migration and agglomeration during fuel cell operations, with a consequent loss of performance [7]. The development of nanostructured thin films [8] with excellent activity and stability demonstrates the potential of Pt extended surfaces. Their conformal morphology minimizes unused platinum, allowing for high ORR-specific activity [9] while preventing electrochemical Ostwald ripening [10]. The preparation of Pt thin films to achieve ultra-low Pt loaded electrodes with high performance and stability is a challenge however, due to the complexity of several fabrication techniques and the difficulty in tailoring thickness and continuity of the deposits [11–13]. For this purpose, electrochemical fabrication methods are often considered, due to their ease of use and overall cost effectiveness [14–16]. Conventional overpotential electrodeposition results in preferential growth at areas with high surface energy, such as step edges and surface defects, leading to incomplete substrate coverage at low thicknesses and non-uniform deposits [17,18]. Surface-limited redox replacement (SLRR) utilizes the monolayer-limited nature of underpotential deposition (UPD) to form an atomic layer of a sacrificial metal, coupled with the galvanic displacement of this sacrificial metal by the deposited metal [19]. This can be repeated over multiple cycles in order to allow for the growth of films of desired thickness [20]. To date, SLRR has been utilized to deposit ultrathin conformal films of Pt using Pb and Cu [21–25] as sacrificial metals. This led to incomplete deposition due to the different oxidation state of the metal ions (two copper (II) or lead (II) exchanged for one platinum (IV)). In this regard, tellurium can be a valuable alternative, as already demonstrated, for the formation of Pt nanostructures further applied in electrocatalysis [26–28], since, based on the stoichiometric relationship of the galvanic replacement reaction (Equation (1)), equivalent molar amounts of Pt atoms are generated while Te ions dissolve in the solution.

$$Pt(IV)Cl_6{}^{2-} + Te(0) + 3H_2O \rightarrow Pt(0) + Te(IV)O_3{}^{2-} + 6Cl^- + 6H^+ \tag{1}$$

The goal of this work was to investigate the deposition of Pt films on a model surface, such as a flat gold substrate, by SLRR, using tellurium as sacrificial metal. After having optimized and maximized the tellurium deposition and the Pt replacement, the Pt layers obtained after various numbers of cycles were characterized by scanning electron microscopy (SEM), coupled with energy dispersive X-ray spectroscopy (EDS), atomic force microscopy (AFM), inductively coupled plasma-mass spectrometry (ICP-MS), and cyclic voltammetry.

2. Materials and Methods

2.1. Solutions and Chemicals

All chemicals were of analytical grade and were used as received. Aqueous solutions were prepared using ACS reagents or higher-grade chemicals (Sigma Aldrich, Inc., St. Louis, MO, USA) and ultrapure water (Milli-Q > 18.2 MΩ cm, Merck KGaA, Darmstadt, Germany). The tellurium oxide solution was prepared with TeO_2 (99.999%) and $HClO_4$ as supporting electrolytes, with concentrations of 10^{-3} M and 0.4 M, respectively. Hexachloroplatinic acid hexahydrate H_2PtCl_6 $6H_2O$ (ACS reagent, $\geq$ 37.50% Pt basis), 10^{-3} M, and $HClO_4$ 0.1 M were used as reagent and supporting electrolytes, respectively. The supporting electrolyte $HClO_4$ 0.1 M was also used as a blank solution. The pH of these solutions was 1.0. Before each series of measurements, solutions were freshly prepared, magnetically stirred, and thoroughly out-gassed with high purity nitrogen gas (N_2) for 30 min, in order to flush out dissolved oxygen.

2.2. Electrochemical Studies

All electrochemical experiments were performed using a Solartron analytical Modulab workstation and carried out using an electrochemical flow cell with three electrodes. An Ag/AgCl (3 M NaCl) served as the reference electrode (AMETEK, Inc., Berwyn, PA, USA) and a platinum wire (Sigma Aldrich, Inc., St. Louis, MO, USA) was used as the counter electrode. The working electrode was a gold rotating disk electrode (RDE) (PINE instruments, Grove City, PA, USA). For the ICP-MS analysis, a melted gold working electrode had been prepared and kept inside a resin, and the electrical connection was assured with a copper wire. For the SEM-EDS and AFM measurements, gold substrates (PHASIS, Inc., Plan-les-Ouates, Geneva, Switzerland) consisting of quartz slides coated with 200 nm thick gold films (99.9%, pure), were used as working electrodes. Under the Au film, a thin (10 nm) Ti under-layer was deposited to improve the adhesion of gold onto quartz.

All electrochemical measurements were performed at room temperature ($\approx$ 25 °C). All the potentials in this paper are referring to the reversible hydrogen electrode (RHE). Pt electrochemical surface area (ECSA) was evaluated using the hydrogen underpotential deposition region by integrating the charge from 0.05 to 0.4 V vs. RHE, after the subtraction of the double layer capacitance and assuming a charge density of 210 $\mu C/cm^2$ for polycrystalline Pt [29].

2.2.1. Preparation of the Gold Electrode

Before the deposition, the gold electrode was polished mechanically with diamond powder (1 μm, 0.3 μm and 0.05 μm), then ultrasonicated in water for 5 min while the gold substrate was pretreated as follows: it was first annealed at 350 °C for 18 h at 10^{-6} Torr in sealed glass tubes and then soaked in hot nitric acid for 5 min.

The electrodes were then treated electrochemically using cyclic voltammetry (CV) in a blank solution. To clean the working electrode, 40 cycles from -0.30 V to 1.65 V vs. RHE with a scan rate of 100 mV/s were performed until a reproducible voltammogram was obtained (Figure S1, Supporting Information, SI).

2.2.2. Underpotential Deposition (UPD) of Te and Te Replacement by Pt

The electrochemical behavior of Te on bare gold and on platinum electrodes was studied from a solution of TeO_2 (Sigma Aldrich 99.99% purity, Inc., St. Louis, MO, USA) 1 mM in $HClO_4$ 0.1 M, using cyclic voltammetry with various potential windows. The underpotential deposition (UPD) was carried out in the potentiostatic mode. Following the Te underpotential deposition on a gold electrode, the Te covered electrodes were rinsed in a blank solution of 0.1 M $HClO_4$ and dipped into a 1 mM H_2PtCl_6 $6H_2O$ + 0.1 M $HClO_4$ solution at open circuit voltage (OCV), in order to form a Pt monolayer via surface-limited redox replacement of tellurium. The Te deposition and its replacement by Pt were progressively optimized, using results from cyclic voltammetry and anodic potentiodynamic scanning.

2.3. Characterization of the Pt Deposits

Several methods were used to provide evidence for and to quantify the presence of Pt on the electrodes. First, X-ray diffraction (XRD) patterns after 5 cycles of Te UPD and Pt replacement were recorded in a Bragg-Brentano configuration using a PANalytical X'Pert diffractometer (PANalytical, Almelo, The Netherlands). The deposited layers were then observed using (i) an Oxford X-Max $50mm^2$ instrument (Oxford Instruments, Abingdon, Oxfordshire, England) equipped with energy-dispersive X-ray spectroscopy analysis (EDS), (ii) and a Bruker Nanoscope Dimension 3100 atomic force microscope (AFM, Bruker, Palaiseau, France). The AFM experiments were performed in the tapping mode under ambient conditions. The cantilever tips were Silicon Point Probe Plus® NCSTR (Nanosensors, Neuchâtel, Switzerland; force constant 6.5 N/m, resonance frequency 157 kHz). All the image treatments were performed using Gwyddion [Czech Metrology Institute, Brno, Czech Republic, open software, http://gwyddion.net/]. Inductively coupled plasma-mass spectrometry (ICP-MS) analysis was used to quantify the amount of platinum deposited. In this case, the deposited films were dissolved in 10 mL of *aqua regia* and the obtained solution was analyzed. In order to obtain a sufficient amount of Pt for ICP-MS analyses, several cycles (3 and 5) of Te UPD and Pt replacement were performed.

Finally, cyclic voltammetry in nitrogen saturated $HClO_4$ 0.1 M was carried out at a scan rate of 40 mV/s, to observe hydrogen UPD on Pt and evaluate from it its electrochemical surface area (ECSA).

3. Results and Discussion

3.1. Underpotential Deposition of Tellurium on Gold

Shown in Figure 1 is the cyclic voltammogram (CV) of the Au electrode immersed in tellurium oxide solution, obtained after several consecutive voltammetry cycles, successively scanned from 1.2 V to various potential limits in the range 0–1.2 V vs. RHE at a scan rate of 40 mV/s. The shape of the CV is comparable to that recorded during underpotential deposition of Te on gold in sulfuric acid solutions by Suggs and Stickney [30].

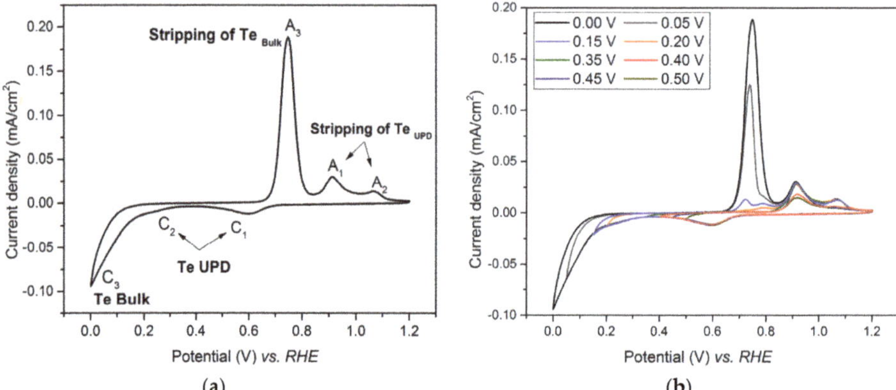

Figure 1. Cyclic voltammetry (CV) of the gold substrate in a solution of 0.4 M $HClO_4$ and 1 mM TeO_2 (pH 1.0) at a potential sweep rate of 40 mV/s. (**a**) CV obtained by scanning the potential from 1.20 V to 0.00V vs. RHE; (**b**) CV obtained by scanning the potential from 1.20 V to progressively lower potential limits (0.50 V, 0.45 V, 0.40 V, 0.35 V, 0.20 V, 0.15 V, 0.05 V and 0.00 V vs. RHE).

The reduction of TeO_2 started at 0.7 V vs. RHE (peak C_1), in agreement with previous work attributing this reduction to the first tellurium UPD peak [31–34]. The second peak (C_2), occurring at the lower potential of 0.4 V vs. RHE, has also been already reported [32,33] and is referred to as the second tellurium UPD, which is only partially resolved from the bulk reduction peak. It is known that the UPD peak shape and potential strongly depend on the crystallographic plane on which it occurs [35–37]. This can explain the slight shift between the UPD potential observed in this work, and those reported in the literature [38].

Detailed CV analysis was performed to identify the location of the two UPD peaks and the peak corresponding to bulk deposition by progressively extending the scanned potential region. The cathodic potential was scanned from 1.2 V vs. RHE to various lower limit values (between 0.50 V and 0.00 V vs. RHE), from which the anodic curves were recorded. The corresponding CVs are displayed in Figure 1b.

A reduction peak C1 and an oxidation peak A1 are seen when the potential was scanned between 1.2 V and 0.5 V vs. RHE. When the potential limit decreases, the intensity of the A1 peak increases and reaches its maximum for a limit value of 0.45 V. This observation suggests that C1 corresponds to a UPD peak for Te, and A1 to its corresponding stripping peak [39]. When the limiting value of the potential further decreases, an additional oxidative peak, A2, is observed. Its intensity increases when the potential limit of the scan decreases, and reaches its maximum when the limit potential is 0.35 V. This suggests the presence of a second UPD peak, C2. When the limit potential value further decreases, an additional reductive peak, C3, appears as well as its corresponding oxidative peak, A3. The intensity of the A1 peak increases as the limit potential value decreases, while the intensity of the peaks A2 and A3 remains constant. The peak C3 should be related to the bulk deposition. From these observations we conclude that the total UPD of Te on gold is reached by using a deposition potential of 0.35 V vs. RHE.

The linear sweep voltammetry (LSV) after UPD Te deposition (after polarization at 0.35 V vs. RHE) was carried out for different durations (Figure S2, Supporting Information, SI). The deposition time was thus optimized from these experiments, and 240 s was found to be the minimum time required to achieve a complete UPD. Beyond this duration, the stripping curves overlaid, indicating that the maximum intensity of the stripping peak had been reached.

The [I-t] curve (Figure S3, Supporting Information, SI) recorded during the deposition at 0.35 V vs. RHE shows that the signal stabilized after 240 s, which is in agreement with the LSV study.

In order to obtain the maximum amount of Te by UPD, and compare with the values reported above (0.35 V and 240 s), a longer deposition duration (300 s) was adopted.

A longer deposition duration (300 s) was adopted in order to obtain the maximum amount of deposited tellurium by UPD. In order to quantify the deposition in these conditions (0.35 V and 300 s), the deposited layer was stripped by applying a potential at 0.35 V and, using the electric charge related to the gold reduction (cf. Supplementary Information), the coverage of the electrode is estimated to be 0.87 monolayers (ML) for total UPD (C_1 and C_2) [36]. This demonstrates that the UPD deposition was successful and that an almost complete film of tellurium on the gold surface was obtained.

3.2. Investigation of Te UPD on Platinum Electrode

The aim here was to perform a multilayer deposition of platinum (by galvanic displacement) onto tellurium deposited by UPD. The possibility of further depositing tellurium onto the formed platinum layer and continuing the successive cycles of UPD/galvanic displacement on it needs to be assessed. For that, a study of Te UPD on a Pt flat electrode was performed and discussed. The same experimental method was used as was for the Au working electrode, but using a Pt electrode instead.

Before the deposition, the Pt electrode was polished mechanically and then treated electrochemically using cyclic voltammetry in a blank solution of 0.1 M perchloric acid. To clean the electrode surface, 40 cycles from −0.05 V to 1.4 V vs. RHE with a scan rate of 100 mV/s were performed until a reproducible voltammogram was obtained (Figure S4, SI).

Figure 2a shows cyclic voltammograms performed on Pt electrode in 1 mM of TeO_2 with a scan rate of 40 mV/s. While C_1/C_2 corresponds to the tellurium UPD, and the reduction of platinum oxide to Pt, and A_1 and A_2 corresponds to their stripping peaks, C_3 corresponds to the bulk reduction of Te and A_3 to its stripping peak.

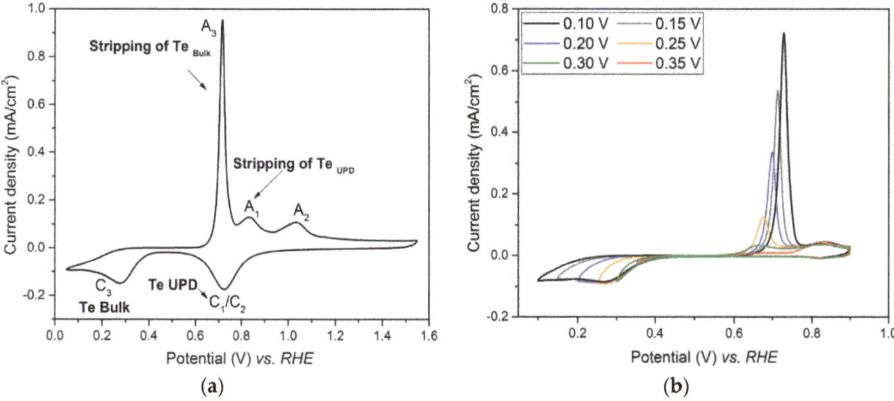

Figure 2. Cyclic voltammograms performed on Pt electrode in 1 mM TeO_2 + 0.4 M $HClO_4$ solution with a scan rate of 40 mV/s. (**a**) potential comprised between 1.60 V and 0.05 V; (**b**) potential from 0.9 V to various cathodic potential limits (0.35 V, 0.30 V, 0.25 V, 0.25 V, 0.15 V and 0.10 V).

Since an overlap occurs between the Te UPD peak, C_1/C_2, and the peak of platinum oxide reduction, the investigation of the Te UPD imposed to avoid high anodic potential (such as 1.6 V vs. RHE as in Figure 2a), which led to the formation of the oxidized species. This can be achieved by limiting the upper potential range to 0.9V vs. RHE, where the platinum oxidation does not occur

(Figure 2b). Indeed, when the potential is scanned from 0.9 V to 0.35 V vs. RHE, only A_1 appears on the anodic curve. When the potential limit is below 0.35 V, the bulk A_3 peak appears. Thus, the underpotential deposition of tellurium on platinum has to be performed between 0.35 V and 0.30 V vs. RHE. The experimental conditions for Te under potential deposition were chosen as E = 0.35 V and t = 450 s. The voltammogram recorded in the range 0.05 V–1.55 V vs. RHE after Te UPD at 0.35 V for 450 s is shown in Figure 3 (full line plot) and compared to that obtained for the Pt electrode in 0.1 M $HClO_4$ (dotted line). The tellurium deposit totally inhibits H adsorption on the Pt surface, as observed in the potential range between 0.40 and 0.05 V vs. RHE.

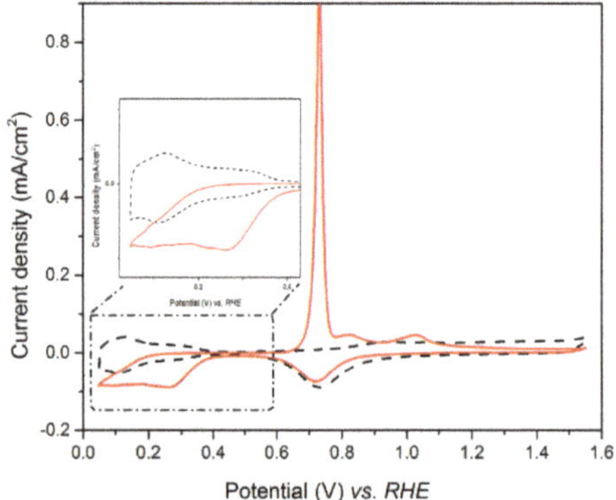

Figure 3. Steady-state voltammetric response of the Pt electrode in 0.1 M $HClO_4$ (dotted line) and first cycle profile after Pt electrode polarization in 1 mM TeO_2 for 450 s at 0.35 V vs. RHE (full line), sweep rate = 40 mV/s.

Using the electric charge related to the platinum reduction, the Te coverage of the electrode is estimated to be 0.86 ML [40]. The UPD deposition of Te on Pt is thus effective.

3.3. Te Replacement by Pt

Following the Te underpotential deposition on bare gold, the Te film was galvanically exchanged with platinum via SLRR. The atomic layer of Te was then oxidized while $PtCl_6^{2-}$ ions were reduced, replacing the Te layer by a Pt layer. The time needed to obtain the highest replacement rate was been determined from the I-t curve (Figure 4), and is 450 s.

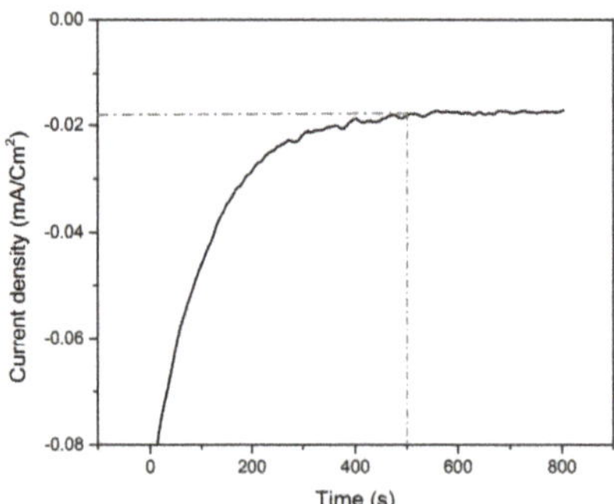

Figure 4. Chronoamperometric curve of the galvanic replacement of Te by Pt at open circuit voltage (OCV).

3.4. Underpotential Deposition of Te onto Pt-Covered Gold Electrode and Subsequent Te Replacements by Pt

The voltammograms obtained on a Pt-covered gold electrode immersed in Te solution after various numbers of replacements (from 1 to 5) are presented in Figure 5. The Au electrode was scanned from 0.45 V to 0.05 V vs. RHE and then up to 0.87 V vs. RHE only, in order to avoid the oxidation of the Pt layer obtained by the previous replacements. One can observe that the intensity of the Te UPD peak increases with the number of replacements and that its potential (0.35 V vs. RHE) is similar to that determined on a platinum bulk electrode (Section 3.2). This latter result is not surprising, since the electrical resistivity of the Pt layer should be very low.

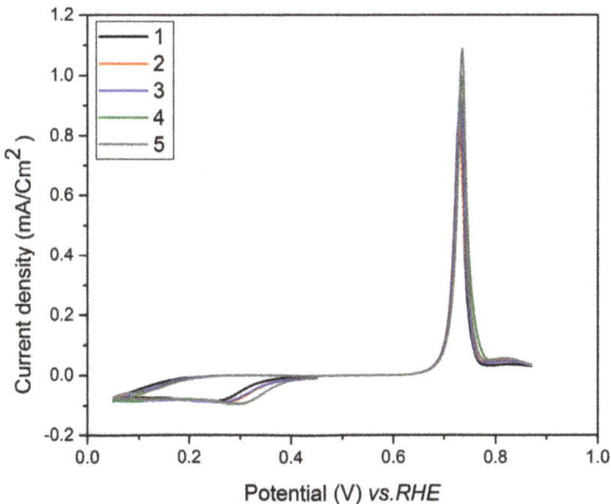

Figure 5. Cyclic voltammograms on Pt-covered Au electrode in 1 mM TeO$_2$ with a scan rate of 40 mV/s (after the number of replacements indicated in the figure).

The I-t curves (Figure 6) recorded during the successive replacements of Te with Pt show that irrespective of the number of replacements, the signal stabilized after 450 s. All these results allow us to define the following experimental conditions: E = 0.35 V and t = 300 s for the Te underpotential deposition; E = OCV and t = 450 s for the Pt replacement.

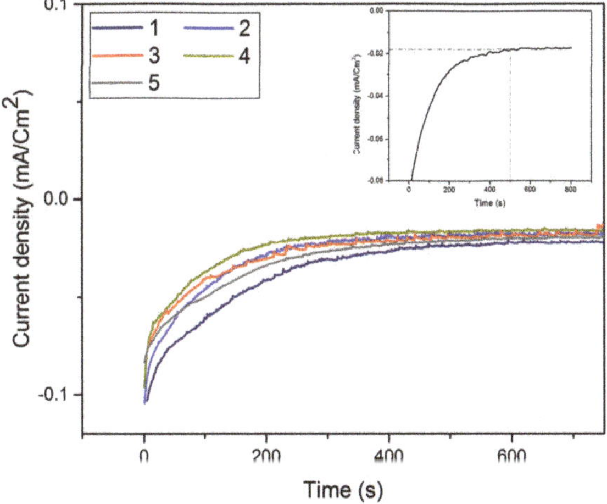

Figure 6. I-t curves recorded at OCV during the Pt replacement of Te layer for various numbers of replacements.

3.5. Characterization of the Pt Film

The success of the Pt deposition was difficult to assess, due to (i) the ultra-low amount formed on the electrode, which was extremely difficult to characterize, (ii) the impossibility of observing the Pt deposit on the gold rotating disk electrode (RDE) directly. RDE was replaced by a 200 nm gold coated quartz slide in a series of experiments of Te deposition and Pt replacement. The Au/quartz slides were then cut in small pieces that could fit the chambers of the different characterization instruments. No information could be obtained from X-ray diffraction, in agreement with the low amount of Pt involved (diffractograms not shown here). Figure 7 shows the SEM image and EDS mappings of the layer after 5 cycles of Te deposition and Pt replacement. They show the presence of Pt distributed over the whole surface of the support. No Te could be detected, which demonstrates its effective replacement by Pt (see the EDS spectra in Figure S5, Supporting Information, SI). The morphology of the layer analyzed by AFM is shown in Figure 8, along with the morphology of the substrate. The latter comprises gold grains of different shapes and dimensions. While the substrate is not completely covered, it appears that the Pt tends to deposit at the gold grain interfaces easier than on the grain surface, leading to a series of contiguous platinum islands formed along the Au-grain interfaces.

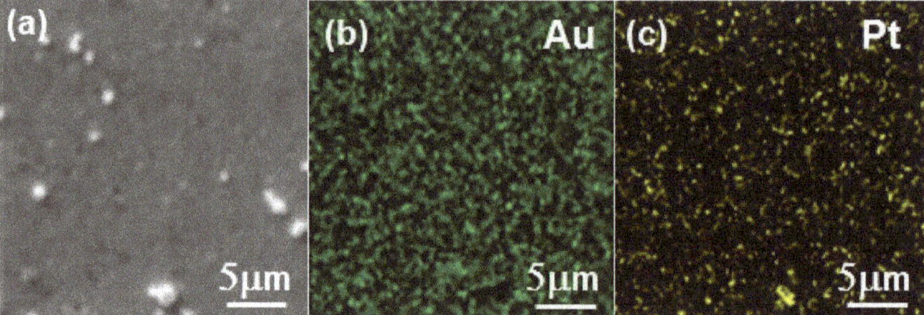

Figure 7. SEM image (**a**) and energy dispersive X-ray spectroscopy (EDS) mapping showing (**b**) Au and (**c**) Pt chemical contrasts for the Pt layer deposited on the gold-coated glass slide after 5 cycles of Te deposition and Pt replacement.

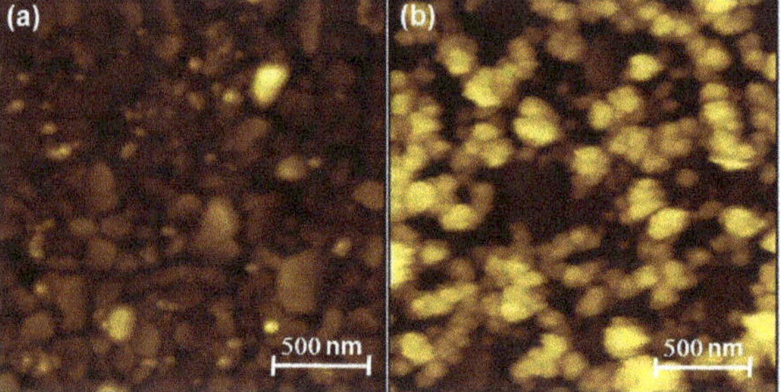

Figure 8. Atomic force microscopy (AFM) micrographs of (**a**) a bare Au-coated glass slide and (**b**) Pt deposit on Au-coated glass slide.

The quantity of deposited Pt has been determined by ICP-MS analysis after 3 and 5 replacement cycles. Taking into account the concentration given by the analyses (0.696 ppb and 3.339 ppb for 3 and 5 cycles, respectively), we conclude that 0.20 ML of Pt were deposited onto the electrode for 3 Te replacements and 0.52 ML for five ones. No trace of tellurium was detected in the analyzed solutions. This allows us to believe that the deposited tellurium has been replaced.

Figure 9 presents a cyclic voltammogram recorded in the range 0.05–1.2 V vs. RHE after five galvanic replacements on the Au electrode. The CV demonstrated the successful deposition of Pt, exhibiting the three characteristic regions of this metal: the hydrogen adsorption/desorption peaks in the potential range between 0.05–0.35 V vs. RHE, the double layer capacitance, and the Pt oxide formation with the corresponding reduction [41]. It should be noticed that the CV presents a certain distortion and asymmetry, probably due to resistive effects, because of the ultra-low amount of Pt and thus difficult electrical connection of the electrode.

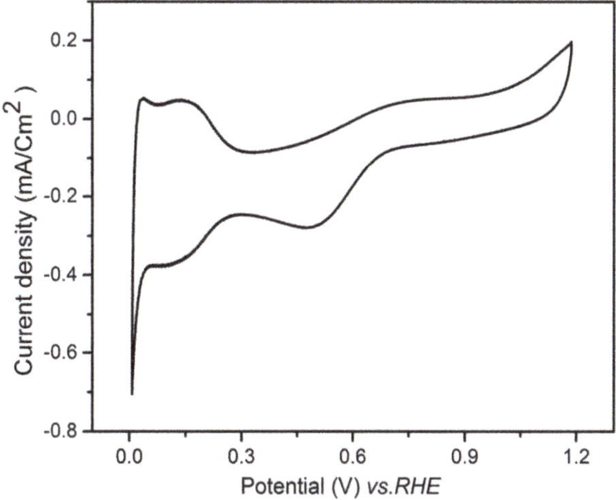

Figure 9. Cyclic voltammogram recorded in 0.1 M $HClO_4$ on Au/Te electrode after five galvanic replacements with Pt.

The electrochemical surface area (ECSA) of the Pt film obtained after 5 replacements was estimated to be 18 m^2/g. This value is in agreement with the ECSA of extended surface catalysts on flat surfaces that in general is lower than that of the corresponding nanoparticles, but gives rise to higher specific activities [42].

The aim of the present work was to investigate the feasibility of ultra-thin Pt deposits by successive cycles of tellurium underpotential deposition and its subsequent replacement with platinum by galvanic displacement. To date, up to five cycles have been achieved, which allowed us to demonstrate the successful replacement of Te by platinum. While the coverage is not complete, platinum deposits along the Au-substrate grains in the form of contiguous islands, a feature that is looked for, since sintering should be energetically unfavourable and the oxide-dissolution mechanism leading to Pt loss should also be reduced. The low ECSA argues for flat Pt islands, again a feature that is looked for, since it should give rise to enhanced specific activities.

On the whole, these first results are encouraging and pave the way to the deposition of ultra-thin Pt layers for utilization as catalysts in electrochemical energy conversion devices. To reach the final goal of a high mass activity, combining high specific activity (presence of active sites) and high ECSA (presence of available surface), the deposition of thin films here reported for a model surface will be extended to nanostructured supports. The mechanical properties such the elastic modulus and the residual stress will then be studied, since they might affect the adhesion of the deposit on the substrate, a major concern when dealing with flexible supports or even free-standing films [43,44].

4. Conclusions

In this work, we studied the deposition of Pt atomic layers onto gold substrate. The deposits were performed by successive cycles, each one consisting in Te underpotential deposition and subsequent Pt replacement of Te by galvanic displacement. The optimal conditions for the underpotential deposition of Te (E = 0.35 V and t = 300 s) on gold and platinum electrodes as well as those for the Pt replacement of Te (OCV for 450 ms) were determined using cyclic voltammetry and anodic potentiodynamic scanning.

The deposited Pt was characterized by scanning electron microscopy coupled with energy dispersive X-ray spectroscopy, atomic force microscopy, inductively coupled plasma-mass spectrometry and cyclic voltammetry. It demonstrated a deposition from 0.2 to 0.5 monolayers of the noble metal, in the form of Pt contiguous islands that form at the substrate Au-grain interfaces preferentially.

The surface-limited redox replacement applied in this work on a model flat gold surface demonstrated to be an effective and straightforward method to obtain platinum ultra-thin deposits. The next step will be its application on electrocatalyst supports used in proton exchange membrane fuel cells, to further validate the possibility of tuning the thickness and ensuring a conformal morphology on more complex structures.

Supplementary Materials: The following are available online at http://www.mdpi.com/2079-4991/8/10/836/s1. Figure S1: Cyclic voltammogram of the Au electrode in 0.1 M HClO4 at a scan rate of 100 mV/s, Figure S2: Linear sweep voltammetry after tellurium UPD deposition for various times of deposition at 0.35 V vs. RHE. (a) peak A1; (b) peak A2, Figure S3: Cyclic voltammogram of the Pt electrode in 0.1 M HClO4 at a scan rate of 100 mV/s, Figure S4: Cyclic voltammogram of the Pt electrode in 0.1 M HClO4 at a scan rate of 100 mV/s, Figure S5: EDS spectra of (a) bare Au-coated glass slide and (b) Pt deposit on Au-coated glass slide. Red arrows indicate the expected position of Te peaks, showing the absence of the element in the layer.

Author Contributions: Formal Analysis, F.H., S.C., A.P. (Annie Pradel), M.-C.R., A.A.P. (Andrea Piarristeguy); Investigation, F.H., M.M.; Writing—Original Draft Preparation, F.H.; Writing—Review & Editing, S.C., A.P. (Annie Pradel), M.-C.R.; Supervision, S.C., A.P. (Annie Pradel), M.-C.R.; Funding Acquisition, S.C.

Funding: This research was funded by the European Research Council under the European Union's Seventh Framework Program (FP/2007-2013)/ERC Grant Agreement No. 306682 (SPINAM).

Acknowledgments: One of the authors (S.C.) acknowledges the support of the French IUF.

Conflicts of Interest: The authors declare no conflict of interest.

References

1. Adzic, R.R.; Zhang, J.; Sasaki, K.; Vukmirovic, M.B.; Shao, M.; Wang, J.X.; Nilekar, A.U.; Mavrikakis, M.; Valerio, J.A.; Uribe, F. Platinum monolayer fuel cell electrocatalysts. *Top. Catal.* **2007**, *46*, 249–262. [CrossRef]

2. Van Der Vliet, D.; Wang, C.; Debe, M.; Atanasoski, R.; Markovic, N.M.; Stamenkovic, V.R. Platinum-alloy nanostructured thin film catalysts for the oxygen reduction reaction. *Electrochim. Acta* **2011**, *56*, 8695–8699. [CrossRef]

3. Zhang, J.; Mo, Y.; Vukmirovic, M.B.; Klie, R.; Sasaki, K.; Adzic, R.R. Platinum Monolayer Electrocatalysts for O_2 Reduction: Pt Monolayer on Pd(111) and on Carbon-Supported Pd Nanoparticles. *J. Phys. Chem. B* **2004**, *108*, 10955–10964. [CrossRef]

4. Adzic, R.R. Platinum monolayer electrocatalysts: Tunable activity, stability, and self-healing properties. *Electrocatalysis* **2012**, *3*, 163–169. [CrossRef]

5. Ercolano, G.; Cavaliere, S.; Rozière, J.; Jones, D.J. Recent developments in electrocatalyst design thrifting noble metals in fuel cells. *Curr. Opin. Electrochem.* **2018**, *9*, 271–277. [CrossRef]

6. Lv, H.; Li, D.; Strmcnik, D.; Paulikas, A.P.; Markovic, N.M.; Stamenkovic, V.R. Recent advances in the design of tailored nanomaterials for efficient oxygen reduction reaction. *Nano Energy* **2016**, *29*, 149–165. [CrossRef]

7. Hartl, K.; Hanzlik, M.; Arenz, M. IL-TEM investigations on the degradation mechanism of Pt/C electrocatalysts with different carbon supports. *Energy Environ. Sci.* **2011**, *4*, 234–238. [CrossRef]

8. Debe, M.K. Electrocatalyst approaches and challenges for automotive fuel cells. *Nature* **2012**, *486*, 43–51. [CrossRef] [PubMed]

9. Papandrew, A.B.; Atkinson, R.W.; Goenaga, G.A.; Wilson, D.L.; Kocha, S.S.; Neyerlin, K.C.; Zack, J.W.; Pivovar, B.S.; Zawodzinski, T.A. Oxygen reduction activity of vapor-grown platinum nanotubes. *ECS Trans.* **2013**, *50*, 1397–1403. [CrossRef]

10. Stephens, I.E.L.; Bondarenko, A.S.; Grønbjerg, U.; Rossmeisl, J.; Chorkendorff, I. Understanding the electrocatalysis of oxygen reduction on platinum and its alloys. *Energy Environ. Sci.* **2012**, *5*, 6744–6762. [CrossRef]

11. Du, Q.; Wu, J.; Yang, H. Pt@Nb-TiO$_2$ catalyst membranes fabricated by electrospinning and atomic layer deposition. *ACS Catal.* **2014**, *4*, 144–151. [CrossRef]

12. Dilonardo, E.; Milella, A.; Cosma, P.; d'Agostino, R.; Palumbo, F. Plasma deposited electrocatalytic films with controlled content of Pt nanoclusters. *Plasma Process. Polym.* **2011**, *8*, 452–458. [CrossRef]

13. Hammer, B.; Norskov, J.K. Theoretical Surface Science and Catalysis—Calculations and Concepts. *Adv. Catal.* **2000**, *45*, 71–129. [CrossRef]

14. Ercolano, G.; Farina, F.; Cavaliere, S.; Jones, D.J.; Rozière, J. Towards ultrathin Pt films on nanofibres by surface-limited electrodeposition for electrocatalytic applications. *J. Mater. Chem. A* **2017**, *5*, 3974–3980. [CrossRef]

15. Alpuche-Aviles, M.A.; Farina, F.; Ercolano, G.; Subedi, P.; Cavaliere, S.; Jones, D.J.; Rozière, J. Electrodeposition of two-dimensional Pt nanostructures on highly oriented pyrolytic graphite (HOPG): The effect of evolved hydrogen and chloride ions. *Nanomaterials* **2018**, *8*, 668. [CrossRef]

16. Farina, F.; Ercolano, G.; Cavaliere, S.; Jones, D.J.; Rozi, J. Surface-Limited Electrodeposition of Continuous Platinum Networks on Highly Ordered Pyrolytic Graphite. *Nanomaterials* **2018**, *8*, 721. [CrossRef] [PubMed]

17. Batina, N.; Kolb, D.M.; Nichols, R.J. In Situ Scanning Tunneling Microscopy Study of the Initial Stages of Bulk Copper Deposition on Au(100): The Rim Effect. *Langmuir* **1992**, *8*, 2572–2576. [CrossRef]

18. Waibel, H.F.; Kleinert, M.; Kibler, L.A.; Kolb, D.M. Initial stages of Pt deposition on Au(111) and Au(100). *Electrochim. Acta* **2002**, *47*, 1461–1467. [CrossRef]

19. Brankovic, S.R.; Wang, J.X.; Adžić, R.R. Metal monolayer deposition by replacement of metal adlayers on electrode surfaces. *Surf. Sci.* **2001**, *474*, L173–L179. [CrossRef]

20. Mrozek, M.F.; Xie, Y.; Weaver, M.J. Surface-enhanced raman scattering on uniform platinum-group overlayers: Preparation by redox replacement of underpotential-deposited metals on gold. *Anal. Chem.* **2001**, *73*, 5953–5960. [CrossRef] [PubMed]

21. Kim, Y.; Kim, J.Y.; Vairavapandian, D.; Stickney, J.L. Platinum Nanofilm Formation by EC-ALE via Redox Replacement of UPD Copper: Studies Using in-Situ Scanning Tunneling Microscopy. *J. Phys. Chem. B* **2006**, *110*, 17998–18006. [CrossRef] [PubMed]

22. Ambrozik, S.; Dimitrov, N. The Deposition of Pt via Electroless Surface Limited Redox Replacement. *Electrochim. Acta* **2015**, *169*, 248–255. [CrossRef]

23. Podlovchenko, B.I.; Maksimov, Y.M.; Maslakov, K.I. Electrocatalytic properties of Au electrodes decorated with Pt submonolayers by galvanic displacement of copper adatoms. *Electrochim. Acta* **2014**, *130*, 351–360. [CrossRef]

24. Gokcen, D.; Bae, S.-E.; Brankovic, S.R. Reaction kinetics of metal deposition via surface limited red-ox replacement of underpotentially deposited metal monolayers. *Electrochim. Acta* **2011**, *56*, 5545–5553. [CrossRef]

25. Papaderakis, A.; Mintsouli, I.; Georgieva, J.; Sotiropoulos, S. Electrocatalysts Prepared by Galvanic Replacement. *Catalysts* **2017**, *7*, 80. [CrossRef]

26. Xu, L.; Liang, H.; Li, H.; Wang, K.; Yang, Y.; Song, L.; Wang, X.; Yu, S. Understanding the stability and reactivity of ultrathin tellurium nanowires in solution: An emerging platform for chemical transformation and material design. *Nano Res.* **2014**, *8*, 1081–1097. [CrossRef]

27. Liu, J.; Huang, W.; Gong, M.; Zhang, M.; Wang, J.; Zheng, J. Ultrathin Hetero-Nanowire-Based Flexible Electronics with Tunable Conductivity. *Adv. Mater.* **2013**, *25*, 5910–5915. [CrossRef] [PubMed]

28. Zuo, Y.; Wu, L.; Cai, K.; Li, T.; Yin, W.; Li, D.; Li, N.; Liu, J.; Han, H. Platinum Dendritic-Flowers Prepared by Tellurium Nanowires Exhibit High Electrocatalytic Activity for Glycerol Oxidation. *ACS Appl. Mater. Interfaces* **2015**, *7*, 17725–17730. [CrossRef] [PubMed]

29. Trasatti, S.; Petrii, O.A. Real surface area measurements in electrochemistry. *J. Electroanal. Chem.* **1992**, *327*, 353–376. [CrossRef]

30. Suggo, D.W.; Stickney, J.L. Studies of the structures formed by the alternated electrodeposition of atomic layers of Cd and Te on the low-index planes of Au. *Surf. Sci.* **1993**, *290*, 362–374. [CrossRef]

31. Andrews, R.W. Determination of tellurium(IV) in perchloric acid by stripping voltammetry with collection. *Anal. Chim. Acta* **1980**, *119*, 47–54. [CrossRef]

32. Mori, E.; Baker, C.K.; Reynolds, J.R.; Rajeshwar, K. Aqueous electrochemistry of tellurium at glassy carbon and gold. A combined voltammetry-oscillating quartz crystal microgravimetry study. *J. Electroanal. Chem.* **1988**, *252*, 441–451. [CrossRef]

33. Bravo, B.G.; Michelhaugh, S.L.; Soriaga, M.P.; Villegas, I.; Suggs, D.W.; Stickney, J.L. Anodic underpotential deposition and cathodic stripping of iodine at polycrystalline and single-crystal gold. Studies by LEED, AES, XPS, and electrochemistry. *J. Phys. Chem.* **1991**, *95*, 5245–5249. [CrossRef]

34. Posey, R.S.; Andrews, R.W. Stripping voltammetry of Tellurium (IV) in 0.1 M perchloric acid at rotating gold disk electrodes. *Anal. Chim. Acta* **1980**, *119*, 55–66. [CrossRef]

35. Solomun, T.; Schardt, B.C.; Rosasco, A.; Wieckoski, J.L.; Stickney, A.T.H. Studies of electrodeposition of Silver in an iodine-pretreated stepped surface: Pt(S)[6 (11 1) × (111)]. *J. Electroanal. Chem.* **1984**, *176*, 309–323. [CrossRef]

36. Gregory, B.W.; Norton, M.L.; Stickney, J.L. Thin-layer electrochemical studies of the underpotential deposition of cadmium and tellurium on polycrystalline Au, Pt and Cu electrodes. *J. Electroanal. Chem.* **1990**, *293*, 85–101. [CrossRef]

37. Stickney, J.L.; Rosasco, S.D.; Schardt, B.C.; Hubbard, A.T. Electrodeposition of silver onto platinum(100) surfaces containing iodineadlattices. Studies by low-energy electron diffraction, Auger spectroscopy, and thermal desorption. *J. Phys. Chem.* **1984**, *88*, 251–258. [CrossRef]

38. Zhu, W.; Yang, J.Y.; Zhou, D.X.; Bao, S.Q.; Fan, X.A.; Duan, X.K. Electrochemical characterization of the underpotential deposition of tellurium on Au electrode. *Electrochim. Acta* **2007**, *52*, 3660–3666. [CrossRef]

39. Sella, C.; Boncorps, P.; Vedel, J. The Electrodeposition Mechanism of CdTe from Acidic Aqueous Solutions. *J. Electrochem. Soc.* **1986**, *133*, 2043–2047. [CrossRef]

40. Zhou, W.P.; Kibler, L.A.; Kolb, D.M. Evidence for a change in valence state for tellurium adsorbed on a Pt (111) electrode. *Electrochim. Acta* **2002**, *47*, 4501–4510. [CrossRef]

41. Zhan, D.; Velmurugan, J.; Mirkin, M.V. Adsorption/Desorption of Hydrogen on Pt Nanoelectrodes: Evidence of Surface Diffusion and Spillover. *J. Am. Chem. Soc.* **2009**, *131*, 14756–14760. [CrossRef] [PubMed]

42. Alia, S.M.; Yan, Y.S.; Pivovar, B.S. Galvanic displacement as a route to highly active and durable extended surface electrocatalysts. *Catal. Sci. Technol.* **2014**, *4*, 3589–3600. [CrossRef]

43. Heremans, P.; Tripathi, A.K.; de Meux, A.d.J.; Smits, E.C.P.; Hou, B.; Pourtois, G.; Gelinck, G.H. Mechanical and Electronic Properties of Thin-Film Transistors on Plastic, and Their Integration in Flexible Electronic Applications. *Adv. Mater.* **2016**, *28*, 4266–4282. [CrossRef] [PubMed]

44. Ghidelli, M.; Sebastiani, M.; Collet, C.; Guillemet, R. Determination of the elastic moduli and residual stresses of freestanding Au-TiW bilayer thin films by nanoindentation. *Mater. Des.* **2016**, *106*, 436–445. [CrossRef]

 *nanomaterials*

Article

Surface-Limited Electrodeposition of Continuous Platinum Networks on Highly Ordered Pyrolytic Graphite

Filippo Farina, Giorgio Ercolano, Sara Cavaliere *, Deborah J. Jones and Jacques Rozière

Institute Charles Gerhardt Montpellier, UMR CNRS 5253, Aggregates Interfaces and Materials for Energy, University of Montpellier, 34095 Montpellier CEDEX 5, France; filippo.farina@umontpellier.fr (F.F.); giorgio.ercolano@umontpellier.fr (G.E.); deborah.jones@umontpellier.fr (D.J.J.); jacques.roziere@umontpellier.fr (J.R.)

* Correspondence: sara.cavaliere@umontpellier.fr; Tel.: +33-467-149-098; Fax: +33-467-143-304

Received: 20 August 2018; Accepted: 11 September 2018; Published: 13 September 2018

Abstract: Continuous thin platinum nanoplatelet networks and thin films were obtained on the flat surface of highly ordered pyrolytic graphite (HOPG) by high overpotential electrodeposition. By increasing the deposition time, the morphology of the Pt deposits can be progressively tuned from isolated nanoplatelets, interconnected nanostructures, and thin large flat islands. The deposition is surface-limited and the thickness of the deposits, equivalent to 5 to 12 Pt monolayers, is not time dependent. The presence of Pt (111) facets is confirmed by High Resolution Transmission Electron Microscopy (HRTEM) and evidence for the early formation of a platinum monolayer is provided by Scanning Transmission Electron Microscopy and Energy Dispersive X-rays Spectroscopy (STEM-EDX) and X-ray Photoelectron Spectroscopy (XPS) analysis. The electroactivity towards the oxygen reduction reaction of the 2D deposits is also assessed, demonstrating their great potential in energy conversion devices where ultra-low loading of Pt via extended surfaces is a reliable strategy.

Keywords: electrodeposition; platinum; highly oriented pyrolytic graphite; 2D growth; thin films

1. Introduction

Carbon-supported platinum electrocatalysts are employed in the electrodes of Proton Exchange Membrane Fuel Cells (PEMFCs), but due to the scarcity of platinum and its cost [1], efforts to reduce the noble metal loading have driven research towards the development of strategies maximising its utilisation [2], including metal@platinum core@shell nanoparticles [3], platinum nanostructures [4] and thin films [5]. Several methods such as Atomic Layer Deposition (ALD) [6,7], Pulsed Laser Deposition (PLD) [8], Surface-Limited Redox Replacement (SLRR) [9], and Magnetron Sputtering (MS) [10] allow the preparation of extended Pt surfaces on various supports including carbon [11].

Electrodeposition of Pt in galvanostatic and potentiostatic modes is conventionally characterized by the 3D growth of platinum, which forms spherical, or flower-like, agglomerates [12–14]. Achieving the formation of thin Pt structures via electrodeposition therefore represents a real challenge. We recently observed that pulsed electrodeposition at high overpotentials can produce crystalline nanoplatelets on carbonaceous surfaces. Thin 2D Pt nanoplatelets were grown on electrospun carbon fibres and on nanotubes developed from their surfaces, and the resulting materials were evaluated as electrocatalysts for the Oxygen Reduction Reaction (ORR) in the cathode of a PEMFC, demonstrating enhanced exploitation of the noble metal [15,16]. Such nanofibres and nanotubes do not lend themselves to in-depth investigations using advanced microscopies, and so to further our understanding of this high overpotential electrodeposition method we turned our attention to the model surface of Highly Ordered Pyrolytic Graphite (HOPG), and herein describe our findings.

The aim is to investigate the process occurring on a flat carbon surface at the nanoscale, leading to the formation of crystalline nanoplatelets, interconnected nanostructures, and thin films. To achieve deposition with a single nucleation step, followed by increasing Pt growth for better insights into the deposit formation process, the electrodeposition approach developed in our previous work was modified to consider only a single potential pulse of varying duration. The choice of HOPG as a model surface is due to its relative flatness as well as to its structural similarity to the graphitized regions of the surface of the carbon nanofibers and nanotubes that inspired this work [15,16]. Furthermore, the possibility of exfoliating HOPG to obtain graphene layers [17], which can be further catalysed with platinum, will open novel perspectives of applications.

While studies of platinum electrodeposition on various carbonaceous surfaces have been reported [18–20], it appears that the process has been investigated mainly for slight overpotential conditions and not at the very high overpotentials (i.e. in the region of strong H_2 evolution) that are the conditions chosen here. It was previously reported that interconnected Pt structures can be obtained by applying an overpotential of -500 mV vs. SCE (-259 SHE) [21,22]. Such a high overpotential could in principle force two-dimensional growth in through underpotential deposition of hydrogen (H_{upd}) on the Pt surface, which prevents or impedes diffusion of the Pt precursor towards the surface of the newly electrodeposited Pt layers [23]. Spontaneous deposition of Pt agglomerates, nanoparticles, and networks on HOPG was induced from water as well as non-aqueous polar solvents [24–26] and was explained in terms of the reduction of Pt by oxidized defect sites on the surface and step-edges [24]. However, the mechanism appears to be more complex, involving not only graphite step-edges, but also hydrogenated sites [27,28]. As a consequence, Pt nuclei may be already present before the application of the overpotential to the electrode [28].

In this paper, we investigate the effect of a high overpotential single-pulse length on the morphology and crystal structure of the Pt deposits on the flat model carbon surface HOPG for better insights on the deposition mechanism and 2D growth of extended metal surfaces using microscopy and surface analysis techniques.

2. Materials and Methods

Highly Oriented Pyrolytic Graphite (HOPG, ZYH grade, 12 mm × 12 mm from Veeco-Bruker, Camarillo, CA, USA) was used as the working electrode in a two-electrode cell, and a 5 cm × 5 cm graphite sheet (Goodfellow) was used as the counter-shorted reference electrode (RE/CE). The working electrode assembly was masked with Kapton® tape (RS Components SAS, Beauvais, France), leaving a 6 mm × 6 mm HOPG surface exposed to the precursor solution, and the HOPG back-plate was connected with a platinum wire to the potentiostat (Bio-Logic® SP300, Bio-Logic SAS, Seyssinet-Pariset, France). Electrodeposition experiments were performed at -3 V vs. RE/CE (measured to be equivalent in the same cell geometry to -1.9 V vs. Ag/AgCl) for various duration times (from 5 to 200 s), followed by 60 s at open circuit voltage (OCV). The deposition solution was a solution of 3 mM hexachloroplatinic acid hexahydrate ($\geq$99.9% trace metal basis) and 0.5 M sodium chloride in MilliQ water (18 MΩ cm). All chemicals were provided by Sigma-Aldrich, St. Louis, MO, USA. An iR correction was performed before each electrodeposition step with the typical electrical resistance being in the range 10–20 Ω; the distance between the electrodes was fixed at 5 cm. Before electrodeposition, the solution was cooled to ~4 °C in an ice bath and purged with nitrogen to remove the dissolved oxygen. During the deposition, the precursor solution was stirred at 900 rpm to remove the hydrogen bubbles that evolve on the surface of the HOPG and to limit the thickness of the precursor diffusion layer. Prior to any characterization, the deposited sample was thoroughly rinsed with MilliQ grade water. Tapping Mode Atomic Force Microscopy (TM-AFM) images were acquired with a Bruker Nanoman (Bruker SAS, Palaiseau, France) driven by Nanoscope 5 electronics. The cantilever tips were Silicon Point Probe Plus® NCSTR (Nanosensors, Neuchâtel, Switzerland; force constant 6.5 N/m, resonance frequency 157 kHz). All the image treatments and the thickness measurements were performed using WSxM 4.0 Beta 8.2 [29] and the resulting data were treated with the Fityk software [30] to correct the background.

Particle size measurements and Fast Fourier Transforms were performed with ImageJ 1.48 v (U. S. National Institutes of Health, Bethesda, MD, USA). High Resolution Transmission Electron Microscopy (HRTEM), Scanning Transmission Electron Microscopy (STEM) micrographs and Energy Dispersive X-rays Spectroscopy (EDX) data were obtained with a JEOL 2200FS (Source: FEG) microscope (JEOL Europe SAS, Croissy-sur-Seine, France) operating at 200 kV and equipped with a CCD camera Gatan USC (16 MP; Gatan, Evry, France). The samples were prepared by carefully peeling away flakes of HOPG and depositing them onto a Cu grid with a drop of silver paste. Scanning electron micrographs were acquired by Field Emission-Scanning Electron Microscopy (FE-SEM) using a Hitachi S-4800 microscope (Hitachi Europe SAS, Velizy, France). The surface composition of the samples was investigated by X-ray Photoelectron Spectroscopy (XPS) on an ESCALAB 250 (Thermo Electron, Villebon Sur Yvette, France). The X-ray excitation was provided by a monochromatic Al K_α (1486.6 eV) source, and the analysed surface area was 400 μm^2. A constant analyser energy mode was used for the electron detection (20 eV pass energy). Detection of the emitted photoelectrons was performed perpendicular to the surface sample. Data quantification was performed on the Avantage software (Thermo Fisher Scientific, Waltham, MA, USA), removing the background signal using the Shirley method. The surface atomic concentrations were determined from photoelectron peaks areas using the atomic sensitivity factors reported by Scofield. Binding energies of all core levels are referenced to the C–C bond of C 1s (284.8 eV). Linear sweep voltammetry (LSV) was performed at 20 mV s^{-1} in O_2 saturated 0.1 M $HClO_4$ on a Bio-Logic® SP300 potentiostat (see above); the working electrode was an HOPG-Pt sample connected to the potentiostat with a Pt wire, while the counter electrode was a graphite rod (Alfa Aesar, Heysam, UK) and reference used was an Ag/AgCl electrode (Fisher Scientific, Illkirch, France; E = 0.197 V vs. NHE). The potentials in the manuscript were reported vs. NHE.

3. Results and Discussion

3.1. TM-AFM Analysis

Tapping Mode Atomic Force Microscopy (TM-AFM) images of electrodeposited Pt samples prepared with different pulse durations are shown in Figure 1, together with their thickness and diameter distributions.

Only separate Pt nanoplatelets were visible for a 5 s deposition time, while thin Pt networks started to clearly form after a 10 s deposition, and progressively covered the surface of the HOPG substrate up to 40 s and, at 200 s, thin large flat Pt islands were fully formed. Larger agglomerates of Pt were also present along all the samples, whatever the pulse duration, possibly formed either during the open circuit voltage (OCV) step of the electrodeposition process or as a result of H_2-induced precipitation [23]. The Pt morphologies resulting from the application of this overpotential method in the range 5–40 s appear to be partially similar to previous reports, but they clearly show that the substrate surface is progressive covered with a network of interconnected flat Pt nanostructures of low thickness [21,22]. An increase of the degree of interconnection of the Pt nanoplatelets is clearly visible with an increase in deposition time. A general trend can be derived for the sample series, however one might take into account a certain intra-sample inhomogeneity.

As shown in Figure 2a and Table 1, the thickness of the Pt nanoplatelets only slightly changed with deposition time. Most of the nanoplatelets did not grow over 4 nm in thickness even at longer deposition times (Table 1), with the average being comprised between 1.0 and 2.5 nm (equivalent to 5 to 12 Pt monolayers). Pt deposits with comparable thickness and diameter were reported for slight overpotential conditions and shorter times as well, but there was no evidence of interconnected or extended surface structures [31].

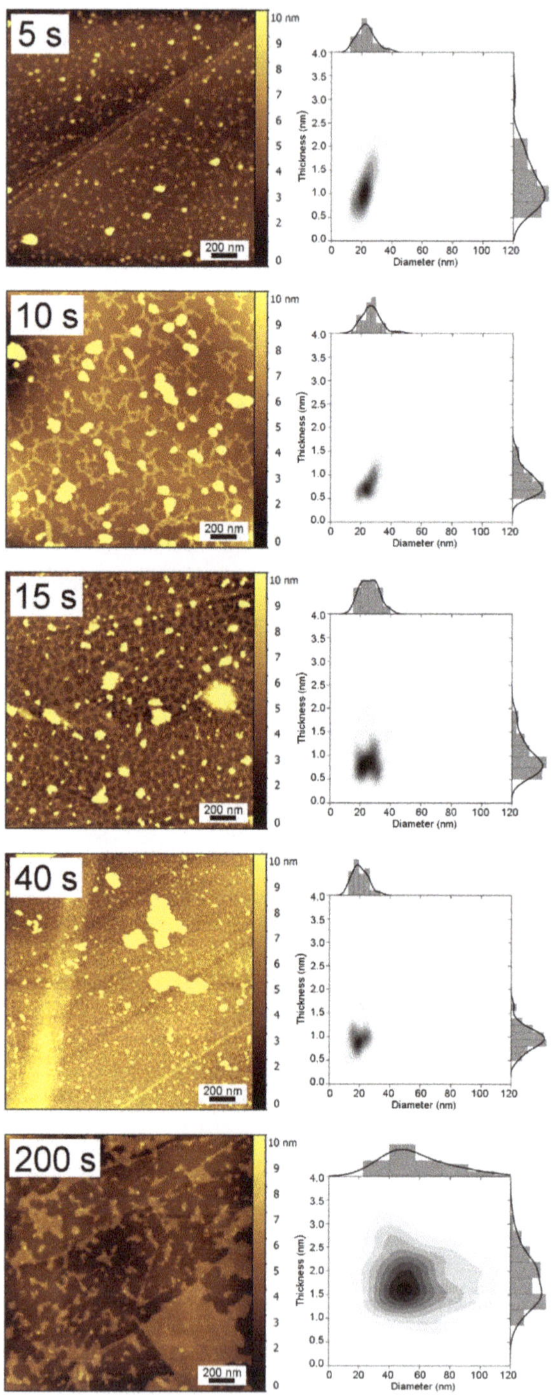

Figure 1. TM-AFM images of Pt on HOPG electrodeposited at different pulse times (**left**, scale bars are 200 nm) with the respective thickness and diameter bivariate distributions (**right**).

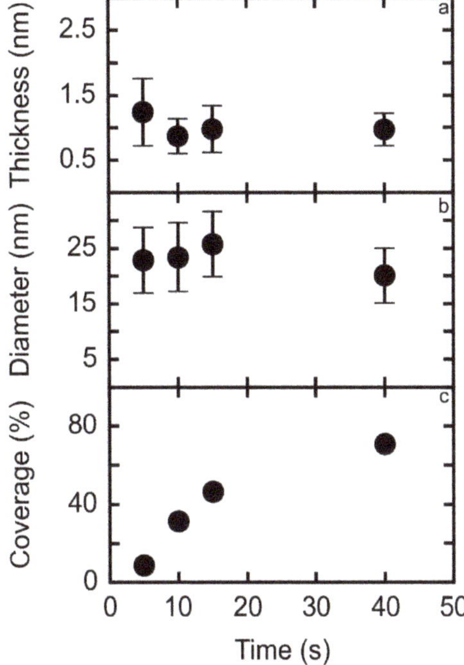

Figure 2. Dependence of average Pt nanoplatelets thickness (with standard deviation) (**a**) average Pt nanoplatelets diameter (with standard deviation). (**b**) and percentage of covered surface. (**c**) on the deposition time. The sample obtained with a 200 s pulse is not included for clarity (see Table 1).

Table 1. Dependence of average thickness and diameter of the electrodeposited Pt nanoplatelets on the deposition time. Surface coverage percentage was calculated on the AFM micrographs shown in Figure 1.

Sample	Thickness (nm)	Diameter (nm)	Surface Coverage (%)
5	1.24 ± 0.52	22.8 ± 6	8.7
10	0.87 ± 0.27	23.4 ± 6	31.4
15	0.98 ± 0.36	25.8 ± 6	46.5
40	0.97 ± 0.25	20.1 ± 5	70.7
200	1.82 ± 0.52	72.0 ± 69	44.9

Similarly, the average nanoplatelet diameter did not vary greatly with the deposition time (Figure 2b) and the size distribution is reasonably similar for all the samples (Figure 1). Although the diameter of the platinum islands is comparable to some reported for other overpotential conditions [32], the present high overpotential conditions lead to a distinct geometry in that the diameter is roughly 20 times the thickness for samples in the 5–40 s time interval, confirming the presence of flat nanoplatelets and thus two-dimensional growth.

The percentage of surface coverage increases with the deposition time (Figure 2c), indicating a surface-limited electrodeposition in the range 5–40 s. The sample prepared with 200 s deposition duration presented different values both for size distribution and dimension ratio due to its clear film-like morphology, dissimilar to the interconnected networks observed at shorter deposition times. It should be noticed that the obtained morphologies and trends are reproducible, while the current densities obtained for Pt electrodeposition can be slightly different from one sample to another. This is due to the different conditions of the HOPG surface exposed by cleavage of the C–C stacking [28].

Different overpotential conditions may affect the aspect of the resulting Pt deposit, and phenomena such as secondary nucleation, surface diffusion, and coalescence should be taken into account to explain the formation of the different structures and morphologies [32]. The supporting electrolyte also takes part in the Pt growth mechanism and in determining the final morphology of the resulting deposit [22,33], its role depending on the concentration, and on the applied potential. In this work, a chloride supporting electrolyte was chosen for compatibility with the Pt precursor (H_2PtCl_6) possibly releasing the same anion. The 0.5 M NaCl solution provided a high concentration of Cl^- anions that may adsorb on the surface of the newly formed Pt structures, preventing further Pt atoms from depositing and growing in a three dimensional way [21,33]. Simultaneous adsorption of H and Cl species on Pt could also occur, possibly synergistically [34,35], or competitively, with the possibility also that at high overpotential only protons are adsorbed [36]. Adsorbed anions may also be partially responsible for the smaller Pt particles migrating from one active site to another or to other Pt surfaces [22,37]. This phenomenon may represent a possible or at least partial explanation for the presence of large Pt agglomerates in these samples. In the very high overpotential conditions used in this work, it is likely that no Cl^- is adsorbed on the Pt surfaces, due to the strong evolution of gaseous H_2. The substrate also plays a role in Pt growth. Indeed, spontaneous electroless deposition on HOPG has been observed from Pt solutions, indicating the presence of metal seeds before the overpotential electrodeposition. The reducing agents for this spontaneous deposition are likely HOPG step edges that get oxidized in solution [28].

3.2. Electron Microscopy Analysis

High resolution transmission electron microscopy (HRTEM) micrographs of a Pt on an HOPG sample electrodeposited using a 15 s pulse were acquired to confirm the chemical nature and morphology of the deposit (Figure 3). The dimensions of the nanoplatelets were around 10 nm. The Fast Fourier Transform (FFT) analysis on selected areas of the TEM image, shown in Figure 3, presents three regions denoted A, B, and C with different d-spacings for the regions where carbon (A) and platinum (B) were clearly distinguishable.

In region A, only a single maximum at 0.21 nm is observed, comparable with the d-spacing of the (011) plane of graphite (0.2031 nm in JCPDS 96-901-2231). Region B clearly corresponds to a Pt platelet with a marked contrast over the carbon background. Indeed, in this region, a d-spacing of 0.23 nm was measured, fully compatible with that of the (111) facet of Pt (0.2299 nm in JCPDS 96-151-2257). In region C, no clear contrast nor evidence for platinum is visible from the TEM image. However, the FFT displays 3 maxima (one of 0.21 nm and two of 0.23 nm) demonstrating the presence of an extremely thin platinum layer, which was further corroborated by use of other techniques. Thus, to seek further evidence for the presence of Pt layers in areas where no obvious contrast was visible by imaging only, Energy Dispersive X-ray Spectroscopy (EDX) spectra on a Scanning Transmission Electron Microscopy (STEM) micrograph were recorded. Similarly to a previous report [38], local elemental analysis was performed on three different morphological features (Pt aggregate, Pt nanoplatelet and apparently bare HOPG) situated in another area of the same sample of Figure 3. The EDX spectra of the three zones highlighted in Figure 4 unquestionably show a Pt signal (Figure 4, inset) also in areas where no Pt nanoplatelets are visible by STEM. To conclude, the electron microscopy results depicted in Figures 3 and 4 provide complementary evidence for the presence of electrodeposited Pt all over the HOPG surface, demonstrating the early stage formation of a continuous ultra-thin electrodeposited Pt layer.

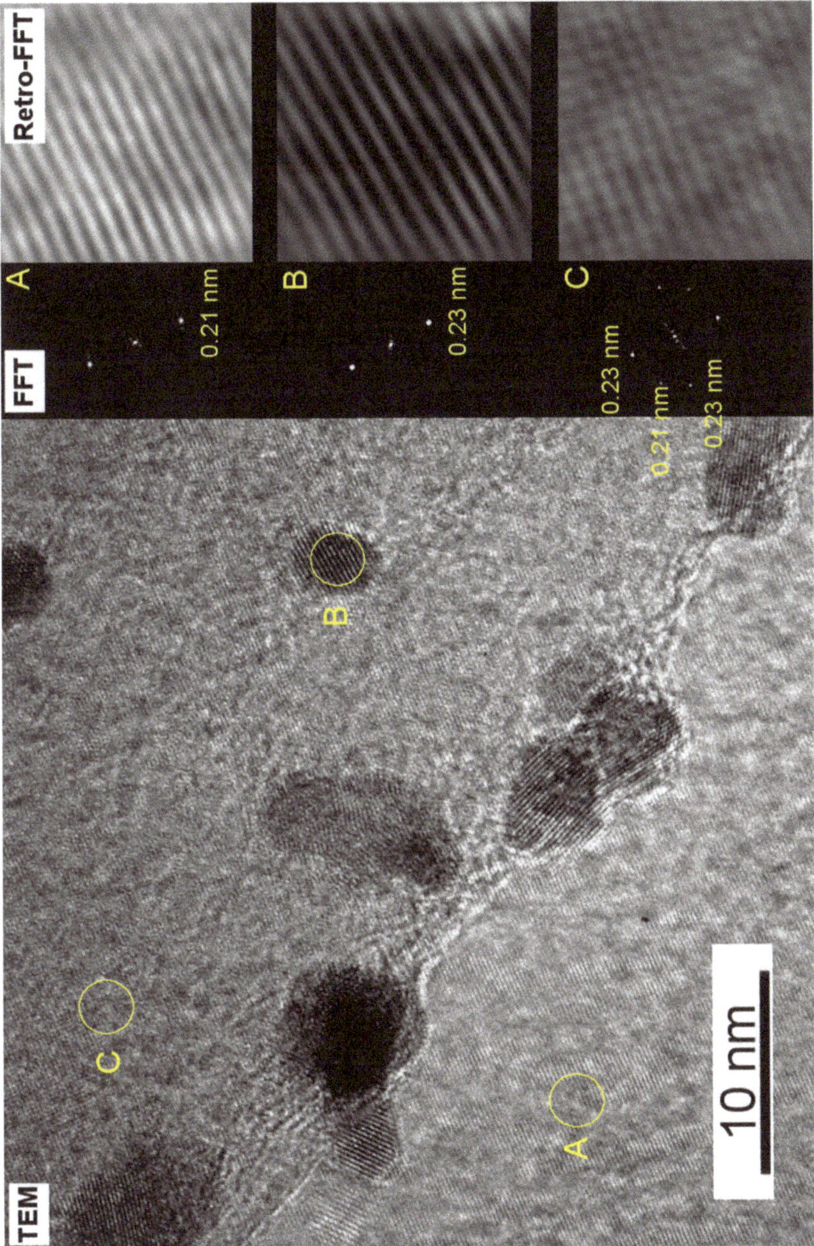

Figure 3. HRTEM micrograph of a slice of HOPG decorated with Pt nanoplatelets (**left**). FFT images of the three selected areas (**centre**) and the derived inverse-FFT (**right**).

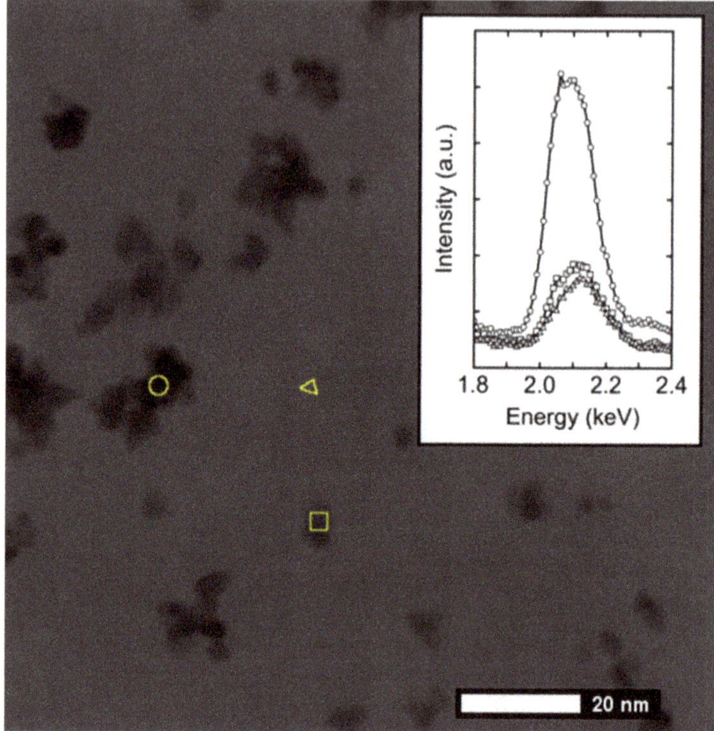

Figure 4. STEM micrograph of a flake of HOPG decorated with Pt aggregates (○) and Pt nanoplatelets (□); the comparison of their EDX spectra with the apparent bare surface of HOPG (Δ) is reported in the inset.

3.3. XPS Characterization

Surface analysis by X-ray Photoelectron Spectroscopy (XPS) was performed on a sample electrodeposited for 200 s both to further endorse the deposition of platinum, as well as to determine its oxidation state (Figure 5). The A bands of the Pt doublet observed on the sample (Pt $4f_{7/2}$ at 71.3 eV, Pt $4f_{5/2}$ at 74.6 eV) correspond to metallic Pt deposited on the HOPG surface. Additional contributions of B bands (72.2 and 75.5 eV respectively) were detected and related to the presence of oxidized forms of Pt, such as PtO_{ads} and/or $Pt(OH)_2$ [39]. The deconvolution of the Pt 4f doublet peaks indicates that Pt^0 is the majority species (73.6% of the total signal, comparable to Pt freshly electrodeposited onto graphite substrates previously reported [40]), rather than ionic Pt, which confirms the predominant metallic nature of the deposit and thus the suitability of the technique to prepare catalytically active Pt thin films.

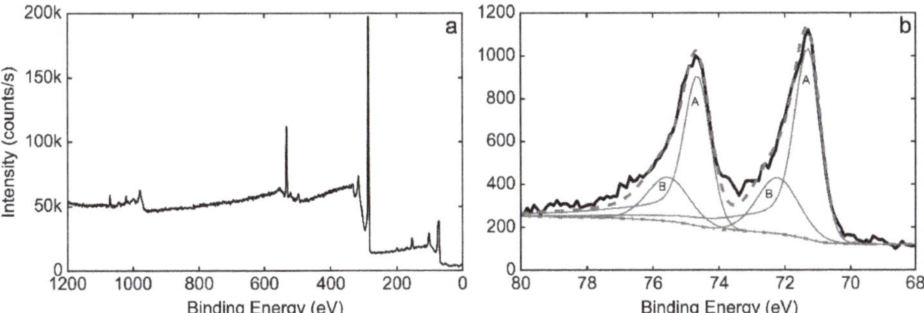

Figure 5. Survey XPS spectrum of a HOPG-Pt sample deposited for 200 s (**a**) and high-resolution spectrum of the Pt 4f region with peaks deconvolution (**b**).

The resulting Pt 4f spectrum is consistent with those reported for Pt films on HOPG with different thicknesses prepared by sputtering [41,42]. It is known that a Pt film thicker than 0.1 Å can already produce an XPS profile with two differentiated peaks, and since the sample prepared in this work also presented some Pt aggregates, the spectrum is the result of the presence of both 2D and 3D structures.

3.4. Electrocatalytic Activity

To assess the catalytic properties of the electrodeposited metal towards the oxygen reduction reaction (ORR), an HOPG-Pt sample deposited for 200 s was analyzed in O_2 saturated 0.1 M $HClO_4$ in a classical three-electrode configuration (scan from 1.2 to 0.3 V vs. NHE). The linear sweep voltammetry (LSV) shown in Figure 6 demonstrates that the deposited Pt^0 is active, with an onset potential of ca 0.98 V vs. NHE. No reliable mass activity values could be derived due to the experimental conditions that are far from steady-state.

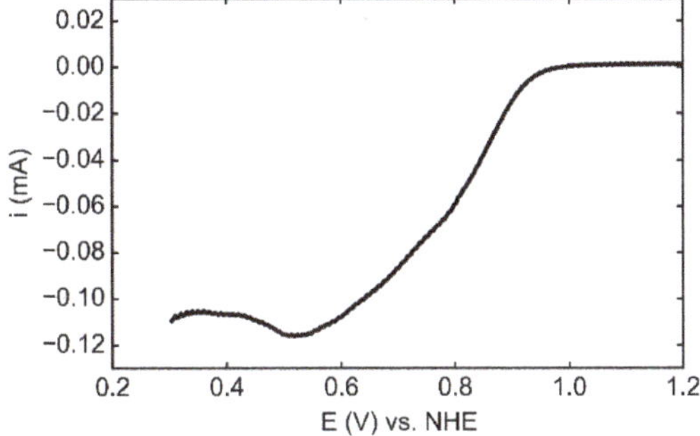

Figure 6. LSV of a HOPG-Pt sample electrodeposited for 200 s in O_2 saturated 0.1 M $HClO_4$ at 20 mV s^{-1}.

4. Conclusions

In conclusion, Pt nanoplatelet networks and Pt thin films were deposited on HOPG by high overpotential electrodeposition with a morphology that depends on pulse duration. The Pt networks formed using 5–40 s pulses comprised of extended islands of average thickness of 1 nm that were fairly homogeneous in diameter (20 times the thickness). Ultra-thin (<2 nm) Pt films are formed at a longer pulse time, 200 s, in which the exposed surfaces are Pt (111) facets. HR-TEM and XPS analysis confirm the metallic nature of the platinum deposit. These greater insights on the morphology of Pt structures deposited at high overpotential could only be obtained on a model carbon surface such as HOPG and will now serve to achieve complete formation of conformal and continuous ultra-thin Pt films on other carbonaceous surfaces (e.g., nanofibers, nanotubes, etc . . .) of applicability in energy conversion, such as the development of ultra-low platinum loaded fuel cell electrodes.

Author Contributions: Conceptualization, S.C., D.J.J. and J.R.; Methodology, G.E., S.C., D.J.J. and J.R.; Formal Analysis, F.F., G.E.; Investigation, F.F., G.E.; Resources, S.C., D.J.J. and J.R.; Writing-Original Draft Preparation, F.F.; Writing-Review & Editing, G.E., S.C., D.J.J. and J.R.; Supervision, S.C., D.J.J. and J.R.; Project Administration, S.C.; Funding Acquisition, S.C.

Funding: The research leading to these results has received funding from the European Research Council (ERC) under the European Union's Seventh Framework Programme (FP/2007-2013)/ERC Grant Agreement SPINAM n. 306682. SC also acknowledges the support of the French IUF.

Acknowledgments: Michel Ramonda (Centrale de Technologie en Micro et nanoélectronique, CTM, University of Montpellier) is acknowledged for AFM image acquisition and helpful discussions.

Conflicts of Interest: The authors declare no conflict of interest.

References

1. Guerrero Moreno, N.; Cisneros Molina, M.; Gervasio, D.; Pérez Robles, J.F. Approaches to polymer electrolyte membrane fuel cells (PEMFCs) and their cost. *Renew. Sustain. Energy Rev.* **2015**, *52*, 897–906. [CrossRef]

2. Ercolano, G.; Cavaliere, S.; Rozière, J.; Jones, D.J. Recent developments in electrocatalyst design thrifting noble metals in fuel cells. *Curr. Opin. Electrochem.* **2018**, *9*, 271–277. [CrossRef]

3. Gan, L.; Heggen, M.; Rudi, S.; Strasser, P. Core−shell compositional fine structures of dealloyed Pt. *Nano Lett.* **2012**, *12*, 5423–5430. [CrossRef] [PubMed]

4. Zhu, C.; Du, D.; Eychmüller, A.; Lin, Y. Engineering ordered and nonordered porous noble metal nanostructures: Synthesis, assembly, and their applications in electrochemistry. *Chem. Rev.* **2015**, *115*, 8896–8943. [CrossRef] [PubMed]

5. Cho, K.Y.; Yeom, Y.S.; Seo, H.Y.; Kumar, P.; Baek, K.-Y.; Yoon, H.G. A facile synthetic route for highly durable mesoporous platinum thin film electrocatalysts based on graphene: morphological and support effects on the oxygen reduction reaction. *J. Mater. Chem. A* **2017**, *5*, 3129–3135. [CrossRef]

6. Lee, H.B.R.; Baeck, S.H.; Jaramillo, T.F.; Bent, S.F. Growth of Pt nanowires by atomic layer deposition on highly ordered pyrolytic graphite. *Nano Lett.* **2013**, *13*, 457–463. [CrossRef] [PubMed]

7. Aaltonen, T.; Ritala, M.; Sajavaara, T.; Keinonen, J.; Leskelä, M. Atomic layer deposition of platinum thin films. *Chem. Mater.* **2003**, *15*, 1924–1928. [CrossRef]

8. Dolbec, R.; El Khakani, M.A. Sub-ppm sensitivity towards carbon monoxide by means of pulsed laser deposited SnO_2:Pt based sensors. *Appl. Phys. Lett.* **2007**, *90*, 173114. [CrossRef]

9. Achari, I.; Ambrozik, S.; Dimitrov, N. Electrochemical atomic layer deposition of Pd ultrathin films by surface limited redox replacement of underpotentially deposited H in a single cell. *J. Phys. Chem. C* **2017**, *121*, 4404–4411. [CrossRef]

10. Gruber, D.; Ponath, N.; Müller, J.; Lindstaedt, F. Sputter-deposited ultra-low catalyst loadings for PEM fuel cells. *J. Power Sources* **2005**, *150*, 67–72. [CrossRef]

11. Su, X.; Zhan, X.; Hinds, B.J. Pt monolayer deposition onto carbon nanotube mattes with high electrochemical activity. *J. Mater. Chem.* **2012**, *22*, 7979. [CrossRef]

12. Whalen, J.J.; Weiland, J.D.; Searson, P.C. Electrochemical deposition of platinum from aqueous ammonium hexachloroplatinate solution. *J. Electrochem. Soc.* **2005**, *152*, C738. [CrossRef]

13. Chen, X.; Li, N.; Eckhard, K.; Stoica, L.; Xia, W.; Assmann, J.; Muhler, M.; Schuhmann, W. Pulsed electrodeposition of Pt nanoclusters on carbon nanotubes modified carbon materials using diffusion restricting viscous electrolytes. *Electrochem. Commun.* **2007**, *9*, 1348–1354. [CrossRef]

14. Ye, F.; Chen, L.; Li, J.; Li, J.; Wang, X. Shape-controlled fabrication of platinum electrocatalyst by pulse electrodeposition. *Electrochem. Commun.* **2008**, *10*, 476–479. [CrossRef]

15. Ercolano, G.; Farina, F.; Cavaliere, S.; Jones, D.J.; Rozière, J. Towards ultrathin Pt films on nanofibres by surface-limited electrodeposition for electrocatalytic applications. *J. Mater. Chem. A* **2017**, *5*, 3974–3980. [CrossRef]

16. Ercolano, G.; Farina, F.; Cavaliere, S.; Jones, D.J.; Rozière, J. Multilayer hierarchical nanofibrillar electrodes with tuneable lacunarity with 2D like Pt deposits for PEMFC. *ECS Trans.* **2017**, *80*, 757–762. [CrossRef]

17. Yivlialin, R.; Bussetti, G.; Brambilla, L.; Castiglioni, C.; Tommasini, M.; Duò, L.; Passoni, M.; Ghidelli, M.; Casari, C.S.; Li Bassi, A. Microscopic analysis of the different perchlorate anions intercalation stages of graphite. *J. Phys. Chem. C* **2017**, *121*, 14246–14253. [CrossRef]

18. Yasin, H.M.; Denuault, G.; Pletcher, D. Studies of the electrodeposition of platinum metal from a hexachloroplatinic acid bath. *J. Electroanal. Chem.* **2009**, *633*, 327–332. [CrossRef]

19. Pang, L.; Zhang, Y.; Liu, S. Monolayer-by-monolayer growth of platinum films on complex carbon fiber paper structure. *Appl. Surf. Sci.* **2017**, *407*, 386–390. [CrossRef]

20. Duarte, M.M.E.; Pilla, A.S.; Sieben, J.M.; Mayer, C.E. Platinum particles electrodeposition on carbon substrates. *Electrochem. Commun.* **2006**, *8*, 159–164. [CrossRef]

21. Lee, I.; Chan, K.-Y.; Lee Phillips, D. Atomic force microscopy of platinum nanoparticles prepared on highly oriented pyrolytic graphite. *Ultramicroscopy* **1998**, *75*, 69–76. [CrossRef]

22. Lee, I.; Chan, K.-Y.; Phillips, D.L. Growth of electrodeposited platinum nanocrystals studied by atomic force microscopy. *Appl. Surf. Sci.* **1998**, *136*, 321–330. [CrossRef]

23. Liu, Y.H.; Gokcen, D.; Bertocci, U.; Moffat, T.P. Self-terminating growth of platinum films by electrochemical deposition. *Science* **2012**, *338*, 1327–1330. [CrossRef] [PubMed]

24. Zoval, J.V.; Lee, J.; Gorer, S.; Penner, R.M. Electrochemical preparation of platinum nanocrystallites with size selectivity on basal plane oriented graphite surfaces. *J. Phys. Chem. B* **1998**, *5647*, 1166–1175. [CrossRef]

25. Shen, P.; Chi, N.; Chan, K.Y.; Phillips, D.L. Platinum nanoparticles spontaneously formed on HOPG. *Appl. Surf. Sci.* **2001**, *172*, 159–166. [CrossRef]

26. Arroyo Gómez, J.J.; García, S.G. Spontaneous deposition of Pt-nanoparticles on HOPG surfaces. *Surf. Interface Anal.* **2015**, *47*, 1127–1131. [CrossRef]

27. Juarez, M.F.; Fuentes, S.; Soldano, G.J.; Avalle, L.; Santos, E. Spontaneous formation of metallic nanostructures on highly oriented pyrolytic graphite (HOPG): An ab initio and experimental study. *Faraday Discuss.* **2014**, *172*, 327–347. [CrossRef] [PubMed]

28. Alpuche-Aviles, M.A.; Farina, F.; Ercolano, G.; Subedi, P.; Cavaliere, S.; Jones, D.J.; Rozière, J. Electrodeposition of two-dimensional Pt nanostructures on highly oriented pyrolytic graphite (HOPG): the effect of evolved hydrogen and chloride ions. *Nanomaterials* **2018**, *8*, 668. [CrossRef]

29. Horcas, I.; Fernández, R.; Gómez-Rodríguez, J.M.; Colchero, J.; Gómez-Herrero, J.; Baro, A.M. WSXM: A software for scanning probe microscopy and a tool for nanotechnology. *Rev. Sci. Instrum.* **2007**, *78*. [CrossRef] [PubMed]

30. Wojdyr, M. Fityk: A general-purpose peak fitting program. *J. Appl. Crystallogr.* **2010**, *43*, 1126–1128. [CrossRef]

31. Lu, G.; Zangari, G. Electrodeposition of platinum nanoparticles on highly oriented pyrolitic graphite: Part II: Morphological characterization by atomic force microscopy. *Electrochim. Acta* **2006**, *51*, 2531–2538. [CrossRef]

32. Ustarroz, J.; Altantzis, T.; Hammons, J.A.; Hubin, A.; Bals, S.; Terryn, H. The role of nanocluster aggregation, coalescence, and recrystallization in the electrochemical deposition of platinum nanostructures. *Chem. Mater.* **2014**, *26*, 2396–2406. [CrossRef]

33. Lu, G.J.; Zangari, G. Electrodeposition of platinum on highly oriented pyrolytic graphite. Part 1: Electrochemical characterization. *J. Phys. Chem. B* **2005**, *109*, 7998–8007. [CrossRef] [PubMed]

34. Li, N.; Lipkowski, J. Chronocoulometric studies of chloride adsorption at the Pt(111) electrode surface. *J. Electroanal. Chem.* **2000**, *491*, 95–102. [CrossRef]

35. Horanyi, G.; Rizmayer, E.Z. A coupled voltammetric and radiometric (voltradiometric) study of the simultaneous adsorption of hydrogen and anions at platinized platinum electrodes. *J. Electroanal. Chem.* **1987**, *218*, 337–340. [CrossRef]

36. Gossenberger, F.; Roman, T.; Groß, A. Hydrogen and halide co-adsorption on Pt(111) in an electrochemical environment: A computational perspective. *Electrochim. Acta* **2016**, *216*, 152–159. [CrossRef]

37. Zhang, Z.C.; Beard, B.C. Agglomeration of Pt particles in the presence of chlorides. *Appl. Catal. A Gen.* **1999**, *188*, 229–240. [CrossRef]

38. Feng, J.; Xiong, W.; Ding, H.; He, B. Hydrogenolysis of glycerol over Pt/C catalyst in combination with alkali metal hydroxides. *Open Chem.* **2016**, *14*, 279–286. [CrossRef]

39. Raynal, F.; Etcheberry, A.; Cavaliere, S.; Noël, V.; Perez, H. Characterization of the unstability of 4-mercaptoaniline capped platinum nanoparticles solution by combining LB technique and X-ray photoelectron spectroscopy. *Appl. Surf. Sci.* **2006**, *252*, 2422–2431. [CrossRef]

40. Spataru, T.; Osiceanu, P.; Marcu, M.; Lete, C.; Munteanu, C.; Spa, N. Functional effects of the deposition substrate on the electrochemical behavior of platinum particles. *Jpn. J. Appl. Phys.* **2012**, *51*, 090119. [CrossRef]

41. Marcus, P.; Hinnen, C. XPS study of the early stages of deposition of Ni, Cu and Pt on HOPG. *Surf. Sci.* **1997**, *392*, 134–142. [CrossRef]

42. Motin, A.M.; Haunold, T.; Bukhtiyarov, A.V.; Bera, A.; Rameshan, C.; Rupprechter, G. Surface science approach to Pt/carbon model catalysts: XPS, STM and microreactor studies. *Appl. Surf. Sci.* **2018**, *440*, 680–687. [CrossRef]

Article

Electrodeposition of Two-Dimensional Pt Nanostructures on Highly Oriented Pyrolytic Graphite (HOPG): The Effect of Evolved Hydrogen and Chloride Ions

Mario A. Alpuche-Aviles [1],*, Filippo Farina [2], Giorgio Ercolano [2], Pradeep Subedi [1], Sara Cavaliere [2],*, Deborah J. Jones [2] and Jacques Rozière [2]

[1] Department of Chemistry, University of Nevada, Reno, NV 89557, USA; psubedi@unr.edu
[2] Institute Charles Gerhardt Montpellier, Laboratory of Aggregates Interfaces and Materials for Energy, University of Montpellier, 34095 Montpellier, France; filippo.farina@umontpellier.fr (F.F.); giorgio.ercolano@umontpellier.fr (G.E.); deborah.jones@umontpellier.fr (D.J.J.); jacques.roziere@umontpellier.fr (J.R.)
* Correspondence: malpuche@unr.edu (M.A.A.-A.); sara.cavaliere@umontpellier.fr (S.C.); Tel.: +1-775-784-4523 (M.A.A.-A.); +33-467-149-098 (S.C.); Fax: +1-775-784-6804 (M.A.A.-A.); +33-467-143-304 (S.C.)

Received: 20 July 2018; Accepted: 23 August 2018; Published: 28 August 2018

Abstract: We discuss the electrodeposition of two-dimensional (2D) Pt-nanostructures on Highly Oriented Pyrolytic Graphite (HOPG) achieved under constant applied potential versus a Pt counter electrode (E_{appl} = ca. -2.2 V vs. NHE, normal hydrogen electrode). The deposition conditions are discussed in terms of the electrochemical behavior of the electrodeposition precursor (H_2PtCl_6). We performed cyclic voltammetry (CV) of the electrochemical Pt deposit on HOPG and on Pt substrates to study the relevant phenomena that affect the morphology of Pt deposition. Under conditions where the Pt deposition occurs and H_2 evolution is occurring at the diffusion-limited rate (-0.3 V vs. NHE), Pt forms larger structures on the surface of HOPG, and the electrodeposition of Pt is not limited by diffusion. This indicates the need for large overpotentials to direct the 2D growth of Pt. Investigation of the possible effect of Cl^- showed that Cl^- deposits on the surface of Pt at low overpotentials, but strips from the surface at potentials more positive than the electrodeposition potential. The CV of Pt on HOPG is a strong function of the nature of the surface. We propose that during immersion of HOPG in the electrodeposition solution (3 mM H_2PtCl_6, 0.5 M NaCl, pH 2.3) Pt islands are formed spontaneously, and these islands drive the growth of the 2D nanostructures. The reducing agents for the spontaneous deposition of Pt from solution are proposed as step edges that get oxidized in the solution. We discuss the possible oxidation reactions for the edge sites.

Keywords: electrodeposition; platinum; highly oriented pyrolytic graphite; 2D growth

---

1. Introduction

The use of Pt is of interest in many renewable energy applications, namely, in the use of technologies that convert chemical energy to electricity, such as proton exchange membrane fuel cells (PEMFCs). Because of the low abundance and high cost of this noble metal, research has centered on the use of Pt nanostructures on a highly conducting and porous support, conventionally carbon black, with the goal of using a minimum loading while preserving high electroactivity towards the oxygen reduction reaction (ORR, the sluggish reaction taking place at the cathode side) and long-term stability. The strategies to minimize the Pt amount call for a shift from monometallic nanoparticles (NPs) to the use of nano-engineered architectures with tailored morphologies and compositions [1,2]. For instance, bi- or tri-metallic particles, alloys [3–6] and successively de-alloyed [7] structures where Pt is associated with other transition metals (e.g., Ni, Co, Cu) have been prepared and demonstrated high ORR activity and durability. Another very promising class of tailored electrocatalysts is represented by core@shell nanostructures with a thin Pt skin covering a transition metal core in 0D (particle-like) [8–10] and 1D (fiber-like) morphologies [11–17]. The advantage of a Pt thin layer morphology on other non-metallic supports (carbon, polymers) has been extensively demonstrated to maximize Pt exploitation by minimizing the contribution of edge/corner sites and to enhance its stability by circumventing any possibility of nanoparticle aggregation [18–20]. In particular, Pt extended surfaces deposited on high aspect-ratio materials such as nanofibers [19,21] and whiskers [22] presented exceptional increase of the ORR specific activity, which was kept after prolonged electrochemical cycling. Among the methods being investigated to produce Pt conformal thin films, atomic layer deposition (ALD), electrochemical atomic layer deposition (EC-ALD), pulsed laser deposition, surface-limited redox replacement (SLRR), galvanic displacement [18,21,23,24] and other vacuum techniques, such as magnetron sputtering are employed [25].

The goal of electrodepositing Pt extended surfaces in a continuous and conformal fashion on the support is complicated by the fact that the 3D growth of this metal is favored with the formation of dendrites [26] and flower-like agglomerates. It was previously reported that it is possible to prepare Pt thin films on flat metallic Au surfaces via a self-limiting electrodeposition process performed at high overpotential [27]. More recently, some of us reported the electrodeposition of thin Pt nanostructures by pulsing the electrodeposition at high overpotentials on carbon fiber webs obtained by electrospinning of precursor solutions [19,28]. The material obtained was determined to be 2D contiguous Pt nanoplatelets. The resulting self-supported nanofibrous electrode (NFE) was demonstrated to be an efficient and stable ORR material. The understanding of the mechanistic aspects that yield the electrodeposition of a Pt thin film on a carbonaceous model surface, such as Highly Oriented Pyrolytic Graphite (HOPG), is crucial for the deposition of ultra-low and continuous coverage of Pt on carbon supports of different morphology and porosity [29].

In this paper, we focus on the study of the electrodeposition step in the first cycle of the electrodeposition sequence (200 s). We address the mechanistic aspects that direct the growth of a 2D film of Pt on HOPG as a model for the deposition of Pt on carbon. The effect of electrolyte concentration and potential on Pt morphology has been recognized for some time [30], and studies include the use of HOPG in mechanistic studies [30–32]. However, up to now, the electrodeposition of Pt has been studied at much lower overpotentials than those used in this work. Penner and coworkers proposed the electrodeposition of Pt nanoparticles with homogeneous size distribution [30]. Simonov et al. [33] revised the conventional model of nucleation and growth [26] and distinguished (i) a primary nucleation of Pt on the substrate, (ii) a secondary nucleation around the Pt deposits, and (iii) the growth phase around the Pt deposits. Here we present evidence that spontaneous deposition of Pt when the HOPG substrate immersed at open circuit, that is, without an external current flowing through the substrate, is a critical initial step. The initial stages of Pt deposition continue to be the subject of investigation because they are thought to control the deposition process. For Pt on HOPG, the spontaneous deposition of Pt has been observed before [30,32,34–36] and although the process is not well understood, it can nonetheless control the size and distribution of the Pt nanostructures deposited during a subsequent potentiostatic step. Zoval et al. first reported this electroless deposition of Pt [30], which led to a decrease in the reproducibility of electrodepositing Pt NPs on HOPG. They used an anodic potential step (ca. 0.8 V vs. NHE for 1 min) to prevent the spontaneous deposition of Pt at open circuit potential (OCP) before the potentiostatic NP deposition. They reported that the oxidizing treatment prevented the spontaneous deposition of Pt at OCP and made the Pt NPs deposited during the subsequent potentiostatic step more homogenous.

Similarly, Lu and Zangari [31] applied an anodic treatment to the substrate to isolate the nucleation and growth during electrodeposition from the electroless deposition of Pt. Later, they addressed the spontaneous deposition of Pt from a 1 mM H_2PtCl_6 solution without additional excess electrolyte and in 0.1–0.2 M HCl. They found that the Pt deposits preferentially on edges although it can also deposit on terraces, regardless of the use of electrolyte. The Pt islands from the spontaneous process lead to dendritic growth of Pt, and thus, they used an oxidation pretreatment (1 V vs. NHE for 2 min) for the HOPG to avoid the spontaneous deposition and to obtain monodispersed Pt NPs [32]. Arroyo-Gómez and García [34] showed that the NPs cover the HOPG surface completely in a solution of 1 mM H_2PtCl_6 + 0.05 M H_2SO_4. Under these conditions, without excess Cl^-, the precursor hydrolyzes quickly [31,33] at RT. Initially, the Pt deposits on edges and at longer times the deposition occurs on the terraces along with larger Pt particles on the surface. After 2 h of immersion the substrate electrocatalytic activity approached that of bulk Pt. The electroless process occurs with [30] or without excess electrolyte [32] (Cl^- [30] or SO_4^{2-} [34]) and in a nonaqueous electrolyte containing chloride (2.5 mM H_2PtCl_6 + 80 mM Et_4NCl in CH_3CN) [35]. In the nonaqueous electrolyte, oxidizing the HOPG surface before immersion in the Pt solution yielded a more uniform NP electrodeposit.

Since the original report, the observation that oxidizing the substrate minimizes the spontaneous deposition of Pt/HOPG has led to the conclusion that a partially oxidized site on the surface is the reducing agent. This reducing agent has been ascribed to defects [30], oxidized [30] and hydrogenated [37] sites on the carbon surface. However, there are no reports on the half oxidation reaction that reduces the Pt precursor. Interestingly, the electrochemical papers lack X-ray photoelectron spectroscopy (XPS) studies of the surface oxidation and, to get insight into the half reactions, we look at the surface energies and composition studies of the spontaneous deposition of Au on HOPG from solutions [38–40]. Here, we propose that edge sites oxidize sequentially to C–OH and C=O groups, based on analogous studies for the deposition of Au on HOPG to produce gas-phase catalysts and our own results that see preferential deposits on the edges, in agreement with previous reports [32,34,35]. We address the effect of the initial electroless deposition of Pt on the formation of 2D nanostructures on HOPG. We point out that in the previous reports, the focus was to electrodeposit NPs, but our goal is the production of a conformal film. Despite evidence that the spontaneous deposition of Pt occurs, we do not see dendritic growth, which indicates that another process limits the growth of the initial deposits. Therefore, we discuss both stages of the Pt deposition, the potentiostatic mode that grows the 2D deposit and the initial electroless deposition.

2. Materials and Methods

Reagents and Materials: HOPG (ZYH grade, 12 × 12 mm from Veeco-Bruker, Camarillo, CA, USA) was used as the working electrode in a two or three electrode cell. The working electrode was freshly exfoliated before every electrochemical experiment and immediately masked with Kapton® tape (RS Components SAS, Beauvais, France). The electrode area exposed to the solution was typically 6 × 6 mm. The backside of the HOPG was taped to a Pt wire or Cu tape, keeping the solution from being exposed to the Pt or Cu. All the reagents used are of analytical grades and used without purification: H_2PtCl_6 (ACS reagent, 37.50% Pt basis) and NaCl were purchased from Sigma-Aldrich (St. Louis, MO, USA). All aqueous solutions were prepared in 18 MΩ·cm water.

Characterization: The samples were characterized with tapping mode atomic force microscopy (TM-AFM). A Bruker Nanoman AFM instrument (Bruker SAS, Palaiseau, France) with a Nanoscope 5 controller was equipped with the following tips: a silicon Point Probe Plus (Nanosensors, Neuchâtel, Switzerland) with 6.5 N/m force constant and f = 157 kHz resonance. The images were processed with Gwyddion version 2.44 (Czech Metrology Institute, Brno, Czech Republic) both for image flattening and thickness measurements.

Electrochemical Measurements: Oxidation currents were positive following the IUPAC convention. Experiments were performed with either a Bio-Logic® SP 300 (Bio-Logic SAS, Seyssinet-Pariset, France) potentiostat (2 channels configuration) or a CH Instruments CHI 760D and 700D workstation (CH Instruments Inc., Austin, Texas, USA). The working electrode was usually an HOPG setup, although some controlled experiments were performed with a 1.6 mm Pt disk electrode. The counter electrode was either a graphite sheet (5 × 5 cm) glassy carbon sheet or a graphite rod (Alfa Aesar, Heysam, UK). We did not use Pt CE in experiments performed in Pt-free solutions. The reference electrode was an Ag/AgCl reference electrode (Fisher Scientific, Illkirch, France; E = 0.197 V vs. NHE) used with a bridge filled with 0.5 M NaCl with the pH adjusted to 2.3, as in the electrodeposition solution. The standard rate constants, k^0 were determined by fitting experimental data to an electrochemical mechanism [41]. The linear sweep voltammograms (LSVs) obtained for the H^+ reduction where simulated with the commercial software DigiElch version 7.FD (ElchSoft, Kleinromstedt, Germany) that allowed us to correct the LSVs for iR_u drop and capacitive charging. The k^0 value was varied until a good fit of the experimental data was obtained using previously reported values of diffusion coefficients of H^+ and H_2 as well as the proton concentration. ($D_{H^+} = 7.1 \times 10^{-5}$ cm^2/s [42] and for H_2, $D_{H_2} = 3.37 \times 10^{-5}$ cm^2/s) [43].

Pt/HOPG Electrode preparation: Full details are given elsewhere for the preparation of the Pt electrode on HOPG and [19,29]. Briefly, the HOPG was exposed to the solution, and the deposition was performed either on a 2- or 3-electrode cell. In the 2-electrode cell, Pt electrodeposition is performed by applying -3 V vs. the counter electrode for 200 s. This potential was applied immediately following an immersion time during which the solution was cooled to 4 °C while bubbled with nitrogen to remove oxygen. The solution was stirred at 900 rpm during electrodeposition to remove hydrogen bubbles evolving from the surface. The electrodeposition of Pt was performed from the following solution: 3 mM H_2PtCl_6, 0.5 M NaCl with the pH adjusted to 2.3 with HCl.

3. Results and Discussion

3.1. Electrodeposition Process Characterization

Figure 1 shows the data of the electrodeposition on HOPG. The electrodeposition was performed in a two-electrode cell, with the first channel of the Bio-Logic potentiostat (floating configuration) with the potential simultaneously monitored with the second potentiostat channel. Initially, the electrode is immersed in the solution and kept under OCP conditions, which is typically around +1 V vs. Ag/AgCl (ca. 0.8 V vs. the normal hydrogen electrode, NHE). After the immersion time, a constant potential of -3 V vs. the CE is applied, which corresponds to ca. -2 V vs. Ag/AgCl (-2.2 V vs. NHE). At these potentials, the current is around 1 µA, which corresponds to ca. 2.8 µA/cm^2. We focus our discussion on two aspects: (1) the initial OCP value and (2) the value of the applied potential (E_{app}) to evaluate the effect of eventual hydrogen formation.

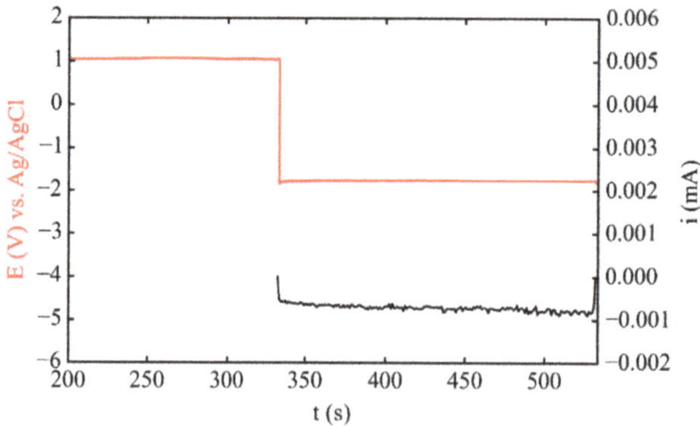

Figure 1. Electrodeposition Pt on Highly Oriented Pyrolytic Graphite (HOPG) (0.36 cm^2) in 3 mM H_2PtCl_6, 0.5 M NaCl, pH 2.3, under N_2 bubbling and 600 rpm stirring. The time between 200 and 320 s corresponds to the sample immersed at OCP = 1 V vs. Ag/AgCl = 0.8 V vs. NHE. At 320 s, E_{app} = -3 V vs. CE, which is ca. -2 V vs. Ag/AgCl and -2.2 V vs. NHE.

3.2. Phase 1: Electrode Immersion

The potential achieved for the HOPG immersed in the Pt solution (Figure 1) is similar to that required for the Pt electrodeposition. Because the solution contains $PtCl_6^{2-}$, the electrodeposition should be a sequential process with the ultimate production of Pt^0 according to the following reactions [44]:

$$PtCl_6^{4-}{}_{(aq)} + 2e \rightarrow PtCl_4^{2-}{}_{(aq)} + 2Cl^-{}_{(aq)} \qquad E^0 = 0.726 \text{ V} \qquad (1)$$

$$PtCl_4^{2-}{}_{(aq)} + 2e \rightarrow Pt + 4Cl^-{}_{(aq)} \qquad E^0 = 0.758 \text{ V} \qquad (2)$$

where all potentials are with respect to NHE. However, the potentials reported in Equations (1) and (2) are derived from complexation schemes; they are normally not observed under usual experimental conditions [45]. The potentials observed in Figure 1 during the incubation period at OCP are similar to those used for Pt deposition, and in fact, in different experiments, we observed potentials more negative than those in Equations (1) and (2). The deposition conditions were investigated by cyclic voltammetry. In Figure 2 are presented the CVs of a 1.6 mm Pt disk electrode immersed blank (without Pt, A) and in the deposition solution (B). The black and red traces correspond to experiments performed with different potential scans at the same scan rate (ν). Concerning the experiments on HOPG in the absence of Pt, the peak around -0.4 V vs. Ag/AgCl (Figure 2A) is attributed to the diffusion limited H^+ reduction to H_2. In the region between -0.4 V and 0.3 V vs. Ag/AgCl, several features in the CV appear when Pt is introduced in the deposition solution. Note that these potentials correspond to -0.2 to 0.5 V vs. NHE, which are positive to the potentials in Equations (1) and (2). Note also that although the step-wise reduction of Pt^{4+} to Pt^0 is expected, the CV shows that the mechanism is more complicated than two consecutive 2-electron processes. Uosaki et al. reported the sequential 2-electron reduction in H_2SO_4 electrolyte with peak potentials observed on Pt(111) in agreement with the values of Equations (1) and (2) [46]. However, in Cl^- electrolytes, the reduction potentials shift to negative values without resolving the two reduction steps on the CV [31]. We will not attempt to discuss here the mechanism of Pt^{4+} reduction, but will use the CVs of the electrode materials in different electrolytes to explain the formation of 2D Pt structures. At negative potential regions (E < 0.7 V vs. NHE), on Pt, the deposition of Pt from one of the $PtCl_x$ complexes is expected.

The CV for HOPG in the Pt deposition solution is a strong function of the nature of the HOPG surface. Figure 3 depicts the linear sweep voltammetry (LSV) of three independently prepared HOPG surfaces in the same electrodeposition solution as the samples shown in Figure 2B. Interestingly, the electrodeposition on HOPG does not display a peak for the diffusion-limited deposition of Pt/HOPG and only the peak for H_2 evolution is present at E = -0.5 V, but no peak for Pt deposition is seen. At potentials more negative than 0.3 V vs. Ag/AgCl, or 0.5 V vs. NHE Pt deposition on HOPG is again apparent. However, the rates of deposition vary widely, and we propose that this is due to the different conditions of the surface exposed by cleavage of the C–C stacking with tape.

We further investigated the electrodeposition of Pt on an HOPG surface by cyclic voltammetry. Figure 4 depicts a series of sequential CVs obtained on the same surface. Note that the scan rate here is twice that in Figure 3, to minimize the time in which the surface is modified during the polarization. Also, each sequential CV was recorded on increasing the potential window (a–e), but no peak characteristic of the deposition of Pt from the 3 mM solution of Pt precursor was observed. After 4 consecutive CVs, the peak for H_2 evolution in the solution is seen (curve e). An anodic peak appears around +1.0 V vs. Ag/AgCl, or 0.8 V vs. NHE, which is consistent with the oxidation of Pt to a $PtCl_x$ complex. The peak intensity increases with the number of CVs, in agreement with a higher amount of Pt present on the surface. This result suggests that at potentials more negative than 0.8 V vs. NHE, the Pt deposit is stable on the HOPG surface. However, the CVs or LSVs do not reach a diffusion-limited rate for the deposition of Pt/HOPG.

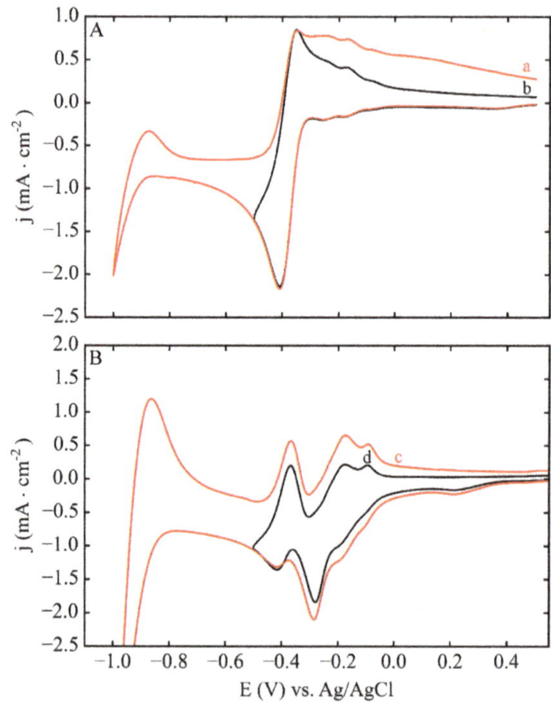

Figure 2. Cyclic voltammetry of a 1.6 mm Pt electrode in (**A**) 0.5 M NaCl pH 2.3 (no Pt) and (**B**) in the same electrolyte in (A) +3 mM H_2PtCl_6. CVs (a,c) and (b,d) were recorded in different potential windows, $\nu = 50$ mV/s.

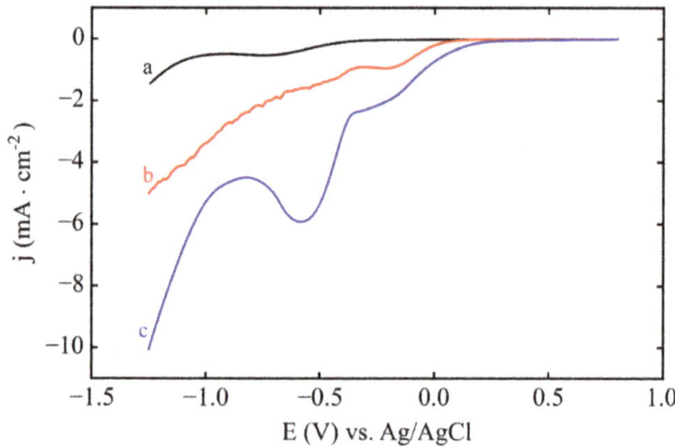

Figure 3. Linear sweep voltammetry of 3 independently prepared HOPG surfaces in 3 mM H_2PtCl_6 0.5 M NaCl pH 2.3 (a–c), $\nu = 100$ mV/s.

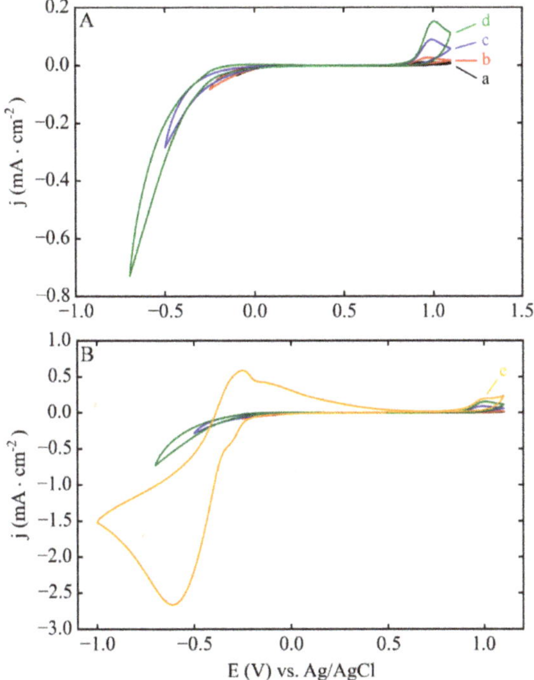

Figure 4. (**A**) Sequence of cyclic voltammograms (CVs) of Pt deposition on the same HOPG surface (a–d), v = 50 mV/s. The CVs increase the potential window in the negative direction. (**B**) Shows the same CVs in (**A**) for comparison with the final CV (e).

The results in Figures 3 and 4 indicate that the Pt deposit is stable on HOPG at potentials that are consistent with the bulk reduction potentials for the complexes in Equations (1) and (2). However, the OCP before deposition indicates that Pt can be formed when the HOPG substrate is immersed in the precursor solution. To investigate this possibility, we immersed a freshly-cleaved HOPG electrode in the electrodeposition solution without an applied potential, and without a connection to an external circuit, and then rinsed it with water. The voltammetry of this electrode in 0.5 M H_2SO_4, and, for comparison, the behavior of HOPG in the same solution, are presented in Figure 5A,B, respectively. The CV of the HOPG exposed to the Pt precursor solution shows the characteristic features for H_2 evolution, which is consistent with the formation of Pt on the surface. Note that the activity is relatively low, since the peaks for hydrogen adsorption and desorption are not well defined. Recently, Arroyo-Gomez and Garcia reported higher electrocatalytic activity for spontaneously deposited Pt on C with electroactivity being a function of time [34]; the highest activity was observed after 2 h of exposure in a solution at room temperature. In the present study, the incubation was performed at low temperatures (4–5 °C), that we propose slow down the deposition of Pt, which has been reported to be a slow process even at room temperature [34].

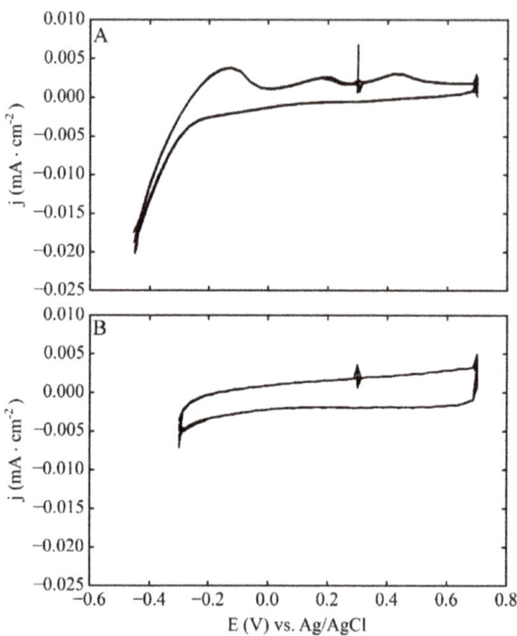

Figure 5. (**A**) CVs of a HOPG electrode in 0.5 M H_2SO_4 exposed to the electrodeposition solution (3 mM H_2PtCl_6, 0.5 M NaCl, pH 2.3) for 200 s. The electrode was immersed but not connected to the external circuit during the exposure to the Pt precursor, then rinsed, dried and characterized in 0.5 M H_2SO_4. (**B**) Behavior of a freshly cleaved HOPG electrode without exposure to the Pt deposition solution.

The AFM images obtained on an identically prepared electrode, exposed to the Pt deposition solution but without a bias for electrodeposition (OCP immersion), are shown in Figure 6. Pt structures have formed on the surface of the HOPG and, interestingly, while they appear to be dispersed on the surface, the features follow the defects on HOPG. Spontaneous deposition on HOPG is not well understood and, while edges and defects on HOPG have been suggested to act as reducing agents [30], there are reports of no preferential deposition around the edges [32]. Our observations support that the defects on HOPG are responsible for the deposition of Pt, (Figure 6D which shows the detail of step edges) which occurs spontaneously once the HOPG is immersed in a Pt-containing solution. Compare with Figure 6E which is the AFM of bare HOPG. The Pt deposition is occurring preferentially on the step/edges of the HOPG substrate and other defects around the surface. One sample was subjected to electrochemical cycling represented in Figure 5, and the AFM of this sample is shown in Figure 6C after electrochemical testing. Figure 6C shows that the Pt structures on the HOPG surface become larger NPs, probably due to the reconstruction of the initial Pt nanostructures during the CVs. This observation suggests that the Pt structures spontaneously deposited on the surface while electrochemically active (Figure 5) are not strongly attached to the HOPG surface because they re-construct (compare Figure 6A,B,D before and after electrochemistry, Figure 6C) and also we observed that the AFM would move some particles during imaging.

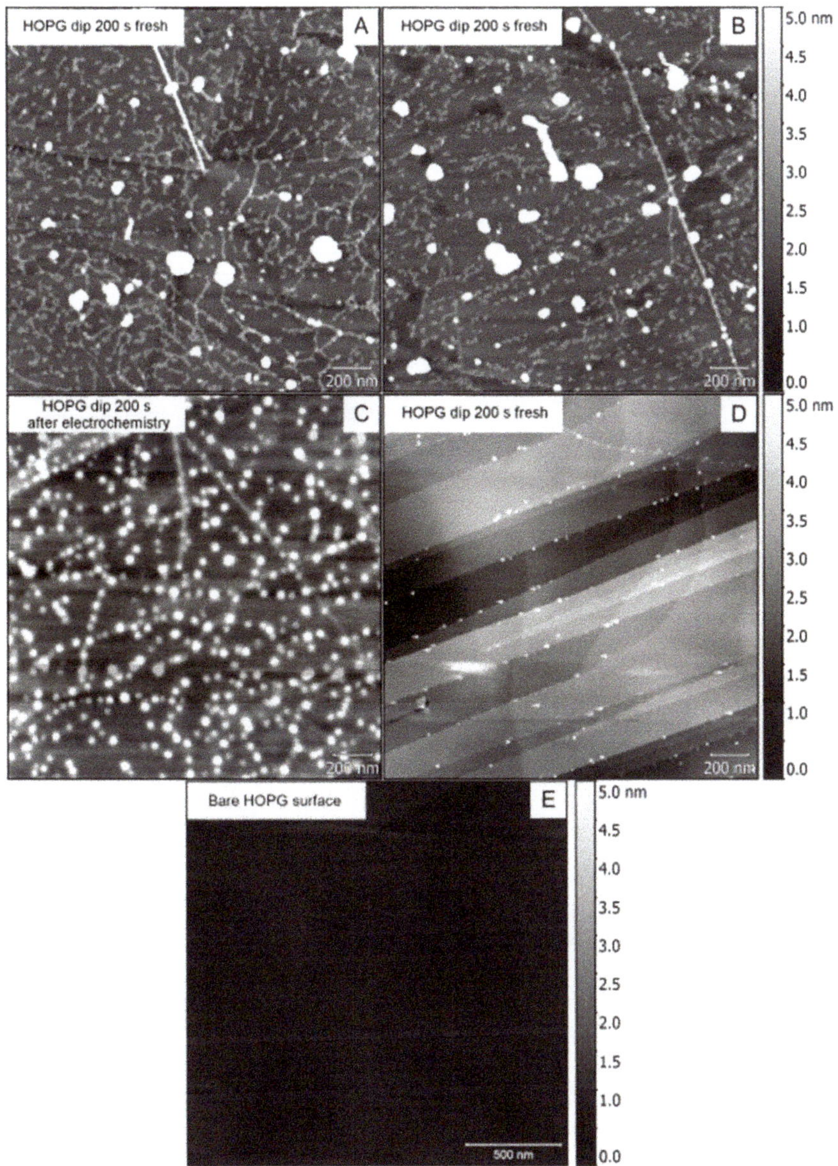

Figure 6. Atomic force microscopy (AFM) images of HOPG surfaces immersed in a Pt deposition solution (3 mM H_2PtCl_6, 0.5 M NaCl, pH 2.3) recorded after the immersion (**A**, **B** and **D** fresh) and after the electrochemical cycles reported in Figure 5 (**C**, after electrochemistry). All the electrodes were prepared by immersion at open circuit potential (OCP), i.e., without applying an external potential. (**E**) Corresponds to a blank HOPG.

The spontaneous Pt deposition that we observed here also occurs under a wide range of conditions [30,32,34–36]. As discussed in the introduction, partially oxidized sites have been proposed as the reducing agent for Pt. Penner and co-workers proposed the oxidation of step edge and other defect sites that could be aldehydes, alcohols and ketons partially oxidized [30]. This assignment was made based on the observation that surface enhanced Raman spectroscopy detect carbonyl functionalities on thermally pitted HOPG substrates. We are not aware of reports on the half oxidation reaction during Pt deposition in the literature, in part because the spontaneous deposition of Pt on C is a point that has received limited attention in the literature. Because the electrochemical papers lack XPS studies of the surface oxidation before and after Pt deposition, we get insight into the half reactions of oxidation from the XPS studies of the adsorption and deposition of Au on HOPG from acidic solutions [38–40]. Because they do not report the redox properties of these surface reactions, we must deduce possible reactions from surface composition studies. XPS showed that immersing HOPG in aqueous HCl (diluted) yields mostly an OH-terminated surface [39] together with a theoretical study to support the experimental finding. Density functional theory (DFT) calculations of the step edges along the HOPG terminated on a C–OH group gave the more stable configuration having the OH group roughly in plane with the graphitic sheet oriented towards the H on the same C=C bond of the armchair edge [39]. Because this configuration has a slightly higher energy than a single graphene sheet (by 0.035 eV/A) the authors proposed that mild oxidation conditions are necessary to produce the C–OH groups on the edges. During heating, the hydroxyl groups oxidize to produce ketone and ester groups on the HOPG surface [39,40]. However, the C–OH groups reduced Au^{3+} to Au^0 from solution while the C–OH oxidize to C=O groups [38]. Therefore, based on the findings by XPS and AFM of Au deposition on HOPG [38–40] along with the AFM studies in this work (Figure 6A,B,D) and previous reports that show preferential deposition of Pt along the step edges [32,34,35], we propose the following oxidation reactions in our precursor solution. Initially, during immersion of HOPG to the precursor solution (3 mM H_2PtCl_6 0.5 M NaCl pH 2.3, which corresponds to dilute HCl), the step edges oxidize to C–OH groups as shown in Scheme 1. To illustrate the reaction, we use a triphenylene to depict the step edges on the HOPG surface that yields an OH and H on the same C=C bond as in the DFT study by Buono et al. [39]. The structure also affords the simplest oxidation case, where the half-reaction is a 2e, $2H^+$ oxidation without the need to create radicals or dangling bonds on the polycyclic structure during the oxidation of the C–H edge sites to C–OH or for the following oxidation of the OH groups.

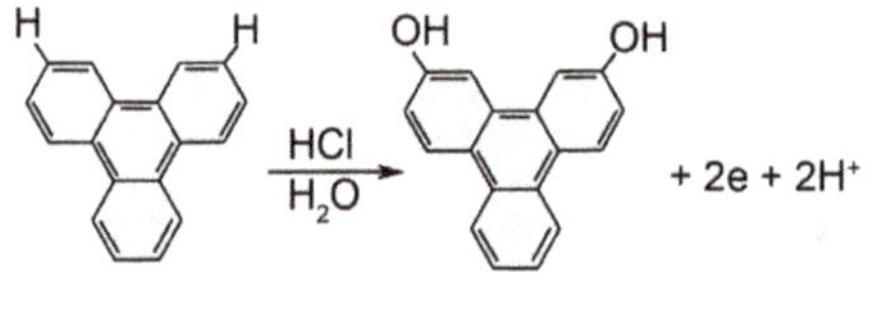

Scheme 1. Proposed oxidation of the HOPG step edges in deposition solution 3 mM H_2PtCl_6 0.5 M NaCl pH 2.3. Triphenylene illustrates a terminal section of the extended HOPG structure. The reaction is deduced from the data in ref [39].

We propose that a second step is the oxidation of the C–OH to C=O. The more likely redox reaction here is the 2e, $2H^+$ oxidation depicted in Scheme 2, with adjacent C–OH groups that can provide enough electrons to reduce the precursor to Pt^0 either by the two sequential, 2e reactions in Schemes 1 and 2 or through the oxidation of adjacent groups.

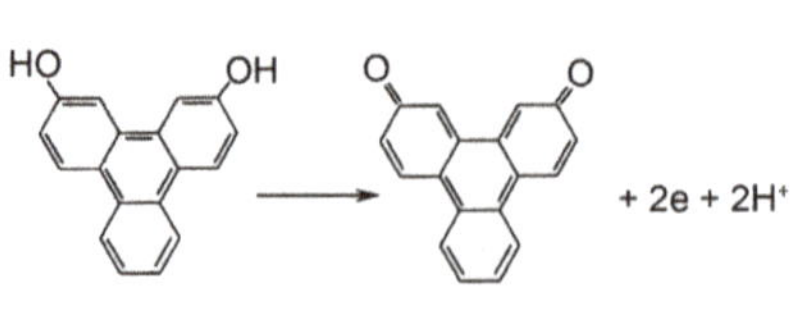

Scheme 2. Proposed oxidation of HOPG step edges as reducing agents for H_2PtCl_6 in the precursor solution. Derived after ref. [38], our results and other reports [32,34,35] that find preferential Pt deposits along the edges.

3.3. Phase 2. Electrodeposition

The results seen on Figures 3 and 4 on the HOPG, compared with those obtained with the Pt electrode (Figure 2), indicate that reduction of $PtCl_x$ complexes is an intrinsically slower process on the surface of HOPG than on Pt. Therefore, the deposition of Pt is expected to occur at a higher rate on Pt structures than on a bare HOPG surface, leading to tridimensional structures. Eventually, at high enough overpotentials, the production of bulk H_2 becomes the predominant electrolytic process in the deposition system. Therefore, we investigated the effect of H_2 on the deposition of the sample. In Figure 7 are shown the LSVs obtained for a freshly cleaved HOPG sample and Pt/HOPG sample, prepared under the conditions of Figure 1. Figure 7A is the data for bare HOPG in the blank electrolyte, note that there is a small peak around −1 V vs. Ag/AgCl, which indicates negligible H^+ reduction activity for HOPG. Figure 7B shows the LSV for Pt/HOPG under the same conditions, and as already pointed out, the peak for the diffusion-limited H_2 evolution is seen around −0.5 V vs. Ag/AgCl. This peak is shifted with respect to that observed for bulk Pt (−0.4 V, Figure 2), because the surface is not completely covered with Pt, and the electrode material behaves with a rate constant lower than that of Pt. For the Pt/HOPG electrode, $k^0 = 6 \times 10^{-3}$ cm/s, as obtained from DigiElch (Elchsoft, Kleinromstedt, Germany) lower than the bulk Pt $k^0 = 0.36$ cm/s [42]. The simulations are shown in Figure 7B (line b) to show the overlap with the experimental data (line a) and were performed based on the previously reported H^+ diffusion coefficient [42] of $D_{H^+} = 7.1 \times 10^{-5}$ cm^2/s for H_2, $D_{H_2} = 3.37 \times 10^{-5}$ cm^2/s [43]. Based on this, a new electrode was prepared with the electrodeposition potential held at −0.5 V vs. Ag/AgCl, which is at the diffusion-limited rate of H_2 evolution. The AFM image of a sample prepared at −0.5 V vs. Ag/AgCl is depicted in Figure 8. The sample prepared under these conditions presents two populations of Pt structures. In addition to platelets 40 nm in diameter and 3–4 nm thick, larger Pt aggregates (ca. 120 × 30 nm) appears, which are not observed when the sample is prepared at larger overpotentials (−2.0 V) [19,29].

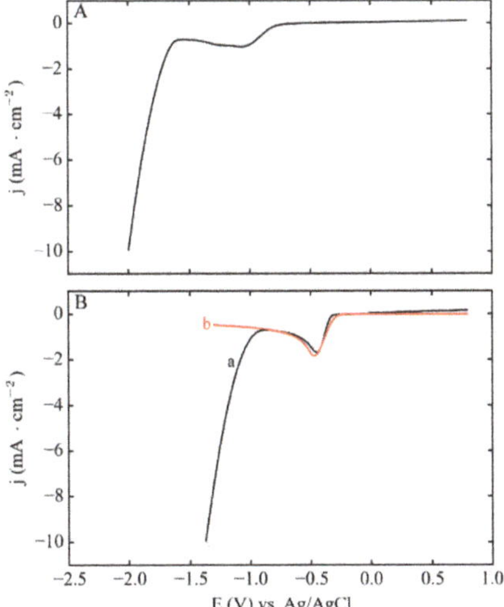

Figure 7. (**A**) Linear sweep voltammograms (LSV) of a bare HOPG electrode in 3 mM H_2PtCl_6, 0.5 M NaCl, pH 2.3, under N_2 and (**B**) For a Pt/HOPG electrode, same conditions as in (**A**). Line (a) shows the experimental data and line (b) the comparison to the digital simulations.

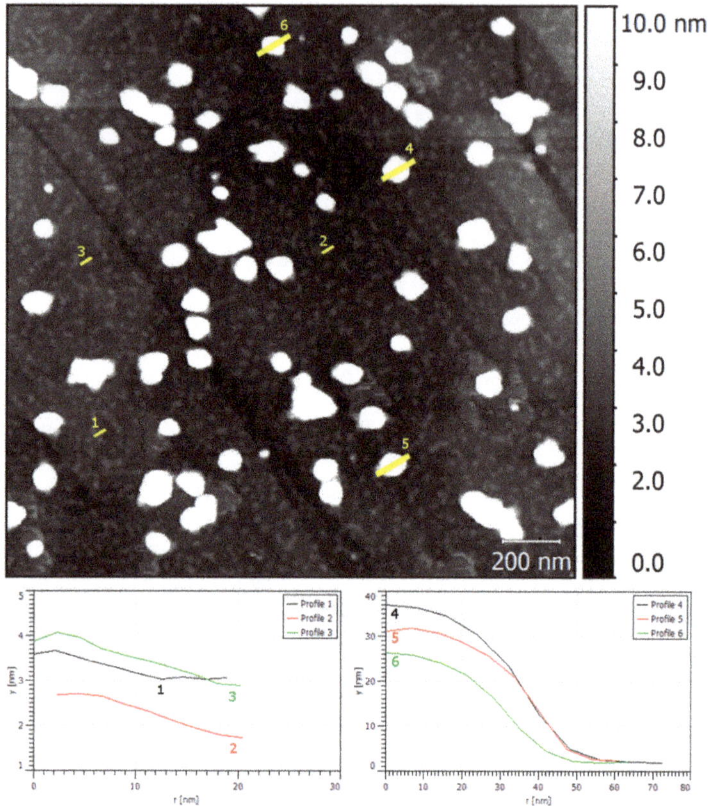

Figure 8. AFM of a sample electrodeposited at -0.5 V vs. Ag/AgCl in 3 mM H_2PtCl_6, 0.5 M NaCl, pH 2.3. Radial profiles of the two population of the Pt features are also reported.

Figure 9 shows the results for the electrodeposition of Pt as a function of time. The current was recorded during deposition, and then, after the Pt deposition was finished, a control experiment was run in the blank solution to study the contribution of H_2 evolution. The corrected current gives an estimate of the amount of Pt deposited under these conditions. Interestingly, the current at short times (t < 50 s) is much lower than expected from the diffusion behavior (note that it is less negative than the calculation based on the reported Pt diffusion coefficient, $D_{Pt}{}^{4+} = 2.2 \times 10^{-5}$ cm^2/s) [31]. These results are consistent with our observation that in the CVs the diffusion-limited peak is not seen on HOPG. At higher deposition times, the current becomes larger than the expected from diffusion-limited behavior. At this point, enough Pt has been deposited on the surface to drive the formation of H_2 bubbles that disturb the solution, therefore, increasing the value of the current. Integrating the current of the Pt deposition provides an estimate for the upper limit of Pt deposition, which corresponds to 40 µg/cm^2. However, under the conditions of high overpotentials (E = -2 V vs. Ag/AgCl), the loading is expected to be much lower based on the size of the Pt features: the Pt NPs in Figure 8 (E = -0.5 V) are much larger than the nanostructures on Figure 10 (E = -2 V) are mostly below 2 nm in high. This sample prepared at high overpotentials shows the nanoplatelets previously described [19,28,29], together with some larger Pt aggregates. The latter are smaller than those seen above obtained at lower overpotentials (Figure 8) (ca. 90 × 20 nm). On the contrary, the nanoplatelets form more extended structures of ca. 200 nm with a 2–6 nm thickness. Therefore, it is likely that H_2 plays a role in limiting the 3D growth of the Pt deposits and favoring the formation of contiguous Pt islands. Based on the work presented here, we propose that the effect of H_2 becomes dominant at larger overpotentials, where water electrolysis is predominant over proton reduction.

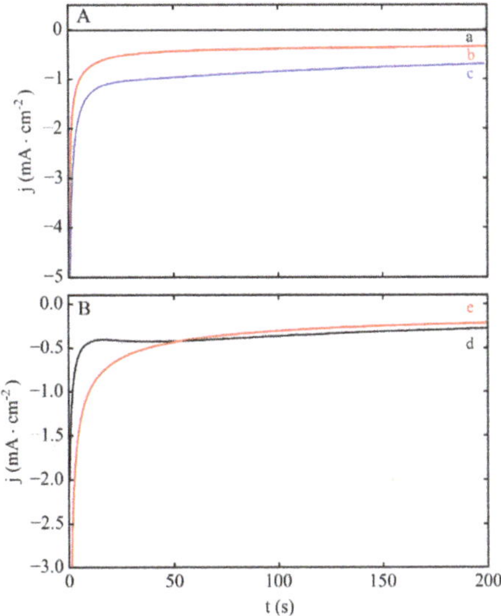

Figure 9. (**A**) Current obtained for the electrodeposition of Pt at E = −0.5 V vs. Ag/AgCl. a: HOPG in blank solution of 0.5 M NaCl, pH 2.3, b: sample deposited with Pt in blank solution, c: Pt deposition in 3 mM H_2PtCl_6, 0.5 M NaCl, pH 2.3. (**B**) Pt deposition current (line d) obtained by subtracting lines (a) and (b) from line (c) in (**A**). The (d) line in (**B**) is the experimental curve, while (e) represents the behavior expected from the Cottrell equation.

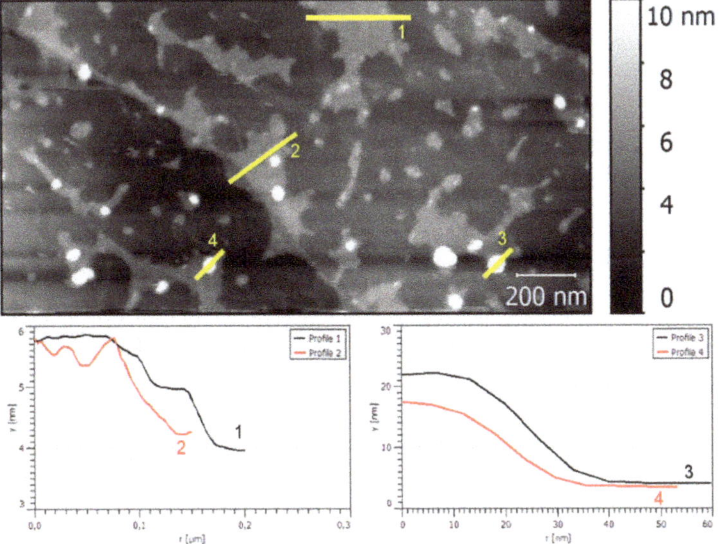

Figure 10. AFM image of a sample electrodeposited at −2 V vs. Ag/AgCl showing the radial profiles measurements of the two populations of deposited structures.

Chloride ions have been proposed to control NP growth [31,32]. A set of CVs of Pt in solutions containing Cl^- is shown in Figure 11 recorded at a scan rate $v = 0.5$ V/s (A, 0.1 M HCl), and are compared with CVs obtained using a solution of the same pH without Cl^- (0.05 M H_2SO_4). The scan rate and the potential window were chosen to show the different peaks, given that the processes involved in the $PtCl_x$ reduction overlap with H_2 evolution. The CVs in HCl show after the formation of $PtCl_x$, several reduction peaks, around 0.8 V, 0.2 and 0 V vs. Ag/AgCl. The last is likely due to H_2 processes because it is also present when using the H_2SO_4 solution. Therefore, the re-dissolution of Cl^- into the solution is proposed to occur in the potential region between 0.8 and 0.2 V vs. Ag/AgCl. These potentials are considerably more positive than those used in the electrodeposition of Pt in this work. However, we should notice that while the stripping of Cl^- is seen at more positive potentials, at negative potentials, Cl^- could still be present in proximity to the Pt surface, as it is expected to be a major component of the double layer. However, the Cl^- will be expected to be weakly bound to the Pt surface under these conditions.

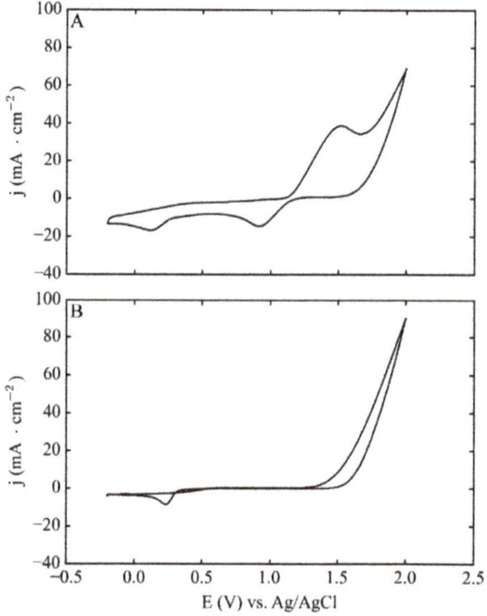

Figure 11. CVs of a 1.6 mm diameter Pt electrode in (**A**) 0.1 M HCl and in (**B**) 0.05 M H_2SO_4, $v = 0.5$ V/s.

4. Conclusions

The electrodeposition conditions for Pt/HOPG were investigated, and it was observed that once the HOPG electrode is immersed in solution, Pt spontaneously deposits on the HOPG surface and forms nanostructures that control the electrodeposition of Pt. This is due to the reduction of $PtCl_x$ complexes being kinetically more favorable on the Pt surface than on the HOPG surface. In fact, the HOPG surface activity towards the reduction of Pt-Cl complexes is a strong function of the nature of the surface characteristics, which at the moment cannot be controlled. The defects on the surface drive the deposition of Pt and that the density of the active sites along the surface (defects, edges, etc.) varies depending on the exfoliation of HOPG. The growth of Pt limited to an island morphology is likely due to H_2 evolution from water reduction, but not from the H^+ reduction. Thus, a different process that occurs at more negative potentials is likely the cause for the limiting Pt growth. From the CVs it is apparent that, besides the reduction of free H^+, the reduction of water also is important at potentials more negative than -1 V. Therefore, it is considered that an intermediate in the reduction of water plays a role in protecting the Pt. These features of the electrodeposition are currently being investigated in our laboratories and will be reported in due time.

Author Contributions: Conceptualization, M.A.A.-A.; Methodology, M.A.A.-A.; Formal Analysis, M.A.A.-A., F.F., P.S.; Investigation, F.F., G.E., P.S.; Resources, M.A.A.-A., S.C., D.J.J. and J.R.; Writing-Original Draft Preparation, M.A.A.-A.; Writing, Review and Editing, S.C., D.J.J. and J.R.; Supervision, M.A.A.-A., S.C., D.J.J. and J.R.; Project Administration, M.A.A.-A. and S.C.; Funding Acquisition, M.A. and S.C.

Funding: The research leading to these results has received funding from the European Research Council (ERC) under the European Union's Seventh Framework Programme (FP/2007-2013)/ERC Grant Agreement SPINAM n. 306682. MAAA acknowledges support for this project by NSF CAREER Award No. CHE-1255387 and from SPINAM during his sabbatical stay at ICGM within the framework of an NSF-ERC program.

Conflicts of Interest: The authors declare no conflict of interest

References

1. Ercolano, G.; Cavaliere, S.; Rozière, J.; Jones, D.J. Recent developments in electrocatalyst design thrifting noble metals in fuel cells. *Curr. Opin. Electrochem.* **2018**, *9*, 271–277. [CrossRef]

2. Lv, H.; Li, D.; Strmcnik, D.; Paulikas, A.P.; Markovic, N.M.; Stamenkovic, V.R. Recent advances in the design of tailored nanomaterials for efficient oxygen reduction reaction. *Nano Energy* **2016**, *29*, 149–165. [CrossRef]

3. Cui, C.; Gan, L.; Heggen, M.; Rudi, S.; Strasser, P. Compositional segregation in shaped Pt alloy nanoparticles and their structural behaviour during electrocatalysis. *Nat. Mater.* **2013**, *12*, 765–771. [CrossRef] [PubMed]

4. Gan, L.; Rudi, S.; Cui, C.; Heggen, M.; Strasser, P. Size-controlled synthesis of sub-10 nm PtNi$_3$ alloy nanoparticles and their unusual volcano-shaped size effect on ORR electrocatalysis. *Small* **2016**, *12*, 3189–3196. [CrossRef] [PubMed]

5. Ma, Y.; Miao, L.; Guo, W.; Yao, X.; Qin, F.; Wang, Z.; Du, H.; Li, J.; Kang, F.; Gan, L. Modulating surface composition and oxygen reduction reaction activities of Pt–Ni octahedral nanoparticles by microwave-enhanced surface diffusion during solvothermal synthesis. *Chem. Mater.* **2018**, *30*, 4355–4360. [CrossRef]

6. Xiong, Y.; Xiao, L.; Yang, Y.; DiSalvo, F.J.; Abruña, H.D. High-loading intermetallic Pt$_3$Co/C core-shell nanoparticles as enhanced activity electrocatalyst towards the oxygen reduction reaction (ORR). *Chem. Mater.* **2018**, *30*, 1532–1539. [CrossRef]

7. Strasser, P.; Koh, S.; Anniyev, T.; Greeley, J.; More, K.; Yu, C.; Liu, Z.; Kaya, S.; Nordlund, D.; Ogasawara, H.; et al. Lattice-strain control of the activity in dealloyed core–shell fuel cell catalysts. *Nat. Chem.* **2010**, *2*, 454–460. [CrossRef] [PubMed]

8. Di Noto, V.; Negro, E.; Polizzi, S.; Vezzù, K.; Toniolo, L.; Cavinato, G. Synthesis, studies and fuel cell performance of "core-shell" electrocatalysts for oxygen reduction reaction based on a PtNi$_x$ carbon nitride "shell" and a pyrolyzed polyketone nanoball "core". *Int. J. Hydrog. Energy* **2014**, *39*, 2812–2827. [CrossRef]

9. Gan, L.; Cui, C.; Rudi, S.; Strasser, P. Core-shell and nanoporous particle architectures and their effect on the activity and stability of Pt ORR electrocatalysts. *Top. Catal.* **2014**, *57*, 236–244. [CrossRef]

10. Gong, H.; Cao, X.; Li, F.; Gong, Y.; Gu, L. PdAuCu nanobranch as self-repairing electrocatalyst for oxygen reduction reaction. *ChemSusChem* **2017**, *10*, 1469–1474. [CrossRef] [PubMed]

11. Bu, L.; Ding, J.; Guo, S.; Zhang, X.; Su, D.; Zhu, X.; Yao, J.; Guo, J.; Lu, G.; Huang, X. A general method for multimetallic platinum alloy nanowires as highly active and stable oxygen reduction catalysts. *Adv. Mater.* **2015**, *27*, 7204–7212. [CrossRef] [PubMed]

12. Bu, L.; Guo, S.; Zhang, X.; Shen, X.; Su, D.; Lu, G.; Zhu, X.; Yao, J.; Guo, J.; Huang, X. Surface engineering of hierarchical platinum-cobalt nanowires for efficient electrocatalysis. *Nat. Commun.* **2016**, *7*, 11850. [CrossRef] [PubMed]

13. Li, M.; Zhao, Z.; Cheng, T.; Fortunelli, A.; Chen, C.-Y.; Yu, R.; Zhang, Q.; Gu, L.; Merinov, B.V.; Lin, Z.; et al. Ultrafine jagged platinum nanowires enable ultrahigh mass activity for the oxygen reduction reaction. *Science* **2016**, *354*, 1414–1419. [CrossRef] [PubMed]

14. Liang, H.-W.; Cao, X.; Zhou, F.; Cui, C.-H.; Zhang, W.-J.; Yu, S.-H. A free-standing Pt-nanowire membrane as a highly stable electrocatalyst for the oxygen reduction reaction. *Adv. Mater.* **2011**, *23*, 1467–1471. [CrossRef] [PubMed]

15. Mao, J.; Chen, W.; He, D.; Wan, J.; Pei, J.; Dong, J.; Wang, Y.; An, P.; Jin, Z.; Xing, W.; et al. Design of ultrathin Pt-Mo-Ni nanowire catalysts for ethanol electrooxidation. *Sci. Adv.* **2017**, *3*, e1603068. [CrossRef] [PubMed]

16. Ruan, L.; Zhu, E.; Chen, Y.; Lin, Z.; Huang, X.; Duan, X.; Huang, Y. Biomimetic synthesis of an ultrathin platinum nanowire network with a high twin density for enhanced electrocatalytic activity and durability. *Angew. Chem. Int. Ed.* **2013**, *52*, 12577–12581. [CrossRef] [PubMed]

17. Wang, D.; Xin, H.L.; Hovden, R.; Wang, H.; Yu, Y.; Muller, D.A.; DiSalvo, F.J.; Abruña, H.D. Structurally ordered intermetallic platinum–cobalt core–shell nanoparticles with enhanced activity and stability as oxygen reduction electrocatalysts. *Nat. Mater.* **2013**, *12*, 81–87. [CrossRef] [PubMed]

18. Alia, S.M.; Pylypenko, S.; Dameron, A.; Neyerlin, K.C.; Kocha, S.S.; Pivovar, B.S. Oxidation of platinum nickel nanowires to improve durability of oxygen-reducing electrocatalysts. *J. Electrochem. Soc.* **2016**, *163*, 296–301. [CrossRef]

19. Ercolano, G.; Farina, F.; Cavaliere, S.; Jones, D.J.; Rozière, J. Towards ultrathin Pt films on nanofibres by surface-limited electrodeposition for electrocatalytic applications. *J. Mater. Chem. A* **2017**, *5*, 3974–3980. [CrossRef]

20. Parsonage, E.E.; Debe, M.K. Nanostructured Electrode Membranes. U.S. Patent US5338430A, 16 August 1994.

21. Alia, S.M.; Ngo, C.; Shulda, S.; Ha, M.A.; Dameron, A.A.; Weker, J.N.; Neyerlin, K.C.; Kocha, S.S.; Pylypenko, S.; Pivovar, B.S. Exceptional oxygen reduction reaction activity and durability of platinum-nickel nanowires through synthesis and post-treatment optimization. *ACS Omega* **2017**, *2*, 1408–1418. [CrossRef]

22. Debe, M.K. Electrocatalyst approaches and challenges for automotive fuel cells. *Nature* **2012**, *486*, 43–51. [CrossRef] [PubMed]

23. Alia, S.M.; Yan, Y.S.; Pivovar, B.S. Galvanic displacement as a route to highly active and durable extended surface electrocatalysts. *Catal. Sci. Technol.* **2014**, *4*, 3589–3600. [CrossRef]

24. Papaderakis, A.; Mintsouli, I.; Georgieva, J.; Sotiropoulos, S. Electrocatalysts prepared by ggalvanic replacement. *Catalysts* **2017**, *7*, 80. [CrossRef]

25. Gruber, D.; Ponath, N.; Müller, J.; Lindstaedt, F. Sputter-deposited ultra-low catalyst loadings for PEM fuel cells. *J. Power Sources* **2005**, *150*, 67–72. [CrossRef]

26. Milchev, A. *Electrocrystallization: Fundamentals of Nucleation and Growth*; Kluwer Academic Publishers: New York, NY, USA, 2002.

27. Liu, Y.; Gokcen, D.; Bertocci, U.; Moffat, T.P. Self-terminating growth of platinum films by electrochemical deposition. *Science* **2012**, *338*, 1327–1330. [CrossRef] [PubMed]

28. Ercolano, G.; Farina, F.; Cavaliere, S.; Jones, D.J.; Rozière, J. Multilayer hierarchical nanofibrillar electrodes with tuneable lacunarity with 2D like Pt deposits for PEMFC. *ECS Trans.* **2017**, *80*, 757–762. [CrossRef]

29. Farina, F.; Ercolano, G.; Cavaliere, S.; Jones, D.J.; Rozière, J. Surface-limited electrodeposition of continuous platinum networks on highly ordered pyrolytic graphite. *Nanomaterials* **2018**. under review.

30. Zoval, J.V.; Lee, J.; Gorer, S.; Penner, R.M. Electrochemical preparation of platinum nanocrystallites with size selectivity on basal plane oriented graphite surfaces. *J. Phys. Chem. B* **1998**, *102*, 1166–1175. [CrossRef]

31. Lu, G.; Zangari, G. Electrodeposition of platinum on highly oriented pyrolyticgraphite. Part I: electrochemical characterization. *J. Phys. Chem. B* **2005**, *109*, 7998–8007. [CrossRef] [PubMed]

32. Lu, G.; Zangari, G. Electrodeposition of platinum nanoparticles on highly oriented pyrolitic graphite: Part II: Morphological characterization by atomic force microscopy. *Electrochim. Acta* **2006**, *51*, 2531–2538. [CrossRef]

33. Simonov, A.N.; Cherstiouk, O.V.; Vassiliev, S.Y.; Zaikovskii, V.I.; Filatov, A.Y.; Rudina, N.A.; Savinova, E.R.; Tsirlina, G.A. Potentiostatic electrodeposition of Pt on GC and on HOPG at low loadings: Analysis of the deposition transients and the structure of Pt deposits. *Electrochim. Acta* **2014**, *150*, 279–289. [CrossRef]

34. Arroyo Gómez, J.J.; García Silvana, G. Spontaneous deposition of Pt-nanoparticles on HOPG surfaces. *Surf. Interface Anal.* **2015**, *47*, 1127–1131. [CrossRef]

35. Shen, P.K.; Chi, N.; Chan, K.Y.; Phillips, D.L. Platinum nanoparticles spontaneously formed on HOPG. *Appl. Surf. Sci.* **2001**, *172*, 159–166. [CrossRef]

36. Quaino, P.M.; Gennero de Chialvo, M.R.; Vela, M.E.; Salvarezza, R.C. Self-assembly of platinum nanowires on HOPG. *J. Argent. Chem. Soc.* **2005**, *93*, 215–224.

37. Juarez, M.F.; Fuentes, S.; Soldano, G.J.; Avalle, L.; Santos, E. Spontaneous formation of metallic nanostructures on highly oriented pyrolytic graphite (HOPG): An ab initio and experimental study. *Faraday Discuss.* **2014**, *172*, 327–347. [CrossRef] [PubMed]

38. Bowden, B.; Davies, M.; Davies, P.R.; Guan, S.; Morgan, D.J.; Roberts, V.; Wotton, D. The deposition of metal nanoparticles on carbon surfaces: The role of specific functional groups. *Faraday Discuss.* **2018**. [CrossRef] [PubMed]

39. Buono, C.; Davies, P.R.; Davies, R.J.; Jones, T.; Kulhavý, J.; Lewis, R.; Morgan, D.J.; Robinson, N.; Willock, D.J. Spectroscopic and atomic force studies of the functionalisation of carbon surfaces: New insights into the role of the surface topography and specific chemical states. *Faraday Discuss.* **2014**, *173*, 257–272. [CrossRef] [PubMed]
40. Burgess, R.; Buono, C.; Davies, P.R.; Davies, R.J.; Legge, T.; Lai, A.; Lewis, R.; Morgan, D.J.; Robinson, N.; Willock, D.J. The functionalisation of graphite surfaces with nitric acid: Identification of functional groups and their effects on gold deposition. *J. Catal.* **2015**, *323*, 10–18. [CrossRef]
41. Bard, A.J.; Faulkner, L.R. *Electrochemical Methods: Fundamentals and Applications*; John Wiley and Sons: New York, NY, USA, 2001; p. 833.
42. Zhou, J.; Zu, Y.; Bard, A.J. Scanning electrochemical microscopy: Part 39. The proton/hydrogen mediator system and its application to the study of the electrocatalysis of hydrogen oxidation. *J. Electroanal. Chem.* **2000**, *491*, 22–29. [CrossRef]
43. Himmelblau, D.M. Diffusion of dissolved gases in liquids. *Chem. Rev.* **1964**, *64*, 527–550. [CrossRef]
44. Colom, F., II. Palladium and platinum. In *Standard Potentials in Aqueous Solution*; Bard, A.J., Parsons, R., Eds.; IUPAC-Marcel Dekker, Inc.: New York, NY, USA, 1985.
45. Ginstrup, O. The redox system platinum(0)/platinum(II)/platinum(IV) with chloro and bromo ligands. *Acta Chem.* Scand. **1972**, *26*, 1527–1541. [CrossRef]
46. Uosaki, K.; Ye, S.; Naohara, H.; Oda, Y.; Haba, T.; Kondo, T. Electrochemical epitaxial growth of a Pt(111) phase on an Au(111) electrode. *J. Phys. Chem. B* **1997**, *101*, 7566–7572. [CrossRef]

 *nanomaterials* 

Article

On the Structure of Ultrathin FeO Films on Ag(111)

Mikołaj Lewandowski [1,*], Tomasz Pabisiak [2,*], Natalia Michalak [3], Zygmunt Miłosz [1], Višnja Babačić [1], Ying Wang [1], Michał Hermanowicz [4], Krisztián Palotás [5], Stefan Jurga [1] and Adam Kiejna [2]

[1] NanoBioMedical Centre, Adam Mickiewicz University, Umultowska 85, 61-614 Poznań, Poland; zmilosz@amu.edu.pl (Z.M.); visbab@amu.edu.pl (V.B.); yinwan@amu.edu.pl (Y.W.); stjurga@amu.edu.pl (S.J.)

[2] Institute of Experimental Physics, University of Wrocław, Pl. M. Borna 9, 50-204 Wrocław, Poland; adam.kiejna@uwr.edu.pl

[3] Institute of Molecular Physics, Polish Academy of Sciences, M. Smoluchowskiego 17, 60-179 Poznań, Poland; michalak@ifmpan.poznan.pl

[4] Institute of Physics, Poznan University of Technology, Piotrowo 3, 60-965 Poznań, Poland; michal.hermanowicz@put.poznan.pl

[5] MTA-SZTE Reaction Kinetics and Surface Chemistry Research Group, University of Szeged, 6720 Szeged, Hungary; kpalotas77@gmail.com

[*] Correspondences: lewandowski@amu.edu.pl (M.L.); tomasz.pabisiak@uwr.edu.pl (T.P.); Tel.: +48-61-829-6717 (M.L.); +48-71-375-9342 (T.P.)

Received: 16 August 2018; Accepted: 9 October 2018; Published: 13 October 2018

Abstract: Ultrathin transition metal oxide films exhibit unique physical and chemical properties not observed for the corresponding bulk oxides. These properties, originating mainly from the limited thickness and the interaction with the support, make those films similar to other supported 2D materials with bulk counterparts, such as transition metal dichalcogenides. Ultrathin iron oxide (FeO) films, for example, were shown to exhibit unique electronic, catalytic and magnetic properties that depend on the type of the used support. Ag(111) has always been considered a promising substrate for FeO growth, as it has the same surface symmetry, only ~5% lattice mismatch, is considered to be weakly-interacting and relatively resistant to oxidation. The reports on the growth and structure of ultrathin FeO films on Ag(111) are scarce and often contradictory to each other. We attempted to shed more light on this system by growing the films using different preparation procedures and studying their structure using scanning tunneling microscopy (STM), low energy electron diffraction (LEED) and X-ray photoelectron spectroscopy (XPS). We observed the formation of a previously unreported Moiré superstructure with 45 Å periodicity, as well as other reconstructed and reconstruction-free surface species. The experimental results obtained by us and other groups indicate that the structure of FeO films on this particular support critically depends on the films' preparation conditions. We also performed density functional theory (DFT) calculations on the structure and properties of a conceptual reconstruction-free FeO film on Ag(111). The results indicate that such a film, if successfully grown, should exhibit tunable thickness-dependent properties, being substrate-influenced in the monolayer regime and free-standing-FeO-like when in the bilayer form.

Keywords: iron oxides; ultrathin films; silver; epitaxial growth; structural characterization; STM; LEED; XPS; DFT; model system

---

1. Introduction

Ultrathin films grown on single crystal supports are believed to constitute a new class of 2D materials with properties governed by low-dimensionality and the interaction with the substrate and, therefore, different from those of the corresponding bulk materials [1]. The substrate- and thickness-mediated effects are particularly pronounced in the case of ultrathin films of insulators or semiconductors grown on conducting supports, where the overlapping of the electronic states of the film and the substrate takes place. Historically, such films were meant to be used as model systems that would allow the use of surface science tools for the studies of materials that are not electrically conductive in the bulk form [2]. However, the unpredicted at the time film-substrate interactions, often manifested by structural reconstructions, such as Moiré superstructures, were found to increase the complexity of those systems and limit their initially intended application. Quite unexpectedly, the unique properties of ultrathin films opened new pathways for the rediscovery of the well-known simple compounds, such as transition metal oxides or dichalcogenides, in their unnatural two-dimensional form, and many reports describing such systems have been published to date.

Monolayer iron oxide (FeO) films can be grown in the close-packed <111> direction on various (111)-oriented and (0001)-oriented metal single crystal supports [3–7]. To date, all such films were found to exhibit lattice mismatch-induced Moiré superstructures which give them unique properties but also make none of them fully-representative as model system imitating bulk oxide material (which would be also interesting to achieve). Among the substrates used, Ag(111) has always been considered a promising candidate for iron oxides growth, as it has the same surface symmetry as FeO(111), only ~5% lattice mismatch (much smaller than the commonly used Pt(111) [3,4]), is considered a weakly-interacting substrate and is relatively resistant to oxidation. There have been several reports on the growth of ultrathin and thin iron oxide films on differently oriented silver single crystal supports [8–19]. An overview of these works is presented in the Supplementary Materials file, Section 1 (SM1). Regarding ultrathin FeO(111) films on Ag(111), most reports were contradictory to each other and lacked detailed structural characterization by atomic-resolution techniques, such as scanning tunneling microscopy (STM). The very first work by Waddil and Ozturk suggested that deposition of submonolayer amounts of iron onto a silver substrate kept at room temperature (RT) and subsequent oxidation results in a (1×1) growth of FeO, i.e. adaptation of the surface lattice constant of FeO (3.04 Å) to Ag (2.89 Å) [8]. The assumption was based on low energy electron diffraction (LEED) and X-ray photoelectron spectroscopy (XPS) results. More recently, Lundgren, Weaver and co-workers grew the films using different procedure, i.e. by depositing iron onto a slightly heated silver substrate and in an oxygen ambient [17–19]. The authors employed STM and observed the formation of a p(9×9) Moiré superstructure with a significantly expanded FeO surface lattice constant (to 3.25 Å) and very small separation between iron and oxygen layers (0.26 Å—meaning that the iron and oxygen atoms are almost in-plane). The superstructure was also visible on the acquired LEED patterns. The conclusions on the structure of FeO(111)/Ag(111) drawn by the two groups differ, however, it has to be underlined that (1) the preparation procedures used by both groups were not the same, (2) the authors of Ref. [8] did not employ STM (or any other atomic-resolution technique) to determine the exact structure of their films and (3) the authors of Refs. [17–19] were clearly stating that their films are not uniform and consist of a mixture of well-defined Moiré-reconstructed FeO (majority phase) and ill-defined "FeO$_x$" species (minority phase).

We try to shed more light on the structure of ultrathin FeO films grown on Ag(111). In our experiments, we applied different preparation procedures and used STM, LEED and XPS to analyze the structure of the fabricated films. Firstly, we deposited iron onto the silver substrate kept at room temperature and post-oxidized it. We found that such procedure leads to the growth of Moiré-reconstructed and structurally ill-defined islands located mainly at the silver step edges. Interestingly, the structural parameters of the observed Moiré superstructure differ significantly from those reported in References [17–19]. Secondly, we modified the procedure by depositing iron onto a heated silver substrate and post-oxidizing. In that way we succeeded to grow uniformly distributed surface structures, some of which were reconstructed and some reconstruction-free. Some of the observed reconstructed regions were virtually identical to those observed following room-temperature iron deposition and oxidation. The nature of the reconstruction-free regions was difficult to identify unambiguously, however, their atomic structure and electronic characteristics seemed to differ from those of clean Ag(111). We also performed density functional theory (DFT) calculations on the structure and properties of a conceptual reconstruction-free FeO/Ag(111). The results indicate that such film, if successfully grown, should exhibit tunable thickness-dependent properties, being substrate-influenced in the monolayer regime and free-standing-FeO-like when in the bilayer form.

2. Experimental and Theoretical Methods

The experiments were performed in an ultra-high vacuum (UHV) system (from Omicron, Taunusstein, Germany) consisting of three inter-connected chambers: The preparation chamber (base pressure: 5×10^{-10} mbar), the scanning probe microscopy chamber (base pressure: 3×10^{-11} mbar) and the load-lock chamber. The preparation chamber hosted a cold cathode sputter gun, single electron beam evaporator and e-beam heating stage. It was also equipped with a LEED instrument and an XPS setup. The scanning probe microscopy chamber was equipped with a variable-temperature STM. All STM measurements were performed in a constant current mode using W tips. The Ag(111) single crystal (purity: 99.999%, polishing accuracy: <0.1°; from MaTeck GmbH, Jülich, Germany) was cleaned by repeated cycles of 1 keV and 0.6 keV Ar^+ (purity: 99.999%; Messer Industriegase GmbH, Bad Soden, Germany) ion sputtering, annealing in O_2 (purity: 99.999%; Messer Industriegase GmbH, Bad Soden, Germany) and under UHV at $T \geq 700$ K. Iron (purity: 99.995%; Alfa Aesar GmbH, Karlsruhe, Germany) was evaporated from a 2 mm rod onto a Ag(111) substrate kept at RT or 500–600 K. The deposition rate, calibrated using STM, was approx. 2 monolayers (MLs) per minute (calibration accuracy: 10%), where 1 ML is defined as the amount of iron that would cover the Ag(111) surface with a closed bcc Fe(110) film. The STM images were processed using Gwyddion (ver. 2.4x, Open Source) [20] and WSxM (ver. 5.0 Develop 8.x, Freeware) [21] computer software. The oxidation of iron was performed by backfilling the preparation chamber with molecular oxygen using a leak valve and heating the sample to 700 K. Following 15–30 min oxidation, the sample was cooled down in oxygen for several minutes. The temperature of the sample was measured using an infrared pyrometer focused on a tantalum holder on which the sample was mounted. The cleanliness of the substrate, as well as the structure of Fe and FeO deposits, were characterized by STM, LEED and XPS. Scanning tunneling spectroscopy (STS) dI/dz experiments were performed using a lock-in technique: A modulation voltage of 60 mV at a frequency of 6777 Hz was applied to the z-piezo of the STM (which resulted in a periodic tip-sample distance change of $+/- 0.1$ nm), the STM bias voltage was set to 50 mV and the curves were acquired locally from the objects of interest. For each measurement, the tip was first retracted from the sample's surface by 1 nm (to attenuate any tip-sample interactions) and then brought to the surface with a speed of 1 nm/s. Each of the presented curves is a result of averaging of 15 similar measured curves and additional smoothing using the locally weighted scatterplot (LOESS) method. The LEED patterns were recorded at various energies ranging from 30 eV to 255 eV. The XPS measurements were performed using a monochromatic $AlK\alpha$ (1486.6 eV) X-ray source and a semispherical electron energy analyzer operating at a pass energy of 50 eV (survey) and 20 eV (regions). The data were calibrated with respect to the Ag $3d_{5/2}$ peak (368.2 eV) [22] and fitted using CasaXPS computer software (ver. 2.3.19PR1.0, Casa Software Ltd, Teignmouth, UK). A linear combination of Gauss and Lorentz functions (the so called Voigt function) and Shirley background subtraction were used for the fittings. Detailed information on all XPS fittings presented in this work is provided in SM2.

The DFT calculations were performed using the Vienna ab initio simulation package (VASP, ver. 5.4.4, Vienna, Austria) [23–25]. The electron-ion core interactions were represented by the projector augmented wave (PAW) potentials [26,27], with Ag $4d^{10}5s^1$, Fe $3d^74s^1$ and O $2s^22p^4$ states considered as the valence states. A plane wave basis set with a kinetic energy cutoff of 400 eV was applied. The exchange-correlation energy was treated at the spin-polarized generalized gradient approximation (GGA) level using the Perdew-Burke-Ernzerhof (PBE) functional [28]. The Brillouin zone was sampled using Γ-centered k-point meshes. A Fermi surface broadening of 0.2 eV was applied to improve convergence of the solutions using the second order Methfessel-Paxton method [29]. To account for the strongly correlated 3d electrons localized on the Fe ions, the Hubbard U correction was applied (GGA+U) within the rotationally invariant approach of Dudarev et al., with the effective parameter $U_{eff} = U - J = 3.0$ eV, where (U,J) = (4,1) eV are the Coulomb and screened exchange parameters, respectively [30]. The applied U_{eff} value is known to provide a satisfactory description of bulk characteristics of FeO [31]. Convergence threshold for the total energy of the studied systems was set to 10^{-6} eV. The lattice parameter of bulk fcc Ag, a = 4.155 Å, calculated using 16×16×16 k-points mesh, was found to agree well with other GGA calculations performed using a similar computational method (e.g. 4.16 Å [32] or 4.14 Å [33]) and overestimated the experimental value, 4.086 Å [34], by less than 1.7%. A silver (111) substrate was modelled with an asymmetric slab consisting of four atomic layers of Ag with a 1×1 or 2×2 surface unit cell. In order to check the appropriateness of the size of the unit cell, additional calculations for larger supercells were also performed. The slabs were separated from their periodic images by a vacuum region of 18 Å. A Γ-centered 16×16×1 k-point mesh was applied for the surface 1×1 unit cell and appropriately reduced for larger cells. The positions of atoms in the bottom two Ag layers were frozen and the atomic positions of the remaining atoms were optimized until the residual Hellman-Feynman forces on atoms were smaller than 0.01 eV/Å. The optimization of the bare surface structure yielded negligibly small expansion of the topmost interplanar separation and a 0.8% contraction of the second interlayer distance. The work function was calculated as the difference between the electrostatic potential energy in the vacuum region and the Fermi energy of the slab. A dipole correction was applied to compensate for the asymmetry of the slab and obtain correct work function values [35]. The calculated work function of the clean Ag(111) surface, 4.49 eV, is very close to the experimental value of 4.46 ± 0.02 eV [36], and the value obtained in other calculations, 4.50 eV [37], using a similar computational method. FeO(111) monolayer was adsorbed on one side of the Ag(111) slab. For the initial structural modelling, the calculated in-plane lattice constants of 2.938 Å for Ag(111) (PBE) and 3.032 Å for FeO(111) (PBE+U) were used [38]. The system was re-optimized after the deposition of a FeO layer within the applied surface unit cell. The electron charges on atoms were calculated using Bader analysis [39,40]. The STM images were simulated using the Tersoff-Hamann method [41].

3. Results and Discussion

3.1. Iron Deposition at Room Temperature and Post-Oxidation

Firstly, we deposited Fe onto a Ag(111) substrate kept at room temperature and post-oxidized it. As prepared iron/silver and iron oxide/silver samples will be further referred to as RT-Fe/Ag(111) and RT-FeO/Ag(111), respectively. Figure 1a presents a large-scale STM image of the 0.5 ML RT-Fe/Ag(111) sample. The deposited iron mainly agglomerated at the silver step edges forming irregularly-shaped nanowires (such growth has also been observed by other authors, see e.g. Reference [42]). In addition, the deposition was found to lead to silver steps erosion and vacancy islands formation (similar substrate etching has been reported for Cu/Ag(111) [43]).

Typical structures observed after oxidation, together with their height profiles, are shown in different magnification in Figure 1b–e. Irregular iron particles mostly changed into well-defined hexagonally-shaped islands (Figure 1b–d). Interestingly, the orientation of the island edges indicated that the <110> crystallographic direction, characteristic for the step edges of (111)-oriented surfaces [44], is not preferred for the as grown FeO islands, while the $\sqrt{3}$ (<112>, <121>, <211> and equivalent) directions are. On most islands, a Moiré superstructure was observed (Figure 1b,c, marked with a red arrow). The atomic resolution image of such an island, shown in Figure 1d, revealed a close-packed atomic arrangement with ~3.2 Å periodicity, while its line profile, presented in Figure 1e, provided information on the height of a typical Moiré-reconstructed island (~2.1 Å), the Moiré period

(45 Å) and the Moiré corrugation (0.4–0.5 Å). Occasionally, a second or even third layer of FeO was observed, with the height of one layer always being around 2.1 Å. The Moiré structure itself was found to run along the $\sqrt{3}$ crystallographic directions of the support (similarly to the island edges). The observed atomic periodicity and height indicated that the islands are indeed FeO(111), as other iron oxide phases (i.e. Fe_3O_4(111) or Fe_2O_3(0001)) would show much larger atomic periodicities in STM (6 Å and 5 Å, respectively [4]), as well as larger heights. The observed height and Moiré corrugation may, of course, depend on the tunneling parameters and tip condition, which are of particular importance when measuring height differences between oxide and metal surfaces. For instance, on the atomic-resolution STM image shown in Figure 1d no Moiré corrugation is visible. The height profile taken along exemplary islands with ill-defined surface structure is presented in Figure 1f. These islands were mainly located at the regular terrace sites of silver, were much higher than the Moiré-reconstructed islands and had a disordered surface structure.

The proposed model for the observed Moiré superstructure is presented in Figure 1g. Based on the Moiré and atomic periodicities, as well as the orientation of the island edges and the Moiré itself, we conclude that the coincidence structure is a 30°-rotated 14×14 FeO on 9×9$\sqrt{3}$ Ag. The numerically determined FeO lattice constant for such a superstructure (3.215 Å) and the Moiré period (45 Å) are in perfect agreement with the experimentally measured values.

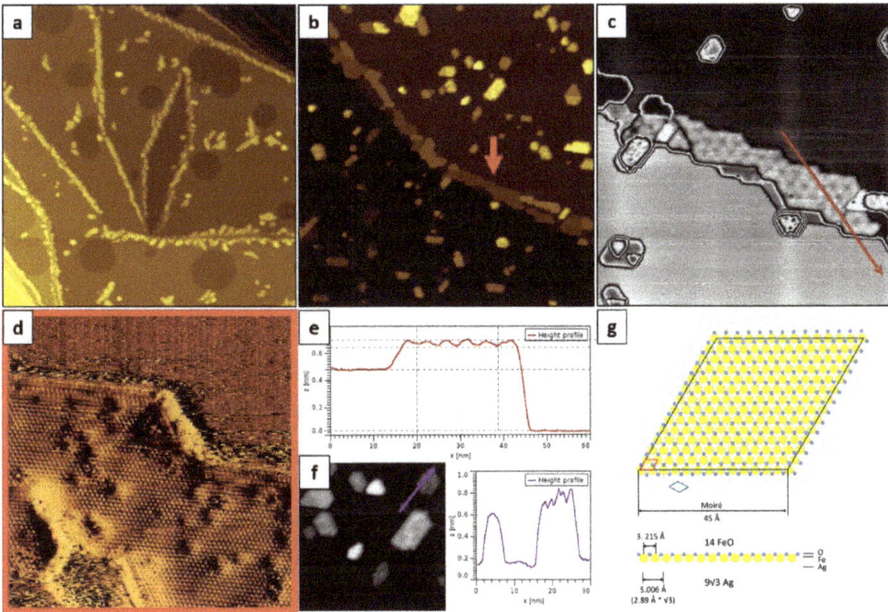

Figure 1. Large-scale topographic scanning tunneling microscopy (STM) images (300 × 300 nm^2) of 0.5 ML Fe deposited onto Ag(111) substrate kept at room temperature (**a**) and subsequently oxidized in 1×10^{-6} mbar O_2 at 700 K (**b**). (**c**) and (**d**) are zoom-in images of (b) (100 × 100 nm^2 and 20 × 20 nm^2, respectively) showing the Moiré superstructure (c) and the atomic structure (d) of the reconstructed island (marked with red arrow in (b)). (**e**) presents the height profile marked in (c). (**f**) shows an STM image (75 × 75 nm^2) of islands with ill-defined structure, as well as an exemplary height profile of such islands (marked on the neighboring STM image). (**g**) presents the proposed model for the observed Moiré superstructure, where the blue rhombus marks the Ag(111) surface cell, the red one the FeO(111) cell and black the Moiré cell (see text for details). Tunneling parameters: +0.7 V (all images), 0.4 nA (a,d,f) and 1.0 nA (b,c).

The most important finding is that the structural parameters of the observed Moiré superstructure differ significantly from those reported by Lundgren, Weaver and co-workers for the Moiré-reconstructed FeO/Ag(111) (26 Moiré period, 3.25 Å atomic period and no Moiré rotation with respect to the support) [17–19]. This contradiction indicates that different preparation procedures may lead to the formation of different well-ordered surface structures that are energetically the most stable for the particular growth conditions. Such behavior was not observed for FeO(111) on any other close-packed single-crystalline substrate.

The LEED pattern obtained for the oxidized sample shown in Figure 1b is presented in Figure 2a. Due to the agglomeration of the material at the silver step edges and a sole fraction of the substrate exposed, the observed pattern could be still roughly interpreted as (1×1). It is worth noting that a (1×1) LEED was also reported by the authors of Ref. [8] following 1–3 MLs Fe deposition at room temperature and post-oxidation. In contrast, a satellite pattern originating from a 26 Å Moiré-reconstructed FeO was reported by other authors following Fe deposition onto a heated substrate and in an oxygen ambient [17–19]. The lack of additional spots on our pattern could be due to small iron oxide coverage (0.5 ML, with a fraction of islands being Moiré-reconstructed and a fraction being structurally ill-defined). Interestingly, a simple simulation of the observed superstructure (Figure 1g) did not produce a clearly-visible Moiré superstructure. Thus, it may be speculated that the experimentally observed corrugation may result from the electronic film-substrate interaction (the lack of visible corrugation on the atomic-resolution image in Figure 1d seems to confirm this theory). The absence of additional spots also confirms the lack of significant amounts of iron oxide phases other than FeO(111), i.e. Fe_3O_4(111) or Fe_2O_3(0001), that would give a (2×2) and $(\sqrt{3}\times\sqrt{3})$R30° LEED patterns, respectively [4,13,15].

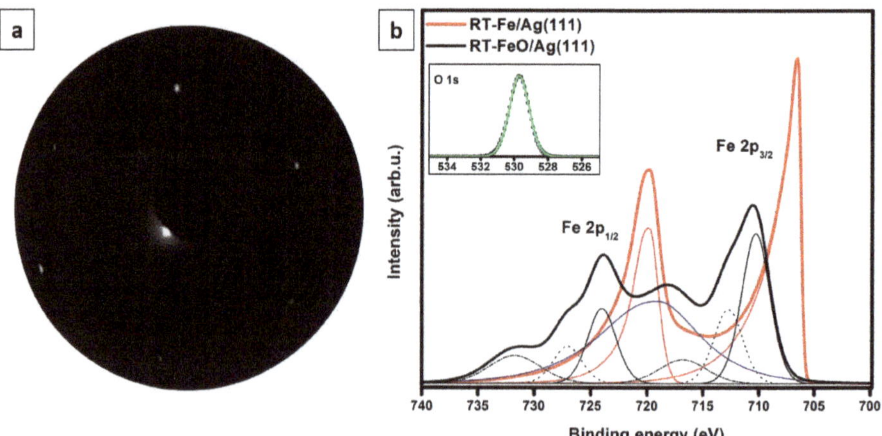

Figure 2. Low energy electron diffraction (LEED) pattern (60 eV) of 0.5 ML FeO film grown by Fe deposition onto Ag(111) substrate kept at room temperature and post-oxidation (**a**). (**b**) shows Fe 2p X-ray photoelectron spectroscopy (XPS) spectra obtained for the 1.0 ML RT-Fe/Ag(111) (red line) and 1.0 ML RT-FeO/Ag(111) (black) samples. The Ag 3s signal, overlapping with the Fe 2p line, is marked in blue. Inset in (**b**) presents the O 1s spectrum (green line) recorded for the oxidized sample.

The XPS data recorded for the 1.0 ML RT-Fe/Ag(111) and 1.0 ML RT-FeO/Ag(111) samples are presented in Figure 2b. The main graph presents the Fe 2p lines obtained for both samples, while the inset shows the O 1s signal obtained for the oxidized sample. A detailed analysis of the Fe 2p data was not trivial, as the region overlaps with the Ag 3s peak centered at around 719 eV.

The Fe 2p signal obtained for the RT-Fe/Ag(111) sample could be fitted with two components centered at around 706.5 eV and 719.9 eV, which are characteristic of metallic iron and result from spin-orbit coupling (Fe $2p_{3/2}$ and $2p_{1/2}$ peaks [45]). The spectrum obtained for the oxidized sample could be fitted with six components—710.3 eV, 712.6 eV, 716.6 eV, 724.0 eV, 727.0 eV and 731.5 eV—which is typical for the FeO phase [46,47]. The 710.3 eV and 724.0 eV peaks, chemically shifted by approx. 4 eV to higher binding energies with respect to metallic iron, result from Fe^{2+} ions in iron oxide. X-ray photoelectron lines of iron oxides, and ionic compounds in general, exhibit broadening due to a multiplet splitting effect (coupling between core holes with unpaired electrons in the outer shell [48–51]) which, in our case, was manifested by the presence of the components centered at around 713.4 eV and 726.9 eV. The 718.8 eV and 731.0 eV peaks on the other hand, correspond to characteristic Fe^{2+} satellite peaks (which appear as a consequence of photoelectrons exciting valence electrons into unoccupied orbitals [46,51,52]). A binding energy difference of nearly 6 eV between the first satellite and the main Fe $2p_{3/2}$ photoemission line and of nearly 7 eV between the second satellite and the main Fe $2p_{1/2}$ line provide clear evidence for the presence of Fe^{2+} iron [46,47,52] (the presence of Fe^{3+} iron would result in a binding energy difference of more than 8 eV [46,52]). Based on the peak positions it may be, therefore, concluded that metallic iron and iron in the Fe^{3+} oxidation state are either not present or present in the amount that falls beyond the detection limit of our XPS system. The oxygen O 1s signal recorded for the oxidized sample could be fitted with a single peak centered around 529.7 eV [49]. The Fe to O ratio, determined from the survey spectrum (not shown) and by taking the elemental photoionization cross-sections [53] into account, was approx. 1:1—as expected for a stoichiometric FeO phase. No other elements, like e.g. carbon, sulfur or other contaminants, could be detected. All these indicated that the observed islands—no matter if well- or ill-defined—consist of Fe^{2+} and O^{2-} ions (however, a contribution of Ag atoms in these structures could not be excluded as well).

3.2. Iron Deposition at High Temperature and Post-Oxidation

Secondly, we modified the preparation procedure by depositing Fe onto a 550 K-heated Ag(111) substrate and post-oxidizing it. The corresponding iron/silver and iron oxide/silver samples will be further referred to as HT-Fe/Ag(111) and HT-FeO/Ag(111) (underlining the fact that iron was deposited at high temperature (HT)). Figure 3a,b presents large-scale STM topography images of the HT-deposited 0.5 ML Fe on Ag(111) before (a) and after (b) oxidation in 1×10^{-6} mbar O_2 at 700 K. In contrast to the observed agglomeration of iron at the silver step edges following RT deposition, the HT deposition was found to lead to the formation of well-defined bcc Fe(110) crystallites uniformly distributed over the silver surface. We explain this difference in the growth mode by partial incorporation of iron into the top silver layer at the initial stage of deposition and, thus, formation of nucleation centers for the growing crystallites. This assumption is based on the observation of small, elongated, rhomboidal particles embedded in the top layer of silver following the deposition of submonolayer iron amounts. The examples of such inclusions are shown in the inset to Figure 3a (marked with blue arrows); they have a height of ~0.6–1.2 Å (approx. half of the height expected for monolayer bcc-Fe, i.e. ~2 Å [54]) and top facets tilted with respect to the substrates' plane—thus suggesting that they are embedded in silver. Such Fe incorporation may appear surprising, taking into account the fact that iron generally does not alloy with silver. However, it has been shown that diffusion of iron into silver bulk is possible at high temperatures [55], therefore, we assume that at temperatures >500 K iron could be incorporated into the top silver layer. For higher iron coverages, mostly multilayer nanocrystals were observed (like the one marked with a green circle in the inset to Figure 3a), with shapes indicating <110>-oriented growth. Interestingly, the fraction of particles embedded in the substrate at the vicinity of the silver step edges was found to modify the steps direction (inset to Figure 3a, orange circle). In contrast to the RT deposition, no substrate etching was observed, which could be due to immediate surface smoothing at 550 K. The topography of the HT-FeO/Ag(111) sample also differed significantly from the RT-FeO/Ag(111): The HT film was inhomogeneous and consisted of different reconstructed and reconstruction-free surface species, most of which were located at the regular terrace sites, more uniformly distributed, had smaller sizes and less defined shapes.

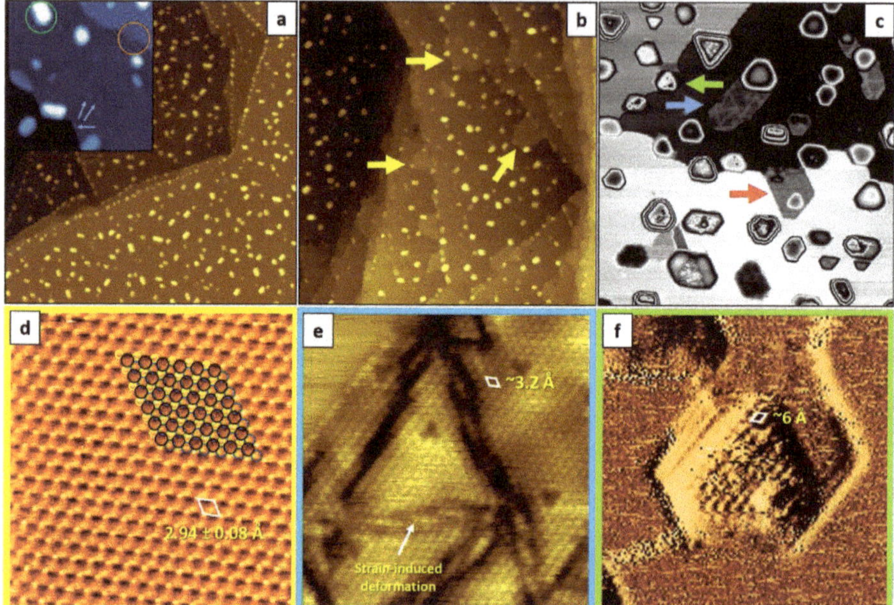

Figure 3. Large-scale topographic STM images (300 × 300 nm^2) of 0.5 ML Fe deposited onto Ag(111) substrate kept at 550 K (**a**) and subsequently oxidized in 1×10^{-6} mbar O$_2$ at 700 K (**b**). The inset in (**a**) (50 × 50 nm^2) shows Fe inclusions in Ag(111) at regular terrace sites (marked with blue arrows) and step edges (orange circle), as well as bcc Fe(110) crystallite (green circle) observed on low-coverage HT-Fe/Ag(111) sample. The yellow arrows in (**b**) show the reconstruction-free islands. (**c**) shows small-scale image (100 × 100 nm^2) with various reconstructed surface regions (marked with green, blue and red arrows). (**d**), (**e**) and (**f**) show atomically-resolved STM images of the structures marked with colors in (b) and (c) (5 × 5 nm^2, 10 × 10 nm^2 and 15 × 15 nm^2, respectively). The unit cells are marked with white rhombuses (see text for details). Tunneling parameters: +0.7 V (a,b,c,e,f, inset in (a)) and +0.1 V (d), 1.0 nA (a) and 0.4 nA (b,c,d,e,f, inset in (a)); (d)—sum of topography and current images, (f)—current image.

Generally, the observed topographic features can be divided into the following categories:

(1) Islands exhibiting a 45 Å Moiré superstructure—virtually identical to the one observed on the RT-FeO/Ag(111) sample—some of which were growing on top of silver terraces and some were embedded into the top silver layer. An example of such an island is marked with a red arrow in Figure 3c;

(2) Strained embedded islands without Moiré. An example of such an island is marked with a blue arrow in Figure 3c. The atomically-resolved image, shown in Figure 3e, reveals an interatomic spacing of ~3.2 Å, which is very similar to the spacing observed on the Moiré-reconstructed islands. Strain-induced deformations, originating most probably from the embedment of the island into the top silver layer, are clearly visible on both STM images. Similar oxide island incorporation into the top silver layer was observed by other authors for MgO(001) and NiO(100) films on Ag(001) [56–58];

(3) Hexagonal islands with large in-plane atomic periodicity (~6 Å) and large heights (~5–15 Å). An example of such an island is marked with a green arrow in Figure 3c. The atomic periodicity, determined from the atomic-resolution STM image shown in Figure 3f, and height indicate that these islands represent the Fe$_3$O$_4$(111) phase [4];

(4) Structurally ill-defined islands—similar to those observed on the RT-FeO/Ag(111) sample;

(5) Atomically-flat islands with no visible signs of lattice mismatch-induced reconstructions. These islands were mostly growing at the lower silver step edges, occasionally being extended in between the particles growing at the regular terrace sites. Examples of such islands are marked with yellow arrows in Figure 3b), S1a and S1b. The height of the islands, ~2.4–2.5 Å, is very similar to the height of a monoatomic step on Ag(111) (~2.35 Å [54]), however, an atomically-resolved STM image, shown in Figure 3d), revealed that they consist of two superimposed atomic sublattices—a lattice of small, sharp protrusions (marked with black circles) and a lattice of a bigger and more diffuse-appearing species (yellow circles). A similarly spaced "lattice" of holes (blue circles) could be also observed. The contributions from different sublattices were visible on both topography and current images, however, with different intensities, therefore, the sum of the two images is presented (for separate topography and current images, see Figures S2a and S2b). The atomic periods within the respective lattices, determined from ~100 line profiles, were found to be 2.94 ± 0.08 Å. The standard deviation, obtained numerically, results from thermal drift and STM image distortion caused by room-temperature imaging at high scanning speed. Notably, the observed atomic structure was locally defect-free and exhibited no out-of-plane rumpling. Even though the height of these reconstruction-free islands and the interatomic distances within them were similar to those of clean Ag(111) (2.89 Å), the appearance of two atomic sublattices was somehow disturbing;

(6) Flat regions in between different types of islands. The height difference between such regions was ~2.35 Å and corresponded to the height of a monoatomic step on Ag(111). Interestingly, at higher oxide coverages (>1 ML), these regions were still present and became difficult to be scanned with STM using negative bias voltages (the tip was crashing the surface). This could indicate that these regions are not purely silver, but were structurally altered by iron deposition and oxidation.

We believe that the large variety of surface structures observed on the HT-FeO/Ag(111) sample compared to the RT-FeO/Ag(111) sample is related to the more uniform distribution of iron prior to oxidation and partial embedment of iron into the top silver layer.

The LEED pattern obtained for the oxidized sample shown in Figure 3b is presented in Figure 4a. Similarly to the RT-FeO/Ag(111) sample, the HT-FeO/Ag(111) sample exhibited a (1×1) LEED with no signatures of structural reconstructions. This may be related to the presence of different surface species, each of which contributed in an insignificant way to the observed pattern.

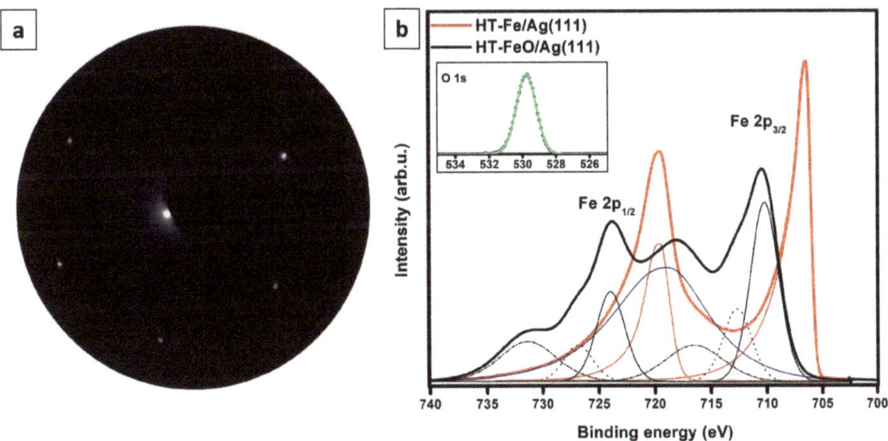

Figure 4. LEED pattern (60 eV) of 0.5 ML FeO film grown by Fe deposition onto Ag(111) substrate kept at 550 K and post-oxidation (**a**). (**b**) shows XPS Fe 2p spectra obtained for the 1.0 ML HT-Fe/Ag(111) (red line) and 1.0 ML HT-FeO/Ag(111) (black) samples. The Ag 3s signal is marked in blue. Inset in (**b**) presents the O 1s spectrum (green line) recorded for the oxidized sample.

The XPS data recorded for the 1.0 ML HT-Fe/Ag(111) and 1.0 ML HT-FeO/Ag(111) samples are presented in Figure 4b. The Fe 2p signal obtained for the HT-Fe/Ag(111) sample could be fitted with two components, centered at around 706.4 eV and 719.6 eV, indicative of the presence of metallic iron. The spectrum obtained after oxidation had six contributions: 710.2 eV, 712.7 eV, 716.5 eV, 724.0 eV, 726.9 eV and 731.3 eV. The 710.2 eV and 724.0 eV peaks were assigned to originate from Fe^{2+} ions in iron oxide and the 712.7 eV and 726.9 eV components from multiplet splitting. The peaks at 716.5 eV and 731.3 eV were the characteristic Fe^{2+} satellite peaks. Similarly to the RT-FeO/Ag(111) sample, the binding energy difference of nearly 6 eV between the main Fe $2p_{3/2}$ photoemission line and the first satellite peak and nearly 7 eV between the main Fe $2p_{1/2}$ line and second satellite peak gave clear evidence for the dominant presence of the Fe^{2+} iron in the studied sample. Metallic iron, iron in the Fe^{3+} oxidation state (originating e.g. from the rarely observed Fe_3O_4 islands) or contaminants—if any—were beyond the detection limit of our XPS instrument. The oxygen O 1s signal recorded for this sample could be fitted with a single peak centered at around 529.8 eV. The Fe to O ratio determined from the survey spectrum (not shown) was approx. 1:1.

The obtained results reveal that slight modification of the oxide preparation procedure may lead to the formation of completely different surface structures: Deposition of Fe onto the Ag(111) substrate held at room temperature and post-oxidation in 1×10^{-6} mbar O_2 at 700 K resulted in the formation of Moiré-reconstructed FeO islands located at the silver step edges, as well as ill-defined structures located at the regular terrace sites of silver, while deposition of iron onto a 550 K-heated substrate and post-oxidation resulted in a wide variety of uniformly distributed reconstructed and reconstruction-free surface species. Even though a slight change in the oxide growth conditions (iron amount, substrate temperature during deposition, oxygen pressure, oxidation temperature) was not critically affecting the structure of FeO films on other close-packed single-crystalline substrates used by different authors so far, the results obtained by us and other groups indicate that on Ag(111) these parameters play a crucial role. The co-existence of reconstructed and reconstruction-free islands with different atomic spacing following high-temperature iron deposition and post-oxidation, points to the high complexity of the Fe-O-Ag system and the need for further studies on this topic. These may be particularly significant with respect to the determination of the nature of the reconstruction-free islands. If, indeed, being Moiré-free FeO(111), such islands/films could constitute a perfect system for model-type electronic, catalytic and magnetic studies. The performed studies also reveal the necessity of using atomic-resolution techniques, such as STM, for the evaluation of the films structure, as XPS and LEED may provide misleading results.

In order to shed more light on the nature of the observed reconstruction-free islands, we performed point STS dI/dz measurements and compared the curves obtained on the islands with those recorded for clean Ag(111) and other single-crystalline substrate: Pt(111). The I(z) spectroscopy resembles the exponential change of the tunneling current with changing tip-sample distance [59]. The dependence can be measured even more precisely by recording the first derivative of the signal, i.e. dI/dz, using the lock-in technique [60,61]. As the slope of the dI/dz curve is proportional to the average work function of the sample and tip, the technique allows distinguishing surfaces with different work function values (assuming a constant work function of the tip during all the measurements). The I(z) method was, for example, successfully applied by other authors to determine local work function variations within the Moiré superstructure of FeO(111)/Pt(111) [62].

The recorded dI/dz curves are shown in Figure 5. The inset presents excerpts from the curves (slopes) plotted in a logarithmic scale. As expected, the slopes of the curves obtained for the probed single crystal surfaces differed significantly, in agreement with their different work function values (the literature values are 4.46–4.50 eV [36,37] and 6.08–6.4 eV [34,63–65] for Ag(111) and Pt(111), respectively). The slope of the curve obtained for the reconstruction-free islands on the HT-FeO/Ag(111) sample was much higher than that of clean Ag(111) and lower than that of Pt(111). The measurements were repeated several times for different sample preparations (and by changing all three samples in a random way to exclude changes in the tip condition in between the measurements) and the trend was always the same. Based on this, it may be concluded that the islands work function value is higher than the work function of silver and lower than the work function of platinum.

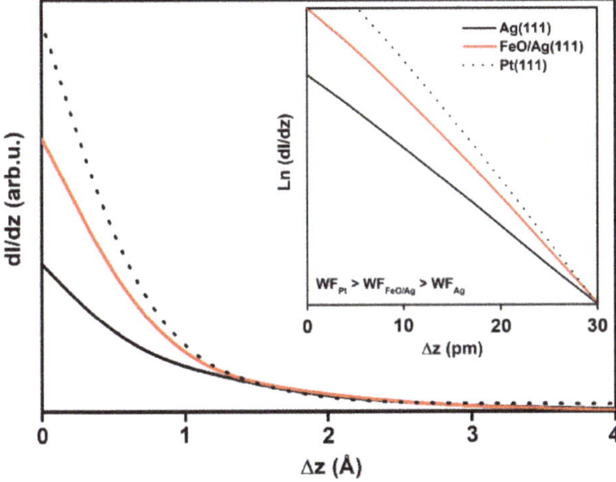

Figure 5. dI/dz curves obtained from clean Ag(111) (black solid curve), clean Pt(111) (black dotted curve) and the reconstruction-free islands grown by Fe deposition onto a Ag(111) substrate kept at 550 K and post-oxidation (red curve). The inset presents the slopes of the curves plotted in a logarithmic scale.

3.3. Calculations on a Conceptual Reconstruction-Free FeO/Ag(111)

In order to determine the structure and properties of a conceptual reconstruction-free FeO/Ag(111), the FeO film was modelled using a 1×1 surface unit cell, thus adopting the lattice constant of the Ag(111) support. This means that during the structural optimization of the systems, the FeO(111) lattice was slightly compressed to match the lattice of Ag(111). Such approach did not allow for any kind of in-plane reconstructions. The results were obtained for the calculated Ag bulk lattice constant of 4.155 Å (PBE) and FeO lattice constant of 4.288 Å (PBE+U). These values differ by 3.1%, while the experimentally-determined lattice parameters of Ag (4.086 Å) and FeO (4.32 Å) differ by 5.7%. Therefore, a similar set of calculations, including structural optimizations of the considered monolayer and bilayer FeO(111) systems, was performed using the experimental lattice constants of Ag and FeO. Importantly, the calculated geometries, presented in SM3, do not show substantial differences with respect to those obtained with theoretically determined lattice constant values.

Several configurations of Fe-terminated and O-terminated FeO(111)/Ag(111) systems were examined, considering both Fe and O atoms placed in fcc or hcp hollows, as well as on top of Ag atoms. The most stable Fe-termination was found to be the one with Fe atoms occupying fcc positions and O atoms in the hcp hollow sites, with an O–Ag distance of 2.45 Å and an Fe–O distance of 0.90 Å. Other Fe-terminations were by at least 10 meV (with respect to the total energy) less favored. The most stable O-termination was again the one with Fe and O atoms sitting in the threefold coordinated fcc and hcp hollow sites of the Ag(111) surface, respectively, with an Fe–Ag distance of 2.35 Å and an O–Fe distance of 0.82 Å. This configuration (presented in Figure 6a) was by 0.79 eV energetically more favored than the Fe-terminated one. It has to be mentioned that for the 1×1 supercell calculations, this configuration was nearly degenerated in energy with the one with Fe in hcp and O in fcc sites (which was less stable by only 2 meV). The calculations performed using a 2×2 supercell confirmed the same preference of adsorption sites with a total energy difference of 6 meV, and those performed using the experimental lattice constant of Ag (4.086 Å) showed that the two configurations differ by 15 meV. Another O-termination, with O atoms in the on-top positions and Fe atoms in either fcc or hcp sites, were by 89 meV less favored. Based on those findings, further calculations were performed for the most energetically-stable O-terminated structure with Fe in the threefold coordinated fcc sites and O in the hcp hollow sites.

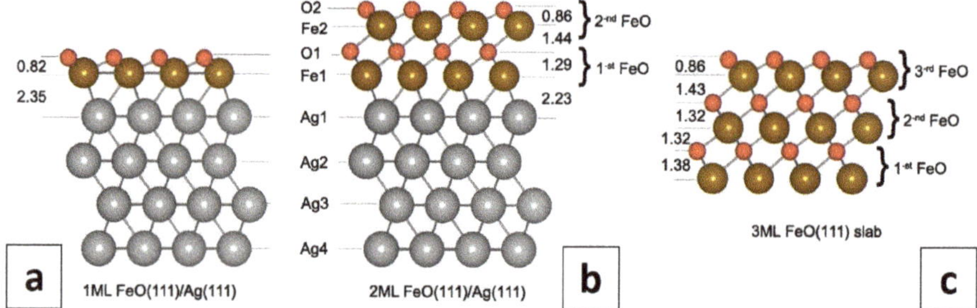

Figure 6. Side views of calculated O-terminated monolayer (**a**) and bilayer (**b**) reconstruction-free FeO(111) films on Ag(111), as well as 3-MLs-thick FeO slab (**c**), determined from PBE+U calculations using a 1×1 surface unit cell with fixed 2.938 Å FeO(111) and Ag(111) in-plane lattice constants (FeO(111) film adopting the geometry of the Ag(111) substrate). All distances are given in Å.

The calculated total height of the O-terminated monolayer FeO(111) film was found to be 3.17 Å. A strongly reduced Fe–O interlayer spacing (0.82 Å) in the FeO(111)/Ag(111) system compared to bulk FeO (1.25 Å) [66] is of similar magnitude to that reported for FeO(111)/Pt(111) [66] and results from a complex stabilization mechanism, i.e. an interplay between structural relaxation, polarity compensation, charge transfer and magnetic ordering [67–70]. The calculated adhesion energy of the FeO(111) monolayer on Ag(111) was found to be 0.91 eV per Fe atom. This is much less than that reported for the FeO(111) monolayer Pt(111) (1.36 eV [18]), but it is doubled compared to the value obtained for FeO(111) on Ag(100) (0.46 eV [18]). The difference in the adhesive bond strength is reflected in the structural properties of the FeO film. The Fe–Ag distance for FeO(111) supported on Ag(111) and Ag(100) is, respectively, 2.35 Å and 2.65 Å, whereas the Fe–Pt distance at the FeO/Pt(111) interface is ~2.7 Å (it varies slightly over different high-symmetry regions of the Moiré supercell) [31]. Interestingly, the reduction of the FeO-substrate bonding distance with the enhanced adhesive energy strength does not lead to a regular change in the distance between oxygen and iron planes in the FeO(111) monolayer (rumpling), which is largest (0.82 Å, Figure 6a) for FeO/Ag(111) and smallest (0.26 Å [18]) for FeO/Ag(100). For the FeO/Pt(111) it is ~0.7 Å [31]. The difference between the values obtained for FeO(111)/Ag(111) (this work) and FeO(111)/Ag(100) [18] may originate from different substrate orientation (Ag(111) vs. Ag(100)) and the computational unit cell used (1×1 vs. p(2×11)) which, in our case, did not allow for any in-plane relaxations. An inspection of the calculated Bader charges (SM4 and Figure S3) on the interface atoms of the FeO(111) monolayer deposited on a Ag(111) substrate showed that adsorption of a FeO film results in the appearance of negative charges on silver (0.17 |e|) and iron (0.04 |e|) atoms, gained at the expense of oxygen (0.24 |e|). This charge transfer may be important with respect to films stabilization. Another stabilizing factor may be the magnetic superstructure. It is worth noting that the applied 1×1 (or a larger 2×2) surface unit cell gives a ferromagnetic (FM) ground state and does not allow an antiferromagnetic (AFM) superstructure, which results from magnetic frustration of antiferromagnetic FeO on the three-fold symmetric oxide layer, to be reproduced [31]. Such a magnetic superstructure is observed, indeed, if a 3×3 or a ($\sqrt{3}\times\sqrt{3}$)R30° surface unit cell is applied, where the number of Fe atoms in a layer is a multiple of three with a 1:2 ratio of Fe atoms with opposite magnetic moments. In that case, the AFM structure is by 0.26 eV per FeO unit more stable than the FM one. However, the AFM alignment of magnetic moments on Fe atoms makes the iron layer distinctly rumpled, with Fe atoms of one direction of magnetization being by around 0.5 Å more distant from the Ag(111) surface than those with opposite magnetization (Figure S4). Such rumpling could be, potentially, experimentally observed below the Néel temperature of FeO (~198 K [71]).

The bilayer FeO(111) film was modelled by placing a second monolayer of FeO(111) on top of the FeO(111)/Ag(111) slab, originally in a flat configuration, with Fe over surface Ag atoms and O atoms in fcc positions over Fe atoms in the first FeO layer (Figure 6b). Upon structural optimization, the lateral positions of Fe and O atoms remained unchanged and the atoms relaxed only vertically. Both FM and AFM stackings of ferromagnetically ordered adjacent FeO layers were considered. The AFM configuration appeared to be by 63 meV more stable than the FM one and, thus, it was adopted in further calculations. The height of the first (interface) monolayer increased to 3.52 Å due to a substantial increase of the O–Fe distance (to 1.29 Å). The thickness of the second (top) monolayer was 2.30 Å, with an O–Fe distance of 0.86 Å. The distance between the iron oxide layers was 1.44 Å and the distance between the Fe atoms in the bottom layer and the silver support was reduced to 2.23 Å. These values suggest strengthening of the adhesive bond between the bilayer FeO(111) and the silver support compared to monolayer FeO film on Ag(111). It has to be mentioned that the results obtained for monolayer and bilayer FeO(111) films did not change with a change of the supercell from 1×1 to 2×2. Notably, the height of the experimentally observed reconstruction-free islands (2.4–2.5 Å) differs significantly from the value calculated for a monolayer FeO on Ag(111) (3.2 Å). In fact, the experimental value is more similar to the height of the calculated second FeO layer on Ag(111) (2.3 Å). It is, of course, possible that the height of the experimentally observed structures, if indeed being FeO, is affected by stabilization mechanisms that were not taken into account in our theoretical considerations. One could also speculate that the experimentally observed structures are the second-layer FeO islands which grow on top of the first-layer FeO islands embedded in the silver substrate. This would make FeO/Ag(111) system similar to MgO(001)/Ag(001) and NiO(100)/Ag(001) [56–58]. On the other hand, the second FeO(111) layer on Pt(111) was reported to undergo strong structural relaxations resulting from polarity compensation [72,73].

To compare the atomic structures, we simulated, based on the calculated geometries, the STM images of 1 ML FeO/Ag(111) and 2 MLs FeO/Ag(111) and set them side by side with the atomic-resolution image obtained experimentally on one of the reconstruction-free islands (Figure 7a–c, respectively). The simulated STM images for 1-ML-thick and 2-MLs-thick films differ by contrast, however, they both seem to fit the experimental image, where the smaller protrusions observed experimentally would correspond to iron atoms, the bigger to oxygen atoms and the holes to the positions of silver atoms.

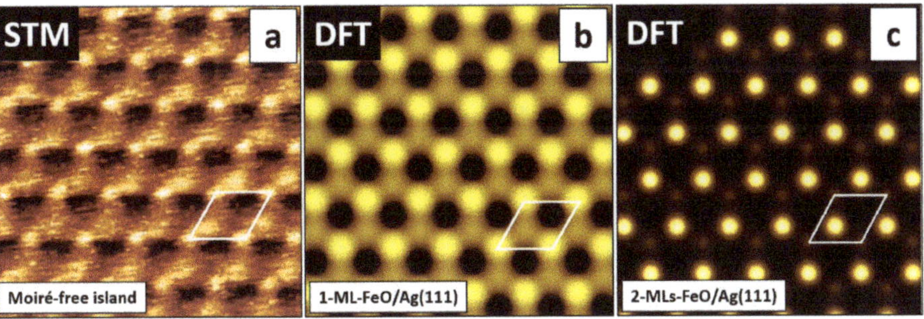

Figure 7. Atomic-resolution STM image (1.73 × 1.73 nm², V = +0.1 V, I = 0.4 nA) obtained on the reconstruction-free island on the HT-FeO/Ag(111) sample (**a**). (**b**) and (**c**) present simulated images of a 1-ML-thick and 2-MLs-thick reconstruction-free FeO(111) films on Ag(111), respectively (see text for details). The simulation is based on density functional theory (DFT) calculations and the Tersoff-Hamann method [41]. The FeO(111)-(1×1) unit cell is marked with a white rhombus.

In order to check the influence of the Ag(111) substrate on the structure of 1-ML-thick and 2-MLs-thick FeO(111) films, we also simulated a free-standing 3-MLs-thick (six atomic layers) FeO(111) slab and compared its parameters with those of Ag(111)-supported films (see Figure 6c and SM4). As the antiferromagnetic configuration of alternating layers was found to be by 2.01 eV more favored than the ferromagnetic one, the further description is given for the AFM phase. The slab is asymmetric due to differently terminated sides which makes the interlayer separations at the O-terminated and Fe-terminated sides, respectively, shrink or expand. Notably, the distances between the inner Fe and O planes (1.32–1.43 Å) were found to be close to the calculated Fe–O spacing in a bilayer FeO(111) on Ag(111) (1.29–1.44 Å) and in bulk FeO (1.25 Å) [66]. The similarity between the Ag(111)-supported bilayer FeO(111) film and a free-standing 3-MLs-thick FeO(111) slab was also reflected in the calculated Bader charges and magnetic moments (SM4). The magnitude of both electron charges and magnetic moments on atoms of the top surface layer of a bilayer film and a free-standing slab is mainly determined by the presence of the surface. In both structures, the O and Fe atoms of the outermost layer lose 0.25 and 0.41–0.43 electrons, respectively. The electron charge gain on Fe1 atoms of the bilayer (0.51 |e|) and Ag1 atoms of the substrate (0.19 |e|) indicate the weak character of the FeO–Ag bonding. The magnetic moments on atoms of the outermost FeO layers of the bilayer and a free-standing 3-MLs-thick slab are nearly the same, which means that they are determined by the presence of both the surface and the underlying FeO monolayer. In contrast, when a monolayer FeO is placed directly on Ag(111), magnetic moments on oxygen atoms (0.39 μB) are induced and the moments on Fe atoms are enhanced compared to "FeO-supported" FeO layer.

The partial densities of electronic states (PDOS) calculated for the FeO(111) monolayer on Ag(111), the second FeO layer from a bilayer FeO(111) film on Ag(111) and the top FeO layer from the 3-MLs-thick FeO slab are presented in Figure 8a–c. The similarity between the electronic structure of all three systems can be seen, especially when comparing the second FeO layer on Ag(111) and a free-standing FeO slab. It suggests that the density of states is mainly determined by the presence of the surface and not by the interaction with the Ag(111) support. However, the presence of the underlying FeO layer is important for the opening of an energy gap of 0.7–0.9 eV in the topmost FeO layer (in the majority spin states; see Figure 8b,c). The Fe-DOS is dominated by the Fe 3d states and the O-DOS by the O 2p states. There is also a great asymmetry in the majority and minority spin states, in particular in those belonging to Fe. The former is almost completely occupied, whereas the latter are mostly empty.

Based on the performed calculations, it may be concluded that the second FeO(111) layer on Ag(111) has a "free-standing FeO" character.

The calculated work function values of a monolayer FeO(111) film on Ag(111) range from 5.87 eV (AFM configuration, $(\sqrt{3} \times \sqrt{3})$R30° unit cell) to 6.56 eV (FM configuration, 1×1 unit cell) (see SM5 for values obtained for different configurations of magnetic arrangements and computational unit cells). Bilayer FeO has an even higher work function value: 6.67 eV. Notably, both values are much higher than the work function of clean Ag(111). The value roughly determined by us experimentally using dI/dz spectroscopy for the reconstruction-free islands seems to be similar to that calculated for monolayer FeO/Ag(111) and lower than the one calculated for bilayer FeO.

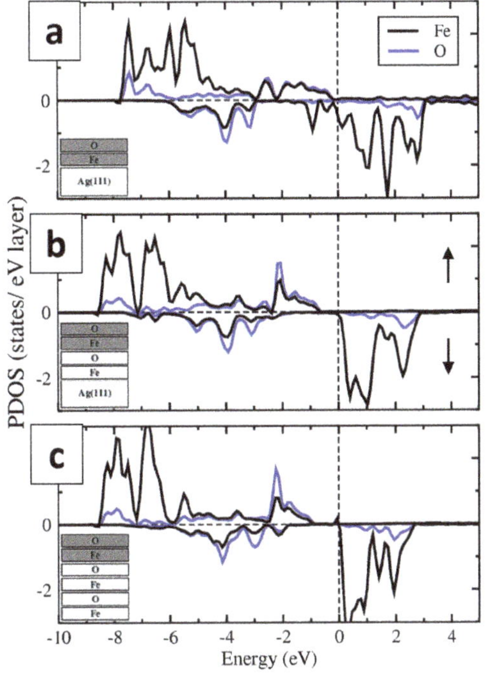

Figure 8. Partial densities of electronic states (PDOS) of a FeO(111) monolayer on Ag(111) (**a**), second FeO(111) layer from a bilayer FeO(111) film on Ag(111) (**b**) and the top FeO layer from a free-standing 3-MLs-thick FeO(111) slab (**c**), resulting from PBE+U calculations.

4. Conclusions

We used STM, LEED and XPS to study ultrathin FeO films grown on Ag(111) using two different procedures: (1) By depositing iron onto a silver substrate kept at room temperature and post-oxidizing it and (2) by depositing iron onto a silver substrate kept at 550 K and post-oxidizing. Following the procedure (1), we observed the formation of islands exhibiting a 45 Å Moiré superstructure. These islands grew mainly at the silver step edges, which is related to iron agglomeration at these surface sites following room-temperature deposition. Procedure (2) was found to lead to the formation of various reconstructed and reconstruction-free surface species, where the reconstruction-free ones grew mainly at the lower silver step edges, while the other species were uniformly distributed across the silver surface. The difference in the growth mode compared to procedure (1) is related to the formation of uniformly distributed bcc Fe(110) crystallites following 550 K deposition and partial embedment of iron/iron oxide in the top silver layer. The results of this and other works on the FeO/Ag(111) indicate the critical role of the growth conditions on the structure of the obtained films. To the best of our knowledge, such dependence has not been observed for FeO on any other close-packed single-crystalline substrate so far. It is now evident that at least two different Moiré superstructures may be formed for FeO/Ag(111), i.e. the 26 Å 8 FeO×9 Ag superstructure ([17–19]) and the 45 Å 14 FeO×9$\sqrt{3}$ Ag (this work), the growth of which is accompanied by the formation of various ill-defined and reconstruction-free surface species. With respect to the reconstruction-free species, their unambiguous compositional identification is not trivial, however, their atomic structure and electronic characteristics seem to differ from those of clean Ag(111), thus it is tempting to assign them to Moiré-free FeO. The performed DFT calculations indicate that a reconstruction-free FeO(111) film on Ag(111), if successfully grown, should exhibit tunable thickness-dependent properties, being substrate-influenced in the monolayer regime and free-standing-FeO-like when in the bilayer form. The latter could constitute a perfect model system for electronic, catalytic and magnetic studies.

Supplementary Materials: The following are available online at http://www.mdpi.com/2079-4991/8/10/828/s1, Overview of the literature data concerning the preparation of iron oxide films on silver single crystal supports, detailed X-ray photoelectron spectroscopy data, geometrical details of the calculated FeO systems, numerically-determined Bader charges, magnetic moments and work function values, additional figures.

Author Contributions: M.L. performed STM, LEED and XPS measurements, data analysis and wrote the manuscript (except for the theoretical part); N.M. and Z.M. assisted in STM, LEED and XPS measurements, as well as data analysis; V.B. and Y.W. performed dI/dz measurements (supervised by M.L.), took part in XPS data analysis (V.B.), creation of the model of the Moiré superstructure (Y.W.) and revision of the manuscript; T.P. performed all numerical calculations (with the contribution of M.H. in the early stage of the work) and analyzed the results together with A.K. who wrote the theoretical part of the manuscript; K.P. contributed to the discussion of simulated STM images; N.M, V.B., Y.W., M.H. and T.P. assisted in the preparation of the figures; S.J., in addition to all the other authors, contributed to the discussion.

Funding: This research was funded by the National Science Centre of Poland, grant No. 2012/05/D/ST3/02855, and the Foundation for Polish Science, grant No. First TEAM/2016-2/14. Computer time was granted by the ICM of the Warsaw University, project No. G44-23.

Acknowledgments: This work was financially supported by the National Science Centre of Poland (SONATA programme, 2013-2017, grant No. 2012/05/D/ST3/02855, MAGNETON project) and the Foundation for Polish Science (First TEAM programme, 2017-2020, grant No. First TEAM/2016-2/14 ("Multifunctional ultrathin Fe(x)O(y), Fe(x)S(y) and Fe(x)N(y) films with unique electronic, catalytic and magnetic properties" project co-financed by the European Union under the European Regional Development Fund)) [74]. Tomasz Pabisiak and Adam Kiejna acknowledge computer time granted by the ICM of the Warsaw University (Project G44-23) [75]. The authors would like to thank Roland Wiesendanger for providing the idea for the studies and for his comments on the early versions of the manuscript (published on arXiv:1608.01376 [cond-mat.mes-hall]).

Conflicts of Interest: The authors declare no conflict of interest.

References

1. Pacchioni, G. Two-Dimensional Oxides: Multifunctional Materials for Advanced Technologies. *Chem. Eur. J.* **2012**, *18*, 10144–10158. [CrossRef] [PubMed]

2. Peden, C.H.F.; Herman, G.S.; Ismagilov, I.Z.; Kay, B.D.; Henderson, M.A.; Kim, Y.-J.; Chambers, S.A. Model catalyst studies with single crystals and epitaxial thin oxide films. *Catal. Today* **1999**, *51*, 513–519. [CrossRef]

3. Vurens, G.H.; Salmeron, M.; Samorjai, G.A. Structure, composition and chemisorption studies of thin ordered iron oxide films on platinum (111). *Surf. Sci.* **1988**, *201*, 129–144. [CrossRef]

4. Weiss, W.; Ranke, W. Surface chemistry and catalysis on well-defined epitaxial iron-oxide layers. *Progr. Surf. Sci.* **2002**, *70*, 1–151. [CrossRef]

5. Zeuthen, H.; Kudernatsch, W.; Peng, G.; Merte, L.R.; Ono, L.K.; Lammich, L.; Bai, Y.; Grabow, L.C.; Mavrikakis, M.; Wendt, S.; et al. Structure of Stoichiometric and Oxygen-Rich Ultrathin FeO(111) Films Grown on Pd(111). *J. Phys. Chem. C* **2013**, *117*, 15155–15163. [CrossRef]

6. Ketteler, G.; Ranke, W. Heteroepitaxial Growth and Nucleation of Iron Oxide Films on Ru(0001). *J. Phys. Chem. B* **2003**, *107*, 4320–4333. [CrossRef]

7. Khan, N.A.; Matranga, C. Nucleation and growth of Fe and FeO nanoparticles and films on Au(111). *Surf. Sci.* **2008**, *602*, 932–942. [CrossRef]

8. Waddill, G.D.; Ozturk, O. Epitaxial growth of iron oxide films on Ag(111). *Surf. Sci.* **2005**, *575*, 35–50. [CrossRef]

9. Lopes, E.L.; Abreu, G.J.P.; Paniago, R.; Soares, E.A.; de Carvalho, V.E.; Pfannes, H.-D. Atomic geometry determination of FeO(001) grown on Ag(001) by low energy electron diffraction. *Surf. Sci.* **2007**, *601*, 1239–1245. [CrossRef]

10. Schlueter, C.; Lübbe, M.; Gigler, A.M.; Moritz, W. Growth of iron oxides on Ag(111)—Reversible Fe$_2$O$_3$/Fe$_3$O$_4$ transformation. *Surf. Sci.* **2011**, *605*, 1986–1993. [CrossRef]

11. Abreu, G.J.P.; Paniago, R.; Pfannes, H.-D. Growth of ultra-thin FeO(100) films on Ag(100): A combined XPS, LEED and CEMS study. *J. Magn. Magn. Mater.* **2014**, *349*, 235–239. [CrossRef]

12. Bruns, D.; Kiesel, I.; Jentsch, S.; Lindemann, S.; Otte, C.; Schemme, T.; Kuschel, T.; Wollschläger, J. Structural analysis of FeO(111)/Ag(001): Undulation of hexagonal oxide monolayers due to square lattice metal substrates. *J. Phys. Condens. Matter* **2014**, *26*, 315001. [CrossRef] [PubMed]

13. Genuzio, F.; Sala, A.; Schmidt, Th.; Menzel, D.; Freund, H.-J. Interconversion of α-Fe$_2$O$_3$ and Fe$_3$O$_4$ Thin Films: Mechanisms, Morphology, and Evidence for Unexpected Substrate Participation. *J. Phys. Chem. C* **2014**, *118*, 29068–29076. [CrossRef]

14. Merte, L.R.; Shipilin, M.; Ataran, S.; Blomberg, S.; Zhang, C.; Mikkelsen, A.; Gustafson, J.; Lundgren, E. Growth of Ultrathin Iron Oxide Films on Ag(100). *J. Phys. Chem. C* **2015**, *119*, 2572–2582. [CrossRef]

15. Genuzio, F.; Sala, A.; Schmidt, Th.; Menzel, D.; Freund, H.-J. Phase transformations in thin iron oxide films: Spectromicroscopic study of velocity and shape of the reaction fronts. *Surf. Sci.* **2016**, *648*, 177–187. [CrossRef]

16. Lamirand, A.D.; Grenier, S.; Langlais, V.; Ramos, A.Y.; Tolentino, H.C.N.; Torelles, X.; De Santis, M. Magnetite epitaxial growth on Ag(001): Selected orientation, seed layer, and interface sharpness. *Surf. Sci.* **2016**, *647*, 33–38. [CrossRef]

17. Mehar, V.; Merte, L.R.; Choi, J.; Shipilin, M.; Lundgren, E.; Weaver, J.F. Adsorption of NO on FeO$_x$ Films Grown on Ag(111). *J. Phys. Chem. C* **2016**, *120*, 9282–9291. [CrossRef]

18. Merte, L.R.; Heard, C.J.; Zhang, F.; Choi, J.; Shipilin, M.; Gustafson, J.; Weaver, J.F.; Grönbeck, H.; Lundgren, E. Tuning the Reactivity of Ultrathin Oxides: NO Adsorption on Monolayer FeO(111). *Angew. Chem. Int. Ed.* **2016**, *55*, 9267–9271. [CrossRef] [PubMed]

19. Shipilin, M.; Lundgren, E.; Gustafson, J.; Zhang, C.; Bertram, F.; Nicklin, C.; Heard, C.J.; Grönbeck, H.; Zhang, F.; Choi, J.; et al. Fe Oxides on Ag Surfaces: Structure and Reactivity. *Top. Catal.* **2017**, *60*, 492–502. [CrossRef]

20. Gwyddion. Available online: http://gwyddion.net (accessed on 9 October 2018).

21. Horcas, I.; Fernandez, R.; Gomez-Rodriguez, J.M.; Colchero, J.; Gomez-Herrero, J.; Baro, A.M. WSXM: A software for scanning probe microscopy and a tool for nanotechnology. *Rev. Sci. Instrum.* **2007**, *78*, 013705. [CrossRef] [PubMed]

22. Hedman, J.; Klasson, M.; Nilsson, R.; Nordling, C.; Sorokina, M.F.; Kljushnikov, O.I.; Nemnonov, S.A.; Trapeznikov, V.A.; Zyryanov, V.G. The Electronic Structure of Some Palladium Alloys Studied by ESCA and X-ray Spectroscopy. *Phys. Scr.* **1971**, *4*, 195–201. [CrossRef]

23. Kresse, G.; Hafner, J. Ab initio molecular dynamics for open-shell transition metals. *Phys. Rev. B* **1993**, *48*, 13115–13118. [CrossRef]

24. Kresse, G.; Furthmüller, J. Efficient iterative schemes for ab initio total-energy calculations using a plane-wave basis set. *Phys. Rev. B* **1996**, *54*, 11169–11186. [CrossRef]

25. Kresse, G.; Furthmüller, J. Efficiency of ab-initio total energy calculations for metals and semiconductors using a plane-wave basis set. *Comput. Mater. Sci.* **1996**, *6*, 15–50. [CrossRef]

26. Blöchl, P.E. Projector augmented-wave method. *Phys. Rev. B* **1994**, *50*, 17953–17979. [CrossRef]

27. Kresse, G.; Joubert, D. From ultrasoft pseudopotentials to the projector augmented-wave method. *Phys. Rev. B* **1999**, *59*, 1758–1775. [CrossRef]

28. Perdew, J.P.; Burke, K.; Ernzerhof, M. Generalized Gradient Approximation Made Simple. *Phys. Rev. Lett.* **1996**, *77*, 3865–3868. [CrossRef] [PubMed]

29. Methfessel, M.; Paxton, A.T. High-precision sampling for Brillouin-zone integration in metals. *Phys. Rev. B* **1989**, *40*, 3616–3621. [CrossRef]

30. Dudarev, S.L.; Botton, G.A.; Savrasov, S.Y.; Humphreys, C.J.; Sutton, A.P. Electron-energy-loss spectra and the structural stability of nickel oxide: An LSDA+U study. *Phys. Rev. B* **1998**, *57*, 1505–1509. [CrossRef]

31. Giordano, L.; Pacchioni, G.; Goniakowski, J.; Nilius, N.; Rienks, E.D.L.; Freund, H.-J. Interplay between structural, magnetic, and electronic properties in a FeO/Pt(111) ultrathin film. *Phys. Rev. B* **2007**, *76*, 075416. [CrossRef]

32. Kokalj, A.; Dal Corso, A.; de Gironcoli, S.; Baroni, S. Adsorption of ethylene on the Ag(001) surface. *Surf. Sci.* **2002**, *507–510*, 62–68. [CrossRef]

33. Müller, M.; Diller, K.; Maurer, R.J.; Reuter, K. Interfacial charge rearrangement and intermolecular interactions: Density-functional theory study of free-base porphine adsorbed on Ag(111) and Cu(111). *J. Chem. Phys.* **2016**, *144*, 024701. [CrossRef] [PubMed]

34. Lide, D.R. *CRC Handbook of Chemistry and Physics*; CRC Press: Boca Raton, FL, USA; London, UK; New York, NY, USA, 2003.

35. Neugebauer, J.; Scheffler, M. Adsorbate-substrate and adsorbate-adsorbate interactions of Na and K adlayers on Al(111). *Phys. Rev. B* **1992**, *46*, 16067–16080. [CrossRef]

36. Chelvayohan, M.; Mee, C.H.B. Work function measurements on (110), (100) and (111) surfaces of silver. *J. Phys. C Solid State Phys.* **1982**, *15*, 2305–2312. [CrossRef]

37. Otálvaro, D.; Veening, T.; Brocks, G. Self-Assembled Monolayer Induced Au(111) and Ag(111) Reconstructions: Work Functions and Interface Dipole Formation. *J. Phys. Chem. C* **2012**, *116*, 7826–7837. [CrossRef]

38. Ossowski, T. Private communication, 2016.

39. Bader, R.F.W. *Atoms in Molecules: A Quantum Theory*; Oxford University Press: Oxford, UK, 1990.

40. Henkelman, G.; Arnaldsson, A.; Jónsson, H. A fast and robust algorithm for Bader decomposition of charge density. *Comput. Mater. Sci.* **2006**, *36*, 354–360. [CrossRef]

41. Tersoff, T.; Hamann, D.R. Theory of the scanning tunneling microscope. *Phys. Rev. B* **1985**, *31*, 805–813. [CrossRef]

42. Buchner, F.; Kellner, I.; Steinrück, H.-P.; Marbach, H. Modification of the Growth of Iron on Ag(111) by Predeposited Organic Monolayers. *Z. Phys. Chem.* **2009**, *223*, 131–144. [CrossRef]

43. Bocquet, F.; Maurel, C.; Roussel, J.-M.; Abel, M.; Koudia, M.; Porte, L. Segregation-mediated capping of Volmer-Weber Cu islands grown onto Ag(111). *Phys. Rev. B* **2005**, *71*, 075405. [CrossRef]

44. Jeong, H.-C.; Williams, E.D. Steps on surfaces: Experiment and theory. *Surf. Sci. Rep.* **1999**, *34*, 171–294. [CrossRef]

45. Graat, P.C.J.; Somers, M.A.J. Simultaneous determination of composition and thickness of thin iron-oxide films from XPS Fe 2p spectra. *Appl. Surf. Sci.* **1996**, *100/101*, 36–40. [CrossRef]

46. Yamashita, T.; Hayes, P. Analysis of XPS spectra of Fe and Fe ions in oxide materials. *Appl. Surf. Sci.* **2008**, *254*, 2441–2449. [CrossRef]

47. García, L.M.; Bernal-Villamil, I.; Oujja, M.; Carrasco, E.; Gargallo-Caballero, R.; Castillejo, M.; Marco, J.F.; Gallego, S.; de la Figuera, J. Unconventional properties of nanometric FeO(111) films on Ru(0001): Stoichiometry and surface structure. *J. Mater. Chem. C* **2016**, *4*, 1850–1859. [CrossRef]

48. Grosvenor, A.P.; Kobe, B.A.; Biesinger, M.C.; McIntyre, N.S. Investigation of multiplet splitting of Fe 2p XPS spectra and bonding in iron compounds. *Surf. Interface Anal.* **2004**, *36*, 1564–1574. [CrossRef]

49. Biesinger, M.C.; Payne, B.P.; Grosvenor, A.P.; Lau, L.W.M.; Gerson, A.R.; Smart, R.S.C. Resolving surface chemical states in XPS analysis of first row transition metals, oxides and hydroxides: Cr, Mn, Fe, Co and Ni. *Appl. Surf. Sci.* **2011**, *257*, 2717–2730. [CrossRef]

50. McIntyre, N.S.; Zetaruk, D.G. X-ray photoelectron spectroscopic studies of iron oxides. *Anal. Chem.* **1977**, *49*, 1521–1529. [CrossRef]

51. Ratner, B.D.; Castner, D.G. *Surface Analysis—The Principal Techniques*, 2nd ed.; Vickerman, J.C., Gilmore, I.S., Eds.; Wiley-VCH: Weinheim, Germany, 2011.

52. Rubio-Zuazo, L.; Chainani, A.; Taguchi, M.; Malterre, D.; Serrano, A.; Castro, G.R. Electronic structure of FeO, γ-Fe_2O_3, and Fe_3O_4 epitaxial films using high-energy spectroscopies. *Phys. Rev. B* **2018**, *97*, 235148. [CrossRef]

53. Scofield, J.H. Hartree-Slater subshell photoionization cross-sections at 1254 and 1487 eV. *J. Electr. Spectrosc. Relat. Phenom.* **1976**, *8*, 129–137. [CrossRef]

54. Davey, W.P. Precision Measurements of the Lattice Constants of Twelve Common Metals. *Phys. Rev.* **1925**, *25*, 753–761. [CrossRef]

55. Longworth, G.; Jain, R. Mossbauer effect study of iron implanted silver alloys as a function of dose and annealing temperature. *J. Phys. F Metal Phys.* **1978**, *8*, 993–1007. [CrossRef]

56. Jaouen, T.; Tricot, S.; Delhaye, G.; Lépine, B.; Sébilleau, D.; Jézéquel, G.; Schieffer, P. Layer-Resolved Study of Mg Atom Incorporation at the MgO/Ag(001) Buried Interface. *Phys. Rev. Lett.* **2013**, *111*, 027601. [CrossRef] [PubMed]

57. Pal, J.; Smerieri, M.; Celasco, E.; Savio, L.; Vattuone, L.; Rocca, M. Morphology of Monolayer MgO Films on Ag(100): Switching from Corrugated Islands to Extended Flat Terraces. *Phys. Rev. Lett.* **2014**, *112*, 126102. [CrossRef] [PubMed]

58. Steurer, W.; Surnev, S.; Fortunelli, A.; Netzer, F.P. Scanning tunneling microscopy imaging of NiO(100)(1 × 1) islands embedded in Ag(100). *Surf. Sci.* **2012**, *606*, 803–807. [CrossRef]

59. Binnig, G.; Rohrer, H.; Gerber, Ch.; Weibel, E. Tunneling through a controllable vacuum gap. *Appl. Phys. Lett.* **1982**, *40*, 178–180. [CrossRef]

60. Schuster, R.; Barth, J.V.; Wintterlin, J.; Behm, R.J.; Ertl, G. Distance dependence and corrugation in barrier-height measurements on metal surfaces. *Ultramicroscopy* **1992**, *42–44*, 533–540. [CrossRef]

61. Yamada, Y.; Sinsarp, A.; Sasaki, M.; Yamamoto, S. Local Tunneling Barrier Height Measurement on Au(111). *Jpn. J. Appl. Phys.* **2003**, *42*, 4898–4900. [CrossRef]

62. Rienks, E.D.L.; Nilius, N.; Rust, H.-P.; Freund, H.-J. Surface potential of a polar oxide film: FeO on Pt(111). *Phys. Rev. B* **2005**, *71*, 241404. [CrossRef]

63. Hulse, J.; Küppers, J.; Wandelt, K.; Ertle, G. UV-photoelectron spectroscopy from xenon adsorbed on heterogeneous metal surfaces. *Appl. Surf. Sci.* **1980**, *6*, 453–463. [CrossRef]

64. Salmeron, M.; Ferrer, S.; Jazzar, M.; Samorjai, G.A. Photoelectron-spectroscopy study of the electronic structure of Au and Ag overlayers on Pt(100), Pt(111), and Pt(997) surfaces. *Phys. Rev. B* **1983**, *28*, 6758–6765. [CrossRef]

65. Derry, G.N.; Ji-Zhong, Z. Work function of Pt(111). *Phys. Rev. B* **1989**, *39*, 1940. [CrossRef]

66. Kim, Y.J.; Westphal, C.; Ynzunza, R.X.; Galloway, H.C.; Salmeron, M.; van Hove, M.A.; Fadley, C.S. Interlayer interactions in epitaxial oxide growth: FeO on Pt(111). *Phys. Rev. B* **1997**, *55*, R13448–R13451. [CrossRef]

67. Noguera, C. Polar oxide surfaces. *J. Phys. Condens. Matter* **2000**, *12*, R367–R410. [CrossRef]

68. Rienks, E.D.L.; Nilius, N.; Giordano, L.; Goniakowski, J.; Pacchioni, G.; Felicissimo, M.P.; Risse, Th.; Rust, H.-P.; Freund, H.-J. Local zero-bias anomaly in tunneling spectra of a transition-metal oxide thin film. *Phys. Rev. B* **2007**, *75*, 205443. [CrossRef]

69. Goniakowski, J.; Finocchi, F.; Noguera, C. Polarity of oxide surfaces and nanostructures. *Rep. Prog. Phys.* **2008**, *71*, 016501. [CrossRef]

70. Goniakowski, J.; Noguera, C. Polarization and rumpling in oxide monolayers deposited on metallic substrates. *Phys. Rev. B* **2009**, *79*, 155433. [CrossRef]

71. McCammon, C.A. Magnetic properties of Fe$_x$O (x > 0.95): Variation of Néel temperature. *J. Magn. Magn. Mater.* **1992**, *104–107*, 1937–1938. [CrossRef]

72. Lu, S.-Z.; Qin, Z.-H.; Cao, G.-Y. Nanostructured double-layer FeO as nanotemplate for tuning adsorption of titanyl phthalocyanine molecules. *Appl. Phys. Lett.* **2014**, *104*, 253104.

73. Lu, S.-Z.; Qin, Z.-H.; Guo, Q.-M.; Cao, G.-Y. Work function mediated by deposition of ultrathin polar FeO on Pt(111). *Appl. Surf. Sci.* **2017**, *392*, 849–853. [CrossRef]

74. The experimental results were partially presented by V. Babačić, Y. Wang and M. Lewandowski at the DPG Spring Meeting 2018, Berlin, Germany, 11–16 March 2018, the NanoTech Poland 2018 Conference, Poznań, Poland, 6–9 June 2018, the AMPERE NMR School 2018, Zakopane, Poland, 10–16 June 2018, 34th European Conference on Surface Science, Aarhus, Denmark, 26–31 August 2018 and the 3rd International Meeting on Materials Science for Energy Related Applications—3IMMSERA, Belgrade, Serbia, 25–26 September 2018.

75. The theoretical results were partially presented by T. Pabisiak and A. Kiejna at the 8th International Workshop on Surface Physics, Trzebnica, Poland, 26–30 June 2017.

 *nanomaterials*

Article

Symmetry-Induced Structuring of Ultrathin FeO and Fe₃O₄ Films on Pt(111) and Ru(0001)

Natalia Michalak [1], Zygmunt Miłosz [2], Gina Peschel [3], Mauricio Prieto [3], Feng Xiong [3,4], Paweł Wojciechowski [1], Thomas Schmidt [3] and Mikołaj Lewandowski [1,2,*]

[1] Institute of Molecular Physics, Polish Academy of Sciences, M. Smoluchowskiego 17, 60-179 Poznań, Poland; michalak@ifmpan.poznan.pl (N.M.); pwojciechowski@ifmpan.poznan.pl (P.W.)

[2] NanoBioMedical Centre, Adam Mickiewicz University, Umultowska 85, 61-614 Poznań, Poland; zmilosz@amu.edu.pl

[3] Fritz-Haber-Institut der Max-Planck-Gesellschaft, Faradayweg 4-6, 14195 Berlin, Germany; peschel@fhi-berlin.mpg.de (G.P.); prieto@fhi-berlin.mpg.de (M.P.); xxf2012@mail.ustc.edu.cn (F.X.); schmidtt@fhi-berlin.mpg.de (T.S.)

[4] Department of Chemical Physics, University of Science and Technology of China, No. 96, JinZhai Road Baohe District, Hefei 230026, China

* Correspondence: lewandowski@amu.edu.pl; Tel.: +48-618-29-6717

Received: 6 August 2018; Accepted: 7 September 2018; Published: 12 September 2018

Abstract: Iron oxide films epitaxially grown on close-packed metal single crystal substrates exhibit nearly-perfect structural order, high catalytic activity (FeO) and room-temperature magnetism (Fe₃O₄). However, the morphology of the films, especially in the ultrathin regime, can be significantly influenced by the crystalline structure of the used support. This work reports an ultra-high vacuum (UHV) low energy electron/synchrotron light-based X-ray photoemission electron microscopy (LEEM/XPEEM) and electron diffraction (μLEED) study of the growth of FeO and Fe₃O₄ on two closed-packed metal single crystal surfaces: Pt(111) and Ru(0001). The results reveal the influence of the mutual orientation of adjacent substrate terraces on the morphology of iron oxide films epitaxially grown on top of them. On fcc Pt(111), which has the same mutual orientation of adjacent monoatomic terraces, FeO(111) grows with the same in-plane orientation on all substrate terraces. For Fe₃O₄(111), one or two orientations are observed depending on the growth conditions. On hcp Ru(0001), the adjacent terraces of which are 'rotated' by 180° with respect to each other, the in-plane orientation of initial FeO(111) and Fe₃O₄(111) crystallites is determined by the orientation of the substrate terrace on which they nucleated. The adaptation of three-fold symmetric iron oxides to three-fold symmetric substrate terraces leads to natural structuring of iron oxide films, i.e., the formation of patch-like magnetite layers on Pt(111) and stripe-like FeO and Fe₃O₄ structures on Ru(0001).

Keywords: iron oxides; FeO; Fe₃O₄; ultrathin films; epitaxial growth; platinum; ruthenium; symmetry; LEEM; LEED; XPEEM

1. Introduction

Iron oxides exhibit unique physical and chemical properties that find applications in various industrial fields. The properties of iron oxide nanostructures, such as nanoparticles or thin films, differ from those of the corresponding bulk oxides, which is related to their limited dimensionality and – in the case of thin films—the interaction with the substrate on which they grow. It has been shown that ultrathin wüstite (FeO) films exhibit superior catalytic activity in the CO oxidation reaction [1, 2], which is related, among other factors, to the strong film–substrate interaction [3]. Ultrathin (few-nanometers-thick) magnetite (Fe₃O₄) films, on the other hand, have been shown to exhibit ferro-

(ferri-)magnetic ordering at room temperature [4–6], with magnetic properties dependent on the morphology of the films influenced by the structure of the used support [7].

Iron oxide films can be grown on various metal single crystal substrates, the two most commonly used of which are cubic Pt(111) [8] and hexagonal Ru(0001) [9]. Even though the surfaces of both of these substrates exhibit close-packed atomic planes with relatively similar lattice spacing, being 2.78 Å for Pt(111) and 2.71 Å for Ru(0001), their bulk crystal structures determine different in-plane orientation of adjacent atomic terraces on stepped surfaces. In the case of fcc platinum, the orientation is identical and results from the ABCABC-stacking. In the case of hcp ruthenium, the terraces are 'rotated' by 180° with respect to each other (when looking at the mutual positions of atoms in the first two atomic layers of each terrace [10]), which is a consequence of the ABAB-stacking. These differences are schematically shown in Figure 1.

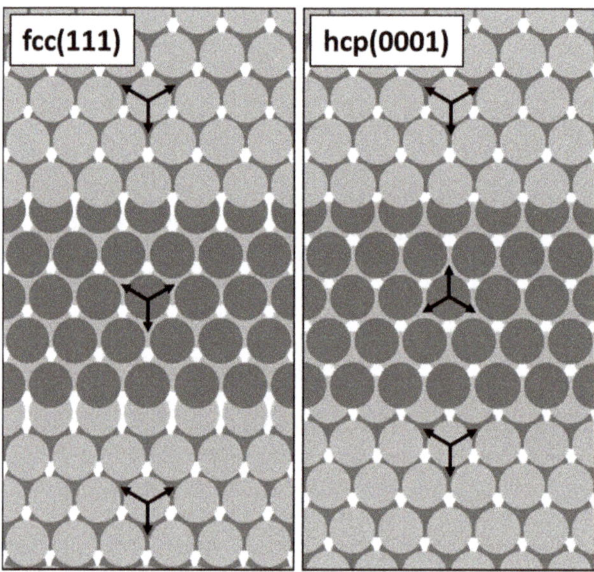

Figure 1. Schematic drawings showing the mutual orientation of adjacent monoatomic terraces on stepped fcc(111) and hcp(0001) surfaces (only the first two atomic layers are shown for each terrace; based on a similar figure published in Reference [10]).

On Pt(111), iron oxide films grow via the so-called Stranski–Krastanov (layer + islands) mode [8]. Firstly, an FeO layer is formed in direct contact with the metal support. Then, three-dimensional Fe_3O_4 islands start to nucleate on top of FeO. These islands may ultimately coalesce and form a closed magnetite film at a total thickness of >100 Å [8]. On Ru(0001), the growth mode depends on the preparation conditions: The use of O_2-assisted Fe deposition onto a heated substrate results, similarly to the case of Pt(111), in the formation of $Fe_3O_4(111)$ islands growing on top of FeO(111) [9]. Iron deposition onto a substrate kept at room temperature and post-oxidation, on the other hand, results in an additional thermodynamically-driven transformation of the FeO(111) film underneath $Fe_3O_4(111)$ islands to magnetite (so that Fe_3O_4 grows directly on Ru(0001) [11]). FeO(111) itself preferably grows as an Fe-O monolayer (ML) on Pt(111) [8] (with the possibility of stabilizing up to 2.5 MLs under certain growth conditions) and as an Fe-O-Fe-O bilayer on Ru(0001) [9] (with the possibility of stabilizing a structurally ill-defined monolayer [11] or a 4 MLs-thick film [9] using certain preparation recipes). The oxide has a rock-salt structure and, when looking along ⟨111⟩ direction, consists of alternately stacked close-packed layers of Fe^{2+} and O^{2-} ions, with iron atoms located at the interstitial sites of the oxygen lattice [8]. Due to the lattice mismatch between the oxide and the support (FeO(111) has an in-plane lattice constant of around 3.1 Å), ultrathin FeO films are characterized by Moiré superstructures the fingerprints of which can be observed in scanning tunneling microscopy (STM) images and low energy electron diffraction (LEED) patterns [8,9]. The coincidence structures responsible for the formation of such superstructures on Pt(111) and Ru(0001) are 8 FeO units on 9 Pt units and 7 FeO units on 8 Ru units (resulting in 25 Å and 21.6 Å Moiré periodicities, respectively). Fe_3O_4(111), on the other hand, has an inverse spinel structure with a mixture of Fe^{2+} Fe^{3+} ions arranged in Kagomé-type and mix-trigonal layers separated by close-packed O^{2-} planes. The interatomic distances within the mix-trigonal lattices equal to the distances between unoccupied sites in the Kagomé lattices and are twice as large as those in the oxygen lattice (approx. 6 Å vs. 3 Å). This gives rise to the characteristic (2×2) LEED pattern and accounts for the atomic spacing seen in the STM images [12].

Low energy electron microscopy (LEEM) is a powerful tool that allows real-time observation of thin films growth with a nanometer-scale resolution. When equipped with an imaging energy analyzer and used with a synchrotron light as an excitation source, it can be operated in the X-ray photoemission electron microscopy (XPEEM) mode which allows obtaining chemical contrast on the acquired images. The instrument also gives the possibility to record micro-spot low energy electron diffraction (μLEED) patterns from the selected sub-micrometer-sized regions. A complete overview of the LEEM-based techniques can be found in [13].

LEEM was used by various research groups for the studies of iron oxide films on Pt(111), Ru(0001) and Ag(111) (see e.g. References [6,11,14–18]). The results provided information on the growth, structure, electronic and magnetic properties of the films at the nanometer-scale. However, none of these studies comprehensively addressed the influence of different substrates' symmetry on the structure of iron oxide islands and films grown on top of them.

This work reports a comparative LEEM/XPEEM and μLEED surface science study of the growth and structure of FeO and Fe_3O_4 films on Pt(111) and Ru(0001). In addition to the well-known dependence of the film morphology on the growth conditions, the results reveal the influence of the substrates' symmetry—more precisely: the mutual orientation of adjacent substrate monoatomic terraces—on the structure of iron oxide islands and films epitaxially grown on top of them. The established procedures allow the preparation of naturally structured wüstite and magnetite layers that may find potential applications in future spintronic devices.

2. Materials and Methods

The experiments were performed using the SMART instrument (built up within a collaboration of several German groups [19,20]) located at the BESSY II synchrotron of the Helmholtz-Zentrum Berlin (HZB) (beamline UE49-PGM-SMART). The instrument is an aberration-corrected and energy-filtered spectro-microscopy system operating under ultra-high vacuum (UHV; base pressure: 1×10^{-10} mbar). It combines real-time electron- and X-ray-based microscopy (LEEM, XPEEM) and electron diffraction (μLEED), being a perfect tool for the studies of the growth and structure of epitaxial thin films (see Refs. [19] and [20] for more details).

Pt(111) and Ru(0001) single crystals (purity 99.999%; from MaTeck GmbH, Jülich, Germany) were cleaned by repeated cycles of 1 keV Ar^+ ions sputtering at room temperature, annealing in 1×10^{-6} mbar O_2 at 800–1000 K (Ar and O_2 99.999% pure; from Westfalen AG, Münster, Germany) and in UHV at 1300 (Pt(111)) or 1450 K (Ru(0001)). The crystalline order and cleanliness of the substrates were monitored with LEEM, μLEED and XPS. Iron (99.995%; Alfa Aesar GmbH, Karlsruhe, Germany) was deposited using a commercial evaporator (Focus GmbH, Hünstetten, Germany) onto the substrates kept at room temperature and post-oxidized in 1×10^{-6} mbar O_2 at 900 K for several minutes. The Fe deposition rate was determined from the obtained LEEM images (1 ML of iron is defined as the amount needed for the formation of a closed FeO(111) film on Pt(111); calibration accuracy: +/−5%). The temperature of the samples was controlled using an infrared pyrometer (LumaSense Technologies, Inc., Santa Clara, CA, USA).

The growth and structure of iron oxide films were characterized by LEEM, XPEEM and μLEED. The energies at which the data were taken are indicated in the captions of the figures. The LEEM-IV data presented in this work were obtained by plotting the intensity of a certain surface region from a series of images recorded at different beam energies, the dark field (DF) LEEM-IV curves from a series of dark field images acquired by mapping a particular diffraction spot, while the XPEEM-IV spectra from a stack of XPEEM images taken at different kinetic energies (the resulting XPEEM-IV curves, calibrated with respect to the valence band, are equivalent to micro-spot X-ray photoelectron spectroscopy (μXPS) spectra). Some sets of spectra were additionally normalized to allow for direct comparisons.

3. Results and Discussion

Figure 2a presents a LEEM image of iron oxide structures grown on Pt(111) by ~1.9 MLs iron deposition under UHV and subsequent oxidation in 1×10^{-6} mbar O_2 at 900 K. For these particular support, iron coverage and oxidation conditions, a formation of a closed FeO(111) film with nucleating Fe_3O_4(111) islands on top, was expected [8]. Interestingly, at certain energies, three different contrasts were observed on the acquired LEEM images, visible as light grey and dark grey hexagonal and irregularly-shaped islands on a bright background (these structures can be seen more clearly in the inset to Figure 2a).

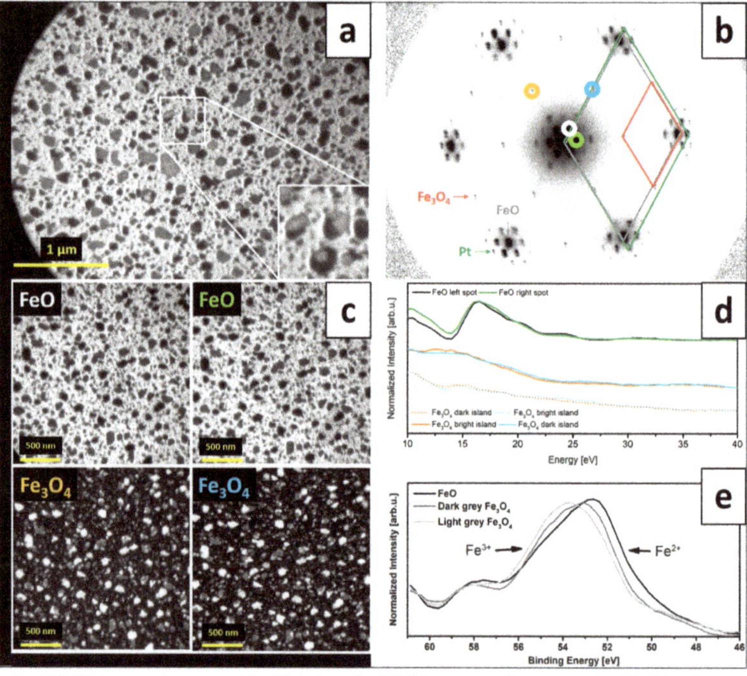

Figure 2. LEEM image (energy: 10 eV) (**a**) and μLEED pattern (energy: 42 eV) (**b**) of iron oxide structures grown on Pt(111) by ~1.9 MLs Fe deposition under UHV at RT and post-oxidation in 1×10^{-6} mbar O_2 at 900 K; the Pt(111), FeO(111) and Fe_3O_4(111) unit cells are marked on the μLEED pattern with green, grey and red rhombuses, respectively; (**c**) presents DF-LEEM images obtained by using the diffraction spots of FeO(111) (imaging energy: 17 eV) and Fe_3O_4(111) (10 eV) (marked in colors in (**b**)), while (**d**) shows the DF-LEEM-IV curves plotted from stacks of different DF-LEEM images taken at different energies (the corresponding pairs of curves were normalized and the curves were offset for clarity); (**e**) presents the XPEEM-IV data (photon energy: 150 eV) plotted for various surface regions that correspond to different iron oxide structures.

The μLEED pattern taken from this surface is shown in Figure 2b. Contributions from several ordered surface structures could be identified: Six main (1 × 1) reflexes were assigned to the substrate Pt(111). Six additional spots, positioned closer to the (0;0) spot and representing a larger interatomic spacing, were assigned to originate from FeO(111). Satellite spots around the Pt(111) spots are known to result from the multiple scattering at the Moiré superstructure [8]. Additional (2 × 2) spots with respect to Pt(111)-(1 × 1) spots were tentatively assigned to originate from Fe_3O_4(111) [8,12]. The number and shape of the Moiré satellite spots indicated the presence of a bilayer (Fe-O-Fe-O) film [21]. The precise assignment of different surface species visible in LEEM was possible thanks to dark field images obtained using the diffraction spots originating from different surface structures (the images are presented in Figure 2c with a color-code that follows the marks on the μLEED pattern in Figure 2b). The results confirmed that the bright background is FeO(111), while the light grey and dark grey islands are Fe_3O_4(111). In general, FeO(111) and Fe_3O_4(111) have a three-fold rotational symmetry, which is the same as the substrate Pt(111), and, therefore, can potentially grow on a three-fold symmetric substrate in two domains rotated with respect to each other by 180°. It was indeed shown in Ref. [22] that FeO(111) can grow in such two domains on Pt(111), where one domain orientation is more favored than the other. However, closed films were found to exhibit only one (more favored) orientation, which is in line with the morphology observed in our experiments. Similarly, for Fe_3O_4(111), which grows on top of FeO(111)/Pt(111), two quantitatively inequivalent in-plane orientations were reported in Refs. [12] and [15] (as could be expected for three-fold symmetric Fe_3O_4(111) growing on three-fold symmetric FeO(111)). However, in our case, dark field imaging of the two neighboring Fe_3O_4(111) diffraction spots (i.e., the (2 × 2) spots with respect to Pt(111)-(1 × 1) spots) did not result in two different image contrasts, thus indicating that all Fe_3O_4 islands have the same in-plane orientation (Figure 2c). As the contrast on the DF-LEEM images may also depend on the beam energy, we recorded DF-LEEM-IV curves for two neighboring μLEED spots of each type, i.e., the FeO-Moiré spots and the Fe_3O_4-(2 × 2) spots (the resulting curves are displayed in Figure 2d and marked in colors corresponding to the rings in Figure 2b). In both cases, the corresponding curves had the same character, which confirmed the same orientation of FeO(111) and Fe_3O_4(111) on all substrate terraces. Notably, at certain energies, the FeO signal could also be observed at the positions of the light grey and dark grey islands. This indicated that FeO was present underneath Fe_3O_4 and that the magnetite islands were relatively thin (thinner than the probing depth of the DF-LEEM). The contrast in LEEM may, of course, also depend on complex interactions between different layers and has to be interpreted with proper caution, however, the observed growth seems reasonable, taking into account the deposited amount of iron and the existing knowledge on the structure of iron oxide films on Pt(111).

The recorded XPEEM data allowed us to plot XPEEM-IV curves from each type of region and determine the compositional differences between the FeO background and two types of Fe_3O_4 islands. The data recorded for the binding energy range where the Fe 3p signals are known to appear are shown in Figure 2e (we used the 3p signals instead of the commonly used 2p due to the specific set-up of the beamline in which lower photon energies result in higher intensity). The spectrum obtained by plotting the intensity from the whole field of view (not shown) revealed a broad and asymmetric peak, indicating superposition of components originating from iron in different oxidation states (i.e., Fe^{2+} and Fe^{3+}, which occur at the low and high energy sides of the spectrum, respectively [17]). The curve taken locally from the regions assigned to FeO(111) shows a pronounced maximum at the low energy side, indicating the expected dominant presence of Fe^{2+} iron in these regions. The spectrum recorded from the dark grey islands, on the other hand, has a contribution of both components, with Fe^{2+} signal being more pronounced than the Fe^{3+}. The plot obtained for the light grey islands is similar but with higher contribution of Fe^{3+} ions. Higher amount of Fe^{3+} in the light grey Fe_3O_4 islands indicated that they were thicker than the dark grey islands, so that less (or no) Fe^{2+} signal was detected from the FeO layer underneath the islands (the amount of Fe^{2+} ions per unit volume is higher in FeO than in Fe_3O_4). In general, the obtained results are in line with other reports on the growth of iron oxide films on Pt(111) following 1-2 MLs Fe deposition under UHV and post-oxidation in 1×10^{-6} mbar O_2 at 900 K, i.e., they reveal the formation of an FeO(111) layer with a mix-valence Fe_3O_4(111) islands of different height nucleating on top of it. The most important observation made was that all the iron oxide structures grown in that way have the same in-plane orientation with respect to the substrate Pt(111).

Deposition of the same amount of iron onto Ru(0001) and post-oxidation resulted in a slightly different sample morphology, as can be seen on the LEEM image in Figure 3a. The μLEED pattern taken from this surface is presented in Figure 3b. Again, LEEM showed bright background with light grey and dark grey islands, while μLEED pattern was a superposition of diffraction spots originating from Ru(0001) ((1 × 1) spots arrangement), FeO(111) ((1 × 1), larger interatomic spacing than that of Ru(0001)), Moiré (satellite spots around the Ru(0001)-(1 × 1) spots) and Fe_3O_4(111) ((2 × 2) spots with respect to the Ru(0001)-(1 × 1) spots) (the pattern can be directly compared with the one published in [14], as they were taken at similar beam energy). DF-LEEM images (Figure 3c) revealed that the bright background is FeO(111) and that the dark islands are Fe_3O_4(111). The amount of Fe_3O_4(111) islands was lower than in the case of Pt(111), however, the average size of the islands was larger. Interestingly, the light grey islands were found not to show any intensity when mapping the FeO diffraction spots and only weak contrast (noise level) at certain energies when mapping the (2 × 2) spots visible in μLEED. In general, the oxidations parameters used should promote, similarly to the case of Pt(111), the growth of bilayer FeO(111) film on Ru(0001) [9] and the appearance of higher order Moiré satellite spots confirmed the presence of such a film in our experiments. However, the amount of deposited iron was not sufficient for the formation of a closed bilayer FeO(111) film fully covering the Ru(0001) substrate. Taking this into account, the light grey islands could be assigned to exposed Ru(0001) and the weak contrast observed when mapping the (2 × 2) LEED spots to oxygen chemisorbed on Ru(0001) (forming the well-known 3O structure [23]).

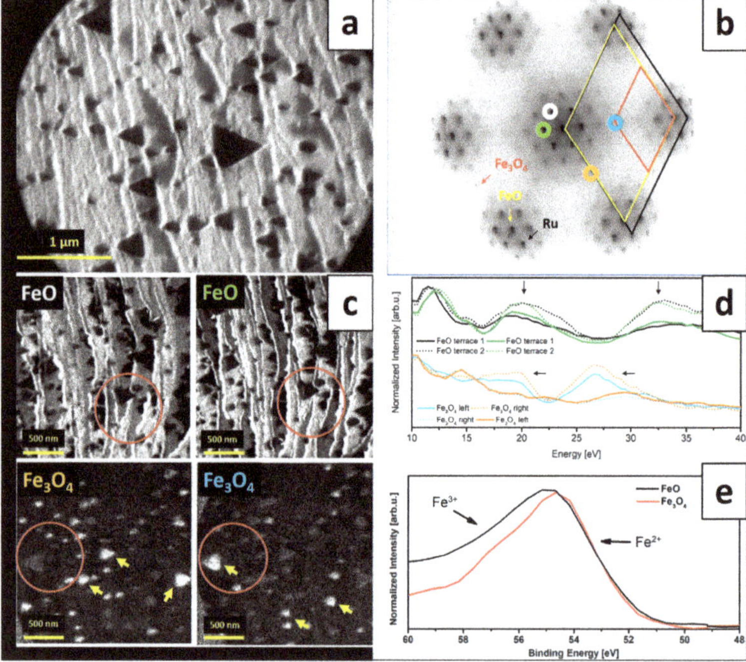

Figure 3. LEEM image (energy: 10 eV) (**a**) and μLEED pattern (energy: 42 eV) (**b**) of iron oxide structures grown on Ru(0001) by ~1.9 MLs Fe deposition under UHV at RT and post-oxidation in 1×10^{-6} mbar O_2 at 900 K; the Ru(0001), FeO(111) and Fe_3O_4(111) unit cells are marked on the μLEED pattern with black, yellow and red rhombuses, respectively; (**c**) presents DF-LEEM images (red circles

mark the same sample position) obtained by using the diffraction spots of FeO(111) (imaging energy: 20 eV) and Fe_3O_4(111) (26 eV; different position than the one at which FeO(111) images were taken) (marked in colors in (**b**)), while (**d**) shows DF-LEEM-IV curves plotted from stacks of DF-LEEM images taken at different energies (the corresponding pairs of curves were normalized and the curves were offset for clarity); (**e**) presents the XPEEM-IV data (photon energy: 150 eV) plotted for various surface regions that correspond to different iron oxide structures.

The observed Fe_3O_4(111) islands had a triangular shape and were similar to those reported in Refs. [6,11,14]. 'Left-' and 'right-oriented' triangles could be observed, indicating two possible in-plane orientations of Fe_3O_4(111) on Ru(0001). Interestingly, within one substrate terrace, only one islands' orientation could be seen (Figure 4) (the only exceptions were the biggest islands that were crossing several terraces—in their case, the borders were probably set by step bunches). Dark field images presented in Figure 3c did not only confirm the 180° rotation of Fe_3O_4(111) on adjacent Ru(0001) terraces (islands with the same orientation, marked with yellow arrows, show much higher intensity when mapping a particular Fe_3O_4 diffraction spot), but also the same rotation of FeO(111) (again, on each substrate terrace only one orientation of FeO(111) was observed). The latter results in the formation of a stripe-like FeO structure on Ru(0001). It has to be mentioned that such a growth mode was already predicted for FeO(111)/Ru(0001) in Ref. [24] based on the six-fold symmetry observed in LEED of few-nanometers-thick FeO(111) films. The 180° rotation is related to the structure of stepped Ru(0001) surfaces, as described in the Introduction (see Figure 1). A single monoatomic Ru(0001) terrace has a three-fold symmetry [10], same as FeO(111) and Fe_3O_4(111), therefore it seems intuitive that the epitaxially growing iron oxide will align to the structure of the Ru(0001) substrate (more specifically: to the structure of a particular substrate terrace on which it grows). The rotation of iron oxides is also evident when looking at the DF-LEEM-IV curves taken locally from FeO(111) on adjacent substrate terraces, as well as from left- and right-oriented Fe_3O_4(111) islands, using the neighboring FeO(111) and Fe_3O_4(111) diffraction spots, respectively (Figure 3d). The DF-LEEM-IV curve obtained for one FeO(111) spot from a particular substrate terrace shows the same character as the curve taken for the other (neighboring) FeO(111) spot from the neighboring terrace. However, the curves taken from the same terrace using different spots are different (the differences are marked with black arrows in Figure 3d). The same holds true for left- and right-oriented Fe_3O_4(111) islands and neighboring Fe_3O_4(111) diffraction spots. Different character of the curves with respect to those obtained for Fe_3O_4(111) on Pt(111) may be due to the fact that the Fe_3O_4 islands on Ru(0001) are much thinner and their I–V characteristics may be differently influenced by the underlying substrate. With this respect, it is important to mention that dark field mapping of FeO(111) spots did not show any signal at the location of Fe_3O_4 islands, thus confirming the expected transformation of FeO(111) underneath Fe_3O_4(111) islands to magnetite [9]. The presence of Fe_3O_4(111)/Ru(0001) interface, different from the Fe_3O_4(111)/FeO(111) one observed on Pt(111), may have a strong influence on the oxide's I–V characteristic.

The analysis of the obtained XPEEM data (Figure 3e) revealed that both FeO(111) and Fe_3O_4(111) structures consist of a mixture of Fe^{2+} and Fe^{3+} ions. The presence of Fe^{+3} ions in magnetite was expected, however, in wüstite it may only be explained by the presence of iron vacancies in the bilayer FeO film [24].

To better visualize the registries between FeO(111), Fe_3O_4(111) and both supports—i.e., Pt(111) and Ru(0001)—we constructed schematic model of the experimentally observed structures and present them in Figure 5. The models take into account the lattice mismatch at different metal-oxide interfaces (FeO(111)/Pt(111), FeO(111)/Ru(0001) and Fe_3O_4(111)/Ru(0001)), as well as the rotation angle reported for FeO(111)/Pt(111) [21].

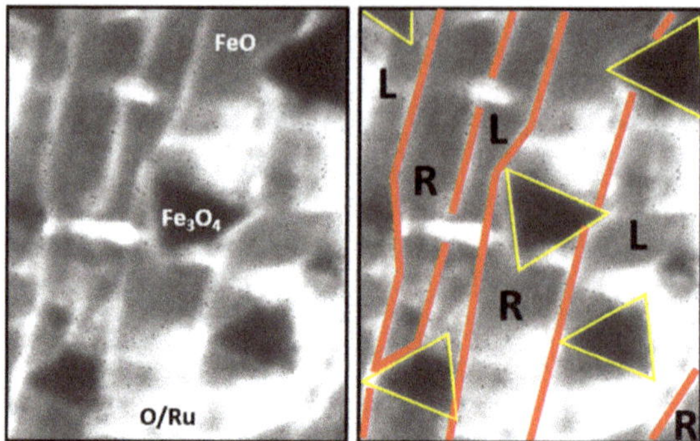

Figure 4. Excerpt from a LEEM image of iron oxides on Ru(0001) showing the 180° rotation of triangular $Fe_3O_4(111)$ islands on adjacent monoatomic terraces of the Ru(0001) substrate. The raw image is shown on the left, while the right hand side shows the same image with marked island contours (in yellow) and substrate monoatomic step edges (in red). "L" and "R" indicate terraces with 'left-oriented' and 'right-oriented' islands, respectively.

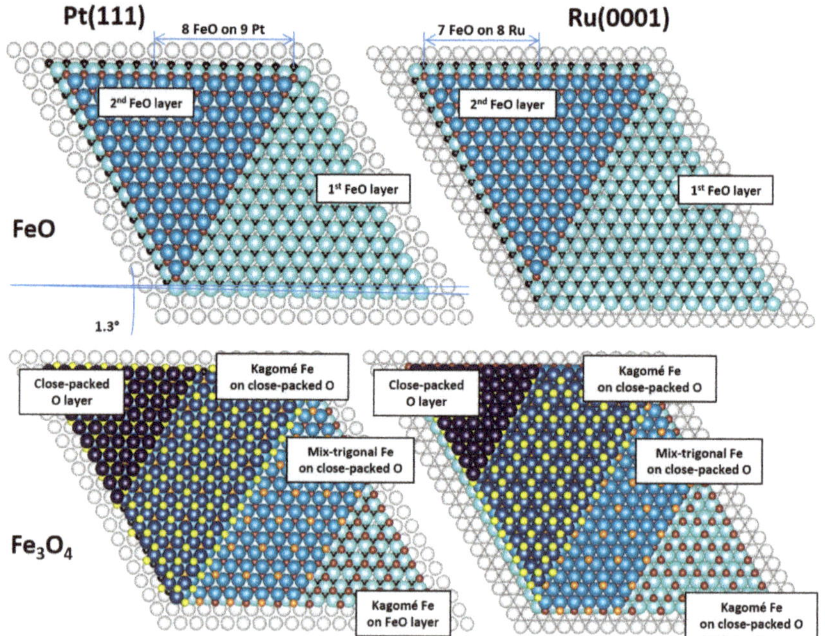

Figure 5. Schematic models of FeO(111) and $Fe_3O_4(111)$ structures on Pt(111) and Ru(0001). The atoms in different iron and oxygen sublattices are partially removed to better visualize their different arrangement within different sublayers. The models were made using VESTA 3.4.3 computer software [25].

The fact that $Fe_3O_4(111)$ islands on both supports exhibit only one in-plane orientation per substrate terrace is believed to be related to (1) the properties of the substrates and (2) the oxidation temperature used. It is well established that the preparation of iron oxide films on Pt(111) and Ru(0001) starts with 1-2 ML Fe deposition and post-oxidation in 1×10^{-6} mbar O_2 at 900–1000 K—the conditions that promote the growth of well-defined FeO(111) wetting layers [8,9]. The oxidation temperature of 900 K favors the formation of bilayer FeO(111) on both supports, which is what we also observe in our experiments. The prolonged oxidation time may, however, result in the transformation of the second FeO(111) layer on Pt(111) to Fe_3O_4(111), so that Fe_3O_4(111) islands grow on top of the first FeO layer [8], and in the transformation of the bilayer FeO(111) film on Ru(0001) to Fe_3O_4(111) (so that Fe_3O_4 grows directly on Ru(0001) [11]). Both transformations may potentially occur at the regular terrace sites (and not at the step edges where Fe_3O_4 crystallites usually nucleate [8,9]), with Fe_3O_4 preserving the same in-plane orientation as the initial FeO. Such growth mode may be driven by various thermodynamic and kinetic factors. Looking at the iron oxides bulk phase diagram [26], the FeO phase should not be thermodynamically stable at the oxidation conditions used in our experiments (1×10^{-6} mbar O_2 and 900 K). However, earlier studies of other authors revealed that in the case of supported iron oxide films the formation of various equilibrium and intermediate surface phases, lying outside of the bulk stability ranges, may be possible [8,9]. On Pt(111), the first FeO layer is stabilized by the strong interaction with the platinum substrate. The second FeO layer can also grow on top of the first layer at certain growth conditions, however, it has a much higher surface free energy and, during prolonged oxidation, undergoes transformation to a more stable Fe_3O_4 phase. The growth of Fe_3O_4 is, in turn, a transition state towards the final transformation to α-Fe_2O_3 (hematite) which, based on the bulk phase diagram, should be the thermodynamically most stable phase at these oxidation conditions. Such transformational order is correlated with an increasing oxidation state of iron (Fe^{2+} in FeO, mixed Fe^{2+}/Fe^{3+} in Fe_3O_4 and Fe^{3+} in α-Fe_2O_3). However, as indicated by the authors of Ref. [26], the complete transformation of Fe_3O_4 to α-Fe_2O_3 would require oxidation times of several hours (unless a higher oxygen pressure is used). It is thus expected that the observed growth order, i.e., first FeO layer $\rightarrow$ second FeO layer $\rightarrow$ transformation of the second FeO layer to Fe_3O_4, is related to the relatively low oxidation temperature used, which allows the initial stabilization of the second FeO layer prior to the onset of the Fe_3O_4 growth (direct growth of Fe_3O_4 islands on the first FeO(111) layer was observed by other authors when the oxidation was carried out at 1000 K [8]), and low oxidation pressure (which makes transformation to α-Fe_2O_3 improbable due to the limited oxidation time). In addition, the amount of iron locally available within the second FeO layer is not sufficient for the formation of Fe_3O_4 islands that would cover a similar surface area—that is why the nucleating magnetite islands grow at the expense of the neighboring second layer FeO regions which are 'eaten up' in the process (the first FeO layer stays intact). The growth of Fe_3O_4 may, therefore, lead to the formation of iron vacancies in the FeO regions, which would explain the appearance of the Fe^{3+} component in our XPEEM-IV data. The growth is thus kinetically limited, as it depends on the iron diffusion rate which is lower at lower oxidation temperatures. On Ru(0001), the situation is slightly different, as the initially stabilized form of FeO, i.e., the bilayer Fe-O-Fe-O film, is more weakly interacting with the substrate than the monolayer Fe-O film on Pt(111) and undergoes complete transformation to Fe_3O_4 (so that the Fe_3O_4 crystallites grow directly on Ru(0001)). The amount of iron locally available for the FeO $\rightarrow$ Fe_3O_4 transformation is twice as large as on Pt(111) and sufficient for the direct transformation. The formation of iron vacancies in the vicinity of the growing islands is also possible, however, not absolutely necessary. The growth is thus not kinetically limited and proceeds more efficiently at the same oxidation conditions.

In order to verify the influence of the substrate and the growth conditions on the structure of thicker iron oxide films on Pt(111) and Ru(0001), we deposited additional ~3.8 MLs of iron (which resulted in a total iron dose of ~5.7 MLs) onto both supports and oxidized in 1×10^{-6} mbar O_2 at 900 K. Such a procedure typically leads to the proceeding growth of Fe_3O_4 [8,9]. The LEEM images obtained for the as prepared samples are shown in Figure 6a,b for iron oxides on Pt(111) and Ru(0001), respectively. The deposited amount of iron was not sufficient for the formation of a closed Fe_3O_4(111) film on Pt(111), as magnetite islands tend to coalesce at a total film thickness of about 100 Å on this particular support [8] (where the height of one Fe_3O_4(111) unit cell is 4.85 Å). Due to this, we expected the surface to consist of a large number of nucleated Fe_3O_4(111) islands with some exposed FeO(111) in between. This morphology can be indeed observed on the LEEM image presented in Figure 6a, which reveals rough surface. The corresponding µLEED pattern, shown in Figure 6c, consists of strong Fe_3O_4 reflexes and weak FeO spots—thus supporting the LEEM data. The additional streaks in between the (2×2) spots are indicative of small Fe_3O_4 domain sizes. On Ru(0001), the deposition–oxidation cycle was expected to promote further transformation of FeO to Fe_3O_4. In fact, the LEEM image shown in Figure 6b exhibits uniform contrast, thus indicating that the surface is fully covered with a structurally-uniform iron oxide layer. The µLEED pattern taken from this surface, presented in Figure 6d, revealed the presence of strong Fe_3O_4 (2×2) reflexes and no diffraction spots originating from FeO. This confirmed that the initial mixed FeO/Fe_3O_4 surface transformed, with the help of the additionally deposited iron, into a uniform Fe_3O_4 layer. This behavior, different from the one observed on Pt(111), is related to the fact that Fe_3O_4 forms much thinner structures on Ru(0001) than on Pt(111) [6] and, therefore, the coalescence of islands may occur for much lower iron dose. To shed more light on the structure of as prepared films, we also performed dark field imaging using the $Fe_3O_4(111)$ spots marked on the diffraction patterns shown in Figure 6c,d, which resulted in the DF-LEEM images shown in Figure 6e,f, respectively. Interestingly, two domains rotated by 180° around the [111] axis were observed on both Pt(111) and Ru(0001), resulting in the formation of a patch-like Fe_3O_4 layer on Pt(111) and stripe-like magnetite structure on Ru(0001).

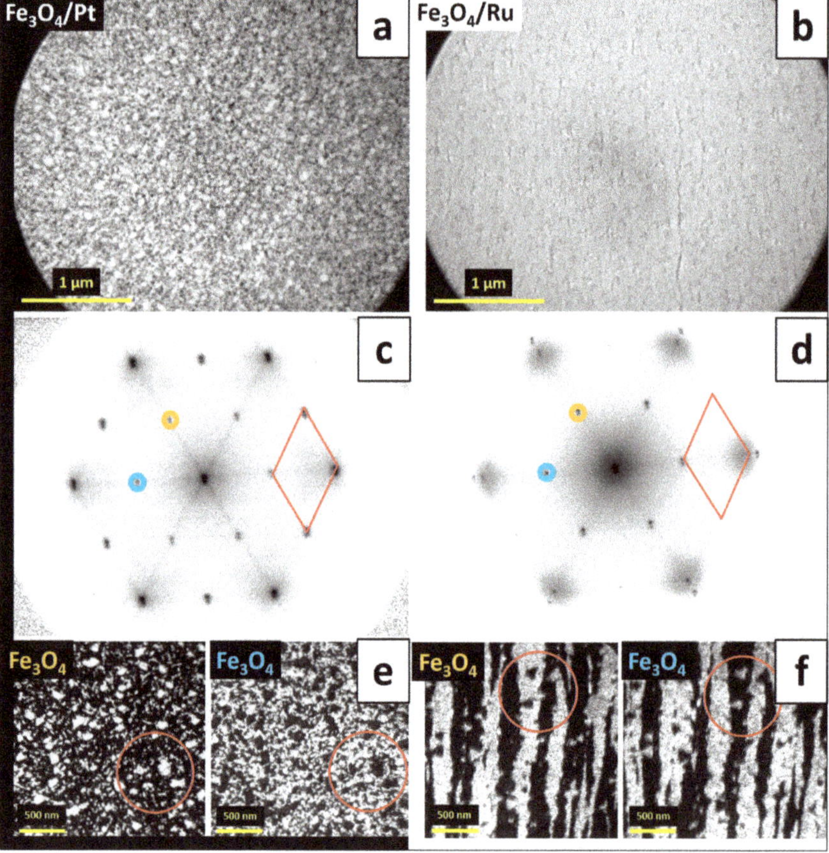

Figure 6. LEEM images (energy: 10 eV) (**a**,**b**) and µLEED patterns (energy: 42 eV) (**c**,**d**) of $Fe_3O_4(111)$ films grown on Pt(111) (**a**,**c**) and Ru(0001) (**b**,**d**) by ~5.7 ML Fe deposition (total dose) under UHV at RT and post-oxidation in 1×10^{-6} mbar O_2 at 900 K; the $Fe_3O_4(111)$ unit cell is marked on µLEED patterns with red rhombus; (**e**,**f**) present DF-LEEM images obtained by mapping the diffraction spots marked with colors in (**c**) and (**d**), respectively (red circles mark the same sample position on respective images) (imaging energies: 15 (**e**) and 20 eV (**f**)).

The appearance of two island orientations on Pt(111) is believed to be related to the presence of newly-nucleated Fe_3O_4 islands that did not form via the transformation of the second FeO layer to Fe_3O_4, but through the oxidation of the additionally deposited iron. Such islands should have higher probability to grow in one of the two possible orientations, even though one orientation seems to be highly dominant independently on the growth conditions (see e.g. Refs. [8,15]). On Ru(0001), the proceeding transformation of a bilayer FeO(111) film to Fe_3O_4(111) results in an iron oxide film with one in-plane orientation per substrate terrace. Only rarely small crystallites with an orientation opposite to that of the terrace on which they grow could be observed (examples of such islands can be seen inside the red circles in Figure 6f). Notably, the stripe-like Fe_3O_4(111) structure on Ru(0001) was also observed after the deposition of additional ~1.9 MLs of iron (total iron dose of ~7.6 MLs). The DF-LEEM images obtained for these films by mapping two neighboring Fe_3O_4(111) diffraction spots are presented in Figure 7a. This indicates that on this particular support and at these oxidation conditions the oxide grows in a Frank–van der Merwe (layer-by-layer) mode, with subsequent layers adopting the structure of the preceding ones. Such a growth mode favors the stripe-like Fe_3O_4 structure also at higher film thicknesses.

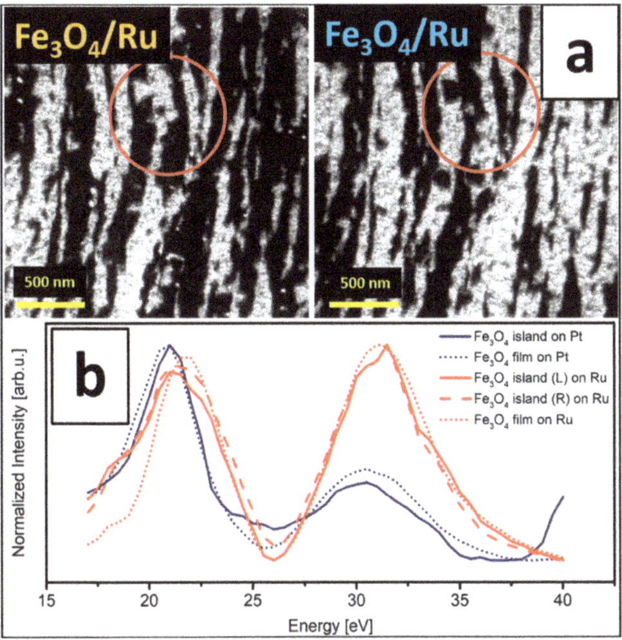

Figure 7. DF-LEEM images of the Fe_3O_4(111) film grown on Ru(0001) by ~7.6 MLs Fe deposition (total dose) under UHV at RT and oxidation in 1×10^{-6} mbar O_2 at 900 K (**a**) (red circles mark the same sample position; imaging energy: 26 eV), obtained by mapping the diffraction spots analogous to those marked in colors in Figure 6d; (**b**) shows LEEM-IV curves obtained for Fe_3O_4 islands and multilayer films on Pt(111) and Ru(0001).

Interestingly, the LEEM-IV characteristics obtained for the initial Fe_3O_4(111) islands and for the multilayer magnetite films on Pt(111) and Ru(0001) were found to show a very similar character within the energy range of 17–40 eV (Figure 7b). The two peaks, centered at around 21 and 31 eV, were observed for all magnetite structures—regardless of the support and magnetite thickness. In particular, the 31 eV peak was explicitly observed for this iron oxide phase and may be used as a LEEM-IV-fingerprint of Fe_3O_4(111). LEEM-IV was already shown to be a reliable method for fingerprinting different iron oxide phases on Pt(111) and Ru(0001), however, usually the energy region of 0–30 eV was used in this respect [11,14,16]. In this region $Fe_3O_4(111)$ shows very similar character to FeO(111) (with only a small difference in peak positions), that is why the peak at ~31 eV, which is observed only for magnetite, seems to be a more proper choice for fingerprinting this particular iron oxide phase.

4. Conclusions

The results revealed the direct influence of the mutual orientation of adjacent monoatomic terraces of Pt(111) and Ru(0001) on the orientation of FeO(111) and $Fe_3O_4(111)$ crystallites epitaxially grown on top of them. The different orientation of terraces originates from different crystal structure (fcc vs. hcp). On cubic Pt(111), which has the same mutual orientation of adjacent monoatomic terraces, FeO(111) grows with the same in-plane orientation on all substrate terraces. For $Fe_3O_4(111)$, one or two orientations can be observed depending on the growth conditions. On Ru(0001), the adjacent terraces of which are 'rotated' by 180° with respect to each other, the in-plane orientation of the initial FeO(111) and $Fe_3O_4(111)$ crystallites is determined by the orientation of the substrate terrace on which they nucleated. Such a growth mode can be explained by the adaptation of three-fold symmetric iron oxides to three-fold symmetric substrate terraces and leads to natural structuring of iron oxide films, i.e., the formation of patch-like magnetite layers on Pt(111) and stripe-like FeO and Fe_3O_4 structures on Ru(0001). Even though the orientation of iron oxide crystallites may also depend on other factors, such as the oxidation temperature or substrate-specific growth mode, the observed symmetry-induced structuring provides a general route for tuning the structure and properties of ultrathin [111]-oriented epitaxial films by switching the substrate from fcc(111) to hcp(0001).

Author Contributions: N.M., Z.M., G.P., M.P., and M.L. prepared the samples and performed the experiments. F.X. assisted in samples preparation. N.M., Z.M., and M.L. analyzed the results and M.L. wrote the manuscript. P.W. contributed to data analysis and figure preparation. T.S., in addition to all the other authors, contributed to the discussion.

Funding: This research was funded by the National Science Centre of Poland, grant nos. 2012/05/D/ST3/02855 and 2016/21/N/ST4/00302, and the Foundation for Polish Science, grant no. First TEAM/2016-2/14. The SMART instrument was financially supported by the Federal German Ministry of Education and Research (BMBF) under the contract 05 KS4WWB/4, as well as by the Max-Planck Society.

Acknowledgments: This work was financially supported by the National Science Centre of Poland (project nos. 2012/05/D/ST3/02855 (Pt(111) part) and 2016/21/N/ST4/00302 (Ru(0001) part) [27]) and the Foundation for Polish Science (First TEAM programme, 2017–2020, grant no. First TEAM/2016-2/14 ("Multifunctional ultrathin Fe(x)O(y), Fe(x)S(y) and Fe(x)N(y) films with unique electronic, catalytic and magnetic properties" project co-financed by the European Union under the European Regional Development Fund)). The SMART instrument was financially supported by the Federal German Ministry of Education and Research (BMBF) under the contract 05 KS4WWB/4, as well as by the Max-Planck Society. We thank the HZB for the allocation of a synchrotron radiation beamtime, Ernst Bauer and Feliks Stobiecki for critical reading of the manuscript and Thomas Vasileiadis for hosting us in Berlin during beamtime measurements. Feng Xiong thanks the China Scholarship Council for financial support.

Conflicts of Interest: The authors declare no conflict of interest.

References

1. Sun, Y.-N.; Qin, Z.-H.; Lewandowski, M.; Carrasco, E.; Sterrer, M.; Shaikhutdinov, S.; Freund, H.-J. Monolayer iron oxide film on platinum promotes low temperature CO oxidation. *J. Catal.* **2009**, *266*, 359–368. [CrossRef]
2. Weng, X.; Zhang, K.; Pan, Q.; Martynova, Y.; Shaikhutdinov, S.; Freund, H.-J. Support effects on CO oxidation on metal-supported ultrathin FeO(111) films. *Chem. Cat. Chem.* **2017**, *9*, 705–712.
3. Giordano, L.; Lewandowski, M.; Groot, I.M.N.; Sun, Y.-N.; Goniakowski, J.; Noguera, C.; Shaikhutdinov, S.; Pacchioni, G.; Freund, H.-J. Oxygen-induced transformations of an FeO(111) film on Pt(111): A combined DFT and STM study. *J. Phys. Chem. C* **2010**, *114*, 21504–21509. [CrossRef]
4. Schedin, F.; Hewitt, L.; Morrall, P.; Petrov, V.N.; Thornton, G.; Case, S.; Thomas, M.F.; Uzdin, V.M. In-plane magnetization of an ultrathin film of $Fe_3O_4(111)$ grown epitaxially on Pt(111). *Phys. Rev. B* **1998**, *58*, R11861. [CrossRef]

5. Morrall, P.; Schedin, F.; Langridge, S.; Bland, J.; Thomas, M.F.; Thornton, G. Magnetic moment in an ultrathin magnetite film. *J. Appl. Phys.* **2003**, *93*, 7960–7962. [CrossRef]

6. Monti, M.; Santos, B.; Mascaraque, A.; Rodríguez de la Fuente, O.; Niño, M.A.; Menteş, T.O.; Locatelli, A.; McCarty, K.F.; Marco, J.F.; de la Figuera, J. Magnetism in nanometer-thick magnetite. *Phys. Rev. B* **2012**, *85*, 020404. [CrossRef]

7. Lewandowski, M.; Miłosz, Z.; Michalak, N.; Ranecki, R.; Sveklo, I.; Kurant, Z.; Maziewski, A.; Mielcarek, S.; Luciński, T.; Jurga, S. Room temperature magnetism of few-nanometers-thick Fe_3O_4(111) films on Pt(111) and Ru(0001) studied in ambient conditions. *Thin Solid Films* **2015**, *591*, 285–288. [CrossRef]

8. Weiss, W.; Ritter, W. Metal oxide heteroepitaxy: Stranski–Krastanov growth for iron oxides on Pt(111). *Phys. Rev. B* **1999**, *59*, 5201. [CrossRef]

9. Ketteler, G.; Ranke, W. Heteroepitaxial growth and nucleation of Iron Oxide films on Ru(0001). *J. Phys. Chem. B* **2003**, *107*, 4320–4333. [CrossRef]

10. de la Figuera, J.; Puerta, J.M.; Cerda, J.I.; El Gabaly, F.; McCarty, K.F. Determining the structure of Ru(0001) from low-energy electron diffraction of a single terrace. *Surf. Sci.* **2006**, *600*, L105–L109. [CrossRef]

11. Santos, B.; Loginova, E.; Mascaraque, A.; Schmid, A.K.; McCarty, K.F.; de la Figuera, J. Structure and magnetism in ultrathin iron oxides characterized by low energy electron microscopy. *J. Phys. Condens. Matter* **2009**, *21*, 314011. [CrossRef] [PubMed]

12. Ritter, M.; Weiss, W. Fe_3O_4(111) surface structure determined by LEED crystallography. *Surf. Sci.* **1999**, *432*, 81–94. [CrossRef]

13. Bauer, E. *Surface Microscopy with Low Energy Electrons*; Springer: New York, NY, USA, 2014; 496p, ISBN 978-1-4939-0934-6.

14. Monti, M.; Santos, B.; Mascaraque, A.; Rodríguez de la Fuente, O.; Niño, M.A.; Menteş, T.O.; Locatelli, A.; McCarty, K.F.; Marco, J.F.; de la Figuera, J. Oxidation pathways in bicomponent ultrathin Iron Oxide films. *J. Phys. Chem. C* **2012**, *116*, 11539–11547. [CrossRef]

15. Sala, A.; Marchetto, H.; Qin, Z.-H.; Shaikhutdinov, S.; Schmidt, Th.; Freund, H.-J. Defects and inhomogeneities in Fe_3O_4(111) thin film growth on Pt(111). *Phys. Rev. B* **2012**, *86*, 155430. [CrossRef]

16. Palacio, I.; Monti, M.; Marco, J.F.; McCarty, K.F.; de la Figuera, J. Initial stages of FeO growth on Ru(0001). *J. Phys. Condens. Matter* **2013**, *25*, 484001. [CrossRef] [PubMed]

17. Genuzio, F.; Sala, A.; Schmidt, Th.; Menzel, D.; Freund, H.-J. Interconversion of α-Fe_2O_3 and Fe_3O_4 thin films: Mechanisms, morphology, and evidence for unexpected substrate participation. *J. Phys. Chem. C* **2014**, *118*, 29068–29076. [CrossRef]

18. Genuzio, F.; Sala, A.; Schmidt, Th.; Menzel, D.; Freund, H.-J. Phase transformations in thin iron oxide films: Spectromicroscopic study of velocity and shape of the reaction fronts. *Surf. Sci.* **2016**, *648*, 177–187. [CrossRef]

19. Fink, R.; Weiss, M.R.; Umbach, E.; Preikszas, D.; Rose, H.; Spehr, R.; Hartel, P.; Engel, W.; Degenhardt, R.; Wichtendahl, R.; et al. SMART: A planned ultrahigh-resolution spectromicroscope for BESSY II. *J. Electron Spectrosc. Relat. Phenom.* **1997**, *84*, 231–250. [CrossRef]

20. Schmidt, Th.; Marchetto, H.; Lévesque, P.L.; Groh, U.; Maier, F.; Preikszas, D.; Hartel, P.; Spehr, R.; Lilienkamp, G.; Engel, W.; et al. Double aberration correction in a low-energy electron microscope. *Ultramicroscopy* **2010**, *110*, 1358–1361. [CrossRef] [PubMed]

21. Ritter, M.; Ranke, W.; Weiss, W. Growth and structure of ultrathin FeO films on Pt(111) studied by STM and LEED. *Phys. Rev. B* **1998**, *57*, 7240–7251. [CrossRef]

22. Zeuthen, H.; Kudernatsch, W.; Merte, L.R.; Ono, L.K.; Lammich, L.; Besenbacher, F.; Wendt, S. Unraveling the edge structures of Platinum(111)-supported ultrathin FeO islands: The influence of oxidation state. *ACS Nano* **2015**, *9*, 573–583. [CrossRef] [PubMed]

23. Kostov, K.L.; Gsell, M.; Jakob, P.; Moritz, T.; Widdra, W.; Menzel, D. Observation of a novel high density 3O(2×2) structure on Ru(001). *Surf. Sci.* **1997**, *394*, L138–L144. [CrossRef]

24. Martin-Garcia, L.; Bernal-Villamil, I.; Oujja, M.; Carrasco, E.; Gargallo-Caballero, R.; Castillejo, M.; Marco, J.F.; Gallego, S.; de la Figuera, J. Unconventional properties of nanometric FeO(111) films on Ru(0001): stoichiometry and surface structure. *J. Mater. Chem. C* **2016**, *4*, 1850–1859. [CrossRef]

25. Momma, K.; Izumi, F. VESTA 3 for three-dimensional visualization of crystal, volumetric and morphology data. *J. Appl. Crystallogr.* **2011**, *44*, 1272–1276. [CrossRef]

26. Ketteler, G.; Weiss, W.; Ranke, W.; Schlögl, R. Bulk and surface phases of iron oxides in an oxygen and water atmosphere at low pressure. *Phys. Chem. Chem. Phys.* **2001**, *3*, 1114–1122. [CrossRef]
27. Michalak, N.; Miłosz, Z.; Peschel, G.; Prieto, M.; Xiong, F.; Schmidt, Th.; Lewandowski, M. Growth of FeO and Fe3O4 films on Pt(111) and Ru(0001) studied with LEEM/XPEEM and micro-LEED. Presented at the NanoTech Poland 2018 Conference, Poznań, Poland, 6–9 June 2018.

 *nanomaterials*

Article

Self-Catalyzed CdTe Wires

Tom Baines [1,*], **Giorgos Papageorgiou** [1,†], **Oliver S. Hutter** [1], **Leon Bowen** [2], **Ken Durose** [1] and **Jonathan D. Major** [1]

1   Stephenson Institute for Renewable Energy, Physics Department, University of Liverpool, Liverpool L69 7XF, UK; g.papageorgiou@swansea.ac.uk (G.P.); O.S.Hutter@liverpool.ac.uk (O.S.H.); dph0kd@liverpool.ac.uk (K.D.); jonmajor@liverpool.ac.uk (J.D.M.)
2   Department of Physics, G.J. Russell Microscopy Facility, Durham University, South Road, Durham DH1 3LE, UK; leon.bowen@durham.ac.uk
*   Correspondence: tbaines@liverpool.ac.uk
†   Current Address: CTF SOLAR GmbH, Zur Wetterwarte 50, 01109 Dresden, Germany.

Received: 23 March 2018; Accepted: 23 April 2018; Published: 25 April 2018

Abstract: CdTe wires have been fabricated via a catalyst free method using the industrially scalable physical vapor deposition technique close space sublimation. Wire growth was shown to be highly dependent on surface roughness and deposition pressure, with only low roughness surfaces being capable of producing wires. Growth of wires is highly (111) oriented and is inferred to occur via a vapor-solid-solid growth mechanism, wherein a CdTe seed particle acts to template the growth. Such seed particles are visible as wire caps and have been characterized via energy dispersive X-ray analysis to establish they are single phase CdTe, hence validating the self-catalysation route. Cathodoluminescence analysis demonstrates that CdTe wires exhibited a much lower level of recombination when compared to a planar CdTe film, which is highly beneficial for semiconductor applications.

Keywords: CdTe; self-catalysed; wires

1. Introduction

Due to their unique properties, semiconductor nano and microwires have attracted a lot of interest for optoelectronic devices. Direct band gap wires offer enhanced performance due to an increased effective surface area and reduced recombination [1–3]. CdTe is a direct semiconductor widely utilized for photovoltaics (PV) as a thin film absorber, owing to its near optimal band gap for PV of 1.5 eV and ease of deposition owing to its comparatively simple phase chemistry [4]. CdTe can be produced by a variety of techniques, such as physical vapor deposition (PVD) [5], chemical vapor deposition (CVD) [6], molecular beam epitaxy (MBE) [7] and solution phase synthesis [8]. It has been suggested that solar cells based on CdTe wires may exhibit a higher performance, due to the advantageous carrier transport properties of the wires. However, the fabrication of CdTe wires is far more challenging than for the planar films. CdTe wires grown by PVD or CVD techniques typically occur via a vapor-liquid-solid (VLS) mechanism necessitating a metal catalyst seed particle to facilitate wire growth [9]. The use of catalysts is problematic when attempting to create device quality material. Au and Bi have been used to successfully synthesize CdTe wires [9,10], however both have been established as deep level recombination centers in CdTe, compromising performance [11,12]. Catalyst-free CdTe wires are therefore preferable to minimize the defect content. Prior work on catalyst free CdTe wires primarily consists of two approaches (i) a solution based method grown via a solution-liquid-solid (SLS) mechanism [13], or (ii) a template assisted electrodeposition route using, for example, an aluminum oxide film as the template [14]. Both of these techniques have their disadvantages; they often require the use of solvents like oleylamine or complex patterning steps for the template layers. Wires produced by solution based methods are also often difficult to incorporate into device structures as many applications require vertically aligned wires projecting from the surface of the substrate [15].

The only prior report of self-catalyzed CdTe wires via PVD was from Wang et al. who reported growth via a thermal chemical method on an ITO substrate via a proposed vapor-solid-solid (VSS) mechanism [16], where the substrate was placed in a alumina vacuum tube furnace and CdTe was deposited at 2×10^{-1} mbar and 700 °C. Nanorods were observed to grow from a single CdTe seed particle formed on the substrate surface. The (111) zincblende crystal surface is more active than the substrate leading to continuous stacking along the CdTe (111) plane resulting in a preferred 1D growth [16]. It is worth noting that we use the term catalysis in this context to infer the CdTe seed particle is the cause for the wire formation, rather than the reaction rate is being increased.

In this work we present self-catalyzed wire growth via a proven scalable deposition route, close space sublimation (CSS) [5] on Mo substrates and as such this work has direct relevance to potential device production [17]. The influence of growth pressure and substrate surface roughness on wire growth were investigated with characterization by scanning electron microscopy (SEM), X-ray diffraction (XRD), energy dispersive X-ray (EDX) spectroscopy and cathodoluminescence (CL).

2. Materials and Methods

A range of substrates were used in this work; uncoated soda-lime glass (SLG), fluorine doped tin oxide (FTO) coated SLG "TEC 6" glass (NSG Ltd., St. Helens, UK) and 0.1 mm Mo foil substrates (Advent, 99.95% pure, Oxford, UK). All substrates were washed with isopropyl alcohol (IPA) and de-ionized (DI) water, then ultrasonically cleaned in DI water prior to deposition. 250 nm (0.5 Ω/sq) Mo films were grown onto the glass substrates via Direct Current (DC) magnetron sputtering at 400 °C using an Ar plasma. 6–8 μm CdTe (Alfa Aesar, 99.99% pure, Lancashire, UK) was deposited via CSS in an N_2 ambient at a variety of pressures and at source and substrate temperatures of 650 °C and 550 °C respectively.

Atomic force microscopy (AFM) was carried out using a Veeco Innova Bruker atomic force microscope (Bruker, CA, USA) in contact mode. XRD measurements were performed using a PANalytical X'pert PRO X-ray diffractometer (PANalytical B. V., Eindhoven, The Netherlands) at room temperature, using CuKα1 line as the X-ray source. SEM images were taken using a JSM-7001F microscope from JEOL with EDX spectrometer (JEOL, Tokyo, Japan). CL spectra was measured with a Hitachi SU-70 SEM (Hitachi, Tokyo, Japan) operating at 12 keV together with a Gatan MonoCL system (Gatan, CA, USA) for CL detection. The pixel dwell time for the panchromatic was 4 s.

3. Results and Discussion

The preferable implementation for CdTe wires in a solar cell structure is via the "substrate" cell structure [17], wherein the CdTe component must be deposited on top of a suitable back contact medium. Typically for substrate CdTe devices the back contact material is Mo as its thermal expansion coefficient is close to that of CdTe and its use is well established for other thin film technologies such as Copper Indium Gallium Selenide (CIGS) and Copper Zinc Tin Sulfide (CZTS) [18,19], although there are some issues with regards to generation of an Ohmic contact [17]. We therefore focused on producing wires on a Mo surface as this is the first step towards wire cell development. We considered two substrate options: Mo coated SLG, the typical route for CIGS/CZTS devices, or growth directly onto Mo foil, which is of interest due to the potential to produce flexible solar cells [18]. Figure 1a,b shows SEM images of CdTe growth on SLG/Mo and Mo foil respectively, using deposition conditions equivalent to that for our standard CdS/CdTe superstrate cell platform [20]. For growth on Mo foil (Figure 1b) we observe what we would classify as a "typical" CdTe thin film growth for these conditions, i.e., complete coverage of the substrate with a grain size 1–5 μm [21]. For growth on the SLG/Mo substrates under identical growth conditions we observed the formation of a field of self-catalysed wires. The wires show an ordered lateral growth with an average diameter of 3.5 ± 0.3 μm, average length of 15.9 ± 3.8 μm and clear evidence of a hexagonal cap to the wires. Prior nucleation and growth studies have identified similar hexagonal CdTe islands to form during the early stages of CSS deposition [22].

From this and the microscopy images we infer a vapor-solid-solid (VSS) growth mode, wherein hexagonal seed crystals are initially formed on the surface then subsequently act as a template for wire growth, following a similar mechanism to the one proposed by Wang et al. [16,23]. Such mechanisms are likely to be highly sensitive to the mobility of adatoms on the surface, which will influence the critical nucleus size [24] and thus the formation of any surface-stable seed crystal, as well as the rate of material flux to feed wire growth. The lack of wire formation for growth on the Mo foil can therefore be attributed to one of two factors either (i) the surface energy of the foil is significantly different compared to SLG/Mo, thus influencing the adatom lifetime on the surface and altering the growth mode [25] or (ii) as the Mo foil is significantly rougher than Mo/SLG (71.98 nm Root Mean Square (RMS) vs. 8.0 nm RMS roughness, Figure S1 in supporting information shows the three dimensional AFM images of the substrates surface) the increased roughness is hampering adatom diffusion disrupting the formation of seed particles.

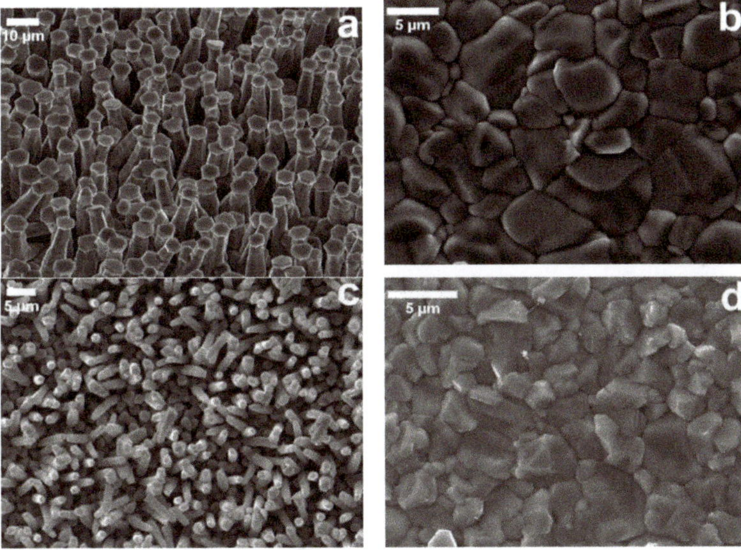

Figure 1. SEM images showing the effect of the substrate on wire growth. CdTe deposition was performed under identical growth conditions on: (**a**) SLG/Mo; (**b**) Mo Foil; (**c**) SLG; and (**d**) SLG/FTO/Mo.

A route to better understand the limiting factor was to utilize identical growth conditions but using uncoated SLG and SLG/FTO/Mo coated substrates. These substrates were selected in particular to separate the influence of the deposition surface from the surface roughness. For deposition on SLG the surface bonding energy is completely different compared to SLG/Mo but the roughness is even lower. FTO coated glass is stable at high temperatures due to its CVD deposition route and has higher roughness than SLG, 16.17 nm RMS (Figure S1), but by coating it with sputtered Mo is offers an identical surface to SLG/Mo only with increased roughness. Via comparison of growth on these substrates it allows us to separate the influence of the Mo layer from the roughness. SEM images for growth on SLG and SLG/FTO/Mo are shown in Figure 1c,d respectively. We observe the formation of wires on the SLG surface but thin-film style growth on the SLG/FTO/Mo surface, hence it is apparent that surface roughness is the controlling factor for wire formation. For the highly smooth Mo-free SLG surface, wires have formed with smaller dimensions than that observed for the

SLG/Mo surface. This again demonstrates the influence of roughness but additionally allows us to rule out the Mo film acting as a catalyst layer and clearly demonstrates the wires are self-catalysed. For the rougher SLG/FTO/Mo surface, growth has reverted to thin films, albeit of slightly smaller grain size than on Mo foil. This suggests that the Mo layer is largely inconsequential to wire formation in contrast to the surface roughness. Wire formation has been shown to be roughness dependent in other materials systems such as ZnO [26,27], but it's crucial role in CdTe wire formation has never previously been established.

In order to determine the level of wire formation control afforded by this self-catalysed route, the influence of CSS deposition pressure was studied as this determines the adatom arrival and re-evaporation rates [28]. Figure 2a–c show SEM images with varying pressure from vacuum (system base pressure i.e., no nitrogen added to ambient) to 30 Torr and to 60 Torr. Deposition was again performed on SLG/Mo to maintain a device relevant stack structure. Layers grown under vacuum exhibit a highly uniform and dense wire array while at 30 Torr the wires become less uniform and are more randomly distributed (11 wires/400 μm^2 and 5 wires/400 μm^2 respectively). It has previously been established for CSS growth of thin film CdTe that higher deposition pressures favor the formation of larger grains owing to a reduction in the adatom arrival rate at the surface and thus the formation of larger critical sized nuclei [21]. Here, for growth at 60 Torr, this has led to the formation of excessively large seed crystals, 12.3 ± 0.9 μm, compared to 3.5 ± 0.3 μm for vacuum conditions, which appear unable to effectively template wire growth. There is some evidence of growth occurring beneath these caps, but true wire growth does not occur. These results indicate that while there is a relatively wide pressure range over which growth will occur, the dimensions and the quality of the wires may be adjusted. This follows from previous work where the formation of seed particles and wires is inversely proportional to the growth pressure due to a reduction in the vapor concentration [16,29].

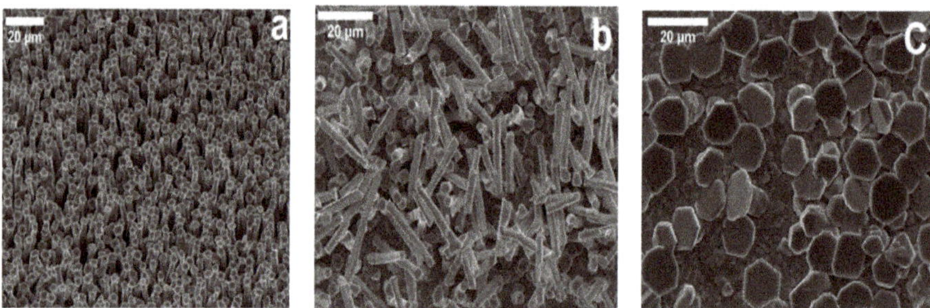

Figure 2. SEM images of CdTe wire growth with varying deposition pressure. (**a**) Vacuum; (**b**) 30 Torr and (**c**) 60 Torr.

The XRD patterns shown in Figure 3 were recorded for wires grown on SLG/Mo at either 30 Torr or under vacuum and compared with a planar film deposited on SLG/FTO/Mo. The growth preference may be determined by calculating the texture coefficients (C_{hkl}) for each diffraction peak using Equation (1).

$$C_{hkl} = \frac{\frac{I_{hkl}}{I_{rhkl}}}{1/n_p \sum_{n_p=1}^{n_p} \frac{I_{hkl}}{I_{rhkl}}} \tag{1}$$

where I_{hkl} is the intensity of a (hkl) diffraction peak, I_{rhkl} is the relative intensity of this diffraction peak for a powder sample and n_p is the number of reflections present in the sample. Then a standard deviation (σ) indicates the extent to which the film deviates from the powder using Equation (2) [30]. A high σ indicates the film is more textured and a low value indicates it is more random [30].

$$\sigma = \sqrt{\sum 1/n_p (C_{hkl} - 1)^2} \tag{2}$$

Table 1 shows the C_{hkl} and σ for all the samples presented in Figure 3. All samples show a typical diffraction pattern for zinc blende CdTe with a preferential (111) orientation however, the degree of preferential orientation varies. Layers grown on SLG/FTO/Mo (Figure 3c), which have no evidence of wire formation, are the most randomly orientated film σ = 1.59. Both wire samples grown on SLG/Mo show an increase in preferred (111) orientation to σ = 1.83 for growth at 30 Torr (Figure 3b) and σ = 2.23 for growth at vacuum (Figure 3a). This increase would indicate that the wires grow in a (111) orientation.

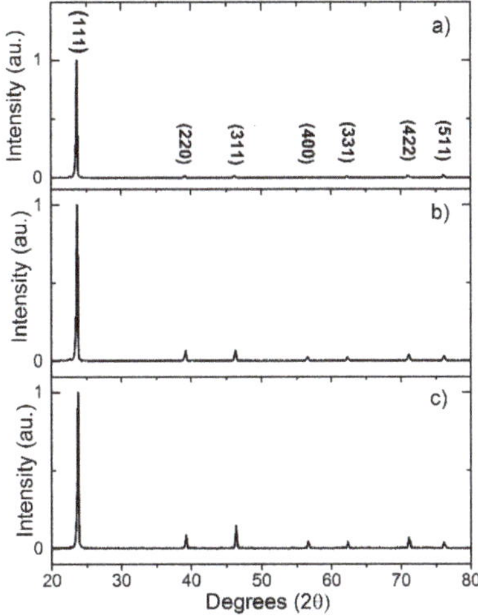

Figure 3. XRD patterns for wire growth on SLG/Mo at (**a**) vacuum and (**b**) 30 Torr; (**c**) Shows XRD pattern of a planar film deposited at 30 Torr on SLG/FTO/Mo.

Table 1. Texture coefficients (C_{hkl}) and standard deviation (σ) calculated for each of the samples shown in Figure 3.

	Under Vacuum SLG/Mo	30 Torr SLG/Mo	30 Torr SLG/FTO/Mo
C_{111}	6.47	5.511	4.90
C_{220}	0.11	0.407	0.43
C_{311}	0.10	0.398	0.70
C_{400}	0.011	0.144	0.23
C_{331}	0.075	0.134	0.21
C_{422}	0.11	0.234	0.35
C_{511}	0.12	0.171	0.19
σ	2.23	1.83	1.59

One possibility was that the wires were being nucleated by some contaminant on the surface. Although the substrates were thoroughly cleaned prior to deposition, outdiffusion of chemical species from the glass can still occur. To verify that contaminants were not nucleating wire growth, EDX was performed at a number of points along the wires (Figure 4). Spectra were taken from the both the cap and shaft of the wire with the chemical composition being compared. The spectra in both cases looks identical, being predominantly Cd and Te with only small additional peaks from the Mo under layer and O. This indicates that the wires are phase pure CdTe with the additional oxygen being detected always being present in CSS deposited CdTe. The EDX and XRD data presented support the inference that the wires are self-catalysed from a CdTe seed particle which subsequently become the cap of the wire [16].

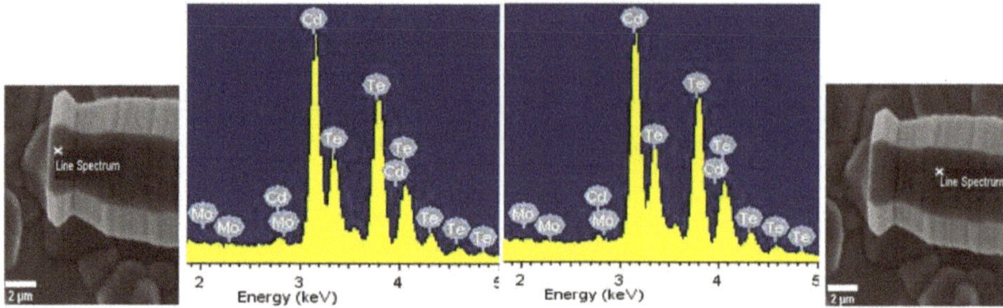

Figure 4. SEM images of the CdTe wires deposited under vacuum on Mo/SLG and EDX spectra at different points of the wires.

It is anticipated that wires should have enhanced carrier transport properties and thus reduced carrier recombination. CL analysis was performed to compare the recombination rates of CdTe wires and thin film. Figure 5a,b show SEM images of the wire and planar layers grown at vacuum on SLG/Mo and SLG/FTO/Mo respectively, with accompanying CL images shown in Figure 5c,d. The CL spectra were collected and normalized in the region of 1.40 to 1.70 eV (Figure 5e). The wires produced a higher signal compared to the planar film indicating the wires possess a reduced level of non-radiative recombination, possibly to a reduction in the number of grain boundaries present in the wires compared to the planar film [31]. This reduced non-radiative recombination makes these ideal for many device applications such as radiation detectors and PV. In order for a direct comparison between the CL signals both were renormalized (Figure 5f) to their respective highest intensity signals, both CL spectra for the wires and planar film show a near band edge transition [11]. The peak for wires is shifted to a slightly higher energy of 1.435 eV compared to 1.419 eV for the planar film, these peaks correspond to a typical donor-acceptor pair transition observed for CdTe [3,5]. The CL spectra for planar films also shows an additional peak at 1.486 eV corresponding to a near band edge transition for CdTe.

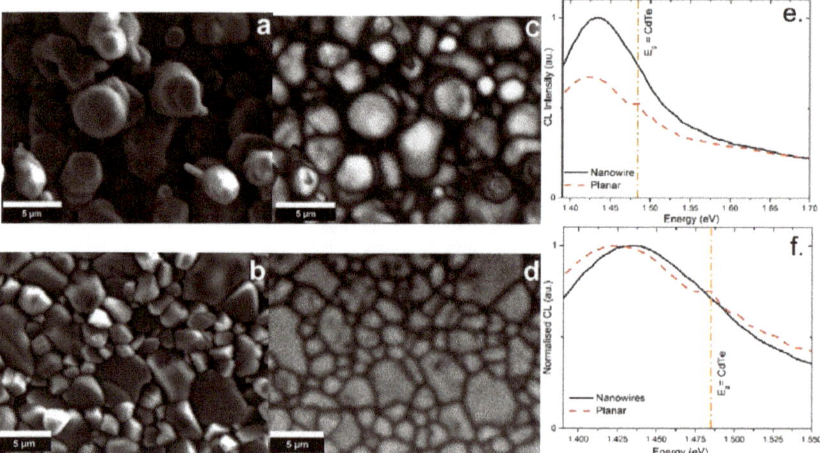

Figure 5. SEM images for CdTe wires (**a**) and the planar (**b**) CdTe film; along with CL images for the wires (**c**) and planar film (**d**). Normalized CL spectra produced for the CdTe wires and CdTe planar film (**e**,**f**). CdTe samples were grown at vacuum. (**e**) Has been normalized with respect to the highest signal, therefore the wire signal was set to 1. In order for the peak shift to be directly analyzed the signals were both normalized to 1 so that the intensities of the signals was neglected (**f**).

4. Conclusions

Self-catalysed CdTe wires have been fabricated in a single step technique using a simple industrially scalable PVD technique (CSS). We have produced samples on a SLG/Mo substrate suitable for cell fabrication and demonstrated wire growth dimensions can be controlled by both the surface roughness and deposition pressure. XRD and EDX results indicate that the wires are preferentially (111) oriented and free from contaminants while CL analysis shows reduced recombination in the wires compared to thin films. This development of catalyst free wires on a useable substrate is important for semiconductor applications as it excludes extrinsic growth templating materials which induce deep defects into the CdTe like Bi_{Te} and Au_{Cd} which are detrimental to performance [12]. The next step will be to incorporate these SLG/Mo/CdTe wire structures into complete solar cell devices.

Supplementary Materials: The following are available online at http://www.mdpi.com/2079-4991/8/5/274/s1, Figure S1 shows the three-dimensional AFM image of the substrates used in this study in order to determine their surface roughness.

Author Contributions: T.B. and J.D.M. conceived and designed the experiments; T.B. performed the experiments; T.B. performed the SEM and XRD measurements; G.P. performed the EDX measurements; O.S.H. performed and analyzed the AFM measurements; L.B. performed the CL measurements; T.B. analyzed the EDX and CL data; K.D. supervised G.P and O.S.H. and provided lab equipment; T.B. and J.D.M. wrote the paper.

Acknowledgments: This work was funded by the UK Engineering and Physical Sciences Research Council grant numbers EP/N014057/1 and EP/J017361/1.

Conflicts of Interest: The authors declare no conflicts of interest.

Data Availability: The data which supports the findings of this work is available from Liverpool's Data Catalogue or from the author.

References

1. Wang, X.; Xu, Y.; Zhu, H.; Liu, R.; Wang, H.; Li, Q. Crystalline Te nanotube and Te nanorods-on-CdTe nanotube arrays on ITO via a ZnO nanorod templating-reaction. *CrystEngComm* **2011**, *13*, 2955–2959. [CrossRef]

2. Ma, Y.; Jian, J.; Wu, R.; Sun, Y.; Li, J. Growth of single-crystalline ultra-long cadmium telluride micron-size wires via thermal evaporation. *Micro Nano Lett.* **2011**, *6*, 596–598. [CrossRef]

3. Di Carlo, V.; Prete, P.; Dubrovskii, V.G.; Berdnikov, Y.; Lovergine, N. CdTe Nanowires by Au-Catalyzed Metalorganic Vapor Phase Epitaxy. *Nano Lett.* **2017**, *17*, 4075–4082. [CrossRef] [PubMed]

4. Chopra, K.L.; Paulson, P.D.; Dutta, V. Thin-film solar cells: An overview. *Prog. Photovolt. Res. Appl.* **2004**, *12*, 69–92. [CrossRef]

5. Consonni, V.; Rey, G.; Bonaim, J.; Karst, N.; Doisneau, B.; Roussel, H.; Renet, S.; Bellet, D. Synthesis and physical properties of ZnO/CdTe core shell nanowires grown by low-cost deposition methods. *Appl. Phys. Lett.* **2011**, *98*, 96–99. [CrossRef]

6. Lee, S.K.C.; Yu, Y.; Perez, O.; Puscas, S.; Kosel, T.H.; Kuno, M. Bismuth-assisted CdSe and CdTe nanowire growth on plastics. *Chem. Mater.* **2010**, *22*, 77–84. [CrossRef]

7. Kret, S.; Szuszkiewicz, W.; Dynowska, E.; Domagala, J.; Aleszkiewicz, M.; Baczewski, L.T.; Petroutchik, A. MBE Growth and Properties of ZnTe- and CdTe-Based Nanowires. *J. Korean Phys. Soc.* **2008**, *53*, 3055–3063.

8. Khalavka, Y.; Sönnichsen, C. Growth of gold tips onto hyperbranched CdTe nanostructures. *Adv. Mater.* **2008**, *20*, 588–591. [CrossRef]

9. Davami, K.; Ghassemi, H.M.; Sun, X.; Yassar, R.S.; Lee, J.S.; Meyyappan, M. In Situ observation of morphological change in CdTe nano-and submicron wires. *Nanotechnology* **2011**, *22*, 435204. [CrossRef] [PubMed]

10. Yang, L.; Wu, R.; Li, J.; Sun, Y.F.; Jian, J.K. CdTe nanosheets and pine-like hyperbranched nanostructures prepared by a modified film technique: Catalyst-assisted vacuum thermal evaporation. *Mater. Lett.* **2011**, *65*, 17 20. [CrossRef]

11. Ruiz, C.M.; Saucedo, E.; Martínez, O.; Bermúdez, V. Hexagonal CdTe-like rods prompted from Bi2Te3 droplets. *J. Phys. Chem. C* **2007**, *111*, 5588–5591. [CrossRef]

12. Wei, S.; Zhang, S.B. Electronic Structures and defect physics of Cd-Based Semiconductors. In Proceedings of the Twenty-Eighth IEEE Photovoltaic Specialists Conference, Anchorage, AK, USA, 15–22 September 2000; IEEE: Piscataway, NJ, USA, 2000; pp. 483–486.
13. Jin, X.; Kruszynska, M.; Parisi, J.; Kolny-Olesiak, J. Catalyst-free synthesis and shape control of CdTe nanowires. *Nano Res.* **2011**, *4*, 824–835. [CrossRef]
14. Zhao, A.W.; Meng, G.W.; Zhang, L.D.; Gao, T.; Sun, S.H.; Pang, Y.T. Electrochemical synthesis of ordered CdTe nanowire arrays. *Appl. Phys. A Mater. Sci. Process.* **2003**, *76*, 537–539. [CrossRef]
15. Neretina, S.; Hughes, R.A.; Britten, J.F.; Sochinskii, N.V.; Preston, J.S.; Mascher, P. Vertically aligned wurtzite CdTe nanowires derived from a catalytically driven growth mode. *Nanotechnology* **2007**, *18*, 275301. [CrossRef]
16. Wang, X.N.; Wang, J.; Zhou, M.J.; Wang, H.; Xiao, X.D.; Li, Q. CdTe nanorods formation via nanoparticle self-assembly by thermal chemistry method. *J. Cryst. Growth* **2010**, *312*, 2310–2314. [CrossRef]
17. Williams, B.L.; Major, J.D.; Bowen, L.; Phillips, L.; Zoppi, G.; Forbes, I.; Durose, K. Challenges and prospects for developing CdS/CdTe substrate solar cells on Mo foils. *Sol. Energy Mater. Sol. Cells* **2014**, *124*, 31–38. [CrossRef]
18. Kranz, L.; Gretener, C.; Perrenoud, J.; Schmitt, R.; Pianezzi, F.; La Mattina, F.; Blösch, P.; Cheah, E.; Chirilă, A.; Fella, C.M.; Hagendorfer, H.; Jäger, T.; Nishiwaki, S.; et al. Doping of polycrystalline CdTe for high-efficiency solar cells on flexible metal foil. *Nat. Commun.* **2013**, *4*, 2306. [CrossRef] [PubMed]
19. Lin, Y.C.; Hsieh, Y.T.; Lai, C.M.; Hsu, H.R. Impact of Mo barrier layer on the formation of $MoSe_2$ in $Cu(In,Ga)Se_2$ solar cells. *J. Alloys Compd.* **2016**, *661*, 168–175. [CrossRef]
20. Major, J.D.; Treharne, R.E.; Phillips, L.J.; Durose, K. A low-cost non-toxic post-growth activation step for CdTe solar cells. *Nature* **2014**, *511*, 334–337. [CrossRef] [PubMed]
21. Major, J.D.; Proskuryakov, Y.Y.; Durose, K.; Zoppi, G.; Forbes, I. Control of grain size in sublimation-grown CdTe, and the improvement in performance of devices with systematically increased grain size. *Sol. Energy Mater. Sol. Cells* **2010**, *94*, 1107–1112. [CrossRef]
22. Major, J.D.; Proskuryakov, Y.Y.; Durose, K.; Green, S. Nucleation of CdTe thin films deposited by close-space sublimation under a nitrogen ambient. *Thin Solid Films* **2007**, *515*, 5828–5832. [CrossRef]
23. Ambrosini, S.; Fanetti, M.; Grillo, V.; Franciosi, A.; Rubini, S. Vapor-liquid-solid and vapor-solid growth of self-catalyzed GaAs nanowires. *AIP Adv.* **2011**, *1*, 042142. [CrossRef]
24. Venables, J.A. *Introduction to Surface and Thin Film Processes*; Cambridge University Press: Cambridge, UK, 2000; ISBN 9780511755651.
25. Venables, J.A.; Spiller, G.D.T.; Hanbucken, M. Nucleation and growth of thin films. *Rep. Prog. Phys.* **1984**, *47*, 399–459. [CrossRef]
26. Chen, S.W.; Wu, J.M. Nucleation mechanisms and their influences on characteristics of ZnO nanorod arrays prepared by a hydrothermal method. *Acta Mater.* **2011**, *59*, 841–847. [CrossRef]
27. Ho, S.T.; Chen, K.C.; Chen, H.A.; Lin, H.Y.; Cheng, C.Y.; Lin, H.N. Catalyst-free surface-roughness-assisted growth of large-scale vertically aligned zinc oxide nanowires by thermal evaporation. *Chem. Mater.* **2007**, *19*, 4083–4086. [CrossRef]
28. Mandl, B.; Stangl, J.; Hilner, E.; Zakharov, A.A.; Hillerich, K.; Dey, A.W.; Samuelson, L.; Bauer, G.; Deppert, K.; Mikkelsen, A. Growth mechanism of self-catalyzed group III-V nanowires. *Nano Lett.* **2010**, *10*, 4443–4449. [CrossRef] [PubMed]
29. Dalal, S.H.; Baptista, D.L.; Teo, K.B.K.; Lacerda, R.G.; Jefferson, D.A.; Milne, W.I. Controllable growth of vertically aligned zinc oxide nanowires using vapour deposition. *Nanotechnology* **2006**, *17*, 4811–4818. [CrossRef]
30. Williams, B.L.; Major, J.D.; Bowen, L.; Keuning, W.; Creatore, M.; Durose, K. A Comparative Study of the Effects of Nontoxic Chloride Treatments on CdTe Solar Cell Microstructure and Stoichiometry. *Adv. Energy Mater.* **2015**, *5*, 1–10. [CrossRef]
31. Mendis, B.G.; Bowen, L. Cathodoluminescence measurement of grain boundary recombination velocity in vapour grown p-CdTe. *J. Phys. Conf. Ser.* **2011**, *326*, 4–8. [CrossRef]

Article

Scalable Fabrication of High-Performance Transparent Conductors Using Graphene Oxide-Stabilized Single-Walled Carbon Nanotube Inks

Linxiang He [1], Chengzhu Liao [2,*] and Sie Chin Tjong [1,*]

[1] Department of Physics, City University of Hong Kong, Tat Chee Avenue, Kowloon, Hong Kong, China; linxiang_he@hotmail.com

[2] Department of Materials Science and Engineering, Southern University of Science and Technology, Shenzhen 518055, China

* Correspondence: liaocz@sustc.edu.cn (C.L.); aptjong@cityu.edu.hk (S.C.T.); Tel.: +86-755-88018761 (C.L.); +852-3442-7702 (S.C.T.)

Received: 2 March 2018; Accepted: 5 April 2018; Published: 7 April 2018

Abstract: Recent development in liquid-phase processing of single-walled carbon nanotubes (SWNTs) has revealed rod-coating as a promising approach for large-scale production of SWNT-based transparent conductors. Of great importance in the ink formulation is the stabilizer having excellent dispersion stability, environmental friendly and tunable rheology in the liquid state, and also can be readily removed to enhance electrical conductivity and mechanical stability. Herein we demonstrate the promise of graphene oxide (GO) as a synergistic stabilizer for SWNTs in water. SWNTs dispersed in GO is formulated into inks with homogeneous nanotube distribution, good wetting and rheological properties, and compatible with industrial rod coating practice. Microwave treatment of rod-coated films can reduce GOs and enhance electro-optical performance. The resultant films offer a sheet resistance of ~80 Ω/sq at 86% transparency, along with good mechanical flexibility. Doping the films with nitric acid can further decrease the sheet resistance to ~25 Ω/sq. Comparing with the films fabricated from typical surfactant-based SWNT inks, our films offer superior adhesion as assessed by the Scotch tape test. This study provides new insight into the selection of suitable stabilizers for functional SWNT inks with strong potential for printed electronics.

Keywords: aqueous dispersion; carbon nanotube; graphene oxide; ink; rod coating; electrical conductivity; optical transmittance; mechanical flexibility

1. Introduction

Recently, there is a strong demand for transparent conductors (TCs) for various optoelectronic devices such as touch screens, solar cells, organic light-emitting diodes (OLEDs), etc. [1–4]. The main TCs used currently are indium-doped tin oxide (ITO). However, indium is very expensive [5], and almost all ITO films are deposited onto glass substrates through a rather slow sputtering technique, resulting in further rises in production costs. The successful development of single-walled carbon nanotubes (SWNTs) in the 1990s has attracted tremendous scientific attention due to their excellent mechanical, electrical and thermal properties [6]. SWNTs can be fabricated into thin films having high optical transparency and electrical conductivity [7], rendering those ideal alternatives for replacing ITO films. Solution-based approaches such as spray coating [8], spin coating [9], Langmuir-Blodgett technique [10], vacuum filtration [11], dip coating and rod coating [12–14], are usually employed to prepare SWNT films. However, as-produced SWNTs generally form agglomerates and are difficult to disperse in water or organic solvents at appropriate concentrations. To effectively disperse them, several strategies, such as chemical modification, surfactants or polymeric dispersant additives, are usually employed [12,13,15–17]. As recognized, the dispersants typically used are electrical insulators and therefore must be removed from the resultant films. This is a difficult or tedious task because dispersants used to debundle SWNTs interact strongly with the nanotubes and are hard to remove from them. Furthermore, the additions and subsequent removal of dispersants increase production costs and degrade the mechanical performance of the films greatly [18].

Graphene oxide (GO), an oxygenated graphene sheet, is a low-cost chemically derived graphene prepared by chemical oxidation of graphite flakes in strong oxidizing solutions [19]. GO consists of hydrophobic domains with a benzene ring structure and hydrophilic sites with hydroxyl, carboxyl and carbonyl groups [20]. This unique feature renders GO acting as an amphiphilic molecule and enables it to form Langmuir-Blodgett films on a water surface [21]. The oxygenated moieties make GO sheets hydrophilic and highly dispersible in water, whereas aromatic domains interact with other aromatic molecules through supramolecular π-π interactions [22]. Using GO to disperse SWNTs has a major benefit: GO can be thermally/chemically reduced to form reduced graphene oxide (rGO) [23], which is electrically conductive, thus enhancing the conductivity of resultant films.

Fabrication of GO-SWNT films have been previously reported by several researchers. For example, Tian et al. carried out a preliminary study on the preparation of transparent GO-SWNT films in which aqueous GO-dispersed SWNT solutions were air-sprayed onto glass substrates [24]. Their GOs were not chemically reduced to rGOs, so the GO-SWNT films exhibited a large sheet resistance of ~4 kΩ/sq at 85% transmittance. Zheng et al. prepared GO-SWNT films by means of a layer-by-layer Langmuir-Blodgett assemble process [25]. The resultant films showed a sheet resistance ranging from 180 Ω/sq to 560 Ω/sq at ~77–86% transparency depending on the number of layers. Very recently, Gorkina et al. fabricated GO-SWNT films using several steps. In the process, SWNT film was first deposited onto a substrate via a dry transfer process, then GO was sprayed on top of SWNT film. GO was thermally reduced at ambient or hydrogen atmosphere and finally reacted with $AuCl_3$ [26]. They reported that the mean sheet resistance of hybrid SWNT/rGO films treated with air and hydrogen was 485 Ω/sq and 264 Ω/sq, respectively, at about 85% transmittance. All these processes are unsuitable for industrial scale-up production. By contrast, coating SWNT inks to plastic substrates with a Mayer rod to form thin films is considered of technological importance. Fluids that can be coated effectively by a Mayer rod is readily adapted to industrial process, allowing rapid roll-to-roll printing at speeds up to 100 m/min on plastic substrates of several meters wide [27]. Moreover, transparent film fabrication by rod coating of SWNT inks is less wasteful, so nearly all the SWNTs are well deposited on the substrate. However, simple SWNT dispersions cannot be used as inks to fabricate thin films, since their rheological properties and wetting behavior are inappropriate for rod coating practice. Therefore, additives are usually incorporated to improve their viscosity, wetting, levelling, etc. However, typical or commercial additives used for modifying rheological properties and/or wetting behavior can induce instability of the dispersions or adversely affect optoelectronic properties of the resulting films. For example, Dan et al. reported that a dual surfactant system, sodium dodecylbenzenesulfonate (SDBS) and Triton X-100 (TX100), can be used to prepare SWNT inks [13]. Li et al. indicated that this mixed surfactants-SWNT system is stable for only a few minutes, making it unsuitable for industrial applications [28]. As such, more efficient strategies are desired to develop large-scale industrial production of these carbon nanostructures.

In a previous study, we reported the use of GO as a dispersant to prepare multi-walled carbon nanotube (MWNT) inks to generate transparent conductive MWNT films [14]. MWNT-based films, however, generally exhibit poor electro-optical performance than the films prepared from SWNTs [29]. This is because every graphene layer of MWNTs contributes to the light absorption, causing the films with strong light absorption [30]. For this reason, the development of carbon nanotube-based TCs has been primarily carried out with SWNTs. Herein, we report and demonstrate large-scale preparation of SWNT films through a Mayer rod coating of aqueous GO-SWNT inks. By converting GO to rGO through microwave irradiation, these films exhibit much improved electro-optical properties, making them very attractive materials for making various optoelectronic devices.

2. Materials and Methods

2.1. Materials and Synthesis

Chemical vapor deposited SWNTs (>95 wt % purity; diameter of 0.8~1.6 nm and length of 5~30 μm) (Figure S1) were supplied by Chengdu Organic Chemicals Co. Ltd., Chengdu, China. Graphite flakes were bought from Sigma-Aldrich, Inc., St. Louis, MO, USA. Graphite oxide was prepared from graphite flakes by means of Hummers process [19]. It was readily exfoliated into monolayer GO sheets in water (Figure S2). All chemical reagents were used as received without further purification.

2.2. rGO-SWNT Films

Aqueous GO dispersions of different concentrations were obtained by stirring graphite oxide solids in water for 4 h, and the mixtures were then sonicated (Tru-Sweep 575 HTAG, Crest Ultrasonics Corporation, Ewing Township, NJ, USA) for 30 min. To such dispersions SWNTs were added, obtaining a concentration of 0.2 mg/mL SWNTs. This concentration was relatively lower than in previous works to ensure full exfoliation of nanotube bundles, since long nanotubes generally caused higher viscosity and decreased the dispersion efficiency. These GO-SWNT dispersions were then tip sonicated (SCIENTZ-II D Ultrasonic Homogenizer) in an ice-water bath for 2 h. This sonication process reduced the size of GO sheets and the diameter of nanotube bundles, resulting in more homogeneous GO-SWNT dispersions. To prepare films, formulated GO-SWNT dispersions were coated onto polyethylene terephthalate (PET) sheets by a Mayer rod. The thickness of the films was controlled by the diameter of a wire coiled on the rod. To convert GO to rGO, the films were firstly exposed to a hydrazine atmosphere for 1 h to partly recover its conductivity, then treated in a microwave oven (Panasonic NN-ST253W, 800 W, Panasonic Corporation, Osaka, Japan) in an argon atmosphere for 2~3 s. This microwave treatment rapidly heated the films and thermally reduced GO to rGO [31,32].

2.3. Material Characterization

The quality of GO-SWNT inks was examined with UV-vis spectroscopy (Agilent Cary 5000 Spectrophotometer, Agilent Technologies Inc., Santa Clara, CA, USA), Raman spectroscopy (Horiba Jobin Yvon LabRAM HR Evolution Micro-Raman spectrometer; excitation wavelength of 532 nm, Horiba Ltd., Kyoto, Japan), transmission electron microscopy (TEM; Philips FEG CM 20, Philips Company, Amsterdam, Netherlands) and zeta potential measurements (Zetaplus, Brookheaven, NY, USA). The contact angle of those inks on PET was obtained with an advanced goniometer (Ramé-Hart Instrument Model 200). The viscosity of the inks was measured using a rheometer (TA Instruments, HR-2 Discovery Hybrid Rheometer, TA Instruments, New Castle, DE, USA). The surface morphology of rGO-SWNT films was observed in both field-emission scanning electron microscope (SEM; Jeol FEG JSM 6335, JEOL Company, Tokyo, Japan) and atomic force microscope (AFM; Veeco Nanoscope V, Veeco Inc., Plainview, NY, USA). We employed X-ray photoelectron spectroscopy (XPS; PHI-5802; Physical Electronics, Chanhassen, MN, USA) with Al K_α source radiation to record chemical bonding states of constituent carbon in the films. The electrical sheet resistance of thin films was determined with a Van der Pauw setup using four electrodes aligned along the circumference of thin films. The light transmittance tests at ambient were performed with a UV-vis spectrophotometer (PerkinElmer Lambda 2S, Perkin Elmer Incorporation, Waltham, MA, USA). For bending cycling, samples were bent to a radius of curvature of 2 mm by sliding around a metal rod. In the meanwhile, electrical resistance was measured up to 2000 cycles. For the film adhesion testing, a strip of tape was applied to the sample and pressed down to uniformly contact the film surface, and then peeled off at ~10 mm/s at a ~45° angle. The sheet resistance and film transmittance before and after the tests were measured to determine the film adhesion properties.

3. Results

3.1. Dispersing SWNTs in Water Using GO

Attaining well-dispersed SWNTs without aggregates is key to ensure successful fabrication of homogeneous films. Due to strong van der Waals interactions amongst SWNTs, they tend to aggregate to form precipitates in water. At sufficient GO, however, SWNTs can be effectively dispersed. We prepared a series of GO-SWNT dispersions with different GO to SWNT mass ratios, and found that homogeneous dispersions can be obtained only when the mass ratio reaches at least 5, at a 0.2 mg/mL SWNT concentration. By diluting with a factor of ~20, these dispersions look like "true solutions" (Figure 1A). UV-vis spectroscopy was then performed to characterize their absorption characteristics (Figure 1B). The band located at ~1000 nm is the S22 interband transition in semiconducting SWNTs, and the M11 peak at ~750 nm relates interband transition in metallic SWNTs. These dispersions are very stable and no visible sediment is found for several months. Zeta potential measurements on these GO-SWNT dispersions agree reasonably with the absorption spectra findings. Pure GO dispersion at 0.1 mg/mL and neutral pH exhibits a zeta potential of 55 ± 5 mV. For the GO-SWNT dispersion with a high SWNTcontent (0.02 mg/mL), zeta potential remains almost unchanged. The results demonstrate that the GO-SWNT dispersion is very stable, and SWNTs are effectively dispersed without changing original electrostatic repulsions among the GO sheets.

Figure 1C,D display Raman spectra of SWNT, GO and the GO-SWNT dispersion. The radial breathing mode (RBM) is a characteristic property of SWNT, i.e., diameter of SWNT is inversely proportional to its RBM peak frequency. Figure 1C reveals the presence of SWNT multi-peaks, so there exists a fairly wide distribution of SWNT diameters. These peaks are significantly suppressed in hybrid GO-SWNT spectrum, demonstrating that GO sheets interact with the nanotubes and severely limit their RBM vibration [33]. From Figure 1D, the G-band at ~1590 cm^{-1} is due to the vibrational modes of sp^2 carbon. The D-band at ~1350 cm^{-1} is Raman active with its value proportional to the amount of amorphous carbon. The G-band of GO is significantly weaker than SWNT, and its D-band is reasonably more intense. The G-band of the GO-SWNT hybrid is dominated by SWNT because no absorption occurs for GO sheets at the excitation wavelength and thus no resonance enhancement [25]. On the other hand, the D-band in the hybrid is disproportionally higher, probably due to the contribution of GO sheets in combination with the effects similar to those displayed by the functionalized nanotubes [24].

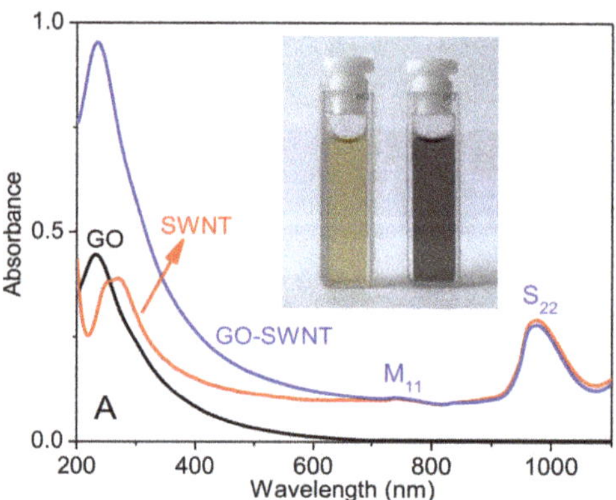

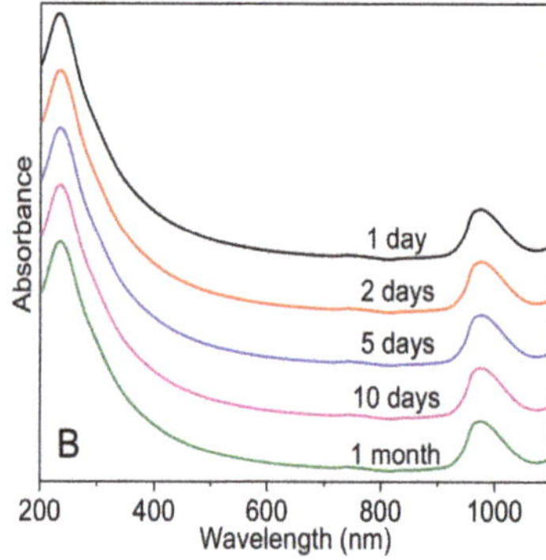

Figure 1. *Cont.*

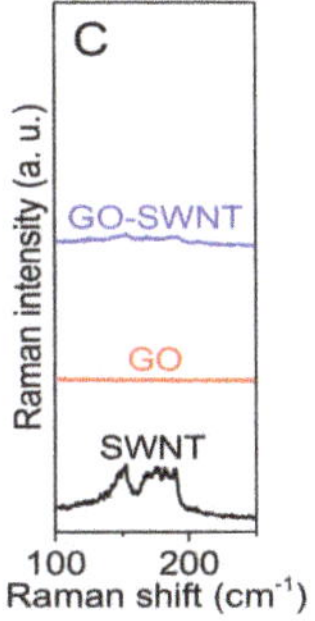
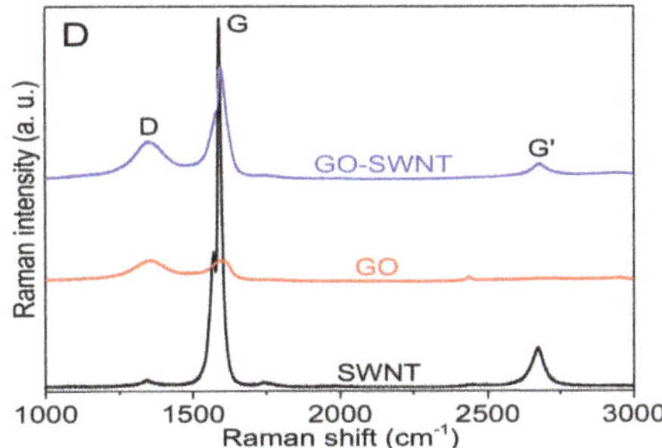

Figure 1. (**A**) UV-vis absorbance of diluted GO and GO-SWNT (weight ratio of 10:1) dispersions. Pure SWNT solution (weakly oxidized by nitric acid) is shown for comparison. The inset photo displays GO-SWNT (left) and GO (right) dispersions. (**B**) Absorbance of GO-SWNT dispersion for different time periods. (**C,D**) Raman profiles of GO-SWNT dispersion (weight ratio of 10:1), GO and SWNT.

The dispersion state of SWNTs was further imaged by TEM as shown in Figure 2. As recognized, pristine SWNTs tended to agglomerate into thick ropes (Figure 2A). In the presence of GO, however, these ropes were exfoliated into thinner ones (~40 nm), and separated from one another by adhering to GO (Figure 2B), showing a strong interaction between GO and SWNT. By increasing the amount of GO, the diameter of the SWNT ropes become thinner. At a mass ratio of 20, these ropes decreased to ~15 nm in diameter (Figure 2C). Due to the huge aspect ratio of SWNTs (length of 5~30 μm and diameter of 0.8~1.6 nm), very strong interactions among them is expected. Therefore, further exfoliation of the SWNT ropes is not possible even in the presence of excessive GO. Figure 2D summarizes the effect of GO concentration on the average diameter of SWNT bundles in the dispersion. These results reveal the successful dispersion of SWNTs with GO. Furthermore, the dispersion is stable during drying and this is crucial for the film fabrication.

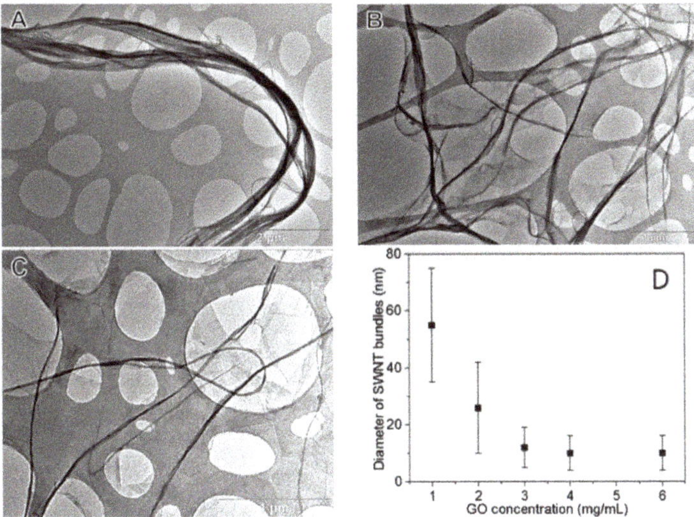

Figure 2. TEM micrographs of (**A**) SWNT bundles in water; (**B**) GO-SWNT dispersion with a GO:SWNT mass ratio of 10:1; (**C**) GO-SWNT dispersion with a GO:SWNT mass ratio of 20:1. The concentration of SWNT in each case is 0.2 mg/mL; (**D**) Effect of GO concentration on the diameter of SWNT bundles in the dispersion. The SWNT content in the ink is 0.2 mg/mL.

Together, these results indicate that GO can effectively debundle SWNTs into individual or thin nanotubes. The energy required for dispersing SWNTs in water is large, because of its huge specific surface area and nonpolar surface behavior, thereby inducing aggregation of SWNTs in water (Figure 2A,B). Two mechanisms can be proposed for good nanotube dispersion with GO. The first one relates to the π-π interaction between GO and SWNT. It is known that GO has lots of π-conjugated aromatic domains within its basal plane, therefore it can interact strongly with SWNTs via π-π attractions [34]. From the literature, drug and dye molecules can adsorb onto the surface of GO through π-π interactions [34]. Besides, it is widely recognized that GOs are strongly hydrated by dispersing in water. When the nanotubes adsorb on GOs, all water molecules attached to the GOs are released, leading to a significant gain in entropy [35]. Thus the crucial factor for the effective dispersion of SWNTs by GOs arises from π-π interactions between the nanotubes and GOs. From the TEM images it is clear that GOs interact with SWNTs by anchoring SWNT bundles. At low GO content (GO:SWNT mass ratio of 10:1), the number of aromatic sites available for anchoring SWNTs is rather low, so thick nanotube bundles tend to form. In the presence of sufficient GOs (GO:SWNT mass ratio of 20:1), these bundles can be effectively exfoliated, resulting in thinner nanotubes. The demonstrated procedure of using GOs to disperse SWNTs in water offers the possibility of low cost and large-scale preparation of excellent SWNT dispersions.

3.2. Wetting Behavior and Rheological Properties of GO-SWNT Dispersions

Rod coating involves the spreading of liquid drops to form a thin liquid layer on the substrate. Contact angle is usually used to evaluate the wettability of a solid surface by a liquid. Large contact angle (larger than 90°) reflects poor wettability. Moreover, the spreading of the liquid causes a significant increase in its surface area, resulting in a steep rise of its surface energy. To minimize the surface energy the wet film would contract its surface accordingly. This often causes film nonuniformities such as contact line recession, dewetting or film rupture. To avoid these nonuniformities (and therefore minimize film contraction), a small surface tension is preferred. The surface tension of GO-SWNT inks and their contact angle with the PET substrate were determined with a goniometer (Figure S3), and were observed to be nearly independent of ink compositions, with $\sigma \sim 70$ mN/m and $\theta \sim 60°$, respectively. The relatively larger contact angle indicated a relatively poor wetting of the GO-SWNT ink on PET. To improve the wetting behavior, we treated the PET substrate with oxygen plasma for ~10 min. This decreased the contact angle to ~10° (Figure S4 and Figure 3B).

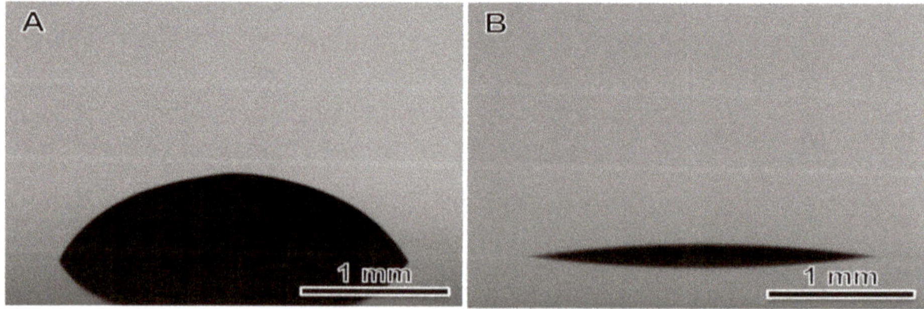

Figure 3. Contact angles for GO-SWNT ink on PET substrate: (**A**) before and (**B**) after plasma treatment. The contact angle is 59.7° and 10.2°, respectively.

Immediately after rod coating, the wet film exhibits a wavy surface in the form of stripes with their interspace distances governed by the diameter of coiled wire. To obtain a flat and smooth surface these stripes must be levelled off. For thin films this leveling process is motivated by the difference in surface tension between the convex and concave regions of wavy liquid. However, the viscous forces play a part in retarding this process. The leveling time is defined as $t_{level} = 3\,\mu r^4 / \sigma h^3$ [36], where μ is the viscosity of the ink, $2\pi r$ is the diameter of coiled wire, and h is the average wet thickness. For liquids like GO-SWNT inks, leveling can therefore be accomplished by reducing the viscosity. During the drying process, dewetting of the wet film may occur, resulting in film nonuniformities as mentioned before. The dewetting time is $t_{dewet} = \mu L / k \sigma \theta^3$ [37], where l is the characteristic length scale of the film and k relates to a fluid property, being about 10^{-3} for primarily water-based system. Therefore, dewetting can be suppressed by reducing the contact angle θ, and/or by increasing the viscosity μ. Since the preparation of a uniform film requires a fast leveling process (small t_{level}) and a slow dewetting process (large t_{dewet}), optimization of the ink viscosity is considered of primary importance.

Inks generally exhibit the non-Newtonian shear-thinning property. At the film formation stage, inks are seriously sheared through the grooves of the rod, thereby flowing at a high-shear rate. During the drying process, they are nearly stationary and flow at a low-shear rate. For a typical rod coating setup, the optimal viscosity value falls in 0.01–1 Pa s range [38]. The characteristic shear rate for the film formation region is ~50 s^{-1} in the present study. Hence, the optimal viscosity value of the GO-SWNT ink should fall in 0.01–1 Pa s range and near 50 s^{-1}. Figure 4 shows that all GO-SWNT inks exhibit a drop in viscosity under shear. This is a thixotropic behavior since the viscosity decreases with shear rate. Such behavior is also displayed by other non-Newtonian fluids such as polymer solutions and biological fluids [39–44]. For GO-SWNT dispersions, this behavior derives from debundle of nanotubes and/or by the increased orientation of nanotubes in the flow direction. Both the SWNT and GO concentration are found to influence the viscosity. Since the nanotubes used in this study have very large aspect ratios, the nanotube concentration would play a dominant role in determining the viscosity [45]. Doubling its concentration from 0.1 mg/mL to 0.2 mg/mL can increase both low-shear viscosity by about one order of magnitude (~2 Pa s at ~0.1 s^{-1}), and high-shear viscosity by about 5-fold (~85 Pa s at ~50 s^{-1}). Increasing the concentration to 0.3 mg/mL would lead to a dramatic increase in low-shear viscosity by about two orders of magnitude (~100 Pa s at ~0.1 s^{-1}), which is high enough to enable the inks to be used for other coating methods like blade coating or screen coating. In contrast, increase the GO concentration from 1 mg/mL to 6 mg/mL (at 0.2 mg/mL SWNT) only slightly increases the viscosity, as shown in Figure 4B. Inks with 0.2 mg/mL SWNTs are found to fall into the optimal viscosity range (0.01–1 Pa s) at a shear rate between 0.1 s^{-1} and 50 s^{-1}, therefore are most suitable for rod coating.

Mayer rod coating of optimized GO-SWNT dispersions generally produces homogeneous defect-free films. The thickness of these films can be controlled by simply choosing coiled rods of different sizes. In this study, the diameter of wire coiled on the rod ranges from 0.2 mm to 2 mm, resulting in wet thicknesses of 20–200 μm and final dry thicknesses of ~10–100 nm.

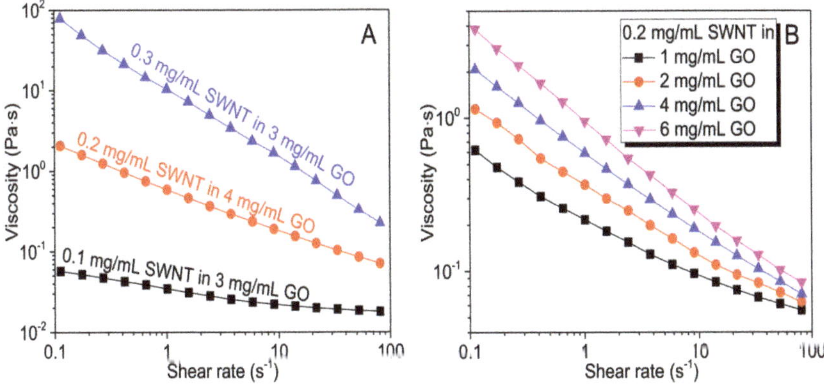

Figure 4. Effects of the concentration of (**A**) SWNT and (**B**) GO on the viscosity of GO-SWNT ink.

Table 1 summarizes different time parameters involved in the fabrication of TCs using GO-SWNT inks with different compositions. t_{level} is determined based on the high-shear viscosity, as the fluid close to the moving rod is highly sheared while being coated. It lies in the order of 10^{-3} s, indicating that levelling occurs immediately after coating to form a flat wet film; t_{dewet} is calculated using the low-shear viscosity, since the wet film is almost stationary during the drying process. It lies in the order of 10^{3} s, which is much larger than t_{dry}, implying that these films can dry before secondary flow occurs. The relationship $t_{level} < t_{dry} < t_{dewet}$ maintains for all these GO-SWNT dispersions, indicating that they can be employed as inks to prepare films with uniform thickness and limited dewetting.

Table 1. Time parameters for fabricating TCs using GO-SWNT inks.

Ink Composition	t_{level} (s)	t_{dry} (s)	t_{dewet} (s)
0.2 mg/mL SWNT + 1 mg/mL GO	8.33×10^{-4}	600	3.32×10^{3}
0.2 mg/mL SWNT + 2 mg/mL GO	1.10×10^{-3}	500	6.16×10^{3}
0.2 mg/mL SWNT + 4 mg/mL GO	1.41×10^{-3}	400	1.12×10^{4}
0.2 mg/mL SWNT + 6 mg/mL GO	1.83×10^{-3}	300	2.05×10^{4}

where t_{level} was calculated using high-shear ($\sim$50 s^{-1}) viscosity; t_{dewet} was determined with low-shear ($\sim$0.1 s^{-1}) viscosity; and t_{dry} was estimated by recording the time taken to cause gelation in the coated films. The diameter of wire coiled on the rod was $\sim$0.6 mm. A length scale (l) of 2 mm ($\sim$10% of width of the coated film) was used for calculating t_{dewet}. The wet coating thickness and wire diameter are assumed to be 1:10 [46].

3.3. Fabrication & Characterization of rGO-SWNT Films

A large GO to SWNT mass ratio ensures a more efficient disentanglement of SWNT bundles, and the relatively large GO concentration renders an ink viscosity appropriate for coating. Rod-coating GO-SWNT ink thus produces uniform, defect-free thin films. SEM images of these films (Figure 5A–C) show that SWNTs are uniformly dispersed and interconnected with each other over a large area, yielding a percolated 2D conducting network throughout the entire film. The very long length of the SWNTs enable the formation of a conductive network at a very low percolation threshold, thus are expected to give a much improved electrical conductivity. With the increase in the GO portion in the ink, SWNT bundles within the film became thinner. This is in good agreement with the TEM observation of SWNT dispersions as shown in Figure 2, implying that the dispersion state of SWNTs is well maintained during the drying process. Even after microwave heating, the distribution of SWNTs is still rather uniform, without any signs of phase separation.

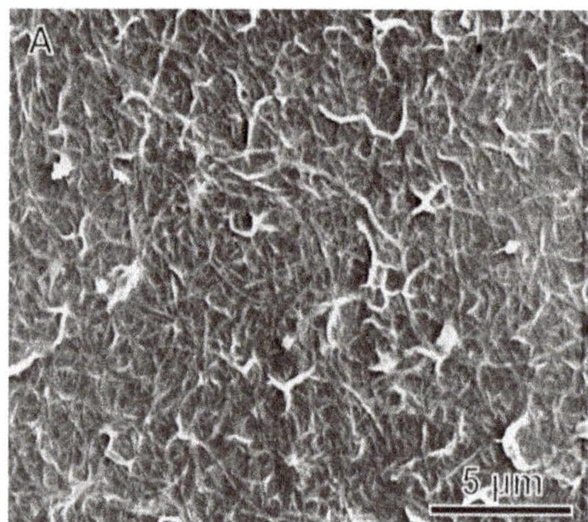
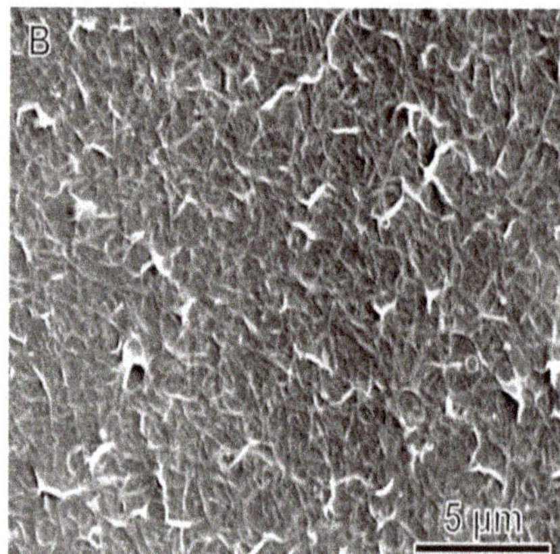

Figure 5. *Cont.*

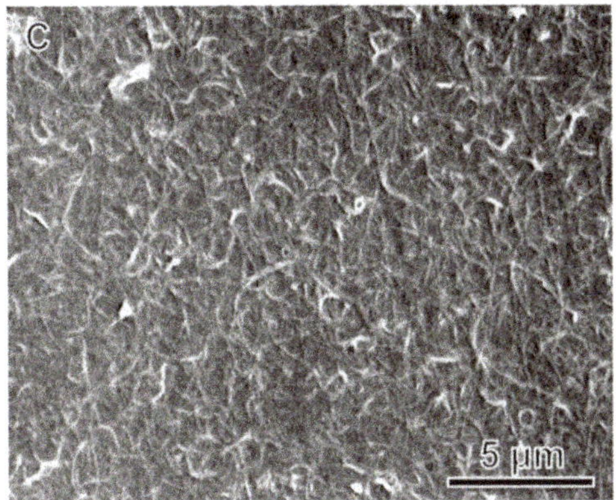

Figure 5. SEM micrographs of rGO-SWNT films prepared from inks with different GO to SWNT mass ratios: (**A**) 5:1, (**B**) 10:1 and (**C**) 20:1.

GO is well known to be nonconductive because of its oxygenated groups. To recover conductivity it must be deoxygenated. Recently, microwave treatment has been shown to be an effective and facile route to remove oxygenated groups from GO and recover its electrical conductivity [32]. Two- to three-second pulses of microwaves can effectively reduce GO into pristine graphene. During the experimental process, we found that microwave treatment of pure GO film (even though it had been treated by hydrazine for ~1 h) did not cause any noticeable change on the film properties, possibly due to the low conductivity of GO, thus leading to poor microwave absorption. In the presence of SWNTs, however, the films heated rapidly to high temperature owing to the strong microwave absorption of SWNTs, which thermally converted GO to rGO. As shown in Figure 6, the resultant rGO-SWNT film contained a negligible amount of oxygen, compared to GO-SWNT film, indicating that the microwave treatment is very efficient in removing the oxygen moieties. The resultant rGO-SWNT films were uniform without disruption or delamination from the substrate.

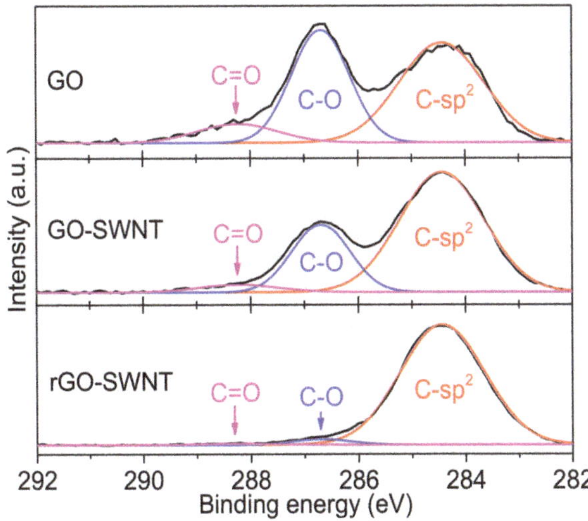

Figure 6. XPS C-1s spectra of GO, GO-SWNT (weight ratio of 10:1) and rGO-SWNT films. Peaks relate to different carbon bonds are indicated.

Figure 7 shows that thin films fabricated from various inks exhibit different electro-optical properties. This can be explained by the electrical contacts between nanostructured elements within the films. As shown in Figure 2, inks with lower GO to SWNT mass ratios have thicker SWNT bundles, leading to limited electrical contacts among the SWNT bundles within the rGO-SWNT films. This results in a relatively larger sheet resistance for these films. Incorporation of more GO leads to exfoliation of SWNT bundles, thus increases the number of electrical contacts within the films and reduces the sheet resistance. AFM images in Figure 8 show that an increase in the GO content in the inks leads to thinner SWNT bundles, rendering the film more efficient in transport charge carriers. Besides, these thinner bundles would result in a more smooth film surface (Figure S5), which is preferred for real applications such as organic photovoltaics or OLEDs. At an even larger GO to SWNT mass ratio (~30), the SWNT bundles would not be further exfoliated (Figure S6A,B and Figure 2G). In this case, the electrical contacts between SWNT bundles would be disturbed by rGO sheets, which account for a large portion of the film are less conductive than SWNTs. This results in an increase in the film resistance.

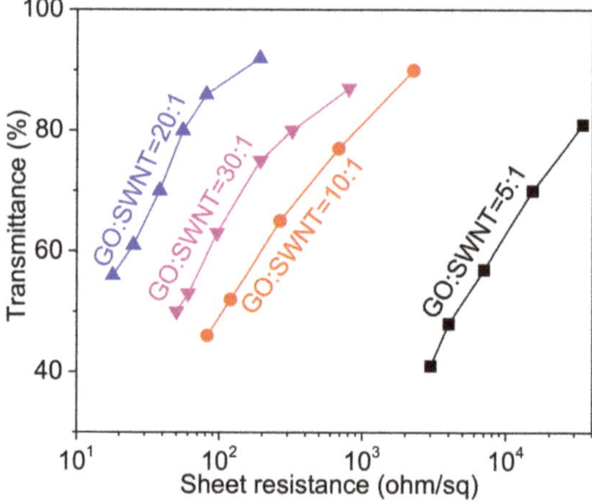

Figure 7. Sheet resistance versus optical transmittance for rGO-SWNT films fabricated from inks with different GO to SWNT mass ratios. The SWNT content in the ink is 0.2 mg/mL.

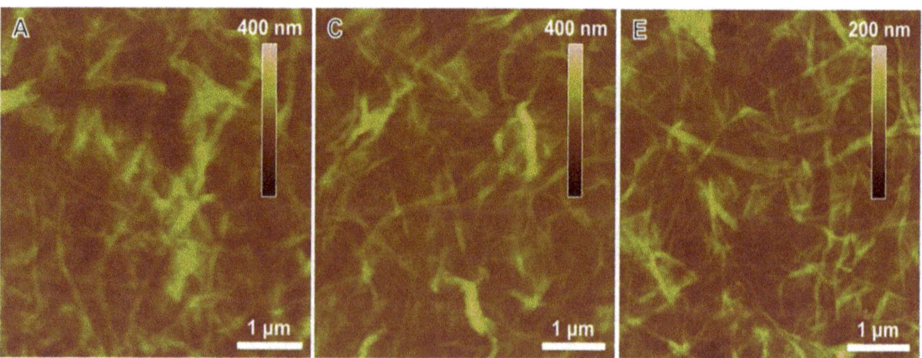

Figure 8. *Cont.*

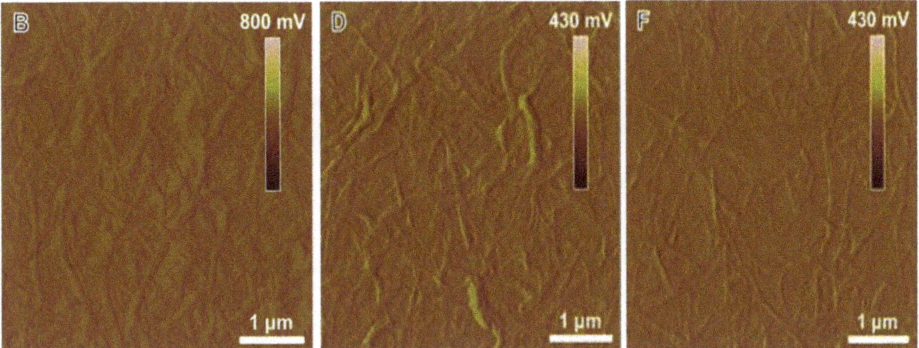

Figure 8. AFM height (**A**, **C** and **E**) and amplitude (**B**, **D** and **F**) microgrqaphs of rGO-SWNT films prepared from inks with different GO to SWNT mass ratios: (**A**) 5:1, (**B**) 10:1 and (**C**) 20:1.

For TCs, the relationship between transmittance and sheet resistance is generally described by [1]:

$$T = (1 + \frac{Z_0}{2R_s}\frac{\sigma_{op}}{\sigma_{dc}})$$

(1)

where Z_0 is the impedance of free space (377 Ω), R_s the sheet resistance, σ_{op} the optical conductivity, and σ_{dc} the DC conductivity. The performance of TCs can be evaluated by a parameter, i.e., figure of merit (FoM) given by σ_{dc}/σ_{op}. FoM offers direct performance comparison between the TCs prepared from various materials or methods over entire range of optical transmittance. Generally, TCs should have low sheet resistance and high optical transmittance, i.e., a large FoM. In Figure 9A we compare the performance of rGO-SWNT film with that of SWNT films prepared using surfactant/polymer as the dispersant. FoMs are determined from the sheet resistance at approximately 75~85% transmittance (Figure 9B). This is because most reported transmittance values of SWNT films fall in this range. It can be seen that the films fabricated from the GO-SWNT system outperform other systems in terms of electro-optical properties. For thin films fabricated from the inks with non-conductive dispersants, the incorporation of dispersants usually results in inferior nanotube-nanotube electrical contacts, giving rise to the resultant film with a large sheet resistance. In contrast, GO sheets are beneficial for the electrical transport of nanotube films. As mentioned before, there exists strong π-π interactions between the GO sheets and SWNTs. After converting to rGO, these conductive sheets serve as nano-"patches" that fixed or repaired SWNT networks by filling their void sites. Additional conducting channels would therefore be created and significantly reduce the film resistance. Another reason for the reduced sheet resistance is mechanical factor in origin. Due to the increased nonpolar nature of rGO compared to GO sheets, the rGO-SWNT interactions should be much stronger than the GO-SWNT interactions. Microwave irradiation of GO-SWNT to yield rGO-SWNT causes film densification. This force of densification is expected to compress SWNT networks, reduce the air gap within the film as well as the nanotube-nanotube junction resistance, and consequently decreases the overall resistance. Since rGO is almost transparent, film transparency of rGO-SWNT remains unchanged as expected. It is noted that the electrical conductivity of rGO-SWNT films can be further enhanced by chemical doping [46]. Simply dipping the films in nitric acid (12M) for 2 min causes a marked decrease in sheet resistance, with a negligible effect on the film transparency (Figure S7A). The average sheet resistance for a ~86% transparent film before doping was ~80 Ω/sq, and decreased to ~25 Ω/sq after doping, demonstrating a three-fold decrease in the equivalent sheet resistance. However, the doping effect does not last very long (Figure S7B), because nitric acid molecules are weakly adsorbed on the nanotube surfaces and can desorb. A poly(3,4-ethylenedioxythiophene):poly-(styrene sulfonate) (PEDOT:PSS) layer has been shown to effectively inhibit the increase of film resistance [47].

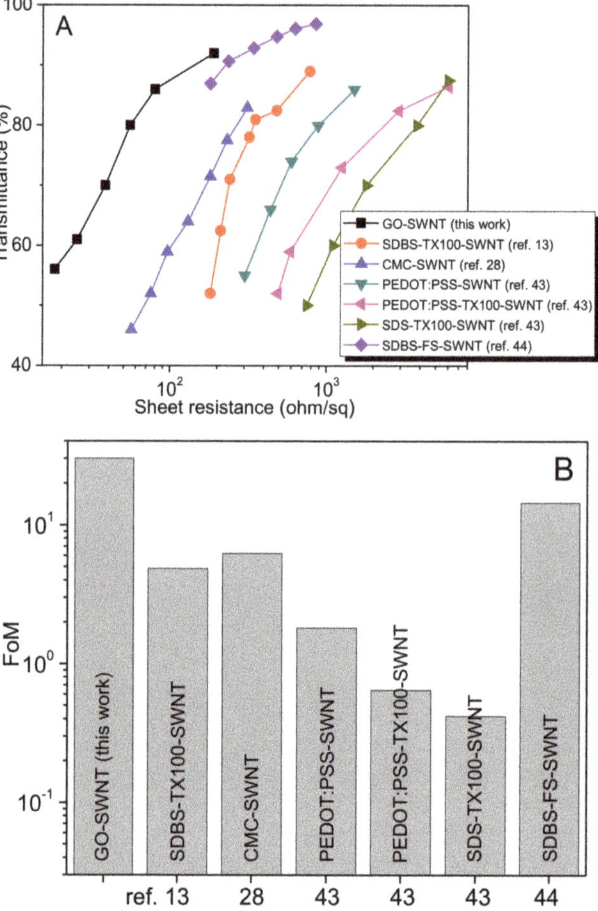

Figure 9. Comparison of electrical properties of rGO-SWNT film with those films contained other dispersing agents. (**A**) Transmittance versus sheet resistance profiles; (**B**) Comparison of figure of merit (FoM) CMC: sodium carboxymethyl cellulose; SDS: sodium dodecyl sulphate; FS: fluorosurfactant (FC-4430).

As mentioned above, we have used Mayer rod coating to deposit rGO-MWNT films onto glass substrates [14]. The rGO-MWNT film exhibits a sheet resistance of 380 Ω/sq at 85% transmittance. Moreover, the rGO-MWNT film has a FOM value of 6. In contrast, the sheet resistance of rGO-SWNT film in this study is ~80 Ω/sq at ~86% transmittance before doping, and reduces to ~25 Ω/sq after doping. The FOM value of rGO-SWNT film is 30 (Figure 9), being much higher than that of rGO-MWNT sample. Apparently, rGO-MWNT films exhibit much higher sheet resistance and poorer electro-optical performance than rGO-MWNT films in this work. This can be attributed to each graphene layer of MWNTs contributes to the light absorption, resulting in strong light absorption of rGO-MWNT films [30]. In addition, rGO-MWNT films are deposited on rigid glass substrates, so their mechanical flexibility cannot be evaluated through mechanical bending test. On the contrary, rGO-SWNT films are coated on flexible PET plastic substrates, so their electro-optical performance due to mechanical bending can be determined accordingly.

In addition to excellent electro-optical properties of rGO-SWNT films, TCs require mechanical durability to withstand mechanical bending of flexible, portable consumer electronics. Figure 10A shows that the electrical resistance of the rGO-SWNT films exhibits negligible change over 2000 bending cycles to a radius of 2 mm, with a minor increase in resistance consistent with previous reports, verifying the utility of this material for devices in which a high tolerance to the twist of flexibility is necessary. While the rGO-SWNT exhibits no degradation, PET substrate exhibits plastic deformation under these conditions, which is the main reason contributing to the resistance increase.

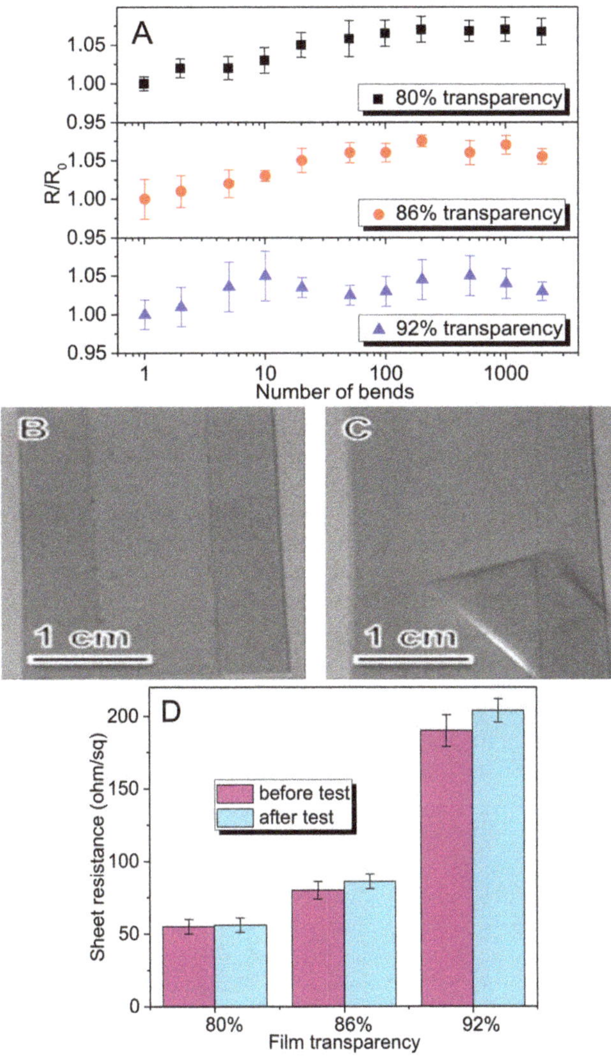

Figure 10. (**A**) Relative sheet resistance vs. benching cycles for rGO-SWNT films. R_0 is the electrical resistance of samples before bending, and R is the electrical resistance that changes with bending cycles. Error bars reflect standard deviation for five measurements, (**B,C**) Mechanical stability of rGO-SWNT/PET film: before and after the Scotch tape test; (**D**) Sheet resistance vs. transparency before and after Scotch tape peeling test.

While bending stability confirms that the material can deform without failure, it does not necessarily imply strong film adhesion to the substrate. It is known that solution-processed CNT film interacts weakly with the polymer substrate, resulting in poor adhesion of the coating. This can cause significant yield loss during device processing. A scotch tape test was therefore performed to assess the substrate adhesion performance of rGO-SWNT films. A strip of tape is applied to the films and peeled off in a controlled manner. Then the tape and sample are inspected for the film damage. The optical pictures of rGO-SWNT film coated on PET substrate before and after tape testing are shown in Figure 10B,C. No significant film damage is found in the rGO-SWNT film after tape testing. Electrical measurements agree reasonably with this result. Specifically, these films exhibit negligible resistance change after tape adhesion test (Figure 10D). We propose that the microwave treatment plays a critical role in enhancing the film adhesion. As reported before, CNT-polymer welding is an important effect caused by microwave heating [48–52]. It usually leads to strong adhesion between the nanotubes and the plastic. Given microwave heating is highly selective, only nanotubes absorb microwave energy and heated accordingly, so it can be carried out without causing significant distortion of underlying PET substrate [49].

It can be concluded that the rGO-SWNT samples exhibit low sheet resistance of ~80 Ω/sq and ~25 Ω/sq respectively at ~86% transmittance before and after nitric acid doping. The rGO-SWNT film also exhibits a high FOM value of 30. The excellent electro-optical properties, coupled with the intrinsic flexibility and good film adhesion property make the rGO-SWNT conductor an excellent or ideal replacement for ITO. In terms of scalable film production and high quality of the products, the method introduced in this study is much superior or better than those films reported in the literature. In Figure S8A we show a film of ~4 $\times$ 5 inch^2 prepared by this method. The film appears to be very uniform without any nanotube aggregates. Moreover, microscopic examinations of different locations of the film also reveal a similar morphology (Figure S8B,C). Therefore, it is expected that the GO-SWNT ink can be well adapted to the coating processes in industries including slot, slide, and roll-to-roll, facilitating a fast and mass production of continuous film with high quality. In addition, the film properties can be fine-tuned by monitoring the compositions and/or the thicknesses of the films. We anticipate that this work can trigger further study of novel dispersants for SWNTs and the better use of resulting SWNT inks with functional properties.

4. Conclusions

We have presented a promising graphene oxide-based approach for preparing carbon nanotube inks for scalable production of carbon-based TCs. This strategy combines scalable, low-cost production of graphene oxide with tunable nanotube ink formulation for a technique compatible with large-scale production, thus providing new opportunities for broad integration of the films in industrial applications. The resulting films exhibit many of the desirable properties such as good film adhesion, sheet resistance of ~80 Ω/sq at 86% transparency, being the best amongst solution-processed carbonaceous films. Moreover, nitric acid doping can further decrease the sheet resistance to ~25 Ω/sq. Finally, microwave treatment induces rapid welding between the film and the substrate, yielding the films resilient to strong mechanical stresses including cyclic bending and adhesion testing. The concurrent realization of these properties in a scalable and adaptable process represents a significant advance for using nanotube inks in manufacturing transparent conductors.

Supplementary Materials: The following are available online at http://www.mdpi.com/2079-4991/8/4/224/s1. Figure S1: TEM micrograph of as-received SWNTs. Figure S2: AFM image of as-synthesized GOs (on freshly cleaved mica). Figure S3: Contact angle results for GO-SWNT inks with different compositions on PET substrates. (A) 1 mg/mL, (B) 2 mg/mL, (C) 3 mg/mL, (D) 4 mg/mL and (E) 6 mg/mL. The SWNT content is 0.2 mg/mL. Contact angles (from left to right) are 59.7°, 54.4°, 59.9°, 56.4° and 55.6°, respectively. Figure S4: Contact angle results for GO-SWNT inks with different compositions on PET substrates after oxygen plasma treatment. (A) 1 mg/mL, (B) 2 mg/mL, (C) 3 mg/mL, (D) 4 mg/mL and (E) 6 mg/mL. The SWNT content is 0.2 mg/mL. Contact angles (from left to right) are 10.2°, 10.5°, 10.7°, 10.7° and 10.6°, respectively. Figure S5: 3D topography of rGO-SWNT films prepared from different GO-SWNT dispersions: (A) 5:1, (B) 10:1 and (C) 20:1. Figure S6: (A) AFM height and (b) amplitude micrographs of rGO-SWNT film prepared from the ink having GO to SWNT mass ratio of

30:1. Figure S7: (A) Effect of nitric acid doping on electro-optical performance of 86% transparent rGO-SWNT films. (B) Stability of nitric acid-doped 86% transparent rGO-SWNT film. Figure S8: (A) Photograph of a GO-SWNT film of ~4 × 5 inch^2 prepared from GO-SWNT ink by rod coating. (B) and (C) Low magnification SEM micrographs of location I and II showing morphological uniformity of the film. The white dots are catalyst impurities.

Acknowledgments: The authors gratefully acknowledged financial support of this work by the National Natural Science Foundation of China (Grant No. 21703096).

Author Contributions: L. He synthesized and fabricated rGO-SWNT films, performed AFM, Raman, XPS, contact angle, viscosity, electrical resistance and optical transmittance measurements. C. Liao carried out optical, SEM and TEM examinations. S.C. Tjong designed the project, analyzed the data and wrote the manuscript.

Conflicts of Interest: The authors declare no conflict of interest.

References

1. He, L.X.; Tjong, S.C. Nanostructured transparent conductive films: Fabrication, characterization and applications. *Mater. Sci. Eng. R Rep.* **2016**, *109*, 1–101. [CrossRef]
2. Artukovic, E.; Kaempgen, M.; Hecht, D.S.; Roth, S.; Grüner, G. Transparent and flexible carbon nanotube transistors. *Nano Lett.* **2005**, *5*, 757–760. [CrossRef] [PubMed]
3. Wang, X.; Zhi, L.J.; Mullen, K. Transparent, conductive graphene electrodes for dye-sensitized solar cells. *Nano Lett.* **2008**, *8*, 323–327. [CrossRef] [PubMed]
4. Hecht, D.S.; Hu, L.; Irvin, G. Emerging transparent electrodes based on thin films of carbon nanotubes, graphene, and metallic nanostructures. *Adv. Mater.* **2011**, *23*, 1482–1513. [CrossRef] [PubMed]
5. Ye, S.R.; Rathmell, A.R.; Chen, Z.F.; Stewart, I.E.; Wiley, B.J. Metal nanowire networks: The next generation of transparent conductors. *Adv. Mater.* **2014**, *26*, 6670–6687. [CrossRef] [PubMed]
6. Iijima, S.; Ichihashi, T. Single-shell carbon nanotubes of 1-nm diameter. *Nature* **1993**, *363*, 603–605. [CrossRef]
7. Hu, L.; Hecht, D.S.; Grüner, G. Carbon nanotube thin films: Fabrication, properties, and applications. *Chem. Rev.* **2010**, *110*, 5790–5844. [CrossRef] [PubMed]
8. Kim, S.; Yim, J.; Wang, X.; Bradley, D.D.; Lee, S.; deMello, J.C. Spin-and spray-deposited single-walled carbon-nanotube electrodes for organic solar cells. *Adv. Funct. Mater.* **2010**, *20*, 2310–2316. [CrossRef]
9. Jo, J.W.; Jung, J.W.; Lee, J.U.; Jo, W.H. Fabrication of highly conductive and transparent thin films from single-walled carbon nanotubes using a new non-ionic surfactant via spin coating. *ACS Nano* **2010**, *4*, 5382–5388. [CrossRef] [PubMed]
10. Kim, Y.; Minami, N.; Zhu, W.; Kazaoui, S.; Azumi, R.; Matsumoto, M. Langmuir-Blodgett films of single-wall carbon nanotubes: Layer-by-layer deposition and in-plane orientation of tubes. *Jpn. J. Appl. Phys.* **2003**, *42*, 7629–7634. [CrossRef]
11. Eda, G.; Lin, Y.Y.; Miller, S.; Chen, C.W.; Su, W.F.; Chhowalla, M. Transparent and conducting electrodes for organic electronics from reduced graphene oxide. *Appl. Phys. Lett.* **2008**, *92*, 233305. [CrossRef]
12. Zhang, D.; Ryu, K.; Liu, X.; Polikarpov, E.; Ly, J.; Tompson, M.E.; Zhou, C. Transparent, conductive, and flexible carbon nanotube films and their application in organic light-emitting diodes. *Nano Lett.* **2006**, *6*, 1880–1886. [CrossRef] [PubMed]
13. Dan, B.; Irvin, G.C.; Pasquali, M. Continuous and scalable fabrication of transparent conducting carbon nanotube films. *ACS Nano* **2009**, *3*, 835–843. [CrossRef] [PubMed]
14. He, L.X.; Tjong, S.C. Aqueous graphene oxide-dispersed carbon nanotubes as inks for the scalable production of all-carbon transparent conductive films. *J. Mater. Chem. C* **2016**, *4*, 7043–7051. [CrossRef]
15. Zhou, M.; Wang, Y.; Zhai, Y.; Zhai, J.; Ren, W.; Wang, F.; Dong, S. Controlled synthesis of large-area and patterned electrochemically reduced graphene oxide films. *Chem. Eur. J.* **2009**, *15*, 6116–6120. [CrossRef] [PubMed]
16. He, L.X.; Tjong, S.C. Low percolation threshold of graphene/polymer composites prepared by solvothermal reduction of graphene oxide in the polymer solution. *Nanoscale Res. Lett.* **2013**, *8*, 132. [CrossRef] [PubMed]
17. He, L.X.; Tjong, S.C. Effect of temperature on electrical conduction behavior of polyvinylidene fluoride nanocomposites with carbon nanotubes and nanofibers. *Curr. Nanosci.* **2010**, *6*, 520–524. [CrossRef]
18. Hu, L.B.; Choi, J.W.; Yang, Y.; Jeong, S.; La Mantia, F.; Cui, L.F.; Cui, Y. Highly conductive paper for energy-storage devices. *Proc. Natl. Acad. Sci. USA* **2009**, *106*, 21490–21494. [CrossRef] [PubMed]
19. Hummers, W.S.; Offeman, R.E. Preparation of graphitic oxide. *J. Am. Chem. Soc.* **1958**, *80*, 1339. [CrossRef]

20. Dreyer, D.R.; Park, S.; Bielawski, C.W.; Ruoff, R.S. The chemistry of graphene oxide. *Chem. Soc. Rev.* **2010**, *39*, 228–240. [CrossRef] [PubMed]

21. Cote, L.J.; Kim, F.; Huang, J.X. Langmuir-Blodgett assembly of graphite oxide single layers. *J. Am. Chem. Soc.* **2009**, *131*, 1043–1049. [CrossRef] [PubMed]

22. Qiu, L.; Yang, X.; Gou, X.; Yang, W.; Ma, Z.F.; Wallace, G.G.; Li, D. Dispersing carbon nanotubes with graphene oxide in water and synergistic effects between graphene derivatives. *Chem. Eur. J.* **2010**, *16*, 10653–10658. [CrossRef] [PubMed]

23. Pei, S.; Cheng, H.M. The reduction of graphene oxide. *Carbon* **2012**, *50*, 3210–3228. [CrossRef]

24. Tian, L.; Meziani, M.J.; Lu, L.; Kong, C.Y.; Cao, L.; Thorne, T.J.; Sun, Y.P. Graphene oxides for homogeneous dispersion of carbon nanotubes. *ACS Appl. Mater. Interfaces* **2010**, *2*, 3217–3222. [CrossRef] [PubMed]

25. Zheng, Q.; Zhang, B.; Lin, X.; Shen, X.; Yousefi, N.; Huang, Z.D.; Li, Z.; Kim, J.K. Highly transparent and conducting ultralarge graphene oxide/single-walled carbon nanotube hybrid films produced by Langmuir-Blodgett assembly. *J. Mater. Chem.* **2012**, *22*, 25072–25082. [CrossRef]

26. Gorkina, A.L.; Tsapenko, A.P.; Gilshteyn, A.P.; Koltsova, T.S.; Larionova, T.V.; Talyzin, A.; Anisimov, A.S.; Anoshkin, I.V.; Kauppinen, E.I.; Tolochko, O.V.; et al. Transparent and conductive hybrid graphene/carbon nanotube films. *Carbon* **2016**, *100*, 501–507. [CrossRef]

27. Hecht, D.S.; Kaner, R.B. Solution-processed transparent electrodes. *MRS Bull.* **2011**, *36*, 749–755. [CrossRef]

28. Li, X.; Gittleson, F.; Carmo, M.; Sekol, R.C.; Taylor, A.D. Scalable fabrication of multifunctional freestanding carbon nanotube/polymer composite thin films for energy conversion. *ACS Nano* **2012**, *6*, 1347–1356. [CrossRef] [PubMed]

29. Geng, H.Z.; Lee, D.S.; Kim, K.K.; Kim, S.J.; Bae, J.J.; Lee, Y.H. Effect of carbon nanotube types in fabricating flexible transparent conducting films. *J. Korean Phys. Soc.* **2008**, *53*, 979–985. [CrossRef]

30. Yu, L.; Shearer, C.; Shapter, J. Recent development of carbon nanotube transparent conductive films. *Chem. Rev.* **2016**, *116*, 13413–13453. [CrossRef] [PubMed]

31. Imholt, T.J.; Dyke, C.A.; Hasslacher, B.; Perez, J.M.; Roberts, J.A.; Scott, J.B.; Wadhawan, A.; Ye, Z.; Tour, J.M. Nanotubes in microwave fields: Light emission, intense heat, outgassing, and reconstruction. *Chem. Mater.* **2003**, *15*, 3969–3970. [CrossRef]

32. Voiry, D.; Yang, J.; Kupferberg, J.; Fullon, R.; Lee, C.; Jeong, H.Y.; Shin, H.S.; Chhowalla, M. High-quality graphene via microwave reduction of solution-exfoliated graphene oxide. *Science* **2016**, *353*, 1413–1416. [CrossRef] [PubMed]

33. Zou, J.H.; Liu, L.W.; Chen, H.; Khondaker, S.I.; McCullough, R.D.; Huo, Q.; Zhai, L. Dispersion of pristine carbon nanotubes using conjugated block copolymers. *Adv. Mater.* **2008**, *20*, 2055–2060. [CrossRef]

34. Zhang, L.; Xia, J.; Zhao, Q.; Liu, L.; Zhang, Z. Functional graphene oxide as a nanocarrier for controlled loading and targeted delivery of mixed anticancer drugs. *Small* **2010**, *6*, 537–544. [CrossRef] [PubMed]

35. Hamedi, M.M.; Hajian, A.; Fall, A.B.; Håkansson, K.; Salajkova, M.; Lundell§, F.; Wågberg, L.; Berglund, L.A. Highly conducting, strong nanocomposites based on nanocellulose-assisted aqueous dispersions of single-wall carbon nanotubes. *ACS Nano* **2014**, *8*, 2467–2476. [CrossRef] [PubMed]

36. Keunings, R.; Bousfield, D. Analysis of surface-tension driven leveling in viscoelastic films. *J. Non-Newton. Fluid Mech.* **1987**, *22*, 219–233. [CrossRef]

37. Redon, C.; Brochard-Wyart, F.; Rondelez, F. Dynamics of dewetting. *Phys. Rev. Lett.* **1991**, *66*, 715–718. [CrossRef] [PubMed]

38. Cohen, E.D.; Gutoff, E.B. *Modern Coating and Drying Technology*; VCH: New York, NY, USA, 1992; pp. 6–7.

39. Nijenhuis, K.; McKinley, G.H.; Spiegelberg, S.; Barnes, H.A.; Aksel, N.; Heymann, L.A.; Odell, J. Thixotropy, rheopexy and yield Stress. In *Springer Handbook of Experimental Fluid Mechanics*; Tropea, C., Yarin, A.L., Foss, J.F., Eds.; Springer: Berlin, Germany, 2007; pp. 661–679.

40. Irgens, F. *Rheology and Non-Newtonian Fluids*; Springer: Berlin, Germany, 2014; pp. 1–16.

41. Meng, Y.Z.; Tjong, S.C. Rheology and morphology of compatibilized polyamide 6 blends containing liquid crystalline copolyesters. *Polymer* **1998**, *39*, 99–107. [CrossRef]

42. Tjong, S.C.; Meng, Y.Z. Morphology and mechanical characteristics of compatibilized polyamide 6-liquid crystalline polymer composites. *Polymer* **1997**, *38*, 4609–4615. [CrossRef]

43. Tjong, S.C.; Liu, S.L.; Li, R.K.Y. Mechanical properties of injection moulded blends of polypropylene with thermotropic liquid crystalline polymer. *J. Mater. Sci.* **1996**, *31*, 479–484. [CrossRef]

44. Meng, Y.Z.; Tjong, S.C.; Hay, A.S.; Wang, S.J. Synthesis and proton conductivities of phosphonic acid containing poly-(arylene ether)s. *J. Polym. Sci. A Polym. Chem.* **2001**, *39*, 3218–3226. [CrossRef]
45. Nicholas, A.; Parra-Vasquez, G.; Stepanek, I.; Davis, V.A.; Moore, V.C.; Haroz, E.H.; Shaver, J.; Hauge, R.H.; Smalley, R.E.; Pasquali, M. Simple length determination of single-walled carbon nanotubes by viscosity measurements in dilute suspensions. *Macromolecules* **2007**, *40*, 4043–4047.
46. Satas, D.; Tracton, A.A. *Coatings Technology Handbook*, 2nd ed.; Marcel Dekker: New York, NY, USA, 2000; p. 132.
47. Jackson, R.; Domercq, B.; Jain, R.; Kippelen, B.; Graham, S. Stability of doped transparent carbon nanotube electrodes. *Adv. Funct. Mater.* **2008**, *18*, 2548–2554. [CrossRef]
48. Zhang, M.; Fang, S.; Zakhidov, A.A.; Lee, S.B.; Aliev, A.E.; Williams, C.D.; Atkinson, K.R.; Baughman, R.H. Strong, transparent, multifunctional, carbon nanotube sheets. *Science* **2005**, *309*, 1215–1219. [CrossRef] [PubMed]
49. Wang, C.Y.; Chen, T.H.; Chang, S.C.; Cheng, S.Y.; Chin, T.S. Strong carbon-nanotube-polymer bonding by microwave irradiation. *Adv. Funct. Mater.* **2007**, *17*, 1979–1983. [CrossRef]
50. Wang, C.Y.; Chen, T.H.; Chang, S.C.; Chin, T.S.; Cheng, S.Y. Flexible field emitter made of carbon nanotubes microwave welded onto polymer substrates. *Appl. Phys. Lett.* **2007**, *90*, 103111. [CrossRef]
51. Shim, H.C.; Kwak, Y.K.; Han, C.S.; Kim, S. Enhancement of adhesion between carbon nanotubes and polymer substrates using microwave irradiation. *Scr. Mater.* **2009**, *61*, 32–35. [CrossRef]
52. Han, J.T.; Kim, D.; Kim, J.S.; Seol, S.K.; Jeong, S.Y.; Jeong, H.J.; Chang, W.S.; Lee, G.W.; Jung, S. Self-passivation of transparent single-walled carbon nanotube films on plastic substrates by microwave-induced rapid nanowelding. *Appl. Phys. Lett.* **2012**, *100*, 163120.

 *nanomaterials*

Article

Optical Study and Experimental Realization of Nanostructured Back Reflectors with Reduced Parasitic Losses for Silicon Thin Film Solar Cells

Zeyu Li [1,2,*], Rusli E [1], Chenjin Lu [1], Ari Bimo Prakoso [1], Martin Foldyna [2], Rasha Khoury [2], Pavel Bulkin [2], Junkang Wang [2], Wanghua Chen [2], Erik Johnson [2] and Pere i Roca Cabarrocas [2]

[1] NOVITAS (Centre of Micro-/Nano-electronics), Nanoelectronics Centre of Excellence, School of Electrical and Electronic Engineering, Nanyang Technological University, 50 Nanyang Avenue, Singapore 639798, Singapore; erusli@e.ntu.edu.sg (R.E.); CLU010@e.ntu.edu.sg (C.L.); ABPRAKOSO@ntu.edu.sg (A.B.P.)

[2] LPICM (Laboratory of Physics of Interfaces and Thin Films), CNRS (Centre national de la recherche scientifique), Ecole Polytechnique, Université Paris Saclay, Palaiseau 91128, France; martin.foldyna@polytechnique.edu (M.F.); rasha.khoury@polytechnique.edu (R.K.); pavel.bulkin@polytechnique.edu (P.B.); junkang.wang@polytechnique.edu (J.W.); wanghua.chen@polytechnique.edu (W.C.); erik.johnson@polytechnique.edu (E.J.); pere.roca@polytechnique.edu (P.iR.C.)

* Correspondence: lize0004@ntu.edu.sg; Tel.: +65-9179-7290

Received: 2 July 2018; Accepted: 14 August 2018; Published: 18 August 2018

Abstract: We study light trapping and parasitic losses in hydrogenated amorphous silicon thin film solar cells fabricated by plasma-enhanced chemical vapor deposition on nanostructured back reflectors. The back reflectors are patterned using polystyrene assisted lithography. By using O_2 plasma etching of the polystyrene spheres, we managed to fabricate hexagonal nanostructured back reflectors. With the help of rigorous modeling, we study the parasitic losses in different back reflectors, non-active layers, and last but not least the light enhancement effect in the silicon absorber layer. Moreover, simulation results have been checked against experimental data. We have demonstrated hexagonal nanostructured amorphous silicon thin film solar cells with a power conversion efficiency of 7.7% and around 34.7% enhancement of the short-circuit current density, compared with planar amorphous silicon thin film solar cells.

Keywords: light trapping; silicon thin film; photovoltaics; polystyrene sphere assisted lithography; nanostructured back reflectors

---

1. Introduction

By applying plasma-enhanced chemical vapor deposition (PECVD), hydrogenated amorphous/ microcrystalline silicon (a/µc-Si:H) tandem solar cells fabricated at low temperatures (around 150 °C) have been researched extensively over the last decade and have achieved over 14% stabilized efficiency [1]. Silicon thin films are promising candidates for future large-scale photovoltaics due to their advantages including lesser material usage, compatible with roll-to-roll processes on flexible substrates, usage of nontoxic and abundant active material, and so on. Effective light trapping mechanisms are of extreme importance in fabricating such high efficient solar cells due to the low absorption coefficient of microcrystalline silicon in the near-infrared region. Nanostructured Ag/ZnO back reflector (BR) substrates are one of the most effective and process compatible approaches for an enhanced light trapping performance in silicon thin film solar cells. A typical BR utilizes metallic or metal/dielectric nanostructures, which produce a far field scattering effect for an extended light traveling path [2–5], hence a higher absorption. We have previously investigated enhanced light scattering of the BRs patterned by using double size polystyrene (PS) assisted lithography [6]. However, due to plasmonic absorption by the metal/dielectric nanostructures, a significant amount of parasitic losses in the BR is also introduced [7]. To address and characterize such losses, Bruggeman Effective Medium Approximation (BEMA) was previously applied by L. R. Dahal et al. to study the parasitic losses due to a flat or "nanoscale roughness" Ag/ZnO and Al/ZnO interface [8]. However, typical metal/dielectric nanostructures will have dimensions larger than a few tens of nanometers. For example, random nanostructured ASAHI U type substrates with a r.m.s. roughness of around 35 nm are often used in silicon thin film solar cells to enhance light trapping [9]. For random Ag/ZnO BR parasitic losses estimation, extra experimental results are needed to set up an accurate BEMA model. To study the periodic metal/dielectric structures, electromagnetic simulations can be used [10,11]. Not only it is a powerful tool for the design, optimization, and interpretation of the light-nanostructures interactions, but also the absorption enhancement effect in the active layer can be systematically studied and different mechanisms of the electric field enhancement in the system can be readily isolated [12].

In this work, we first systematically studied optical losses in flat Ag/ZnO BRs and random nanostructured Ag/ZnO ASAHI BRs. With different ZnO thickness as the extra variable, we assessed the simulation accuracy of the proposed BEMA model by experimental results. The origin of the parasitic losses in random ASAHI nanostructured Ag/ZnO BRs is concluded. We have used effective medium theory only to describe the effect of random surface roughness of the BRs. After that, by using PS sphere assisted lithography [13], hexagonal nanostructured Ag/ZnO BRs were fabricated. Light scattering properties, as well as optical losses in such BRs were compared with random nanostructured ASAHI BRs. With the help of high frequency electromagnetic field simulator (HFSS), we further investigated the corresponding optical losses in each layer separately. Last but not least, a full device was built upon the BR in the simulation model. Different light scattering modes and each layer's parasitic losses in the final device were extracted from the simulation results. Photo-generated current was calculated using the AM1.5G solar spectrum. Based on the optical model, we fabricated and characterized silicon thin film solar cells. Experimental results are presented for comparison and verification.

2. Experimental Details

2.1. PS Assisted Lithography for Ag Nanostructures on Glass and Hexagonal Back Reflectors

The PS spheres used in this work were a monodisperse suspension with 5 wt. % in water, obtained from microparticles Gmbh (Berlin, Germany) with nominal diameters of 607 ± 15 nm. We used methanol as the spreading agent, mixed in 1:1 volume ratio with the pristine PS solution. Prior to the PS sphere monolayer transfer, Corning glasses were cleaned with acetone, isopropanol (IPA), and de-ionized (DI) water in an ultrasonic bath, followed by a 10 min UV-Ozone treatment to keep the surface hydrophilic. We used the floating transfer technique for PS spheres monolayer deposition on Corning glass. After that, we used an electron cyclotron resonance (ECR) oxygen plasma etching process for a uniform size reduction of the PS spheres [13]. For the oxygen plasma etching process, we used 40 sccm of O_2, a pressure of 2.7×10^{-3} mbar, and 160 s of etching time at 500 W of plasma power. After thermal evaporation of 100 nm Ag, the substrates were immersed for 20 min in a low power ultrasonic bath in toluene for a complete removal of remaining PS spheres. Another 10 min IPA and DI water ultrasonic bath was applied for a complete removal of any organic solvent residue. Overall fabrication process and SEM image of the hexagonal back reflector are shown in Figure 1.

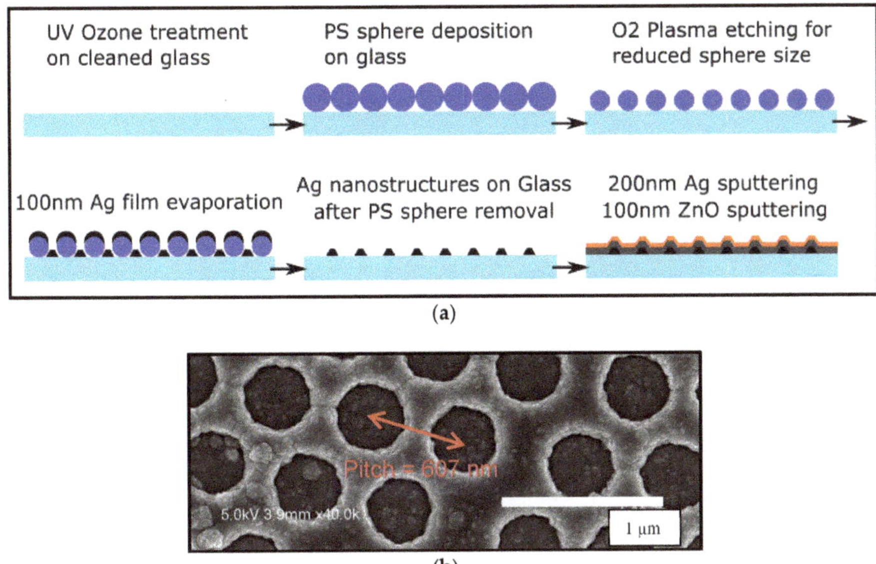

(a)

(b)

Figure 1. (**a**) Fabrication process and (**b**) Top view SEM image of hexagonal back reflectors (pitch = 607 nm) taken after the final step in Figure 1a.

2.2. Solar Cell Fabrication and Characterization

Amorphous silicon based p-i-n solar cells were deposited in a radio frequency (RF) PECVD system [14]. First, a 25 nm thick n-type microcrystalline silicon oxide layer was deposited on the BR [15]. Without breaking the vacuum, ~230 nm of intrinsic amorphous silicon layer was deposited using SiH4 plasma at 150 °C, 2 W RF power, and 30 sccm SiH_4 flow rate, followed by a 25 nm thick p-type amorphous silicon layer. Lastly, 80 nm of indium tin oxide (ITO) was deposited to form the top contacts and to define the cell size of 0.126 cm^2. Such a fabrication order is often referred to as an n-i-p solar cell [16]. A schematic image of the cross section of the device and its corresponding top view SEM image acquired using a Hitachi S-4700 scanning electron microscope is shown in Figure 2.

The current density-voltage (J-V) characteristics of the solar cells were measured under AM1.5G illumination with a commercial solar simulator (Oriel AAA, Irvine, CA, USA), calibrated using a crystalline Si reference cell. Total and diffused reflectance were measured using a Perkin-Elmer Lambda 950 spectrophotometer (London, UK) from 350 nm to 1100 nm wavelength range with a 150 mm integrating sphere and an InGaAs detector.

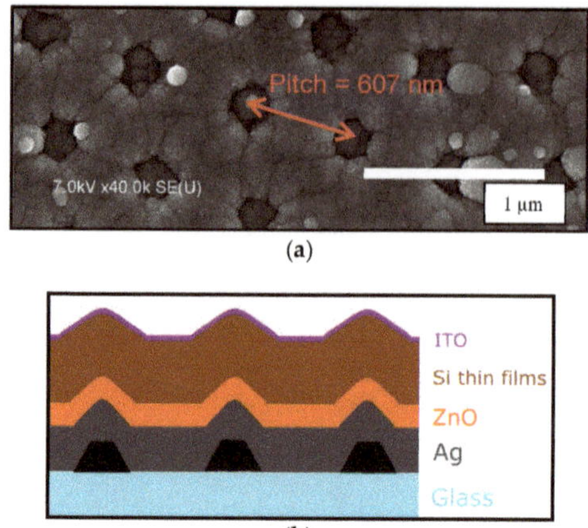

(a)

(b)

Figure 2. (a) Top view SEM image of the complete solar cell and (b) Schematic cross section of a hexagonal back reflector solar cell.

3. Results and Discussion

In this section, optical reflectance measurements on a series of flat Ag/ZnO BRs with different ZnO thickness, ASAHI Ag/ZnO BRs, and modeling results based on BEMA will be discussed first. Then, both experimental and HFSS simulation results of the hexagonal Ag/ZnO nanostructured BRs will be presented and compared for absorption mechanism analysis. After that, HFSS simulation results with an active silicon layer on the BR will be presented and analyzed. Last but not least, J-V, external quantum efficiency (EQE), and optical absorption ($1 - R_{total} - T_{total}$) results of the solar cells fabricated on a hexagonal Ag/ZnO nanostructured BR will be compared with the devices made on flat Ag/ZnO BR and on flat ZnO/Corning glass substrates.

3.1. Optical Characterization of Flat and ASAHI Back Reflectors

Before we proceed to study the optical properties of our nanostructured BRs, reference flat Ag/ZnO on glass substrates were fabricated and reflectance data were measured. Figure 3 shows the reflectance data from sputtered ZnO (0 to 160 nm) on 200 nm sputtered Ag on the glass. Figure 4a shows the reflectance data of ASAHI BR with 0, 50, and 100 nm of ZnO. Figure 4b shows the BEMA simulation model we used. Figure 4c,d show the simulated results for various thicknesses of the rough interface layer L2.

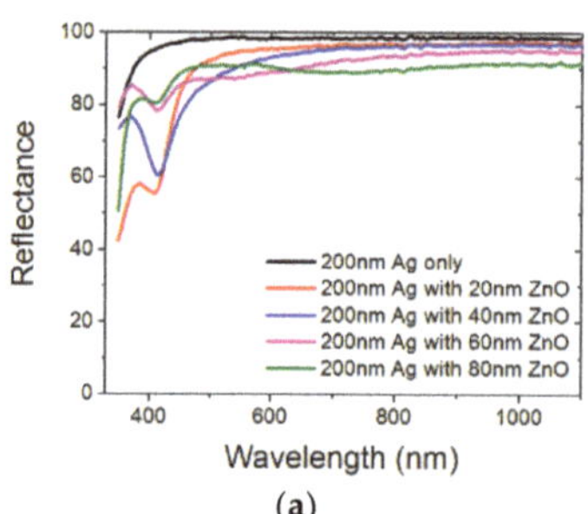

(a)

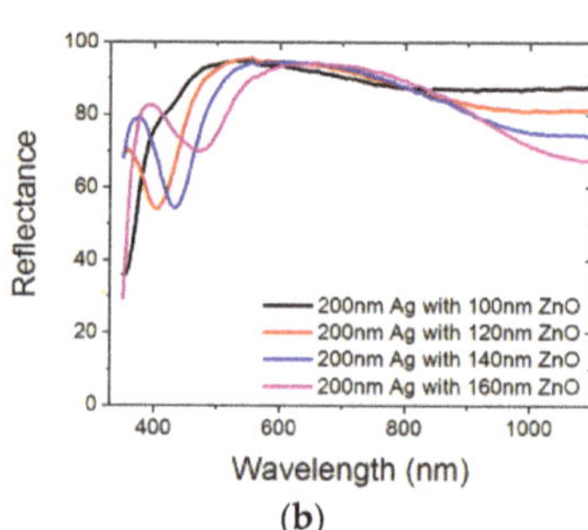

(b)

Figure 3. Total reflectance measured on a glass substrate coated with 200 nm Ag and various thicknesses of ZnO (0–160 nm). (a) 0–80 nm and (b) 100–160 nm.

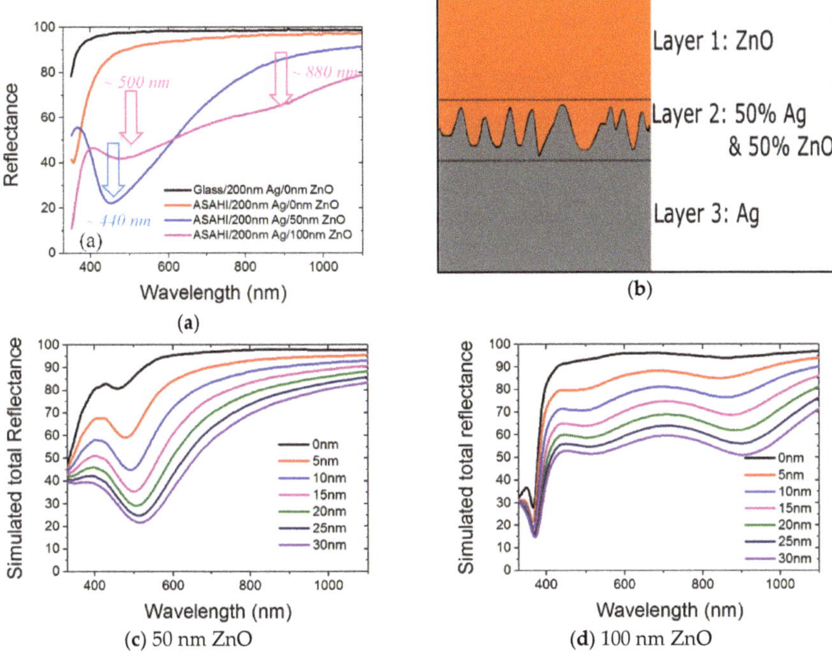

Figure 4. (**a**) Total reflectance measured on ASAHI substrate coated with 200 nm Ag and with 0, 50, and 100 nm of ZnO; (**b**) Bruggeman effective medium approximation model applied for the Ag/ZnO back reflector (BR) study; (**c**) and (**d**) Simulated total reflectance of 50 and 100 nm ZnO (layer 1) with 0 to 30 nm of Ag/ZnO interface (layer 2) thickness.

3.1.1. Total Reflectance from Flat and ASAHI Back Reflectors

As expected, a pure 200 nm of Ag film on glass provides a highly reflective surface from the 350 nm to 1100 nm wavelength range, with minimum absorption at shorter wavelength. In Figure 3a, we observed that when a thin ZnO was deposited on top of Ag, with its increasing thickness from 20 nm to 100 nm, the local reflectance minimum at 415~420 nm diminishes. D. Sainju et al. [17] concluded that the rough Ag/ZnO interface introduces localized plasmonic absorption at around 2.9 eV due to the nano-protrusions on the Ag surface, which exactly corresponds to local reflectance minimum at 420 nm observed in Figure 3a. With increasing ZnO thickness from 20 nm to 80 nm, such a local reflectance minimum becomes less significant. A previous study has reported that refractive index of sputtered ZnO thin film are different when deposited on different substrates [18]. E. Moulin et al. concluded that for thin film silicon solar cells with a Ag BR structure, dielectric material with a lower refractive index between Ag and active silicon thin films results in lower optical losses in BRs [19]. From [20], the author argues that when an interference minimum in reflectance matches the plasmon energy of the rough Ag/ZnO interface, this will lead to a strong coupling of the energy from the optical field into the plasmon mode. With the increased film thickness, the interference minimum shifted away from 420 nm due to Fabry-Perot resonance. Such a shift has led to a decreased plasmon absorption, hence a decreased local reflectance minimum at 420 nm is observed.

The shifts of the interference minimum due to Fabry-Perot resonance are easier to observe in Figure 3b. From Figure 3b, when a thicker ZnO was deposited on top of Ag, with the increasing thickness from 120 nm to 140 nm to 160 nm, reflectance minimum shifts from 400 nm to 430 nm to

470 nm. By using the refractive index of crystalline ZnO on metal from reference [18], we can calculate the condition of destructive interference using

$$\lambda\_destructive = (2 \times n\_ZnO \times t\_ZnO \times \cos(7°))/(m - 1/2) \qquad (1)$$

where $m = 1$ for the first and $m = 2$ for the second destructive interference respectively. At 120 nm ZnO thickness, by using n_ZnO = 2.46 at 400 nm wavelength [18] (where we suspect 400 nm is the 2nd destructive inference), the calculated λ_2nd_destructive at $m = 2$ is 390 nm. This calculation matches the observation of the red shifting of the reflectance minima with increasing ZnO thickness. The difference between the calculated reflectance minima and our experimental data might be due to the difference in the ZnO refractive index when deposited on different substrates; the metal substrate used in reference 14 is Pt, while we are using Ag in our case. The first destructive interference wavelength λ_1st_destructive is calculated to be 1010 nm, by using n_ZnO = 2.12 at 900 nm wavelength [18], $m = 1$. This further explains the reduced reflectance for the wavelength range from 900 nm and beyond in Figure 3b. Other than above, the increased free carrier absorption of a thicker ZnO layer will also play a role in the overall optical performance [21]. From the experimental data, to avoid parasitic absorption by the BRs in the longer wavelength range, 100 nm of ZnO might be more desirable when compared with thicker ZnO cases, as the thickness of ZnO determines the absorption behavior of flat Ag/ZnO BRs. Special care shall be taken at ZnO thickness when BRs are nanostructured.

With the above understanding of flat Ag/ZnO BRs performance, the reflectance results from ASAHI BRs are presented in the next section. In Figure 4a, 200 nm of Ag sputtered on rough ASAHI substrate shows a lower reflectance compared with 200 nm of Ag sputtered on flat glass substrate. This is due to a higher plasmonic absorption on a rough Ag surface [22]. With 50 nm of sputtered ZnO on top, significant absorption is observed for all wavelengths with local reflectance minimum at around 440 nm. With 100 nm of sputtered ZnO on top, reflectance further reduces from the 600 to 1100 nm wavelength range but partially recovers from the 400 to 600 nm range. Two local minima at around 500 and 880 nm have been observed. Such large differences in optical performance between flat BR and ASAHI BR shall be fully understood before we proceed to study the more complex hexagonal nanostructured Ag/ZnO BRs. Other than the total reflectance properties, the effect of nanostructure morphology on diffused reflectance will also be studied and presented in later sections.

3.1.2. Bruggeman Effective Medium Approximation Simulation

In this section we use the Bruggeman Effective Medium Approximation (BEMA) method to estimate the optical performance of the random nanostructured ASAHI Ag/ZnO BRs. BEMA is a fast simulation method for the macroscopic properties of composite materials, which we use to simulate Ag and ZnO mixture in BRs. With a 35 nm r.m.s. roughness on ASAHI substrate surface [9], one would expect less than 35 nm surface roughness at Ag/ZnO interface with added smoothing effect of the sputtered Ag layer [23]. To estimate the Ag/ZnO interface roughness and to correlate between interface roughness and the optical performance, BEMA simulation is used to study the random nanostructure ASAHI Ag/ZnO BRs. Figure 4b shows the BEMA simulation model we used. We designed our simulation model such that Ag/ZnO interface roughness is modelled as a thin film layer (L2) composed of 50% Ag and 50% ZnO, in between the pure ZnO layer (L1) and pure 200nm Ag layer (L3). By assuming a uniform Ag surface roughness and also a conformal coating of the ZnO on top, 50% to 50% composition ratio of the two materials and 0% void should be a good approximation to model the interface roughness layer. Figure 4c,d show the simulated reflectance with 50 and 100 nm of ZnO layer respectively. Interface roughness, that is, the thickness of L2, was changed from 0 to 30 nm in steps of 5 nm.

With 50 nm of ZnO, comparing between the experimental results in Figure 4a and simulation results in Figure 4c, we can see that in terms of the absolute magnitude (~20%), the reflectance from ASAHI BRs is close to the case of simulated 25 nm interface roughness. However, in terms of the position of the local reflectance minima, the reflectance from ASAHI BRs is close to the case of simulated 10 nm interface roughness. Similarly, in the case of 100 nm ZnO, a simulated 25 nm interface roughness shows closer results in absolute magnitude, and a simulated 10 nm interface roughness shows a closer result to the experimental data in position of the local reflectance minima. The above observations can be explained considering that with increasing Ag/ZnO interface layer thicknesses, the effective overall ZnO thickness will be increased. The red shift of the reflectance minima is due to the increased ZnO thickness.

From the simulation results above, we can observe that ZnO thickness (L1) determines the shape of the reflectance curve, while Ag/ZnO interface thickness (L2) determines the magnitude of the reflectance value. We suggest that top ZnO (L1) governs the interference behavior of the incident light. Ag/ZnO interface (L2) on one hand could absorb light which reduces the overall reflectance, and on another hand could be effectively treated as an antireflection layer in between ZnO and bulk Ag. Such an antireflection effect causes more light to be transmitted into bulk Ag (L3), hence more parasitic absorption. From Figure 4c,d, with a uniform increasing L2 thickness, we observe a uniform reduction in reflectance for the wavelength range around 1000 nm. However, in the wavelength range around 500 nm, the reduction of reflectance is large when L2 is thin (15% reduction from 5 to 10 nm) but small when L2 is thick (5% reduction from 25 to 30 nm). This can be explained by the antireflection effect, which saturates with increasing L2 thickness. The saturation in absorption loss of the BRs has been reported by J. Springer et al. [24], while we propose a new possibility to explain such behavior using BEMA approximation.

Our model predicts that such BR has an interface thickness of 25 nm. Surprisingly if we compare the simulated 5 nm interface roughness (100 nm ZnO layer 1) with 100 nm ZnO on 200 nm Ag reflectance data in Figure 3, we can also observe a similar reflectance magnitude and local minima position. Hence, from the perspective of our simulation results, the nominally flat Ag/ZnO BR also has a 5 nm Ag/ZnO interface thickness. The ZnO/air surface roughness has been investigated as well, but no significant changes in the reflectance behavior were observed.

3.2. Experiment and Simulation Results on Hexagonal Ag/ZnO Nanostructured Back Reflectors

In previous sections, we have presented the study on the flat and random nanostructured ASAHI BRs. Their optical loss mechanisms were investigated and correlated with the thickness of ZnO layer as well as Ag/ZnO interface layer. In this part, hexagonal nanostructured BRs are studied. Figure 5 shows the top view of the BRs with different O_2 plasma etching times (see Figure 1a for the process flow).

BEMA is not appropriate to describe the optical behavior of our hexagonal nanostructured BRs. For such BRs, with a semi-periodic arrangement of the metal/dielectric nanostructures and close-to-visible wavelength inter-nanostructure distance, we expect more plasmonic modes to be present in the system, hence higher optical losses in the hexagonal nanostructured BRs compared with random ASAHI BRs. However, a higher optical loss of the nanostructured BR might not necessarily mean a lower quality. For the nanostructured BRs, light scattering capability is another figure of merit that might be more crucial for one to take into consideration. The diffused reflectance is an indirect way to characterize the scattering capability of the BRs [25]. To further investigate, not only the total reflectance but also the diffused reflectance data were collected.

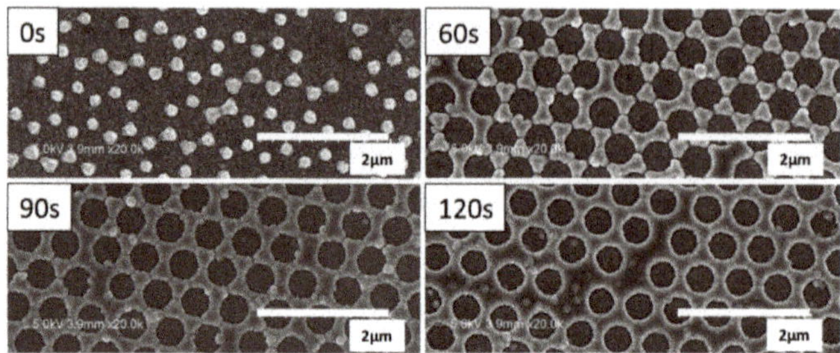

Figure 5. SEM images of hexagonal Ag/ZnO back reflectors obtained with 0, 60, 90, and 120 s of O_2 plasma treatment of polystyrene spheres (See Figure 1a for process flow).

3.2.1. Total and Diffused Reflectance of Hexagonal Ag/ZnO Nanostructured Back Reflectors

Figure 6a,b show the total and diffused reflectance of the hexagonal nanostructured BRs and ASAHI BR with 100 nm ZnO (see Figure 5 for SEM images). Similar optical performance of the different BRs is observed. Comparing with ASAHI BR, we observe higher total reflectance as well as higher diffuse reflectance from around 370 nm to around 570 nm range. This shows that for the above range, hexagonal nanostructured BRs not only have a lower optical loss but also have a stronger light scattering capability. However, from around 570 nm to 840 nm range, which is more critical for light trapping in silicon thin film solar cells, the total reflectance of hexagonal nanostructured BRs is significantly lower than that of the ASAHI BR, especially for the BRs fabricated using O_2 plasma etching. This indicates that, other than Ag/ZnO interface roughness, additional absorptions are present in the hexagonal nanostructured BRs. Due to such significant additional absorptions, the amount of light that is diffused is reduced as well. Hence, for around 570 nm to 700 nm, the diffuse reflectance values of hexagonal nanostructured BRs are lower than ASAHI BR. For the 60 s hexagonal nanostructured BR case, from 700 to 900 nm range, a similar or even higher diffuse reflectance compared to ASAHI BR is obtained (Figure 6b), regardless of the observed higher optical losses on the 60 s hexagonal nanostructured BR. What's more, with the increasing etching time, we notice a decrease in the diffused reflectance, and a red shift in the total reflectance. To understand the source of the reduced reflectance at different local minimum and the effect of the etching time for the losses in total reflectance, the HFSS model was designed and simulation results were extracted and compared with the experiment results.

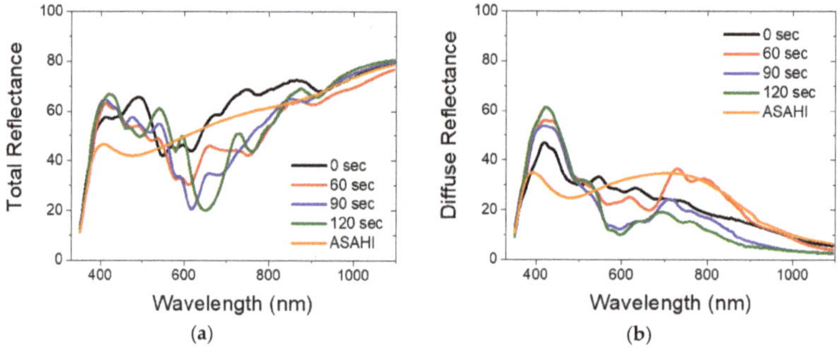

Figure 6. (**a**) Total and (**b**) diffused reflectance of hexagonal Ag/ZnO BRs and ASAHI BR with 100 nm ZnO.

3.2.2. Simulated Absorption of Hexagonal Ag/ZnO Nanostructured Back Reflectors

Figure 7a,b shows the unit cell in the simulation model and corresponding cross sectional view of the BRs we designed in HFSS. Due to the periodic boundary conditions placed onto the six sides of the unit cell, an infinite large area is simulated by repeating the unit cells next to each other. Such hexagonal nanostructure has a fixed periodicity of 600 nm (P), which is the diameter of the PS spheres we used during fabrication. The opening of the nanostructures (*D*) is set to be either 595 nm or 400 nm. At *D* = 595 nm, the opening of the nanostructure is close to the initial diameter of the PS sphere. We use such a condition for the approximation of 0 s etched BR performance estimation. For the 120 s etched BR case, the opening diameter is measured from the SEM image and approximately to be 400 nm. Hence *D* = 400 nm is used for approximation of the 120 s etched BR performance estimation. The ZnO thickness t_1 was set at 100 nm, below Ag hexagonal nanostructures thickness t_2 was set at 100 nm. Bottom bulk Ag thickness was set at 200 nm.

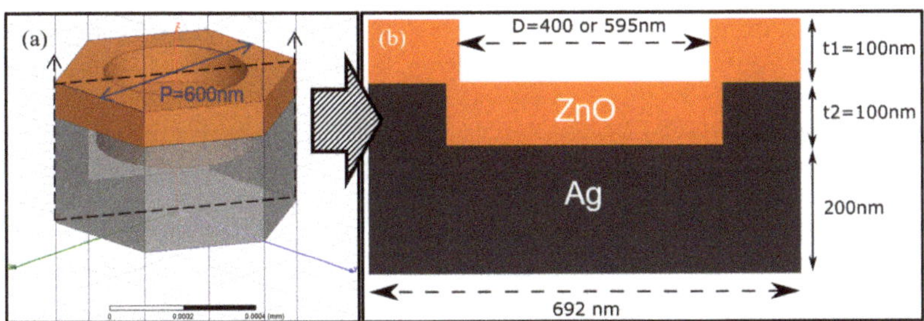

Figure 7. Simulated hexagonal nanostructured back reflector; (**a**) HFSS unit cell in the simulation model and (**b**) corresponding cross sectional view.

Figure 8 shows the comparison of the measured absorptance (1 − *R*) for 0 and 120 s etched BRs and simulated absorption in each layer for 595 and 400 nm opening models. Other than the absorption by ZnO at its optical band gap around 360 nm, four distinctive absorption peaks (a–d highlighted in yellow) are observed both from experiment and simulation results. The peak marked a in Figure 8 is attributed to the ZnO bulk absorption [18]. The peak b is attributed to the Ag plasmonic resonance absorption, which shifts with the opening size. However, this peak did not contribute to overall BR absorption significantly in the experiment. This might be due to the fact that in real devices, the semi-periodic nanostructures did not produce a strong plasmonic effect as compared with the simulation case where they are rigorously periodic. Peaks c and d near 500–600 nm are mainly due to ZnO absorption which was enhanced by the nanostructured Ag below [26]. A consistent red shift of the experimental results with respect to the simulations is observed. This might be due to the minor difference in ZnO film thickness between experiment and simulation model.

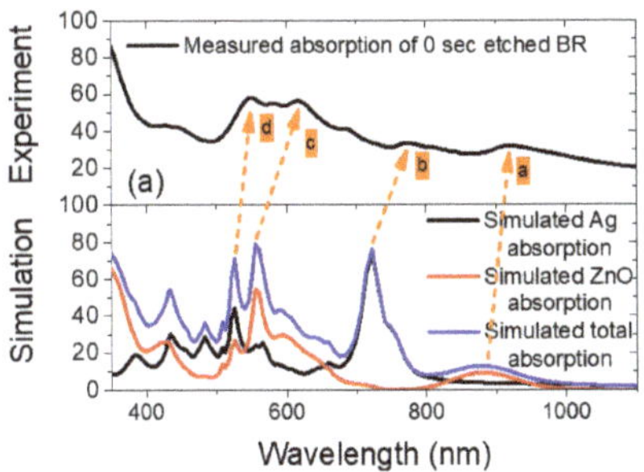

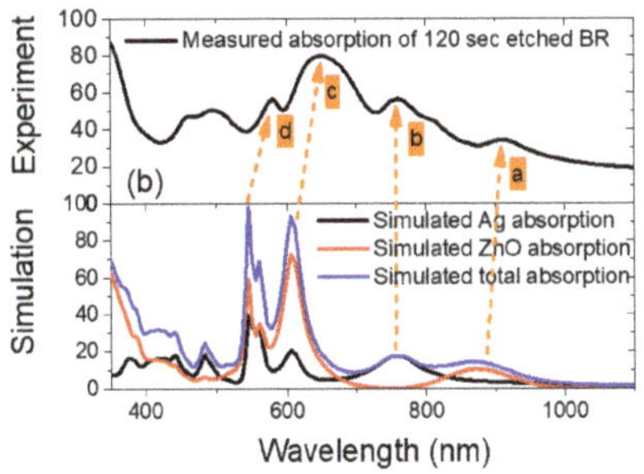

Figure 8. (**a**) 0 s etched BR measured absorption (top) compared with a simulated absorption under 595 nm opening (*D* = 595 nm) simulation case; (**b**) 120 s etched BR measured absorption (top) compared with simulated absorption under 400 nm opening (*D* = 400 nm) simulation case.

With the help of HFSS simulation, we managed to assign each of the optical losses in the periodical hexagonal nanostructured Ag/ZnO BR. Other than the rough Ag/ZnO interface absorption as modelled by the BEMA method, ZnO absorption enhanced by the bottom Ag nanostructure also contributes significantly to the BR absorption. The significant background absorption in NIR region for the experimental results, which differs by about 20% simulation, are attributed to Ag plasmonic absorption due to rough surfaces.

3.3. Experiment and Simulation Results on Hexagonal Ag/ZnO Back Reflector Solar Cells

To investigate the potential parasitic losses from the BRs in solar cell devices as well as to understand different light scattering modes that appeared due to such nanostructures, 25 nm n-type amorphous silicon, 230 nm intrinsic amorphous silicon, 25 nm p-type amorphous silicon, as well as 80 nm top ITO are added on top of the 400 nm opening BR. The 400 nm opening of the BR and final 240 nm opening of the finished device is measured from SEM images in Figures 1b and 2a respectively. This model simulates the 120 s O_2 plasma etched BR and its corresponding solar cell. Due to additional layers added in the model, parasitic losses will be introduced not only by the BRs, but also from ITO and the recombination in highly n/p doped layers. Indeed, those layers will also absorb light but will not contribute to photo-current generation. Figure 9a shows the unit cell and the corresponding cross section of the simulation model of a hexagonal BR solar cell. The overall absorption and individual layers' absorption are simulated.

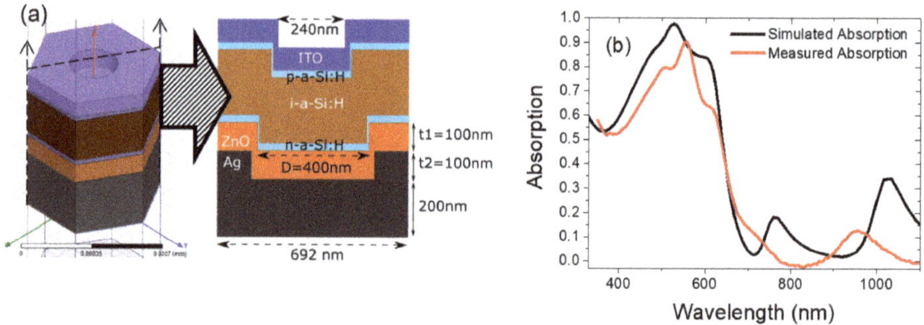

Figure 9. (**a**) High frequency electromagnetic field simulator (HFSS) simulation model for hexagonal BR solar cell; (**b**) Simulated and measured absorption of the plannar Ag/ZnO BR solar cell.

To validate our simulation results, simulated overall absorptance, and measured absorptance (1–measured reflectance) of the flat Ag/ZnO BR solar cell device are plotted in Figure 9b. We can observe similar absorption trend between simulation and experiment. Again experimental results are slightly red shifted, due to the minor difference in film thickness (including ITO, ZnO, and silicon thin films) between experiment and simulation model. Additional contribution to increased absorptance in measured data might be reduced reflectance due to surface roughness, which were not been included in the simulation. In the next section, we present the simulation results of hexagonal nanostructured Ag/ZnO BR solar cells.

3.3.1. Parasitic Losses and Absorption Enhancement Analysis

To analyze the parasitic losses in the different layers of the n-i-p solar cell and the absorption enhancement in the intrinsic silicon layer, the base line for us to compare is the simulated planar device. Based on the difference between planar and nanostructured solar cells, solid conclusions on the parasitic losses, absorption enhancement, and light scattering modes can be drawn. Figure 10a–d shows the simulated ITO, p-type silicon, n-type silicon, and BR absorption for both planar and nanostructured solar cells. From the results we can see that after the solar cells are nanostructured, ITO absorption is greatly enhanced, p layer absorption is reduced, and n layer absorption is reduced from 500 to 650 nm but enhanced from 650 to 720 nm range. For the Ag and ZnO layers, our results show that in planar BR solar cell, Ag almost does not absorb any light and minimum absorption around 620 nm is observed for ZnO. For the nanostructured BR solar cell, ZnO absorption is reduced and red shifted to 690 nm. Two peaks at 700 and 760 nm are observed for Ag. These two absorption peaks will contribute to parasitic absorption as well. Unlike the above results where high absorption is observed for the BRs in air medium, when silicon active layers are deposited on top, BR parasitic absorption is less important as compared to the top ITO parasitic absorption.

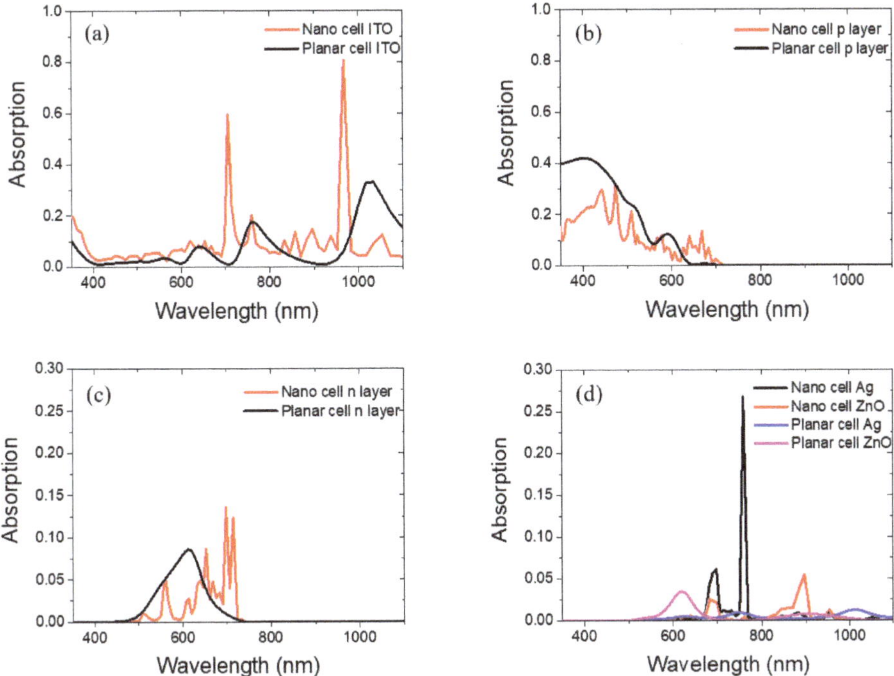

Figure 10. (**a**) Indium tin oxide (ITO) absorption; (**b**) p layer absorption; (**c**) n layer absorption; (**d**) Ag/ZnO BR absorption.

To study the absorption enhancement effect, intrinsic amorphous silicon absorption is plotted in Figure 11a for planar and nanostructured cases. We can observe that the absorption in the intrinsic layer for the nanostructured solar cell is significantly enhanced not only in the short wavelength range (350–520 nm) but also for longer wavelength range (550–720 nm). With referring to the formula below [27] and by using 1.71 eV band gap for amorphous silicon and the AM1.5G solar spectrum, we calculated the generated photo-current density to be 18.3 mA/cm^2 and 15.0 mA/cm^2 for a nanostructured and planar solar cell respectively. In the formula below, $I(\lambda)$ is the spectral irradiance of the standard AM1.5G solar spectrum, $A(\lambda)$ is the absorption in the active absorber layer, and $E(\lambda)$ is the energy of the photon at its corresponding wavelength:

$$J_{SC} = \int_{300\ nm\ to\ 1200\ nm} \{[E_g \times I(\lambda) \times A(\lambda)]/E(\lambda)\}\ d\lambda \qquad (?)$$

To characterize the light resonance mode for the nanostructured solar cell, we selected 670 nm where a sharp absorption enhancement peak is observed in Figure 11a and we compared the electric field distribution in the nanostructured and planar solar cell. Figure 11b shows the electric field distribution at 670 nm incident wavelength. From Figure 11b, we can clearly see the light trapping effect of the nanostructured solar cell. Multiple "hot spots" are induced. The generation of those "hot spots", that is, high electric field strength regions, is due to interferences and waveguide modes. Strong resonance modes are induced inside intrinsic layer. Hence the absorption in the intrinsic layer is strongly enhanced.

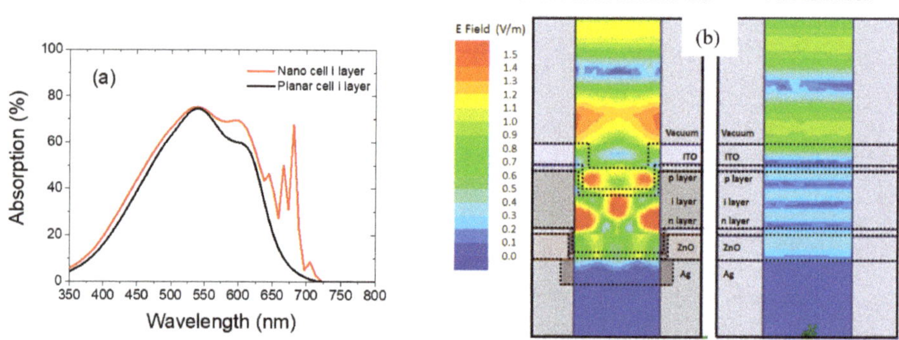

Figure 11. (**a**) Simulated intrinsic amorphous silicon absorption; (**b**) E field distribution of nanostructured and planar solar cell at 670 nm.

3.3.2. Experimental Results

Last but not least, we fabricated solar cells using conventional 1 µm ZnO on glass substrates, Ag/ZnO flat BR substrates, and hexagonal nanostructured Ag/ZnO BR substrates. The experimental results are compared and analyzed. Figure 12a shows the *J-V* characteristics and a table summarizing their parameters. Figure 12b shows the EQE and optical absorption data (1–total reflectance) of the planar Ag/ZnO BR and hexagonal nanostructured Ag/ZnO BR solar cell.

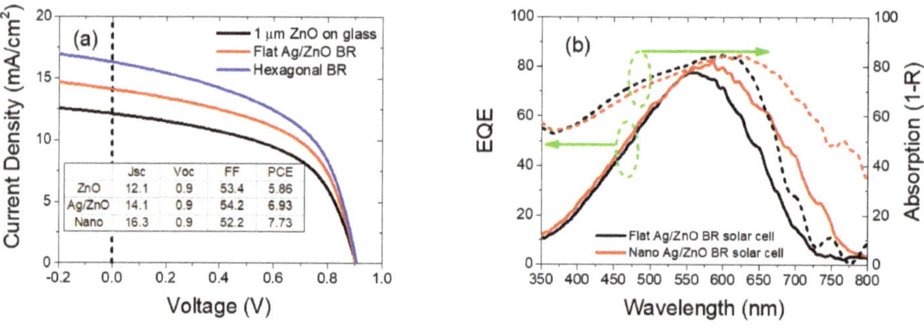

Figure 12. (**a**) Density-voltage (*J-V*) and solar cell parameters of different solar cells; (**b**) external quantum efficiency (EQE) and absorption (1 – R_total) of nanostructured and planar Ag/ZnO BR solar cells.

From Figure 12a, we observe an increasing short-circuit current density. With the incorporation of flat Ag/ZnO BR, 2.0 mA/cm^2 of short-circuit current density are gained compared to the transparent ZnO on a glass substrate. The current collection is enhanced by 16.5%. With the incorporation of hexagonally nanostructured Ag/ZnO BR, not only is the open-circuit voltage not compromised but also the short-circuit current density is further enhanced by 2.2 mA/cm^2. The total Jsc enhancement compared to the reference transparent flat ZnO substrate is as high as 34.7%. From EQE and absorption results in Figure 12b, we observed the enhanced light absorption from 550 to 750 nm wavelength range. This indicates an improved light collection efficiency by incorporating the nanostructured BR into the flat BR device, which leads to an enhanced short-circuit current. Notably, for the flat Ag/ZnO BR substrate we find good agreement between the measured short-circuit current density in the experiment and the calculated short-circuit current by simulation (6% difference). However, simulation predicted an 18.3 mA/cm^2 short-circuit current density for the hexagonal nanostructured solar cell, while experimentally we only obtained 16.3 mA/cm^2 (10.9% difference). Such difference is expected, as in our ideal simulation model, electrical resistivity, defects at the n/i interface, i/p interface, as well as bulk defects in the intrinsic layer are not considered and the collection efficiency of charge carriers is assumed to be 100%. Furthermore, the nanostructured solar cells have a larger contact area compared to the planar solar cells. Hence we would expect more interface defects, which leads to a larger simulation to experiment difference for the nanostructured solar cell than the flat solar cell. From Figure 12b, for the EQE of our real devices, resonance modes are broadened due to the semi-periodic arrangement of the nanostructures and a smooth and continuous light trapping effect from 550 to 750 nm is observed. While in simulated intrinsic silicon absorption, strong resonances at specific wavelengths contribute significantly to the collected current. This is due to the periodicity of the feature being rigorously achieved during the calculation, hence strong light trapping effects are observable only at specific wavelengths. Paetzold et al. [28] reported a series of systematic study and proved that disordered periodic are superior than the rigorous periodic nanostructured BRs for silicon thin film solar cell. While in our study, the reasons why the experimental results are not better than the simulation results are mainly due to the imperfect and unrealistic simulation model which excluded electrical losses. Even though we cannot conclude that our semi periodic nanostructured BRs are better than the rigorous periodic nanostructured BRs, nevertheless, the method used in this work is a good demonstration to achieve a quasi-periodic nanostructures using cost effective PS sphere assisted lithography. The distorter that contribute to the quasi-periodic nanostructures in our work is originated from the deviation of close-packed hexagonal assembly of PS spheres, as well as deviation from oxygen plasma etching of the PS materials.

4. Conclusions

In this work, we have demonstrated a novel hexagonal nanostructured design and fabrication process for light trapping in silicon thin film solar cells. The hexagonal nanostructured back reflectors and corresponding solar cells were modeled by the finite element method and verified against experimental data. The parasitic losses in the back reflector mainly come from the rough Ag/ZnO interface plasmonic absorption and enhanced ZnO absorption by Ag nanostructures. The high frequency electromagnetic field simulator was proven to be an effective tool for modeling our hexagonal nanostructured back reflectors and solar cells. Simulation results show similar trends with experimental data. From simulation, parasitic losses in our nanostructured solar cells mainly come from the ITO top contact layer, while back reflector parasitic losses are less significant. Calculated photo generated current values are close to the experimental short-circuit current, with 6.0% and 10.9% differences for the planar and nanostructured solar cells respectively. Such a difference is attributed to certain parameters that were not considered in the simulation, such as electrical resistivity, defects in the bulk and interfaces. Furthermore, our fabrication approach is not limited to a-Si:H silicon thin film solar cells, but can also be applied to μc-Si:H silicon thin film solar cells. As a low cost, robust, and process compatible method, these nanostructured BR strategies are an effective candidate for improved light trapping performance in silicon thin film solar cells.

Author Contributions: Conceptualization, M.F., J.W., W.C., and P.iR.C.; Data curation, Z.L., R.K., and P.B.; Formal analysis, Z.L.; Funding acquisition, R.E. and P.iR.C.; Investigation, Z.L., C.L., M.F., and P.iR.C.; Methodology, M.F.; Resources, R.E., R.K., P.B., W.C., E.J., and P.iR.C.; Software, C.L. and A.B.P.; Supervision, R.E. and P.iR.C.; Validation, M.F., J.W., W.C., E.J., and P.iR.C.; Writing original draft, Z.L.; Writing review and editing, R.E., C.L., A.B.P., M.F., R.K., P.B., J.W., W.C., E.J., and P.iR.C.

Acknowledgments: The authors thank Jérôme Charliac and Jacqueline Tran, LPICM, CNRS, Ecole Polytechnique, for their help and support with this work.

Conflicts of Interest: The authors declare no conflict of interest.

References

1. Sai, H.; Matsui, T.; Matsubara, K. Stabilized 14.0%-efficient triple-junction thin-film silicon solar cell. *Appl. Phys. Lett.* **2016**, *109*, 183506.

2. Chen, X.; Jia, B.; Saha, J.K.; Cai, B.; Stokes, N.; Qiao, Q.; Wang, Y.; Shi, Z.; Gu, M. Broadband enhancement in thin-film amorphous silicon solar cells enabled by nucleated silver nanoparticles. *Nano Lett.* **2012**, *12*, 2187–2192. [CrossRef] [PubMed]

3. Khazaka, R.; Moulin, E.; Boccard, M.; Garcia, L.; Hänni, S.; Haug, F.J.; Meillaud, F.; Ballif, C. Silver versus white sheet as a back reflector for microcrystalline silicon solar cells deposited on LPCVD-ZnO electrodes of various textures. *Prog. Photovolt. Res. Appl.* **2015**, *23*, 1182–1189. [CrossRef]

4. Bhattacharya, J.; Chakravarty, N.; Pattnaik, S.; Slafer, W.D.; Biswas, R.; Dalal, V.L. A photonic-plasmonic structure for enhancing light absorption in thin film solar cells. *Appl. Phys. Lett.* **2011**, *99*, 131114. [CrossRef]

5. Smeets, M.; Smirnov, V.; Meier, M.; Bittkau, K.; Carius, R.; Rau, U.; Paetzold, U.W. On the geometry of plasmonic reflection grating back contacts for light trapping in prototype amorphous silicon thin-film solar cells. *J. Photonics Energy* **2014**, *5*, 057004. [CrossRef]

6. Li, Z.; Rusli, E.; Foldyna, M.; Wang, J.; Chen, W.; Prakoso, A.B.; Lu, C.; Cabarrocas, P.R. Nanostructured back reflectors produced using polystyrene assisted lithography for enhanced light trapping in silicon thin film solar cells. *Sol. Energy* **2018**, *167*, 108–115. [CrossRef]

7. Jang, Y.H.; Jang, Y.J.; Kim, S.; Quan, L.N.; Chung, K.; Kim, D.H. Plasmonic Solar Cells: From Rational Design to Mechanism Overview. *Chem. Rev.* **2016**, *116*, 14982–15034. [CrossRef] [PubMed]

8. Dahal, L.R.; Sainju, D.; Podraza, N.J.; Marsillac, S.; Collins, R.W. Real time spectroscopic ellipsometry of Ag/ZnO and Al/ZnO interfaces for back-reflectors in thin film Si:H photovoltaics. *Thin Solid Films* **2011**, *519*, 2682–2687. [CrossRef]

9. Wienke, J.; van der Zanden, B.; Tijssen, M.; Zeman, M. "Performance of spray-deposited ZnO:In layers as front electrodes in thin-film silicon solar cells. *Sol. Energy Mater. Sol. Cells* **2008**, *92*, 884–890. [CrossRef]

10. Paetzold, U.W.; Moulin, E.; Pieters, B.E.; Carius, R.; Rau, U. Design of nanostructured plasmonic back contacts for thin-film silicon solar cells. *Opt. Express* **2011**, *19*, A1219–A1230. [CrossRef] [PubMed]

11. Zhang, C.; Guney, D.O.; Pearce, J.M. Plasmonic enhancement of amorphous silicon solar photovoltaic cells with hexagonal silver arrays made with nanosphere lithography. *Mater. Res. Express* **2016**, *3*, 10. [CrossRef]

12. Isabella, O.; Solntsev, S.; Caratelli, D.; Zeman, M. 3-D optical modeling of thin-film silicon solar cells on diffraction gratings. *Prog. Photovolt. Res. Appl.* **2013**, *21*, 94–108. [CrossRef]

13. Khoury, R.; Li, Z.; Bulkin, P.; Roca i Cabarrocas, P.; Johnson, E.V. Detailed study of Electron Cyclotron Resonance Oxygen Plasma Etching of Polystyrene Nanosphere Arrays. Unpublished work.

14. Roca i Cabarrocas, P.; Chevrier, B.J.; Huc, J.; Lloret, A.; Parey, Y.J.; Schmitt, M.J.P. A Fully Automated Hot-wall Multiplasma monochamber Reactor for Thin Film Deposition. *J. Vac. Sci. Technol. A* **1991**, *9*, 2331–2341. [CrossRef]

15. Abolmasov, S.N.; Woo, H.; Planques, R.; Holovský, J.; Johnson, E.V.; Purkrt, A.; Roca i Cabarrocas, P. Substrate and p-layer effects on polymorphous silicon solar cells. *Eur. Phys. J. Photovolt.* **2014**, *5*, 55206. [CrossRef]

16. Kim, K.H.; Kasouit, S.; Johnson, E.V.; Cabarrocas, P.R.I. Substrate versus superstrate configuration for stable thin film silicon solar cells. *Sol. Energy Mater. Sol. Cells* **2013**, *119*, 124–128. [CrossRef]

17. Sainju, D. Spectroscopic Ellipsometry Studies of Ag and ZnO Thin Films and Their Interfaces for Thin Film Photovoltaics. Ph.D. Thesis, University of Toledo, Toledo, USA, 2015.

18. Dejneka, A.; Aulika, I.; Makarova, M.V.; Hubicka, Z.; Churpita, A.; Chvostova, D.; Jastrabik, L.; Trepakov, V.A. Optical Spectra and Direct Optical Transitions in Amorphous and Crystalline ZnO Thin Films and Powders. *J. Electrochem. Soc.* **2010**, *157*, G67–G70. [CrossRef]

19. Moulin, E.; Paetzold, U.W.; Kirchhoff, J.; Bauer, A.; Carius, R. Study of detached back reflector designs for thin-film silicon solar cells. *Phys. Status Solidi Rapid Res. Lett.* **2012**, *6*, 65–67. [CrossRef]

20. Sainju, D.; Van Den Oever, P.J.; Podraza, N.J.; Syed, M.; Stoke, J.A.; Chen, J.; Yang, X.; Deng, X.; Collins, R.W. Origin of optical losses in Ag/ZnO back-reflectors for thin film Si photovoltaics. In Proceedings of the Conference Record of the 2006 IEEE 4th World Conference on Photovoltaic Energy Conversion, Waikoloa, HI, USA, 7–12 May 2006; pp. 1732–1735.

21. Owen, J.I. *Growth, Etching, and Stability of Sputtered ZnO:Al for Thin-Film Silicon Solar Cells*; Forschungszentrum Jülich: Jülich, Germany, 2011.

22. McPeak, K.M.; Jayanti, S.V.; Kress, S.J.P.; Meyer, S.; Iotti, S.; Rossinelli, A.; Norris, D.J. Plasmonic films can easily be better: Rules and recipes. *Am. Chem. Soc. Photonics* **2015**, *2*, 326–333. [CrossRef] [PubMed]

23. Jovanov, V.; Xu, X.; Shrestha, S.; Schulte, M.; Hüpkes, J.; Zeman, M.; Knipp, D. Influence of interface morphologies on amorphous silicon thin film solar cells prepared on randomly textured substrates. *Sol. Energy Mater. Sol. Cells* **2013**, *112*, 182–189. [CrossRef]

24. Springer, J.; Poruba, A.; Müllerova, L.; Vanecek, M.; Kluth, O.; Rech, B. Absorption loss at nanorough silver back reflector of thin-film silicon solar cells. *J. Appl. Phys.* **2004**, *95*, 1427–1429. [CrossRef]

25. Mendes, M.J.; Morawiec, S.; Simone, F.; Priolo, F.; Crupi, I. Colloidal plasmonic back reflectors for light trapping in solar cells. *Nanoscale* **2014**, *6*, 4796–4805. [CrossRef] [PubMed]

26. Ko, Y.H.; Yu, J.S. Optical absorption enhancement of embedded Ag nanoparticles with ZnO nanorod arrays. *Phys. Status Solidi* **2011**, *208*, 2778–2782. [CrossRef]

27. Tiedje, T.; Yablonovitch, E.; Cody, G.D.; Brooks, B.G. Limiting Efficiency of Silicon Solar Cells. *IEEE Trans. Electron Devices* **1984**, *31*, 711–716. [CrossRef]

28. Paetzold, U.W.; Smeets, M.; Meier, M.; Bittkau, K.; Merdzhanova, T.; Smirnov, V.; Michaelis, D.; Waechter, C.; Carius, R.; Rau, U. Disorder improves nanophotonic light trapping in thin-film solar cells. *Appl. Phys. Lett.* **2014**, *104*, 131102. [CrossRef]

Article

Influence of InAlN Nanospiral Structures on the Behavior of Reflected Light Polarization

Yu-Hung Kuo, Roger Magnusson, Elena Alexandra Serban, Per Sandström, Lars Hultman, Kenneth Järrendahl, Jens Birch and Ching-Lien Hsiao *

Thin Film Physics Division, Department of Physics, Chemistry, and Biology (IFM), Linköping University, SE-581 83 Linköping, Sweden; kuoknightly@gmail.com (Y.-H.K.); roger.magnusson@liu.se (R.M.); alese81@ifm.liu.se (E.A.S.); per.o.sandstrom@liu.se (P.S.); larhu@ifm.liu.se (L.H.); kenneth.jarrendahl@liu.se (K.J.); jbh@ifm.liu.se (J.B.)
* Correspondence: hcl@ifm.liu.se

Received: 14 February 2018; Accepted: 9 March 2018; Published: 12 March 2018

Abstract: The influence of structural configurations of indium aluminum nitride (InAlN) nanospirals, grown by reactive magnetron sputter epitaxy, on the transformation of light polarization are investigated in terms of varying structural chirality, growth temperatures, titanium nitride (TiN) seed (buffer) layer thickness, nanospiral thickness, and pitch. The handedness of reflected circularly polarized light in the ultraviolet–visible region corresponding to the chirality of nanospirals is demonstrated. A high degree of circular polarization (P_c) value of 0.75 is obtained from a sample consisting of 1.2 μm InAlN nanospirals grown at 650 °C. A film-like structure is formed at temperatures lower than 450 °C. At growth temperatures higher than 750 °C, less than 0.1 In-content is incorporated into the InAlN nanospirals. Both cases reveal very low P_c. A red shift of wavelength at P_c peak is found with increasing nanospiral pitch in the range of 200–300 nm. The P_c decreases to 0.37 for two-turn nanospirals with total length of 0.7 μm, attributed to insufficient constructive interference. A branch-like structure appears on the surface when the nanospirals are grown longer than 1.2 μm, which yields a low P_c around 0.5, caused by the excessive scattering of incident light.

Keywords: InAlN; nanospiral; metamaterial; sputtering; chirality

---

1. Introduction

Chirality-induced polarization effect in the cuticle of scarab beetles, such as *Cetonia aurata* and *Chrysina argenteola*, is well-known to reflect light with brilliant color and a high degree of circular polarization [1–3]. The reflected circularly polarized light from an incident unpolarized light toward this biological structure is due to its exo-skeleton, which consists of chitin-based layers, each progressively rotated by a small twist angle θ with respect to the previous one. Such a naturally helicoidal structure inspires studies of the optical polarization in synthetic chiral nanospirals or a twistedly layer-stacking helicoidal structure for applications such as circular polarizers, bandpass filters, and handedness converters. These are used for, e.g., fiber optical communication, three-dimensional (3D) glasses, and autostereoscopy [4–7].

Efforts have been devoted to the fabrication and exploration of various chiral structures and materials by methods including glancing angle deposition (GLAD) [8,9], direct laser writing (DLW) [10,11], and holographic lithography (HL) [12]. Different materials can be used to fabricate chiral structures for such polarization-sensitive optical materials, including oxides and fluorides [13,14], gold [11], liquid crystals [15], and silicon nitride [16]

GLAD is the most common technique for the deposition of dielectric chiral materials in the form of either self-assembled or periodically helical columns, including spirals. The mechanism of producing circularly polarized light is described by the interaction between light and the periodic structure, induced by a Bragg phenomenon [17,18]. Accordingly, chirality may be tailored to transform unpolarized incident light to left- or right-handed circularly-polarized light. Large-pitch and long nanostructures with a suite of periods up to several micrometers is required to produce a high degree of circular polarization. This is, however, challenging when using the GLAD technique since broadening and branching of the columns often occur as they grow thicker [13,14].

The DLW method, on the other hand, uses the multi-photon absorption and focused-point laser light with a high magnification objective to write directly into a photoresist. However, at the sharp focal point the intensity of the laser light may lead to local polymerization in areas over one hundred nanometers. This limits the sample size to tens of micrometers, and is thus not economical, both in terms of material usage and process time.

Holographic lithography (HL) is another method, which combining holography and photo-induced polymerization techniques to produce uniform periodic as well as quasi-periodic large 3D structures in photoresist in the optical range [12]. However, this technique is limited to patterning arrayed features or uniformly distributed periodic patterns only. Hence, for the fabrication of arbitrarily shaped patterns, other techniques are required.

Most recently, curved-lattice epitaxial growth (CLEG) [19,20] was demonstrated to produce unique chiral structures by tailoring the nanospiral geometry and internal grading composition of indium aluminum nitride (InAlN) semiconductors. These were demonstrated to have a very high degree of circular polarization (P_c) in ultraviolet and visible ranges [20–22]. A remarkable advantage of CLEG nanorods is their intrinsic curvature, which is fundamentally different from bent rods; bent rods result from a sharp interface between the two different materials with different lattice parameters (such as AlN: 3.11Å and InN: 3.54Å) and coefficients of thermal expansions grown together side-by-side rather than a gradual compositional change. Thus, the latter has a very large internal stress and strain due to lattice mismatch and strained lattices at the interfacial boundary between two materials. In contrast, CLEG curved nanorods have negligible internal stress and strain due to their graded lattice constant and form an epitaxial structure with internal chirality. Furthermore, the lateral compositional gradient can lead to lateral gradients in optical properties. However, detailed information with regard to the effect of CLEG nanospiral structural configurations on the transformation of light polarization is insufficient, which motivates the present study.

The goal of this study is to gain more information about how different growth conditions (nanospiral structural arrangement, shape, geometric confinement, and chemical composition) affect the relationship between the exhibiting circular polarization state and its corresponding wavelength of light reflection. InAlN nanospirals were thus grown by reactive magnetron sputter epitaxy (MSE) with varying external chirality, growth temperatures, TiN seed thickness, nanospiral thickness, and pitch. The optical properties of the resultant nanospirals were analyzed by Mueller matrix spectroscopic ellipsometry (MMSE) and the CompleteEASE software [21,22]. The main P_c peak wavelength and band of reflected light were found to be highly correlated with the pitch, length, turn, and composition of the grown InAlN nanospirals. In addition, the possible mechanisms of optical response behavior were discussed.

2. Experimental Details

An ultra-high-vacuum (UHV) MSE deposition chamber equipped with two 50 mm-diameter and two 75 mm-diameter targets was utilized to grow InAlN alloys [20,23,24]. The chamber was evacuated to a base pressure of $<3 \times 10^{-9}$ Torr with a combination of turbomolecular and mechanical pumps. High-purity 75 mm-diameter Al (99.999%) and 50 mm-diameter In (99.999%) targets were used to co-sputter ternary InAlN under pure nitrogen ambient, supplied as pure nitrogen gas (99.999999%) achieved through a getter purifier. Prior to InAlN growth, a titanium nitride (TiN) seed layer was grown on *c*-plane-oriented sapphire substrates at 850 °C. Typical dc-magnetron powers provided for In, Al, and Ti targets were of 10, 300, and 250 W, respectively. During the sputtering process, a negative substrate bias of 30 V was applied to the sample holder to enhance growth by low-energy ion assistance. Reference [23] A temporal control of substrate rotating angles and the azimuthal orientation of the inclined deposition fluxes were used to tailor the twist-angle θ of segments and chirality of the nanospirals, respectively. Here, $\theta = 90°$ so that four segments completes one turn. More details of CLEG nanospiral growth can be found elsewhere [20].

The composition, crystal structure, and growth plane of the sample surface were determined by θ–2θ scan X-ray diffraction (XRD) using a Philips PW1820 powder diffractometer with a Cu-K$_\alpha$ X-ray lab source. The morphology and surface geometry of the samples were characterized by a LEO-1550 field-emission scanning electron microscope (FE-SEM).

All optical measurements were done in the spectral range of 245–1000 nm at incident angles of 25° using a Mueller matrix spectroscopic ellipsometer (MMSE) produced by J.A. Woollam Co., Inc. (Lincoln, NE, USA) The CompleteEASE software was used to analyze the data measured by MMSE [21,22]. The light polarization state in MMSE measurement was obtained through $\mathbf{S}_o = \mathbf{M}\mathbf{S}_i$ [11,21], where $\mathbf{S}_o = [S_{o0}, S_{o1}, S_{o2}, S_{o3}]^T$ and $\mathbf{S}_i = [S_{i0}, S_{i1}, S_{i2}, S_{i3}]^T$ are the polarization states of outgoing and incident lights, respectively, described with the Stokes vector, and $\mathbf{M}$ is the full 16-element Mueller matrix. For an unpolarized incident light, $\mathbf{S}_i = [1, 0, 0, 0]^T$, the outgoing Stokes vector, will be the same as the first column in the normalized Mueller matrix, $\mathbf{S}_o = [1, m_{21}, m_{31}, m_{41}]^T (m_{11} = 1)$. An important aspect of the nanospirals in this report is their chirality and potential for transforming unpolarized light into circularly polarized light upon reflection. Thus, the total degree of circular polarization $P_C = \frac{S_3}{S_0}$ of the reflected light is presented.

3. Results and Discussion

3.1. Nanospiral Chirality

Figure 1a,b show side-view SEM images of right- and left-handed InAlN nanospirals, respectively, grown on sapphire substrates assisted with a 120-nm TiN seed layer under identical conditions except for the rotational sense. These chiral nanospirals consisted of four complete turns with a designed pitch of 250 nm, as shown in the images. However, the first pitch is slightly shorter and yields thinner spirals than those in the subsequent pitches, which is attributed to incubation from the self-induced formation of nuclei followed by a coalescence and coarsening process until steady-state growth is achieved in the axial direction [25–27]. The crystalline structure of these nanospirals characterized by $\theta/2\theta$ scan XRD are shown in Figure 1c. Reflections of InAlN(0002), TiN(111), and Al$_2$O$_3$(0006) are exclusively present in this scan, located at 35.1, 36.9, and 42.7°, respectively. The SEM and XRD results indicate that these InAlN chiral nanospirals grow preferentially along the c-axis and have the same In content of 0.19, as calculated by Vegard's law without taking strain effect into consideration [23]. The circular polarization measurements of the samples are performed by MMSE. The opposite chirality of the P_c curves measured from two samples with different handedness can be clearly seen in Figure 1d. The mirroring shape of the two curves is proof of high reproducibility in the MSE growth process. Though the values of P_c are slightly different between the samples; the high P_c of around 0.75 at 417 nm indicates that the CLEG InAlN chiral nanospirals are very promising for use as high performance chiral optical elements. It should be noted that the high P_c values are an indication that these advanced structures of nanospirals, with complex internal crystal structures instead of homogenous layers [20], cannot be fully modeled using an ideal stratified 90° twisted structure, since that would suggest suppression of the chiral Bragg peaks. While the full modeling of the structures is beyond the scope of the present work, it is the topic of an extensive study, including a larger set of rotation schemes, which is under way.

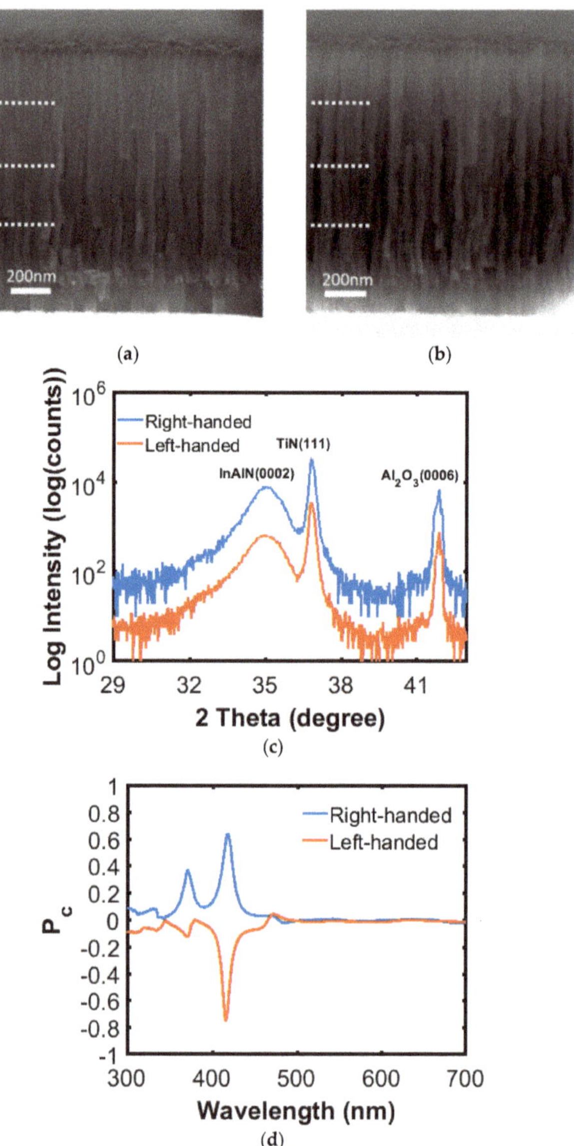

Figure 1. SEM images of (**a**) right-handed and (**b**) left-handed indium aluminum nitride (InAlN) nanospiral film; (**c**) X-ray diffraction (XRD) patterns of right- and left-handed nanospirals; (**d**) P_c versus wavelength of the light reflected from nanospirals with opposite handedness. The negative and positive P_c is referred to left- and right-handed circularly polarized light, respectively.

3.2. InAlN Growth Temperature

Figure 2 shows plan-view SEM images of InAlN grown at temperatures of 450, 550, 650, and 750 °C. Except for the temperature, all samples were grown under the same controlled conditions where left-handed nanospirals where formed by making four complete turns with a transient substrate rotation angle θ of 90° between segments. All TiN seed layers had a thickness of 60 nm.

Figure 2. Plan-view SEM images of InAlN nanospirals grown at (**a**) 750 °C; (**b**) 650 °C; (**c**) 550 °C; and (**d**) 450 °C, respectively. All images use the same scale bar as Figure 2a.

At growth temperatures of 650 and 750 °C, well-separated hexagonal nanospirals are formed, having a well-defined diameter with little deviation as can be seen in Figure 2a,b. When reducing the temperature to 550 °C, the top of the nanospirals becomes tapered with the diameter varying widely from 10 to 50 nm, as shown in Figure 2c. At temperatures lower than 450 °C, the InAlN initially grows with nanospiral geometry (not shown here), but then coalesces, hence forming a thin-solid film-like nanostructure with size bigger than 200 nm, as shown in Figure 2d. In addition, the thickness of the film-like nanostructure is thinner than that of the well-separated nanospirals. Since the volume of the deposited material is the same, a higher growth rate is expected for the well-separated nanostructures. However, a trend of decreasing nanospiral total thickness from 1.19 to 0.85 μm with increasing temperature from 550 to 750 °C is observed, which is attributed to a lower In incorporation rate at a higher growth temperature [26,27].

The InAlN growth temperature affects not only the morphology, but also the nanospiral material's optical properties. Figure 3 shows a plot of P_c versus wavelength for five different temperatures. Distinct P_c peaks can be seen when the growth temperature is in the range of 550–750 °C. The highest $|P_c|$ of approximately ~0.75 is obtained for the sample grown at 650 °C. $|P_c|$ decreases to 0.5 for the 550 °C sample, and more dramatically to 0.1 at 750 °C. Almost no P_c peak can be identified for the samples grown at temperatures lower than 450 °C. As seen in the SEM images, the sample grown at 450 °C loses the separated nanospiral geometry owing to coalescence, which removes the conditions for constructive Bragg reflection [17,18]. However, the spiral form is retained for the 750 °C sample but its In-content, as determined by Vegard's law, is found to be around 0.09. This is substantially less than 0.17 ± 0.02 for samples grown at 650 °C or lower, owing to less efficient In incorporation during high temperature growth [26–28]. Since lower In incorporation results in a less lateral compositional gradient in the nanospirals, the birefringence contributed from the internal compositional gradient becomes less effective on P_c [20–22]. This consequence of low P_c may indicate that the high $|P_c|$ obtained from the 650 °C sample is attributed to both external nanospiral geometry and internal compositional gradient [20]. In addition, the wavelength λ for the $|P_c|$ peak value versus growth temperature is plotted in the inset of Figure 3. The variation of λ shows a trend of shifting towards a shorter wavelength with increasing growth temperature, which can be attributed to a decreased pitch of the nanospirals (due to reduced In incorporation), and is in agreement with the circular

Bragg reflection condition $\lambda = p\bar{n}$, where p is the pitch of nanospirals and $\bar{n}$ is the effective refractive index [17,18].

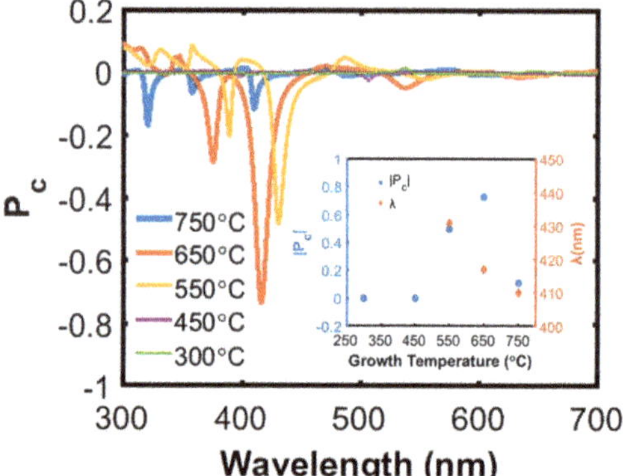

Figure 3. Effect of InAlN nanospiral growth temperature on P_c with respect to wavelength. The inset shows $|P_c|$ and the corresponding peak wavelength versus growth temperature.

3.3. TiN Seed Layer

In this section, we studied the effect of the TiN seed layer thickness on $|P_c|$ and the corresponding peak wavelength. The thickness of the seed layer was varied from 16 to 120 nm, while the InAlN nanospiral growth was kept constant using a growth temperature of 650 °C. As can be seen in Figure 4, a large increase of $|P_c|$ from 0.37 to 0.69 is obtained with increasing TiN thickness from 16 to 30 nm. A further increase of thickness to 120 nm gives slightly increased $|P_c|$ to 0.75. Also, the peak wavelength λ changes and shifts to longer wavelengths with increasing buffer thickness but remains constant when the buffer is thicker than 60 nm. Since a high $|P_c|$ results from a strong constructive Bragg reflection, we can speculate that a thicker and thus more reflecting TiN buffer layer minimizes the depolarization and incoherency occurring in a thinner and semi-transparent buffer layer. TiN is reported to be almost opaque for thicknesses larger than 160 nm [29], but already at a thickness of 60 nm most of the light should be reflected from the TiN buffer layer. Together with the observation that the peak wavelength shifts to higher values, we conclude that a larger pitch of well-defined InAlN nanospirals is promoted by the thicker buffer layers.

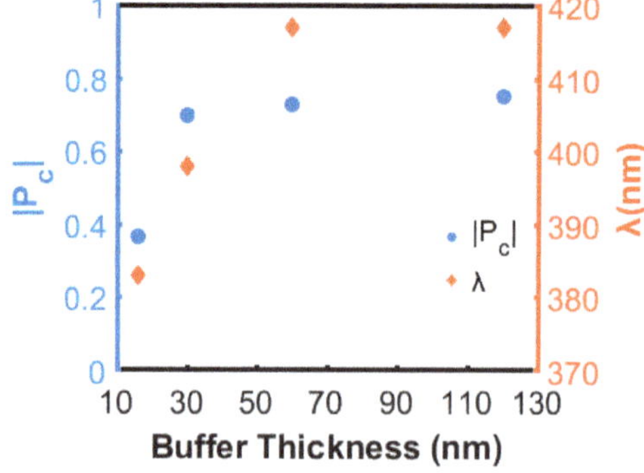

Figure 4. Plot of $|P_c|$ and corresponding peak wavelength as a function of titanium nitride (TiN) buffer thickness.

3.4. Total Thickness and Pitch of Nanospirals

This set of nanospirals was deposited on top of 60 nm-thick TiN buffer, with little difference in P_c compared to the 120 nm-thick buffer. For the total thickness series, we conducted 1-, 2-, 4-, 6- and 8-turn nanospiral growth experiments, with a designed pitch of 250 nm. All films were formed using a growth temperature of 650 °C. We observed that both the diameter and total thickness of the nanospirals increase with growth time, but the diameter change is less pronounced when the total thickness is smaller than 1 μm, as can be seen in Figure 5a. From side-view SEM images (not shown here), the surface of 6- and 8-turn nanospirals grew rougher and more uneven compared to 4-turn nanospirals, and a branch-like structure was formed for 8-turn nanospirals. The spiral geometry becomes rod-like with some branches owing to the formation of new nucleation sites on the surface. Figure 5b reveals trends in $|P_c|$ and λ as a function of total thickness. A dramatic increase of $|P_c|$ from 0 to 0.75 is obtained by adding more turns to the nanospirals, as expected since constructive interference enhances with the number of periods in the structure. However, $|P_c|$ can be reduced if the periodic structure deteriorates, as suggested by the results obtained from 6- and 8-turn nanopirals. In addition, λ seems to have a small red-shift trend towards longer wavelengths with growth time. Detailed examination of the present nanospiral growth rate indicated an increase over time, with a resulting increase of pitch. Since the sputtering power was fixed for all depositions, we observed an increase of Al cathode voltage correlated with a decrease of current with time. A higher cathode voltage may offer a higher sputtering yield, resulting in a higher growth rate [30].

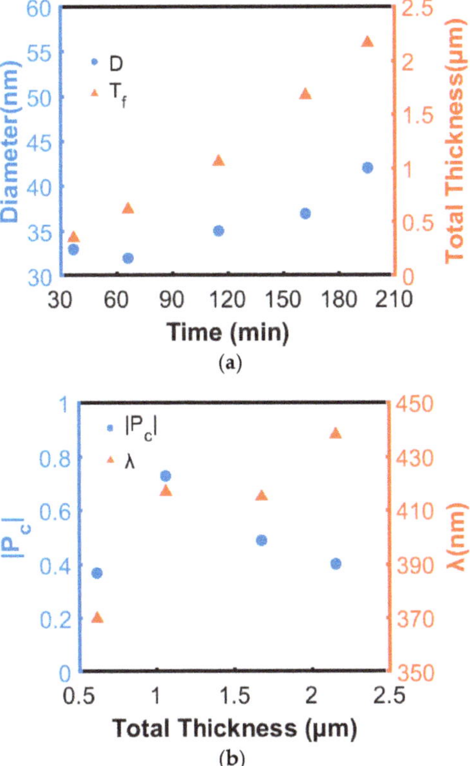

Figure 5. (**a**) Trend of nanospiral diameter (D) and thickness (T_f) with growth time; (**b**) $|P_c|$ and λ for different total thicknesses.

Hence, we further studied the effect of pitch on $|P_c|$ and λ. A series of samples were grown at 650 °C with different pitches, while total thickness was kept constant at around 1 µm. As can be seen in Figure 6a, λ shifts to longer wavelength with increasing pitch. The $|P_c|$ values are lower for shorter and longer pitch. For shorter pitch, λ is expected to fall into the UV region where the InAlN material has a stronger absorption and thus reduces the reflected light. If the pitch is too long, the film contains fewer periods, leading weaker constructive interference. The peak wavelength position is expected to vary linearly with the pitch according to the Bragg condition. A plot of λ versus pitch analyzed from a larger set of samples, regardless of variation in growth parameters, is shown in Figure 6b. The peak wavelengths in the range of 350–470 nm have high consistency with their corresponding pitch size ranges from 170–320 nm. This monotonic increase of wavelength with pitch is evidence that the circular polarization is connected to the constructive Bragg reflection, as expected. Hence, precise control of the pitch of the grown nanospirals is important for the design of chiral-optical response elements.

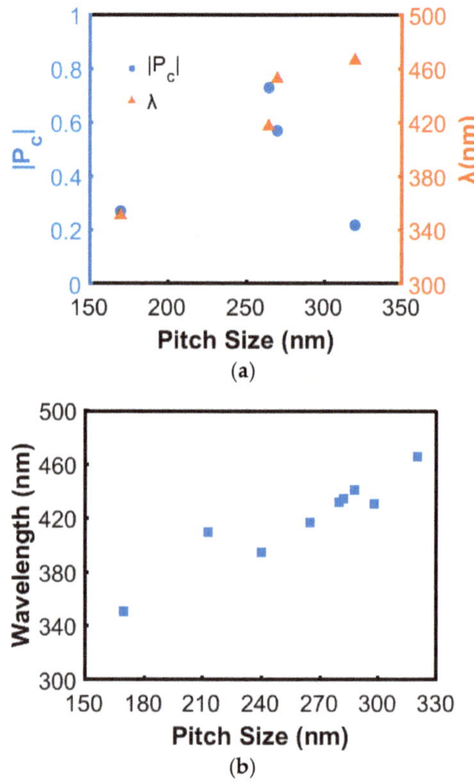

Figure 6. (a) The experimental results in P_c and λ for different pitch sizes while the total length was kept constant at around 1 um; (b) A plot of the relationship between pitch and peak wavelength summarized from all experiments.

4. Conclusions

This study reports on the analysis of the polarization behavior of light reflected from InAlN nanospirals grown on c-plane Al_2O_3 substrates with a TiN buffer layer by reactive MSE with various structural configurations. The degree of circular polarization of the reflected light and the corresponding wavelength range were found to be highly dependent on the buffer thickness, pitch, and morphology of nanospirals. The handedness of the reflected light was also demonstrated to follow the structural chirality. A TiN buffer layer thicker than 30 nm was needed for sufficient reflection. In addition, a film-like structure formed at temperatures lower than 450 °C. When the growth temperature was raised to 750 °C, less than 0.1 In-content incorporated into the InAlN nanospirals. Both cases revealed very low $|P_c|$ values. A pitch ranging from 200 to 300 nm gave promising $|P_c|$ higher than 0.5 in the UV to blue region. A transformation of unpolarized to circularly polarized light followed from the circular Bragg condition. Control over the structural configuration and morphology was also demonstrated to be a determinant for obtaining high $|P_c|$ at a designed peak wavelength.

Acknowledgments: This work was supported by the Swedish Research Council (VR) under grants Nos. 621-2012-4420, 621-2013-5360, 621-2013-4018. The Swedish Government Strategic Research Area in Materials Science on Functional Materials at Linköping University (Faculty Grant SFO-Mat-LiU 2009-00971) and the Centre in Nano science and technology (CeNano) at Linköping University are also acknowledged for financial support.

Author Contributions: All authors conceived and designed the experiments; Y.-H.K. and C.-L.H. performed the growth of InAlN nanospirals; Y.-H.K., E.A.S., L.H. and C.-L.H. characterized sample morphology; R.M., Y.-H.K. and K.J. performed MMSE and analyzed the data together with L.H., J.B. and C.L.-H.; Y.-H.K., L.H. J.B., and C.-L.H. analyzed XRD data; Y.-H.K. and C.-L.H. wrote the manuscript with revision by R.M., E.A.S., P.S., K.J., L.H. and J.B. All authors agreed on the final version of the manuscript.

Conflicts of Interest: The authors declare no conflict of interest.

References

1. Arwin, H.; Magnusson, R.; Landin, J.; Järrendahl, K. Chirality-induced polarization effects in the cuticle of scarab beetles: 100 years after Michelson. *Philos. Mag.* **2012**, *92*, 1583–1599. [CrossRef]
2. Sharma, V.; Crne, M.; Park, J.O.; Srinivasarao, M. Structural origin of circularly polarized iridescence in Jewwled Beetles. *Science* **2009**, *325*, 449–451. [CrossRef] [PubMed]
3. Parker, A.R.; Townley, H.E. Biomimetics of photonic nanostructures. *Nat. Photonics* **2007**, *2*, 347–353. [CrossRef] [PubMed]
4. Hodgkinson, I.J.; Lakhtakia, A.; Wu, Q.H. Experimental realization of sculptured-thin-film polarization discriminatory light-handedness inverters. *Opt. Eng.* **2000**, *39*, 2831–2834.
5. Wu, Q.; Lakhtakia, A.; Hodgkinson, I.J. Circular polarization filters made of chiral sculptured thin films: Experimental and simulation results. *Opt. Eng.* **2000**, *39*, 1863–1868. [CrossRef]
6. Agrawal, G.P.; Radic, S. Phase-shifted fiber Bragg gratings and their application for wavelength demultiplexing. *IEEE Photonics Technol. Lett.* **1994**, *6*, 995–997. [CrossRef]
7. Kim, Y.; Hong, K.; Yeom, J.; Hong, J.; Jung, J.-H.; Lee, Y.W.; Park, J.-H.; Le, B. A frontal projection-type three-dimensional display. *Opt. Exp.* **2012**, *20*, 20130–20138. [CrossRef] [PubMed]
8. Krause, K.M.; Brett, M.J. Spatially graded nanostructured chiral films as tunable circular polarizers. *Adv. Funct. Mater.* **2008**, *18*, 3111–3118. [CrossRef]
9. Hawkeye, M.M.; Brett, M.J. Glancing angle deposition: Fabrication, properties, and applications of micro- and nanostructured thin films. *J. Vac. Sci. Technol. A* **2007**, *25*, 1317–1335. [CrossRef]
10. Deubel, M.; von Freymann, G.; Wegener, M.; Pereira, S.; Busch, K.; Soukoulis, C.M. Direct laser writing of three-dimensional photonic-crystal templates for telecommunications. *Nat. Mater.* **2004**, *3*, 444–447. [CrossRef] [PubMed]
11. Gansel, J.K.; Thiel, M.; Rill, M.S.; Decker, M.B.; Saile, K.; Freymann, V.G.; Linden, S.; Wegener, M. Gold helix photonic metamaterial as broadband circular polarizer. *Science* **2009**, *325*, 1513–1515. [CrossRef] [PubMed]
12. Pang, Y.K.; Lee, J.C.W.; Lee, H.F.; Tam, W.Y.; Chan, C.T.; Sheng, P. Chiral microstructures (spirals) fabrication by holographic lithography. *Opt. Exp.* **2005**, *13*, 7615–7620. [CrossRef]
13. Hodgkinson, I.; Wu, Q.H. Inorganic Chiral Optical Materials. *Adv. Mater.* **2001**, *13*, 889–897. [CrossRef]
14. Young, N.O.; Kowal, J. Optically active fluorite films. *Nature* **1959**, *183*, 104–105. [CrossRef]
15. Hwang, J.; Song, M.H.; Park, B.; Nishimura, S.; Toyooka, T.; Wu, J.W.; Takanishi, Y.; Ishikawa, K.; Takezoe, H. Electro-tunable optical diode based on photonic bandgap liquid-crystal heterojunctions. *Nat. Mater.* **2005**, *4*, 383–387. [CrossRef] [PubMed]
16. Zhang, W.; Potts, A.; Bagnall, D.M. Giant optical activity in dielectric planar metamaterials with two-dimensional chirality. *J. Opt. A Pure Appl. Opt.* **2006**, *8*, 878–890. [CrossRef]
17. Robble, K.; Brett, M.J.; Lakhtakia, I. Chiral sculptured thin films. *Nature* **1996**, *384*, 616. [CrossRef]

18. Valyukh, S.; Arwin, H.; Birch, J.; Järrendahl, K. Bragg reflection from periodic helicoidal media with laterally graded refractive index. *Opt. Mater.* **2017**, *72*, 334–340. [CrossRef]

19. Radnóczi, G.Z.; Seppänen, T.; Pécz, B.; Hultman, L.; Birch, J. Growth of highly curved $Al_xIn_{1-x}N$ nanocrystals. *Phys. Stat. Sol.* **2005**, *202*, R76–R78. [CrossRef]

20. Hsiao, C.-L.; Magnusson, R.; Palisaitis, J.; Sandström, P.; Persson, P.O.Å.; Valyukh, S.; Hultman, L.; Järrendahl, K.; Birch, J. Curved-Lattice Epitaxial Growth of $In_{1-x}Al_xN$ Nanospirals with Tailored Chirality. *Nano Lett.* **2015**, *15*, 294–300. [CrossRef] [PubMed]

21. Magnusson, R.; Hsiao, C.L.; Birch, J.; Arwin, H.; Järrendahl, K. Chiral nanostructures producing near circular polarization. *Opt. Mater. Exp.* **2014**, *4*, 1389–1403. [CrossRef]

22. Magnusson, R.; Birch, J.; Sandström, P.; Hsiao, C.L.; Arwin, H.; Järrendahl, K. Optical Mueller Matrix Modeling of Chiral $Al_xIn_{1-x}N$ Nanospirals. *Thin Solid Films* **2014**, *571*, 447–452. [CrossRef]

23. Hsiao, C.L.; Palisaitis, J.; Junaid, M.; Persson, P.O.Å.; Jensen, J.; Zhao, Q.-X.; Chen, L.C.; Chen, K.H.; Birch, J. Room-temperature heteroepitaxy of single-phase $Al_{1-x}In_xN$ films with full composition range on isostructural wurtzite substrates. *Thin Solid Films* **2012**, *524*, 113–120. [CrossRef]

24. Hsiao, C.L.; Palisaitis, J.; Junaid, M.; Chen, R.S.; Persson, P.O.Å.; Sandström, P.; Holtz, P.O.; Hultman, L.; Birch, J. Spontaneous formation of AlInN core-shell nanorod arrays by ultrahigh-vacuum magnetron sputter epitaxy. *Appl. Phys. Exp.* **2011**, *4*, 115002. [CrossRef]

25. Palisaitis, J.; Hsiao, C.-L.; Hultman, L.; Birch, J.; Persson, P.O.Å. Core-shell formation in self-induced InAlN nanorods. *Nanotechnology* **2017**, *7*, 015303. [CrossRef] [PubMed]

26. Hsiao, C.L.; Palisaitis, J.; Persson, P.O.Å.; Junaid, M.; Sandström, P.; Hultman, L.; Birch, J. Nucleation and core-shell formation mechanism of self-induced $In_xAl_{1-x}N$ core-shell nanorods grown on sapphire substrates by magnetron sputter epitaxy. *Vacuum* **2016**, *131*, 39–43. [CrossRef]

27. Serban, E.A.; Persson, P.O.Å.; Poenaru, I.; Junaid, M.; Hultman, L.; Birch, J.; Hsiao, C.L. Structural and compositional evolutions of $In_xAl_{1-x}N$ core-shell nanorods grown on Si(111) substrates by reactive magnetron sputter epitaxy. *Nanotechnology* **2015**, *26*, 215602. [CrossRef] [PubMed]

28. Palisaitis, J.; Hsiao, C.-L.; Hultman, L.; Birch, J.; Persson, P.O.Å. Direct observation of spinodal decomposition phenomena in InAlN alloy during in-situ STEM heating. *Sci. Rep.* **2017**, *7*, 44390. [CrossRef] [PubMed]

29. Jeyachandran, Y.L.; Narayandass, S.K.; Mangalaraj, D.; Areva, S.; Mielczarski, J.A. Properties of titanium nitride films prepared by direct current magnetron sputtering. *Mater. Sci. Eng. A* **2007**, *445–446*, 223–236. [CrossRef]

30. Martin, P.M. *Handbook of Deposition Technologies for Films and Coatings: Science, Application and Technology*, 3rd ed.; William Andrew: Norwich, UK, 2010; Chapter 5; pp. 253–296.

 *nanomaterials*

Article

Linear Birefringent Films of Cellulose Nanocrystals Produced by Dip-Coating

Arturo Mendoza-Galván [1,2,*]**, Tania Tejeda-Galán** [1]**, Amos B. Domínguez-Gómez** [1]**,
Reina Araceli Mauricio-Sánchez** [1]**, Kenneth Järrendahl** [2] **and Hans Arwin** [2]

[1] Cinvestav-Querétaro, Libramiento Norponiente 2000, 76230 Querétaro, Mexico;
tania.tejeda@cinvestav.mx (T.T.-G.); bdominguez@cinvestav.mx (A.B.D.-G.);
amauricio@cinvestav.mx (R.A.M.-S.)
[2] Materials Optics, Department of Physics, Chemistry and Biology, Linköping University,
SE-58183, Linköping, Sweden; kenneth.jarrendahl@liu.se (K.J.); hans.arwin@liu.se (H.A.)
* Correspondence: amendoza@cinvestav.mx or arturo.mendoza@liu.se; Tel.: +52-442-211-9922

Received: 16 November 2018; Accepted: 22 December 2018; Published: 31 December 2018

Abstract: Transparent films of cellulose nanocrystals (CNC) are prepared by dip-coating on glass substrates from aqueous suspensions of hydrolyzed filter paper. Dragging forces acting during films' deposition promote a preferential alignment of the rod-shaped CNC. Films that are 2.8 and 6.0 μm in thickness show retardance effects, as evidenced by placing them between a linearly polarized light source and a linear polarizer sheet in the extinction configuration. Transmission Mueller matrix spectroscopic ellipsometry measurements at normal incidence as a function of sample rotation were used to characterize polarization properties. A differential decomposition of the Mueller matrix reveals linear birefringence as the unique polarization parameter. These results show a promising way for obtaining CNC birefringent films by a simple and controllable method.

Keywords: nanostructured films; birefringence; nanocrystalline cellulose; Mueller matrix

1. Introduction

Cellulose is the most abundant renewable biopolymer on earth. Its polymeric chain of D-anhydroglucopyranose units, through a hierarchical arrangement, leads to a fibrous macroscopic structure with a semicrystalline character [1]. The extraction of cellulose nanocrystals (CNC) from cellulose fibrils through controlled, sulfuric acid-catalyzed degradation was reported more than half a century ago [2]. Since then, the needle-shaped CNC and their functionalization capabilities have found applications in diverse fields [3–5]. Besides removal of the amorphous regions, the sulfuric acid hydrolysis of cellulose fibrils functionalizes the CNC surface, resulting in negatively charged sulphate groups. The electrostatic interaction between charged CNC leads to stable aqueous suspensions. Decades ago, it was discovered that above a critical concentration, the CNC self-assemble in a chiral nematic liquid crystalline phase [6]. The slow drying of these aqueous suspensions produces films with helicoidal ordering that reflect left-handed polarized light [7,8]. Circular dichroism and circular birefringence are the characteristic polarization properties of chiral CNC films [8]. The spectral location and strength of this so-called circular Bragg reflection depend on the value and distribution of the helicoidal pitch, as well as on the birefringence of CNC.

The linear birefringence of cellulose-based fibers is known only at some wavelengths [9,10]. On the other hand, the birefringence of cellulose derivatives in film form has been studied extensively [11]. In recent times, interest in investigating the birefringence of CNC has increased [12–20]. Particularly, methods for the fabrication of birefringent CNC films have been focused on producing an effective alignment of CNC along a preferential direction. Spin-coating [15] and shear-ordering [16–20] methods have been reported. The development of new methods to fabricate birefringent CNC films will contribute to a better understanding of their fundamental properties and open new opportunities for novel applications. Specifically, thinking in optical biomimetics, they could be used as retarders sandwiched between chiral layers also made from cellulose nanocrystals. With an adequate thickness, the retardation could be tuned to manipulate left- and right-handed circular polarization in a specific spectral range.

In this work, we show that a simple and non-expensive dip-coating technique is suitable to prepare transparent and birefringent CNC films from aqueous suspensions. For a complete characterization of the polarization properties of the films, Mueller-matrix transmission spectroscopic ellipsometry is used. A unique capability of this approach is that the depolarization introduced by the sample, that is, how much the sample affects the degree of polarization of incident light, is also provided. Furthermore, as CNC in aqueous suspension tend to self-assemble in a chiral nematic liquid crystal phase, circular dichroism and circular birefringence could be expected. Therefore, a differential (logarithmic) decomposition of the Mueller matrix data was performed to determine the basic polarization properties of the dip-coated CNC films, including both linear and circular birefringence, as well as dichroism.

2. Materials and Methods

Aqueous suspensions of CNC were obtained following procedures as reported in the literature, but with slight variations [17,21,22]. Filter paper (Whatman 40) was grinded in a coffee mill by four cycles, 35 s each. The milled paper was hydrolyzed at 60 °C for 50 min under vigorous stirring using 64 wt% sulfuric acid at a ratio of 8.75 mL per 1 g of filter paper. To stop the hydrolysis, the CNC suspension was diluted with cold water (10 times the volume of the acid solution) and allowed to settle for two weeks. The clear top layer was decanted, and the remaining cloudy layer was subject to three cycles of centrifugation (9000 rpm for 10 min) and washing with water to remove water-soluble cellulose materials. The thick white suspension was dialyzed against water for three days. The initial concentration of the CNC suspension was 6.5 wt%, and it was diluted with water to attain a concentration of 5.7 wt%, suitable for dip-coating. Ultrasonic dispersion was not applied. Glass slides 25 × 75 mm (Corning 2947) washed with detergent were used as substrates. CNC films were then produced by dip-coating at withdrawal speeds of 10 and 20 cm/min by using a home-made apparatus. Subsequently, the samples were vertically placed and allowed to dry at room temperature for about 3–4 h.

Transmission Mueller-matrix measurements were performed with a dual rotating compensator ellipsometer (RC2, J. A. Woollam Co., Inc., Lincoln, NE, USA) at normal incidence in the wavelength (λ) range 210–1690 nm. Data are here presented versus photon energy given by $E = hc/\lambda$, where h is Planck's constant and c the vacuum speed of light. A motorized sample rotator was used to measure at rotation angles between 0° and 360° in steps of 5°. Transmittance irradiance measurements at normal incidence in the spectral range of 250–840 nm were performed with a FilmTek 3000 system (SCI, Inc., Carlsbad, CA, USA). Atomic force microscopy (AFM) images in tapping mode were acquired with an Innova system (Bruker, Madison, WI, USA). Complementary characterization included X-ray diffraction data (Rigaku/Dmax2100, Austin, TX, USA), attenuated total reflection (ATR) infrared spectroscopy measurements (Spectrum GX system/Perkin Elmer Inc., Waltham, MA, USA), and cross-sectional scanning electron microscopy (SEM) (Phillips XL 30 system, North Billerica, MA, USA). To avoid charging during acquisition of SEM images, a thin layer of gold was deposited on the sample using Ar as carrier gas, 20 μA, 5×10^{-2} torr for 30 s (Denton Vacuum desk V, Moorestown, NJ, USA).

3. Results and Discussion

3.1. Formation of Nanostructured Films and Optical Performance

At the concentration of the CNC which was used (5.7 wt%), the coexistence of isotropic and anisotropic phases in the suspension was expected, as reported for hydrolyzed filter paper at similar conditions [21,22]. Therefore, we hypothesized that by using the dip-coating technique, the drag of draining forces acting during the removal of the substrate could align the nanocrystals in the isotropic phase and partially unwind the cholesteric order. As the substrate leaves the suspension, the draining of water produces an increase of CNC concentration in the entrained suspension, which limits the mobility of CNC. At the withdrawal speeds used, the coating process took about 15–30 s. During the drying stage, the draining of remaining water in the films promoted further alignment of CNC for about 30 min. The film then reached a gel-like state, where the CNC were frozen in a partial nematic ordering. At this stage, the film loses water by evaporation for about 2–3 h. Notice the difference in time scale between the dip-coating process and the evaporation-induced self-assembly of chiral films, as the latter takes a few days.

Figure 1 shows AFM images of the surfaces of the dip-coated CNC films. A preferential alignment of the CNC is observed, and thus, our hypothesis is confirmed. X-ray diffraction data (see Figure A1 in Appendix A) corroborated that the films retain the monoclinic crystalline structure of cellulose Iβ. ATR infrared spectroscopy measurements (see Figure A2 in Appendix A) evidenced that the molecular integrity of the cellulose was retained as well. The thicknesses determined from cross-sectional SEM images (see Figure A3 in Appendix A) were 2.8 ± 0.1 and 6.0 ± 0.5 μm for 10 and 20 cm/min withdrawal speeds, respectively.

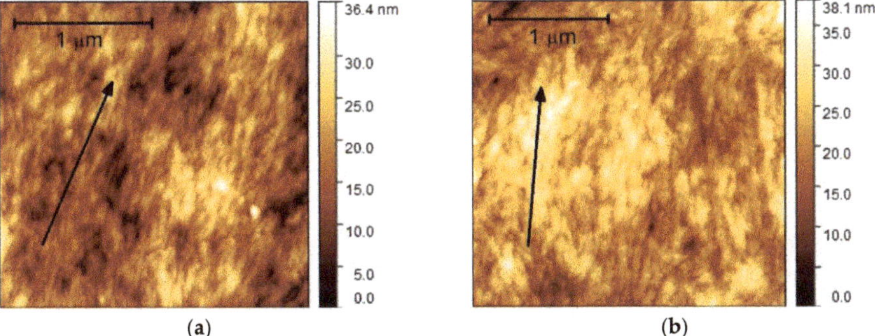

(a)   **(b)**

Figure 1. Atomic force microscopy (AFM) images of dip-coated cellulose nanocrystal (CNC) films on glass substrates produced at withdrawal speeds of (**a**) 10 and (**b**) 20 cm/min. The arrows illustrate the preferential ordering of CNC.

The transmittance (*T*) spectra at normal incidence of unpolarized light from the cellulose side of the dip-coated CNC films are shown in Figure 2. The spectra correspond to the average of three measurements performed at different regions on the films. The standard deviation of the three measurements were plotted as error bars. As can be seen, the films show good transparency with *T* values above 80% in the visible spectral range. Interference oscillations are missing due to the low optical contrast at the cellulose–glass interface because the refractive indices of cellulose ~1.54 [10] and the glass substrate ~1.52 are similar. The decrease in *T* at wavelengths shorter than 350 nm is due to absorption of the glass substrate.

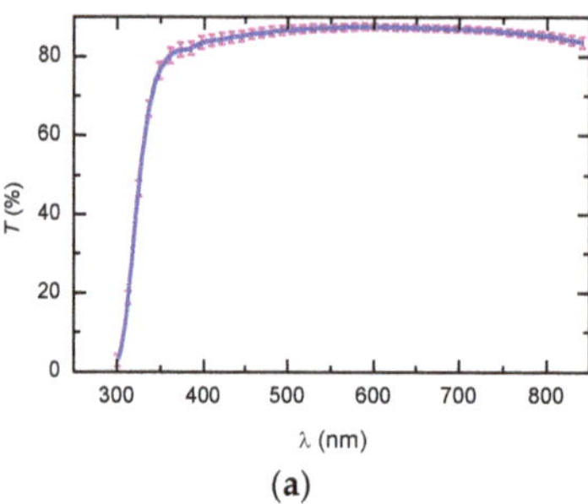 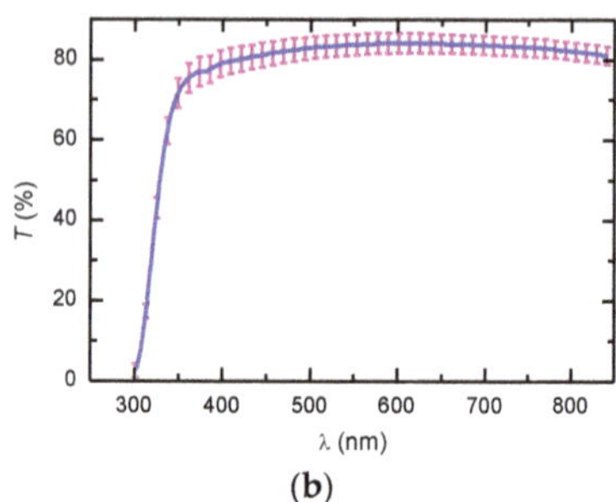

(a)   **(b)**

Figure 2. Average transmittance of unpolarized light of dip-coated CNC films on glass substrates at withdrawal speeds of **(a)** 10 and **(b)** 20 cm/min. Error bars are the standard deviation of three spectra.

Qualitative and simple evidence of the birefringence of a sample is to place it between crossed polarizers. This can be done by using a liquid crystal display (LCD) as a source of linearly polarized light and a linear polarizer sheet. Figure 3 shows pictures of the assembly used. The plane of polarization of light coming from the LCD screen was at about +45° from the horizontal direction, and the polarizer sheet was set in the extinction configuration (see dark regions around the samples). As can be seen, when the polarization plane of the light coming from the LCD is perpendicular to the withdrawal direction (indicated with the arrows), the sample looks dark (sample orientation −45°). This means that the linear polarization is not affected by the sample, and the outcoming light is cancelled out by the polarizer sheet. On the other hand, by rotating the samples to the horizontal (H) and vertical (V) orientations, the image on the LCD screen can clearly be seen. In these configurations, which are the so-called maximum transmittance, the polarization state of the linearly polarized incident light is altered and, in general, exits the sample with elliptical polarization. Thus, the in-plane anisotropy in the films is evident, and its origin resides in the preferential ordering of CNC and in the intrinsic anisotropy of cellulose, as discussed in Section 3.4. Another salient feature is the homogeneity over large areas of the samples. However, some inhomogeneities can be seen at the edges of the substrates due to accumulated material by border effects. The draining also accumulates material at the bottom part of the substrates.

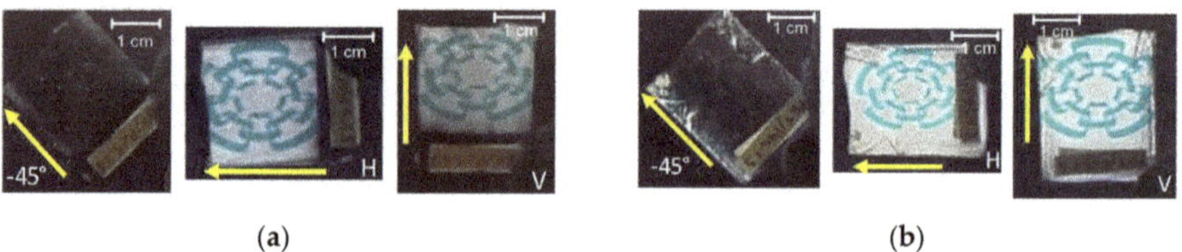

(a)   **(b)**

Figure 3. Pictures of dip-coated CNC films on glass substrates at withdrawal speeds of **(a)** 10 and **(b)** 20 cm/min placed between an LCD monitor and a polarizer in the extinction configuration without a sample. The arrows indicate the withdrawal direction. Sample orientation at +45° looks like −45°.

3.2. Mueller Matrix Data Analysis

The 4 × 4 Mueller matrix (**M**) with elements m_{ij} ($i, j = 1 \ldots 4$) provides a full description of the polarizing and depolarizing properties of a sample [23]. It relates the Stokes vectors of the incident ($\mathbf{S}_i$) and transmitted ($\mathbf{S}_t$) light beams by:

$$\mathbf{S}_t = \mathbf{M}\mathbf{S}_i. \tag{1}$$

In Equation (1), the Stokes vectors of the incident and transmitted light beams (assuming propagation along the z-axis) are expressed in terms of a set of six irradiances:

$$\mathbf{S} = \begin{bmatrix} I_x + I_y \\ I_x - I_y \\ I_{+45°} - I_{-45°} \\ I_R - I_L \end{bmatrix}, \tag{2}$$

where I_x, I_y, $I_{+45°}$, and $I_{-45°}$ correspond to linear polarization along the coordinate axes x and y, and at $+45°$ and at $-45°$ from the x-axis, whereas I_R and I_L correspond to right- and left-handed circularly polarized light, respectively. In this work we used Mueller matrices, which were normalized to total transmittance for unpolarized light, i.e., the first element in the first row of $\mathbf{M}$. Thus, we have $m_{11} = 1$, and other elements have values in the range $[-1,1]$. For the measurements, the samples were placed with the withdrawal direction nearly parallel to the y-axis.

Figure 4 shows polar contour maps of the normal-incidence Mueller matrix transmission data of dip-coated CNC films. The radial and angular coordinates correspond to the photon energy (in eV) and sample rotation angle (ϕ), respectively. Data above 3.75 eV (below λ = 330 nm) were omitted because the glass substrate strongly absorbs the ultraviolet range (T lower than 50% in Figure 2). First, we observed that the elements of the first row and first column were close to zero, whereas the other elements show a richer structure. Second, we found that $\mathbf{M}$ was not block diagonal, which it would be for an isotropic sample or for uniaxially samples with the optic axis aligned with the z-direction. In Figure 4, the non-zero off-diagonal elements m_{42} and m_{24} thus provide evidence of in-plane anisotropy. Other characteristics of $\mathbf{M}$ are related to the ϕ-dependence: (i) m_{44} is invariant; (ii) m_{24}, m_{42}, m_{34}, and m_{43} are periodic with period 180°; (iii) a 90° periodicity is observed in m_{22}, m_{23}, m_{32}, and m_{33}; (iv) rotational shift relationships noted are $m_{22}(\phi) = m_{33}(\phi - \pm 45°)$, $m_{34}(\phi) = m_{24}(\phi - 45°)$, and $m_{43}(\phi) = m_{42}(\phi + 45°)$. Other symmetries are also observed: $m_{23} \cong m_{32}$, $m_{24} = -m_{42}$, and $m_{34} = -m_{43}$. Furthermore, we noticed that increases in film thickness resulted in larger variations of the m_{ij} values. As an example, the diagonal elements vary between 0 and 1 for the 2.8 µm-thick film, whereas the variation is between -1 and 1 in the case of the 6 µm-thick film. Indeed, the contour maps in Figure 4a look like a zoom of the central part of the corresponding ones in Figure 4b.

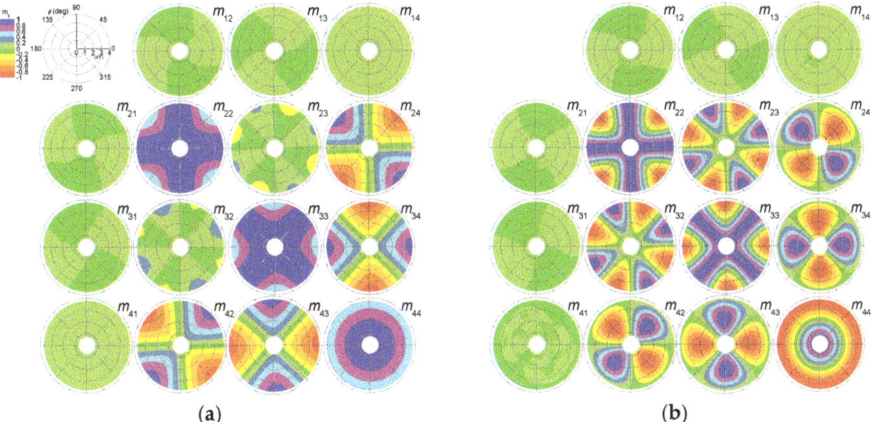

Figure 4. Polar contour maps of normal-incidence Mueller matrix transmission measurements for dip-coated films produced at withdrawal speeds of (**a**) 10 cm/min and (**b**) 20 cm/min.

All the observations listed in the previous paragraph on the structure of **M** in Figure 4 come from the CNC films, because the Mueller matrix of the glass substrate measured at different rotation angles corresponds to the 4 × 4 unity matrix (see Figure A4 in Appendix A). Furthermore, **M** in Figure 4 qualitatively agrees with the Mueller matrix $\mathbf{M_R}$ of an ideal linear retarder plate in which the fast axis is horizontal, and is given by [23]:

$$\mathbf{M_R} = \begin{bmatrix} 1 & 0 & 0 & 0 \\ 0 & \cos^2 2\phi + \cos\delta\sin^2 2\phi & \sin 2\phi\cos 2\phi(1-\cos\delta) & -\sin\delta\sin 2\phi \\ 0 & \sin 2\phi\cos 2\phi(1-\cos\delta) & \sin^2 2\phi + \cos\delta\cos^2 2\phi & \sin\delta\cos 2\phi \\ 0 & \sin\delta\sin 2\phi & -\sin\delta\cos 2\phi & \cos\delta \end{bmatrix}, \tag{3}$$

where $\delta = 2\pi(n_y - n_x)d/\lambda$ is the retardance, d the sample thickness, and n_x and n_y are the refractive indices for polarization along the x- and y-axes, respectively. However, as the matrices in Figure 4 were experimentally determined, it is necessary to investigate how much they deviate from those of ideal polarizing systems. This can be done by determining the depolarizance (D) of the system, which is given by [24]:

$$D = 1 - \left[\frac{1}{3}\left(\frac{tr(\mathbf{M}^{\mathrm{T}}\mathbf{M})}{m_{11}^2} - 1 \right) \right]^{1/2}, \tag{4}$$

where tr and T stand for trace and transpose, respectively. It holds that $D = 0$ corresponds to the Mueller matrix of an ideal non-depolarizing system, and $D = 1$ to an ideal depolarizer. In Figure 5, D is shown for selected values of ϕ. As can be seen, the experimental Mueller matrices of the CNC films correspond very closely to those of ideal systems. This means that for totally polarized incident light, the transmitted beam emerges completely polarized.

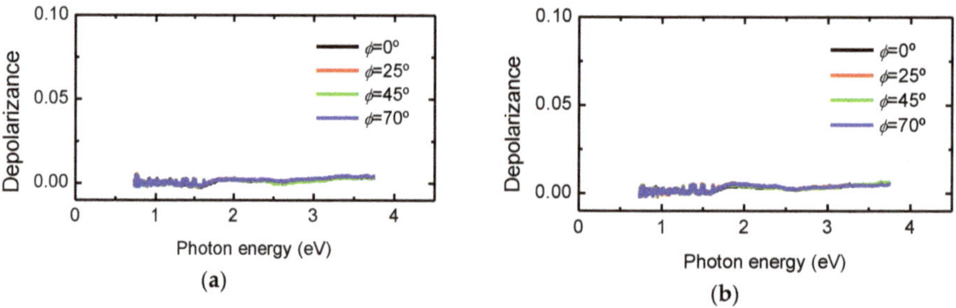

Figure 5. Depolarizance of normal-incidence Mueller matrices for dip-coated films produced at withdrawal speeds of (**a**) 10 cm/min and (**b**) 20 cm/min at selected rotation angles.

3.3. Differential Decomposition of Mueller Matrices

To quantitatively determine the polarization properties of the dip-coated CNC films, we used differential (logarithmic) decomposition of the measured Mueller matrices. This decomposition establishes that **M** and its spatial variation along the direction of wave propagation z are related as $d\mathbf{M}/dz = \mathbf{mM}$, where **m** is the differential matrix [25–28]. For homogeneous media, **m** is independent of z and direct integration gives $\mathbf{L} = \ln\mathbf{M}$, where $\mathbf{L} = \mathbf{m}d$ with d the sample thickness. The matrix **L** is split into $\mathbf{L} = \mathbf{L_m} + \mathbf{L_u}$, where $\mathbf{L_m}$ and $\mathbf{L_u}$ are G-antisymmetric and G-symmetric matrices, respectively, given by $\mathbf{L_m} = (\mathbf{L} - \mathbf{GL}^{\mathrm{T}}\mathbf{G})/2$ and $\mathbf{L_u} = (\mathbf{L} + \mathbf{GL}^{\mathrm{T}}\mathbf{G})/2$, where $\mathbf{G} = \mathrm{diag}[1, -1, -1, -1]$. Since the dip-coated CNC films are non-depolarizing, it holds $\mathbf{L_u} = 0$. $\mathbf{L_m}$ contains the six elementary polarization properties, which are given by [26]:

$$\mathbf{L_m} = \begin{bmatrix} 0 & LD & LD' & CD \\ LD & 0 & CB & -LB' \\ LD' & -CB & 0 & LB \\ CD & LB' & -LB & 0 \end{bmatrix}, \tag{5}$$

where LB (LD) and LB' (LD') are the linear birefringence (dichroism) along the x-y and $\pm 45°$ axes, respectively, whereas CD is the circular dichroism and CB is the circular birefringence.

Figure 6 shows $\mathbf{L_m}$ corresponding to the experimental Mueller matrices in Figure 4. For the sake of clarity, data are only shown for $0° \leq \phi \leq 180°$ in steps of $10°$. For both samples, only LB and LB' differ from zero, and both show a nearly linear dependence with photon energy. The small deviation from linearity implies a small dispersion in birefringence, as will be discussed later. Furthermore, LB increases with withdrawal speed (i.e., film thickness). The variation with rotation angle is illustrated in LB and LB' panels in Figure 6b, and is analyzed in more detail in Section 3.4. The fact that CD and CB are zero indicates the absence of the chiral phase in the films.

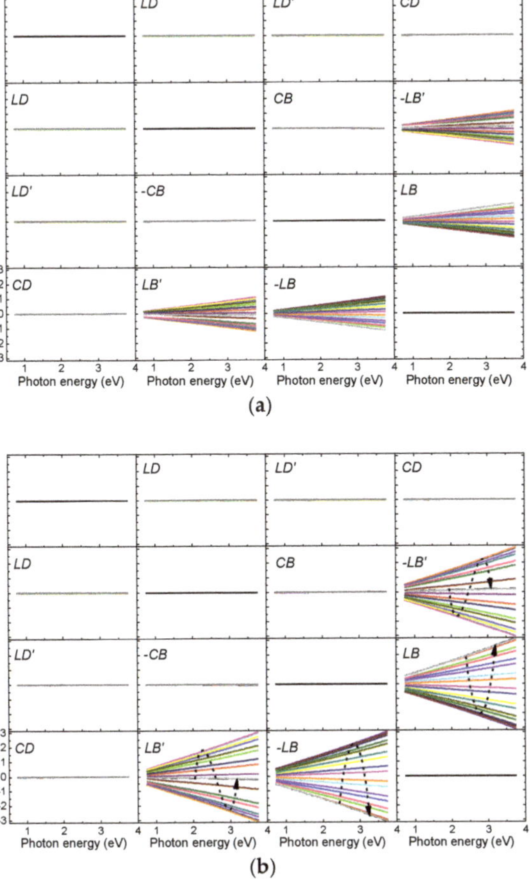

Figure 6. Differential decomposition of the Mueller matrices in Figure 4 for dip-coated films produced at withdrawal speeds of (**a**) 10 cm/min and (**b**) 20 cm/min. For clarity, data are presented only in the $0° \leq \phi \leq 180°$ range in steps of $10°$. The ϕ-variation is indicated with the dashed arrows in panels LB and LB' of (**b**).

3.4. Birefringence of Dip-Coated CNC Films

An insightful view of $LB(\phi)$ and $LB'(\phi)$ determined from the differential decomposition is provided when they are plotted in polar contour maps, as shown in Figure 7a where the radial coordinate corresponds to photon energy and the polar angle to sample azimuth ϕ. From Figure 7a, the obvious relationship $LB'(\phi) = LB(\phi - 45°)$ is clear. Therefore, among the six basic polarization properties in Equation (5), only linear birefringence characterizes the dip-coated CNC films. The ϕ-dependence of LB (and LB') can be expressed as [29], $LB(\phi) = |LB| \cos 2(\phi - \phi_0)$ and $LB'(\phi) = |LB| \sin 2(\phi - \phi_0)$, respectively, where ϕ_0 is the azimuth offset between the y-axis and the optical axis of the sample. Values of $\phi_0 = -2.2$ and $-1.9°$ were determined from the measurements at $\phi = 0°$ of the laboratory frame, as $\tan 2\phi_0 = -LB'(0)/LB(0)$. Since the sample birefringence $|LB| = 2\pi|n_y - n_x|d/\lambda$, due to the preferential ordering of CNC the effective birefringence $\langle \Delta n \rangle = \langle n_y - n_x \rangle$ can be easily obtained and is shown in Figure 7b. The values determined for $\langle \Delta n \rangle$ are about half of those reported for cotton fibers [9,10].

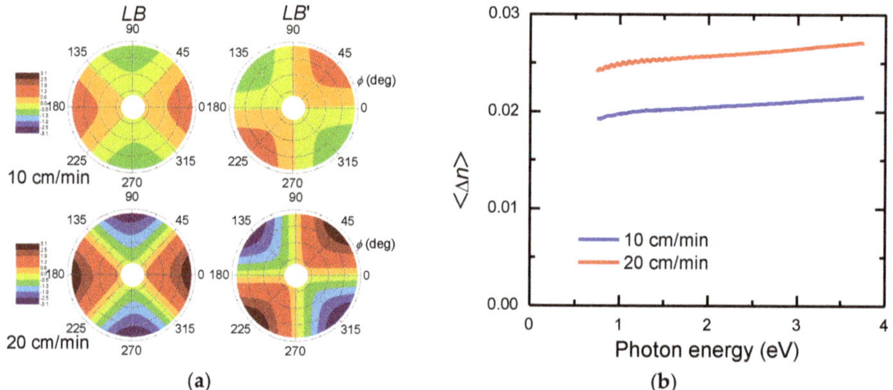

Figure 7. (**a**) Polar contour maps of linear birefringence of dip-coated CNC films at withdrawal speeds of 10 cm/min and 20 cm/min. The radial scale (photon energy) is the same as in Figure 4. (**b**) Effective birefringence <Δn> = <$n_e - n_o$> of the dip-coated CNC films.

According to the monoclinic crystalline structure of cellulose Iβ, its dielectric tensor corresponds to a biaxial crystal given as $\text{diag}[\varepsilon_1, \varepsilon_2, \varepsilon_3]$ in the principal axes frame. Considering that the crystallographic c-axis corresponding to the polymer chain direction (crystallite length) is orthogonal to a and b axes, we have $\varepsilon_3 = \varepsilon_c$ but ε_1 and ε_2 are not necessarily oriented along the a and b axes; besides, their orientation might be wavelength-dependent. Therefore, $n_y = n_e$ would be mainly related to ε_c and $n_x = n_o$ to both ε_1 and ε_2, where n_e and n_o are the extraordinary and ordinary refractive indices, respectively. In the ideal case of perfect alignment and full packing, it is expected that $n_e^2 = \varepsilon_c$ and $n_o^2 = (\varepsilon_1 + \varepsilon_2)/2$, and we obtain the effective birefringence $\Delta n = n_y - n_x = n_e - n_o$. In the present case where the CNC comprising the films are not completely aligned, the structural birefringence will also come into play. In this case, the films could be envisaged as a two-component composite of aligned (anisotropic) and non-aligned (isotropic) phases. Thus, the effective birefringence would be a function of the volume fraction of the anisotropic phase. For the films studied in this work, 0.4–0.5 is a rough estimate of the anisotropic phase volume fraction, assuming a linear variation. Of course, the ordering of CNC crystals in the films depends on the various parameters intervening for their fabrication. In this regard, two main stages are identified: Preparation of the CNC aqueous suspension and films deposition. The physicochemical properties of the CNC suspension largely depend on parameters like hydrolysis temperature, hydrolysis time, acid ionic strength, among others. On the other hand, for films deposition, the withdrawal speed, CNC concentration, and drying conditions are of main importance. The large number of parameters open possibilities to fabricating CNC films with tailored birefringence, and is subject to ongoing investigation which will be reported elsewhere.

4. Conclusions

Birefringent cellulose nanocrystal films were obtained from aqueous suspensions of hydrolyzed filter paper at 60 °C for 50 min. The films prepared were transparent and non-depolarizing. The linear birefringence (*LB*) of the films was determined by a differential decomposition of Mueller matrices, and showed a nearly linear dependence with photon energy. The effective birefringence $<\Delta n>$ was due to the preferential ordering of CNC, and had average values in the visible range of 0.021 and 0.026 for two different film thicknesses. The absence of the chiral phase in the films was ascribed to the short drying time. The dip-coating method is thus shown to be a suitable, non-expensive, and promising way to fabricate birefringent films of cellulose nanocrystals.

Author Contributions: Conceptualization, all authors; project administration, A.M.-G.; methodology, A.M.-G., H.A., K.J.; investigation and validation, T.T.-G., R.A.M.-S., A.B.D.-G.; formal analysis, A.M.-G., A.B.D.-G., T.T.-G., R.A.M.-S.; writing—original draft preparation, A.M.-G.; writing—review and editing, A.M.-G., H.A., K.J. All authors commented, reviewed and gave final approval for publication.

Funding: This research received no external funding.

Acknowledgments: T.T.-G. and A.B.D.-G. acknowledge the scholarship from Concayt-Mexico for master (454210) and doctoral (463797) studies, respectively. A.M.-G. acknowledges the scholarship from Conacyt (2018-000007-01EXTV-00169) to spend a sabbatical leave at Linköping University. K.J. acknowledges the Swedish Government Strategic Research Area in Materials Science on Advanced Functional Materials at Linköping University (Faculty Grant SFO-Mat-Liu No. 2009-000971). The technical assistance of Ana K. S. Rocha-Robledo is acknowledged for acquisition of AFM images. LIDTRA (Conacyt grants N-295261 and LN-254119) is acknowledged for SEM images. The Knut and Alice Wallenberg foundation, the Swedish Research Council, and the Carl Tryggers foundation are acknowledged for financial support.

Conflicts of Interest: The authors declare no conflict of interest.

Appendix A

This Appendix contains additional figures supporting the results presented in the main text. Figure A1 shows X-ray diffraction data of the filter paper used as the source of cellulose and the dip-coated CNC films at withdrawal speeds of 10 and 20 cm/min. The Miller indices are those corresponding to the monoclinic structure of cellulose Iβ, with lattice parameters $a = 7.8$ Å, $b = 7.9$ Å, $c = 10.1$ Å, $\beta = 95.15°$ and space group P2$_1$ [30–32]. The differences in peaks intensities of the data in Figure A1a are due to film thickness as can be seen in Figure A1b when the data are normalized with respect to the most intense (200) peak. It can be noticed that the processing does not affected the crystalline quality of cellulose. The crystallite width evaluated with the Scherrer formula was 6 nm which is typical of cotton cellulose [32].

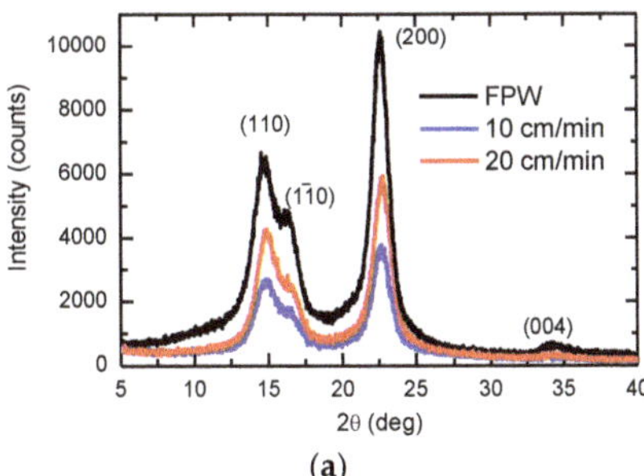
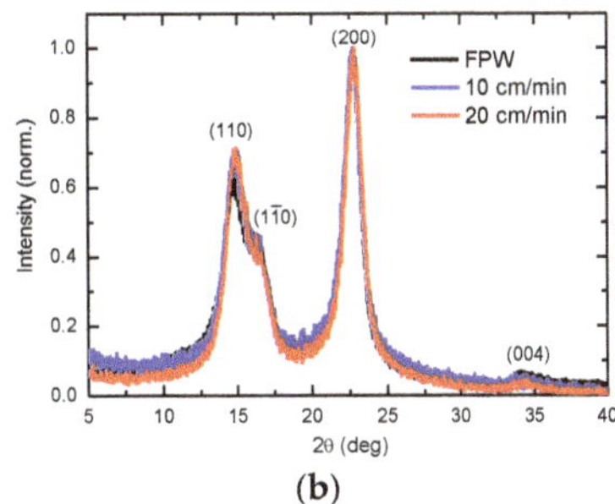

Figure A1. X-ray diffraction patterns of filter paper (FPW) and dip-coated CNC films at withdrawal speeds of 10 and 20 cm/min: (**a**) raw data (**b**) normalized to the (200) peak.

Figure A2 shows FTIR spectra of the filter paper used as the source of cellulose and dip-coated CNC films at withdrawal speeds of 10 and 20 cm/min. The spectra were normalized respect to the band at 1030 cm^{-1}. As can be seen, the spectra of dip-coated CNC films and filter paper are similar indicating that processing does not affect the molecular integrity. The bands of functional groups in cellulose molecule are labeled as well as the characteristic bands identifying cellulose Iβ [32,33].

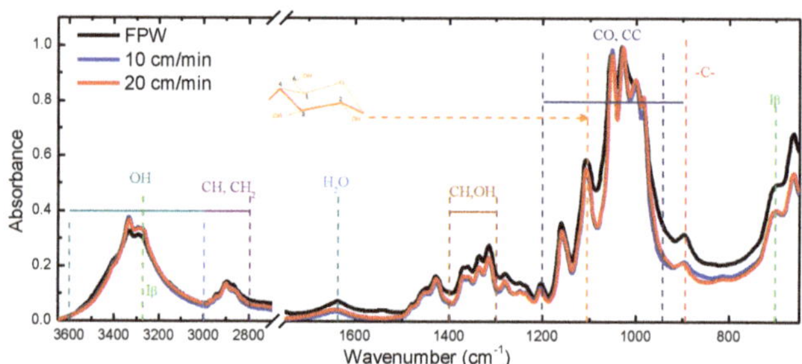

Figure A2. ATR infrared absorption spectra of filter paper (FPW) and dip-coated CNC films at withdrawal speeds of 10 and 20 cm/min

SEM images used to determine the film thicknesses of the dip-coated CNC films at withdrawal speeds of 10 and 20 cm/min are shown in Figure A3.

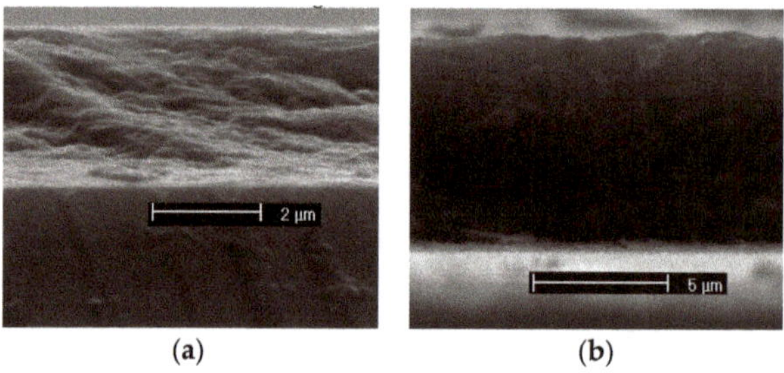

(a) (b)

Figure A3. Cross-section SEM images of dip-coated CNC films at (a) 10 and (b) 20 cm/min.

The normal incidence transmission Mueller matrices of a bare glass substrate measured at different rotation angles are shown in Figure A4. It can be noticed that they correspond to the 4 × 4 unity matrix.

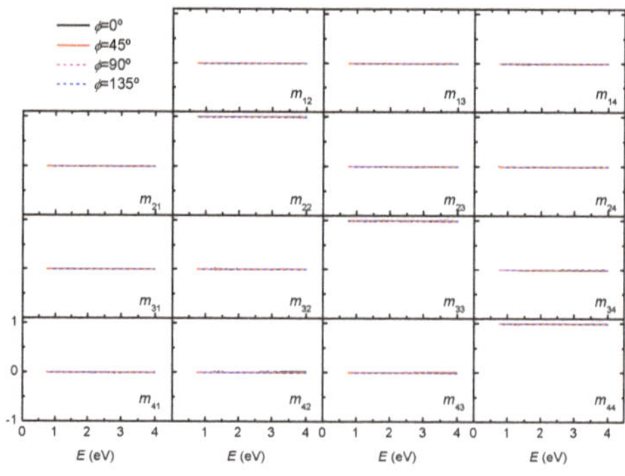

Figure A4. Mueller matrix of glass substrate measured at four rotation angles.

References

1. Wertz, J.-L.; Bédué, O.; Mercier, J.P. *Cellulose Science and Technology*, 1st ed.; EPFL: Lausanne, Switzerland, 2010; ISBN 9782940222414.
2. Rånby, B.G. The colloidal properties of cellulose micelles. *Discuss. Faraday Soc.* **1951**, *11*, 158–164. [CrossRef]
3. Hamad, W.Y. *Cellulose Nanocrystals: Properties, Production and Applications*, 1st ed.; Wiley: Chichester, UK, 2017; ISBN 9781119968160.
4. Kontturi, E.; Laaksonen, P.; Linder, M.B.; Gröschel, A.H.; Rojas, O.J.; Ikkala, O. Advanced materials through assembly of nanocelluloses. *Adv. Mater.* **2018**, *30*, 1703779. [CrossRef] [PubMed]
5. Almeida, A.P.C.; Canejo, J.P.; Fernandes, S.N.; Echeverria, C.; Almeida, P.L.; Godinho, M.H. Cellulose-based biomimetics and their applications. *Adv. Mater.* **2018**, *30*, 1703655. [CrossRef] [PubMed]
6. Revol, J.-F.; Bradford, H.; Giasson, J.; Marchessault, R.H.; Gray, D.G. Helicoidal self-ordering of cellulose microfibrils in aqueous suspension. *Int. J. Biol. Macromol.* **1992**, *14*, 170–172. [CrossRef]
7. Parker, R.M.; Guidetti, G.; Williams, C.A.; Zhao, T.; Narkevicius, A.; Vignolini, S.; Frka-Petesic, B. The self-assembly of cellulose nanocrystals: Hierarchical design of visual appearance. *Adv. Mater.* **2018**, *30*, 1704477. [CrossRef] [PubMed]
8. Mendoza-Galván, A.; Muñoz-Pineda, E.; Ribeiro, S.J.L.; Santos, M.V.; Järrendahl, K.; Arwin, H. Mueller matrix spectroscopic ellipsometry study of chiral nanocrystalline cellulose films. *J. Opt.* **2018**, *20*, 024001. [CrossRef]
9. Iyer, K.R.K.; Neelakanthan, P.; Radhakrishnan, T. Birefringence of native cellulosic fibers I: Untreated cotton and ramie. *J. Polym. Sci.* **1968**, *6*, 1747–1758. [CrossRef]
10. Ganster, J.; Fink, H.-P. Physical constants of cellulose. In *Polymer Handbook*, 4th ed.; Brandrup, J., Immergut, E.H., Grulke, E.A., Eds.; Wiley: New York, NY, USA, 1999; Volume 2, pp. V/135–V/157. ISBN 9780471166283.
11. Yamaguchi, M.; Manaf, M.E.A.; Songsurang, K.; Nobukawa, S. Material design of retardation films with extraordinary wavelength dispersion of orientation birefringence: A review. *Cellulose* **2012**, *19*, 601–613. [CrossRef]
12. Chindawong, C.; Johannsmann, D. An anisotropic ink based on crystalline nanocellulose: Potential applications in security printing. *J. Appl. Polym. Sci.* **2014**, *131*, 41063. [CrossRef]
13. Frka-Petesic, B.; Sugiyama, J.; Kimura, S.; Chanzy, H.; Maret, G. Negative diamagnetic anisotropy and birefringence of cellulose nanocrystals. *Macromolecules* **2015**, *48*, 8844–8857. [CrossRef]
14. Kim, D.H.; Song, Y.S. Anisotropic optical film embedded with cellulose nanowhisker. *Carbohydr. Polym.* **2015**, *130*, 448–454. [CrossRef] [PubMed]
15. Cranston, E.D.; Gray, D.G. Birefringence in spin-coated films containing cellulose nanocrystals. *Colloids Surf. A* **2008**, *325*, 44–51. [CrossRef]
16. Diaz, J.A.; Wu, X.; Martini, A.; Youngblood, J.P.; Moon, R.J. Thermal expansion of self-organized and shear-oriented cellulose nanocrystal films. *Biomacromolecules* **2013**, *14*, 2900–2908. [CrossRef] [PubMed]
17. Haywood, A.D.; Davis, V.A. Effects of liquid crystalline and shear alignment on the optical properties of cellulose nanocrystal films. *Cellulose* **2017**, *24*, 705–716. [CrossRef]
18. Chowdhury, R.A.; Peng, S.X.; Youngblood, J. Improved order parameter (alignment) determination in cellulose nanocrystal (CNC) films by a simple optical birefringence method. *Cellulose* **2017**, *24*, 1957–1970. [CrossRef]
19. Chowdhury, R.A.; Clarkson, C.; Youngblood, J. Continuous roll-to-roll fabrication of transparent cellulose nanocrystal (CNC) coatings with controlled anisotropy. *Cellulose* **2018**, *25*, 1769–1781. [CrossRef]
20. Sanchez-Botero, L.; Dimov, A.V.; Li, R.; Smilgies, D.-M.; Hinestroza, J.P. In situ and real-time studies, via synchrotron X-ray scattering, of the orientational order of cellulose nanocrystals during solution shearing. *Langmuir* **2018**, *34*, 5263–5272. [CrossRef]
21. Dong, X.M.; Kimura, T.; Revol, J.-F.; Gray, D.G. Effects of ionic strength on the isotropic-chiral nematic phase transition of suspensions of cellulose crystallites. *Langmuir* **1996**, *12*, 2076–2082. [CrossRef]
22. Dong, X.M.; Revol, J.-F.; Gray, D.G. Effect of microcrystallite preparation conditions on the formation of colloid crystals of cellulose. *Cellulose* **1998**, *5*, 19–32. [CrossRef]
23. Goldstein, D.H. *Polarized Light*, 3rd ed.; CRC Press: Boca Raton, FL, USA, 2010; ISBN 978143930413.
24. Gil, J.J. Components of purity of a Mueller matrix. *J. Opt. Soc. Am. A* **2011**, *28*, 1578–1585. [CrossRef]

25. Azzam, R.M.A. Propagation of partially polarized light through anisotropic media with or without depolarization: A differential 4 × 4 matrix calculus. *J. Opt. Soc. Am.* **1978**, *68*, 1756–1767. [CrossRef]

26. Ossikovski, R. Differential matrix formalism for depolarizing anisotropic media. *Opt. Lett.* **2011**, *36*, 2330–2332. [CrossRef] [PubMed]

27. Ossikovski, R.; De Martino, A. Differential Mueller matrix of a depolarizing homogeneous medium and its relation to the Mueller matrix logarithm. *J. Opt. Soc. Am. A* **2015**, *32*, 343–348. [CrossRef] [PubMed]

28. Arteaga, O.; Kahr, B. Characterization of homogeneous depolarizing media based on Mueller matrix differential decomposition. *Opt. Lett.* **2013**, *38*, 1134–1136. [CrossRef] [PubMed]

29. Arteaga, O. On the existence of Jones birefringence and Jones dichroism. *Opt. Lett.* **2010**, *35*, 1359–1360. [CrossRef] [PubMed]

30. Morais, J.P.S.; Rosa, M.F.; Filho, M.M.S.; Nascimento, L.D.; Nascimento, D.M.; Cassales, A.R. Extraction and characterization of nanocellulose structures from raw cotton linter. *Carbohydr. Polym.* **2013**, *91*, 229–235. [CrossRef]

31. Nam, S.; French, A.D.; Condon, B.D.; Concha, M. Segal crystallinity index revisited by the simulation of X-ray diffraction patterns of cotton cellulose Iβ and cellulose II. *Carbohydr. Polym.* **2016**, *135*, 1–9. [CrossRef]

32. Oh, S.Y.; Yoo, D.I.; Shin, Y.; Kim, H.C.; Kim, H.Y.; Chung, Y.S.; Park, W.H.; Youk, J.H. Crystalline structure analysis of cellulose treated with sodium hydroxide and carbon dioxide by means of X-ray diffraction and FTIR spectroscopy. *Carbohydr. Res.* **2005**, *340*, 2376–2391. [CrossRef]

33. Oh, S.Y.; Yoo, D.I.; Shin, Y.; Seo, G. FTIR analysis of cellulose treated with sodium hydroxide and carbon dioxide. *Carbohydr. Res.* **2005**, *340*, 417–428. [CrossRef]

 *nanomaterials*

Article

Interfacial Model and Characterization for Nanoscale ReB$_2$/TaN Multilayers at Desired Modulation Period and Ratios: First-Principles Calculations and Experimental Investigations

Shangxiao Jin [1,2] and Dejun Li [1,2,*]

[1] College of Physics and Materials Science, Tianjin Normal University, Tianjin 300387, China; jinshangxiao@126.com

[2] Tianjin International Joint Research Centre of Surface Technology for Energy Storage Materials, Tianjin 300387, China

* Correspondence: dejunli@tjnu.edu.cn; Tel.: +86-22-2376-6519; Fax: +86-22-2376-6519

Received: 7 May 2018; Accepted: 6 June 2018; Published: 10 June 2018

Abstract: The interfacial structure of ReB$_2$/TaN multilayers at varied modulation periods (Λ) and modulation ratios (t_{ReB2}:t_{TaN}) was investigated using key experiments combined with first-principles calculations. A maximum hardness of 38.7 GPa occurred at Λ = 10 nm and t_{ReB2}:t_{TaN} = 1:1. The fine nanocrystalline structure with small grain sizes remained stable for individual layers at Λ= 10 nm and t_{ReB2}:t_{TaN} = 1:1. The calculation of the interfacial structure model and interfacial energy was performed using the first principles to advance the in-depth understanding of the relationship between the mechanical properties, residual stresses, and the interfacial structure. The B-Ta interfacial configuration was calculated to have the highest adsorption energy and the lowest interfacial energy. The interfacial energy and adsorption energy at different t_{ReB2}:t_{TaN} followed the same trend as that of the residual stress. The 9ReB$_2$/21TaN interfacial structure in the B-Ta interfacial configuration was found to be the most stable interface in which the highest adsorption energy and the lowest interfacial energy were obtained. The chemical bonding between the neighboring B atom and the Ta atom in the interfaces showed both covalency and iconicity, which provided a theoretical interpretation of the relationship between the residual stress and the stable interfacial structure of the ReB$_2$/TaN multilayer.

Keywords: ReB$_2$/TaN multilayers; modulation structure; first-principles calculation; interfacial model; adsorption energy; interfacial energy

---

1. Introduction

Hard and wear-resistant coatings are increasingly used to reduce the material losses or to increase the lifetime of tools and machine parts. Thin-film structures consisting of alternating nanoscale multilayers have been attractive subjects in the area of protective coatings due to their extraordinary properties, such as their enormous hardness, which cannot be obtained in uniform bulk materials or in monolithic coatings of the constituent materials. These multilayers are usually produced with physical vapor deposition (PVD), e.g., reactive magnetron sputtering [1]. Recently, the superhardness of rhenium diboride (ReB$_2$) has been discovered to possess a maximum hardness of 55.5 GPa [2]. Researchers developed this superhard material through the optimization of two parameters: the high valence electron density (Re has the second highest valence electron density of all the transition metals) and the bond covalency (B, C, and N form the strongest covalent bonds). The obtained ReB$_2$ films were found to be superhard, as the intrinsic film hardness value (52 GPa) was close to that of bulk ReB$_2$ [3].

Transition-metal diborides, such as TiB_2, have also been studied extensively due to their high hardness and thermochemical stability [4,5]. In this paper, ReB_2 is introduced as a multilayer system and is combined with tantalum nitride (TaN) to synthetize ReB_2/TaN multilayers. We choose the TaN layer as a component of the multilayers because TaN and Ta have been extensively used in coating systems due to their good diffusion barrier properties and their relatively stable structure [6–12]. In addition, TaN is effective in forming nano-crystalline characteristics in multilayers [13–16]. In our previous work, we performed a preliminary study of a ReB_2/TaN multilayer coating. It was found that the modulation period can control the mechanical properties of the ReB_2/TaN multilayers and induce the highest hardness of 28 GPa and modulus of 345.9 GPa with a lower residual stress [17]. However, no simulation and calculation results of the interfacial model and hard mechanism were studied in order to advance the in-depth understanding of the relationship between the mechanical properties and the multilayered structure.

In most multilayers, for example, TiN/VN and Cu/Ni [18,19], the individual layers have the same crystal structure. In this work, therefore, TaN is chosen to achieve an isostructure with ReB_2. Theoretical works suggest that three kinds of models have been used for the strengthening mechanism in multilayers. The first is the analytical model that is based on the Hall–Petch powder law $\sigma \sim \Lambda^{1/2}$, where Λ is the modulated period thickness [19–23]. The strength/hardness versus the period fits to this law at nanometer length scales, since the dislocation pileups can be treated as a continuum. At the sub-micrometer length scale, too few dislocations reside in the pile-ups to be treated as a continuum, and the Hall–Petch relation must be modified. However, the dislocation pile-up-based models are incapable of describing the strength behavior at decreasing layer thicknesses at a few of tens of nanometers. The second type of model is based on the confined layer slip mechanism that involves the glide of the single Orowan-type loops that are bounded by two interfaces [24–26]. This builds on earlier works of a similar confined layer slip mechanism in plastic yielding thin films on substrates. The third kind of model involves atomistic simulations of dislocation transmission across interfaces that provide an upper bound estimation of the interface barrier strength. In our previous work, we also conducted a preliminary theoretical study of ReB_2/TaN multilayers, in which only the interface of the TaN(100)/ReB_2(001) was chosen to calculate the interfacial structure, which showed that the B-N interface had a stronger covalent bonding [27].

The aforementioned investigations focus exclusively on the dislocation generation and motion mechanisms. However, an additional complication for isostructural materials, which is also typically miscible, is that the interface broadening during deposition can decrease the hardness, making the interpretation of the hardness data more difficult [28]. The miscible, isostructural nanolayers that are discussed above are known to interdiffuse, and thereby to lose their enhanced hardness. Along with atomistic simulations of the interface, this paper presents new data and more in-depth interpretation on the hardness enhancement of ReB_2/TaN multilayers.

In this work, in addition to studying the experimental changes in the mechanical properties, modulation period, and modulation ratio, we particularly focused on the hardness mechanism and the relationship between the mechanical properties, residual stress, and interfacial structure using first-principles calculations that are based on density functional theory (DFT). For the first time (to the authors knowledge), we calculated and found the contribution of the interfacial energy, adsorption energy, and the interfacial stability to the residual stress release by controlling the appropriate interfacial structure.

2. Materials and Methods

2.1. Experimental Methods

To evaluate the influence of the layer thickness on the results, two series of ReB_2/TaN multilayers at different bilayers (modulation period, Λ and modulation ratio, $t_{ReB2}{:}t_{TaN}$) as well as monolithic coatings were synthesized using an Radio frequency (RF) magnetron sputtering system, which has been described previously in detail [29]. RF magnetron sputtering is a process that RF power is applied to magnetron sputter source to control sputter rate of the target. The biggest advantage of RF power supply is sputtering insulator targets, such as ReB_2 and TaN. The substrates were deposited 7 cm away from two water-cooled magnetron sputter sources with 99.95% stoichiometric ReB_2 and TaN targets in Ar (99.99%) at a total pressure of 0.5 Pa. Silicon (100) wafers were used as the substrate materials, which were chemically cleaned in an ultrasonic agitator in acetone and absolute alcohol, before being mounted in the vacuum chamber. Prior to deposition, the substrates were sputter cleaned for 10 min in pure argon plasma at -500 V. Different Λ and t_{ReB2}:t_{TaN} were achieved by controlling a computer-driven shutter. To enhance the adhesion between the multilayers and the substrate, an approximately 30 nm-thick Ta buffer layer was deposited on the Si substrates, after which the TaN and ReB_2 layers were deposited alternately. In the process of deposition, the sputtering power was 50 W for ReB_2 and 110 W for TaN, and a negative bias of 80 V was applied to the substrate.

The cross-section of the sample was examined with a field-emission scanning electron microscopy (SEM, SU8010, Hitachi, Japan). X-ray diffraction (XRD) and X-ray reflection (XRR) scans were performed with Cu K_α (40 kV, 20 mA, λ = 1.54056 Å) radiation in a D8A diffractometer (Bruker, Karlsruhe, Germany). The scans were performed while keeping X and ψ fixed and varying θ at intervals of 0.02° (the symbols have their usual meanings [30]). The modulation period Λ was determined by matching the XRR peak positions. The layer thickness ratio was then adjusted so that the relative peak intensities matched. Nanoindentation and nanoscratch measurements on the as-deposited multilayers were performed at room temperature using a Nano indenter system (XP, Agilent, CA, USA). The hardness and elastic modulus of the multilayers as a continuous function of depth from a single indentation were obtained using the continuous stiffness measurement (CSM) technique in the nanoindentation test. The triangular Berkovich diamond indenter tip was calibrated using fused silica [24,31]. Each sample was indented ten times, at a maximum load of 40 mN, which yielded typical maximum depths of 400 nm. The maximum load in the scratch test was up to 100 mN in order to measure the fracture resistance. Then, a post-scan was performed to measure the profile of the scratch surface. The variation in the chemical composition and the element chemical bonding states of the multilayers were analyzed using X-ray photoelectron spectrometer (XPS, PHI 5300, Kanagawa, Japan).

2.2. Theoretical

To reveal the layer's impact and hard mechanism in the multilayers, first-principles that was based on density functional theory (DFT) [32,33] was used to calculate the optimal electronic structure, according to the measured structure. The exchange correlation functional was treated using the generalized gradient approximation (GGA) with the Perdew–Burke–Ernzerhof (PBE) [34]. The interactions between ionic and valance electrons were described with the ultrasoft pseudopotential [35]. The k-point sampling and kinetic energy cutoff convergence have been tested for all calculated surfaces and interfaces. The theoretical calculation of all the surface and interfaces is done after the convergence test. The detailed theoretical calculation results of all the surface and interfaces will be given in the results and discussion section.

3. Results and Discussion

3.1. Microstructure Characterizations

The X-ray reflection (XRR) pattern that is shown in Figure 1a,b give the modulation information of the ReB_2/TaN multilayers at two different modulation period (Λ), and it is used to accurately calculate Λ value. The reflection peaks of different orders n in the XRR spectra occur at 2θ positions given by the modified Bragg's law [29]:

$$\mathrm{Sin}^2\theta = \left(n\lambda / 2\Lambda \right)^2 + 2\delta \tag{1}$$

where λ is the X-ray wavelength (1.5 Å), δ is related to the average reflective index, and Λ is the modulation period of a multilayer (bilayer period). A straight line is fitted according to spots of $\sin^2\theta$ vs. n^2 to determine δ and Λ, while the error of each spot was determined using the internal θ. Using Formula (1), 9.2 and 21 nm-Λ are obtained from the linear regressions of the $\sin^2\theta$ versus n^2 plots shown in Figure 1a and 1b, which are consistent with our design before the experiment. The strong multiple superlattice reflections that are resulting from the large difference in the X-ray scattering factors of the periodic layers indicate the existence of well-defined layered structures between two individual layers in the multilayer system, which are a beneficial characteristic for the hardness enhancement [24,29,31,36].

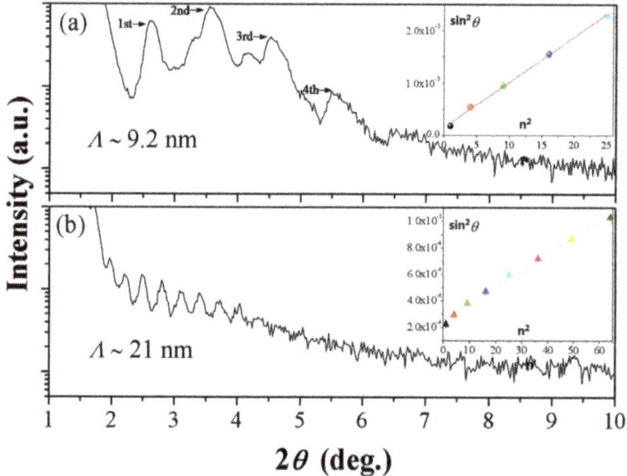

Figure 1. X-ray reflection (XRR) patterns of rhenium diboride (ReB$_2$)/tantalum nitride (TaN) multilayers at (**a**) $\Lambda\sim8$ nm, $t_{\text{ReB2}}{:}t_{\text{TaN}} = 1{:}2$; and (**b**) $\Lambda\sim21$ nm, $t_{\text{ReB2}}{:}t_{\text{TaN}} = 1{:}1$. The insets are the linear least squares fit of $\sin^2\theta$ vs. n^2.

Figure 2a,b show the high-resolution cross-sectional scanning electron microscopy (SEM) images at low and high magnifications from identical ReB$_2$/TaN samples. ReB$_2$/TaN samples, as measured by the X-ray reflection (XRR), are 418 nm thickness. An approximately 31-nm-thick Ta buffer layer is observed. The coating replays the multilayered nanostructure at $\Lambda\sim10$ nm and clearly indicates planar interfaces along the growth direction. The observed 10-nm-thick Λ is near the calculated 9.2 nm, as shown in Figure 1.

The XPS depth profiles of this multilayer are shown in Figure 2c. The periodic variation of the concentrations of Re and Ta as the main elements throughout the thickness gives direct evidence of the multilayered modulation structure in our design. The formation of alternating ReB$_2$ and TaN at the nanoscale is also confirmed. The elemental composition of Re:Ta = 1:2 within the bilayer thickness is nearly equal to the modulation ratio of $t_{\text{ReB2}}{:}t_{\text{TaN}} = 3.4{:}6.6$, as observed in Figure 2b. Figure 2c is artificially corresponded with Figure 2b to suggest that Re shows a darker contrast. The typical cross-sectional SEM images of Figure 2b $\Lambda\sim10$ nm and Figure 2d $\Lambda\sim20$ nm are taken from identical ReB$_2$/TaN samples that were used in the XRR measurements of Figure 1 and directly show a well-defined composition modulation and multilayered structure, which is in agreement with the XRR results above. The SEM results prove that the layered structures are defined well in broad nanolayers. The intermixing crossing interfaces can be observed in Figure 2a,b, due to the lack of sharp interfaces.

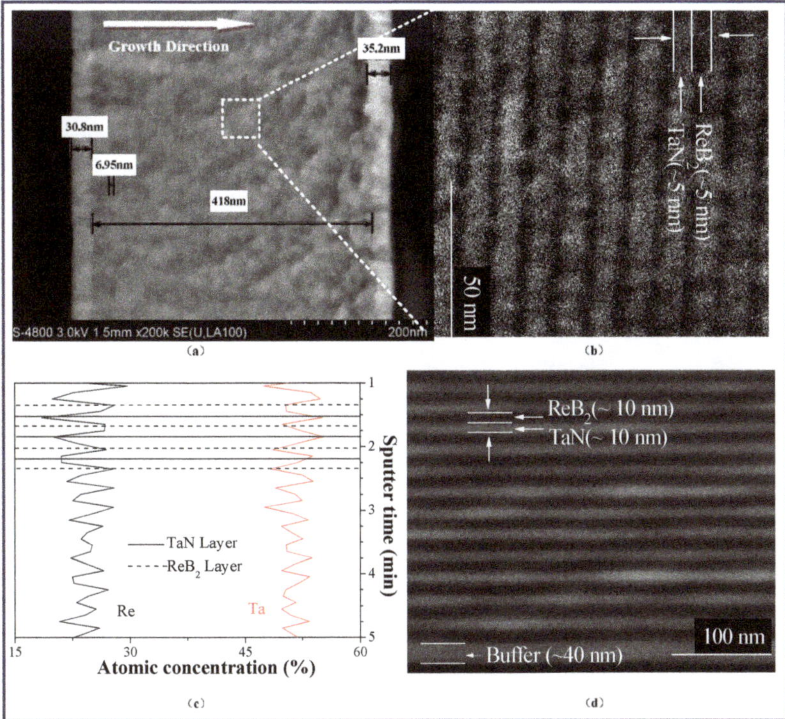

Figure 2. Cross-sectional scanning electron microscopy (SEM) images of ReB$_2$/TaN multilayer at (**a**) Λ~7 nm at low magnification; (**b**) 7 nm, t_{ReB2}:t_{TaN} = 1:2 at high magnification; (**c**) X-ray photoelectron spectrometer (XPS) depth profile of ReB$_2$/TaN multilayer at Λ~7 nm, t_{ReB2}:t_{TaN} = 1:2; (**d**) 20 nm, t_{ReB2}:t_{TaN} = 1:1 at high magnification.

3.2. Mechanical Properties

Figure 3 indicates the regularity of hardness and elastic modulus fluctuation versus Λ and t_{ReB2}:t_{TaN} for the ReB$_2$/TaN multilayers. To compare the multilayers, this figure also shows the hardness and elastic modulus values of monolithic ReB$_2$ and TaN coatings synthesized under identical deposition conditions. The error bars are drawn using the standard deviation that was calculated from the 10 indents. As we can see, the results agree well with the change regulations of other types of the multilayers [6,13–19,23,31,36]. For the ReB$_2$/TaN multilayers, the hardness reaches a maximum value of 38.7 GPa at Λ = 10 nm and t_{ReB2}:t_{TaN} = 1:1, and then decreases as Λ increases or t_{ReB2}:t_{TaN} decreases further.

The rule-of-mixtures hardness of the ReB$_2$/TaN multilayers can be calculated using [18]:

$$H_{ReB_2/TaN} = H_{ReB_2} t_{ReB_2} / \Lambda + H_{TaN} t_{TaN} / \Lambda \qquad (2)$$

The calculated rule-of-mixtures hardness of the ReB$_2$/TaN multilayers at Λ = 10 nm and t_{ReB2}:t_{TaN} = 1:1 is 17.6 GPa. Whereas, the measured hardness for this multilayer is up to 38.7 GPa, which is 120% higher than the rule-of-mixtures value. Detailed analyses of the deformation mechanisms and the interfacial model will be presented in Section 3.3.

Despite the high hardness, the performance of the multilayers depends on many other factors, including the residual stress, friction, fracture toughness, adhesive ability, etc. The reduction of residual stress in the multilayers is a key factor affecting their industrial applications. The residual stress σ is calculated applying the Stoney formula [37] and using the substrate curvature that is determined from a surface profiler:

$$\sigma = -\frac{E_s t_s^2}{6t_c(1-v_s)R} \tag{3}$$

where E_s (131 GPa), t_s (0.0005 cm) and v_s (0.28) are, respectively, the elastic modulus, thickness, and Poisson's ratio of the substrate; t_c is the coating thickness; and, R is the radius of curvature of the multilayer coated substrate. In this test, we choose a scan length of 2 cm for the R measurement. Although, in our study, a precise residual stress is difficult to determine using the available instruments, and this result reflects the reducing trend of the residual stress when compared with the monolithic layer.

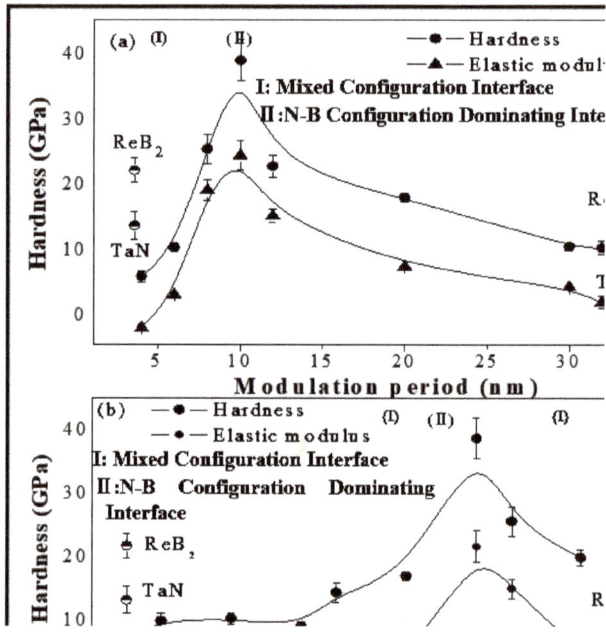

Figure 3. Hardness and Elastic modulus of ReB$_2$/TaN multilayers vs. (**a**) Λ; (**b**) $t_{\text{ReB2}}{:}t_{\text{TaN}}$.

Figure 4 indicates the residual stress, i.e., compressive stress, at different Λ and $t_{\text{ReB2}}{:}t_{\text{TaN}}$ for the ReB$_2$/TaN multilayers. To compare monolithic coatings with the multilayers, the residual stresses of individual ReB$_2$ and TaN coatings that are synthesized under identical deposition conditions are also shown in this figure. The error bars are drawn using the standard deviation, as calculated from the 10 indents. Nearly all of the multilayers exhibit lower residual stress than the average value of the monolithic ReB$_2$ and TaN coatings. The residual stress of the multilayers reaches the lowest value at $\Lambda\sim10$ nm and $t_{\text{ReB2}}{:}t_{\text{TaN}} = 1{:}3$. We believe that periodic insertion of TaN into ReB$_2$ layers suppresses the grain growth, which releases stress that is built up in the ReB$_2$ layers.

The elastic modulus of the coatings is another important factor to obtain good wear resistance. The H^3/E^2 ratio is a strong indicator of the coating's resistance to plastic deformation [38,39]. Table 1 listed these values of the multilayers with different parameters. The resistance plastic deformation of our hardest ReB$_2$/TaN multilayer is obviously improved ($H^3/E^2 = 25.8\%$) due to high hardness and a relatively low elastic modulus when compared with that of the other multilayers and monolithic ReB$_2$ and TaN coatings. Note that another important advantage of the multilayer nanostructured coatings is that one can fabricate superhard materials with identical hardnesses but different values of the plastic deformation (H^3/E^2). This means that superhard multilayers can be produced with different combinations of elastic and plastic properties, which provides a wide choice of multilayers for various specific tasks. Thus, the service life of multilayers can be enhanced while using multilayers that better fit the application substrates, such as steel, in terms of minimizing the internal stresses that occur at the coating/substrate interface.

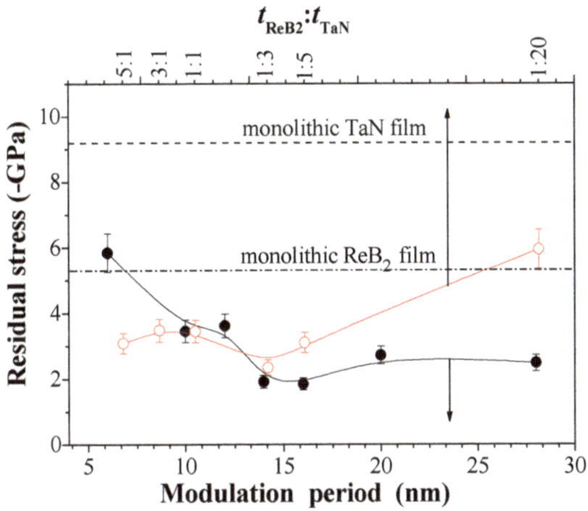

Figure 4. Residual stresses of ReB$_2$/TaN multilayers vs. Λ, $t_{ReB2}{:}t_{TaN}$.

Table 1. The values of H^3/E^2 for the multilayers (H^3/E^2 of monolithic ReB$_2$ and TaN coatings is 11.9% and 4.5%).

Λ (nm)	$t_{ReB2}{:}t_{TaN}$	H^3/E^2 (%)
4	1:1	4.35
10	1:1	25.8
30	1:1	4.09
10	5:1	4.89
10	1:2	11.5

Figure 5a-d show the results of the scratch test at different $t_{ZrB2}{:}t_{AlN}$ and Λ, reflecting the fracture resistance of the ReB$_2$/TaN multilayers. The post-scan curves are always above the scratch scan curves in test due to the plastic recovery after scratching. The normal load corresponding to the point in which the scratch scan profile shows an abrupt change is the critical fracture load L_c that can characterize the adhesion strength of the multilayers. The scratch scan profiles of all the multilayers indicate an abrupt increase point in the scratch depth, except the $t_{ReB2}{:}t_{TaN}$ = 1:1 and Λ = 10 nm sample. This means that the hardest multilayer at $t_{ReB2}{:}t_{TaN}$ = 1:1 and Λ = 10 nm has the highest fracture resistance and adhesion strength between the multilayer and the substrate, which fits the plastic deformation (H^3/E^2) results in Table 1 well. We believe that the improved fracture resistance appears to be directly related to a lower compressive stress, higher hardness, and strong plastic recovery of the coating with a multilayered structure.

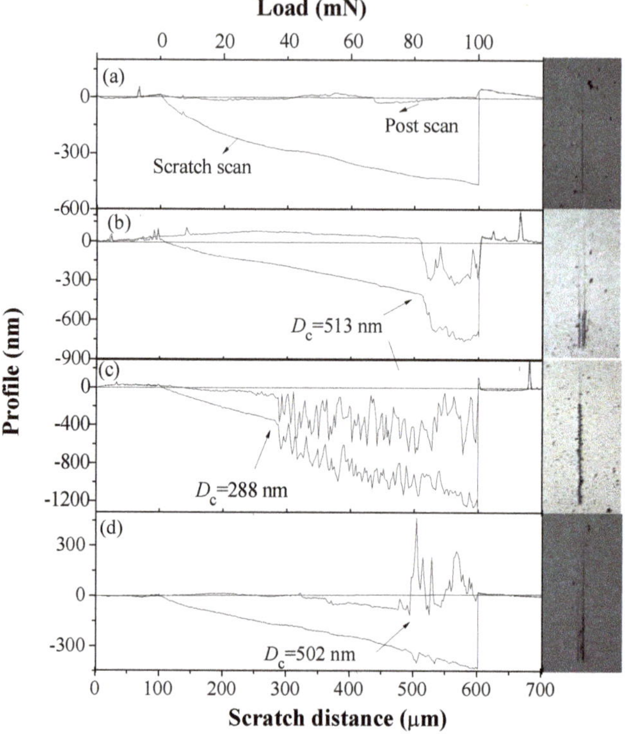

Figure 5. Surface profiles of the scratch scan, post scan, and scratch tracks on ReB$_2$/TaN multilayers at different t_{ZrB2}:t_{AlN} and Λ, (**a**) 1:1, 10 nm; (**b**) 1:4, 10 nm; (**c**) 2:1, 10 nm; and (**d**) 1:1, 28 nm.

3.3. Hard Mechanism

For the nano-indentations, the hardness is normally defined as the ratio of the maximum applied load divided by the corresponding projected contact area, i.e., $H = \frac{P_{max}}{A_C}$, where H, P_{max}, and A_C are the hardness, the maximum applied load, and the projected contact area at the maximum applied load, respectively. However, in this case, several additional observations concerning the behavior of the supperlattice materials can be seen from the load versus displacement curves. This is done by examining different combinations of the curves of the two epitaxially grown ReB$_2$/TaN multilayers and their monolithic components.

A comparison of the load versus displacement curves for ReB$_2$, TaN, and their multilayer at shallow indentation depths and the resulting integral curves are shown in Figure 6a,b, respectively. When the same displacement is pressed into the surface, the multilayer requires a larger load than two monolithic coatings, indicating that the multilayer is harder than both of the monolithic coatings. Figure 6c shows the typical loading-unloading sequences for the multilayers. The highest hardness occurs at Λ~10 nm and t_{ReB2}:t_{TaN} = 1:1, which agrees with the measured results that are mentioned above in Figure 3. The comparison in Figure 6b also indicates that ReB$_2$ and TaN initially follow the same loading pattern as the multilayer, but start to deviate from it at approximately 5 nm. Since the coherent interface repeat periods for multilayers ($\Lambda/2$~5 nm) appear at 5 nm first, this phenomenon provides direct evidence that the behavior of the interface as a barrier to dislocation motion begins to affect the deformation of multilayer, leading to an increase in the hardness.

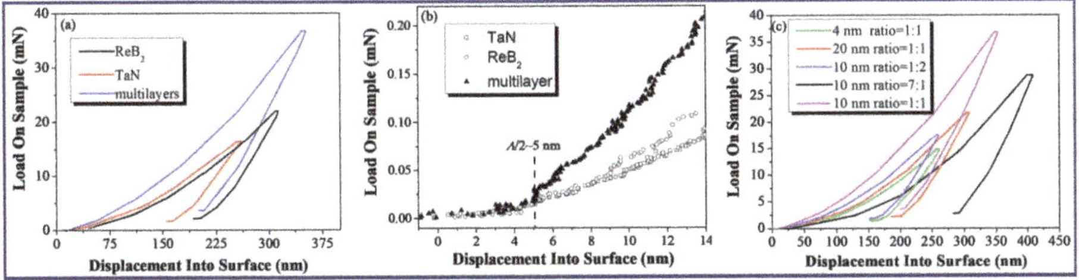

Figure 6. (**a**) Comparison of load *vs* displacement data of ReB$_2$/TaN multilayer (Λ~10 nm) with monolithic ReB$_2$ and TaN coatings; (**b**) Comparison of load vs. displacement data for ReB$_2$/TaN multilayer (Λ~10 nm) with monolithic ReB$_2$ and TaN coatings at shallow indentation depths; and, (**c**) Comparison of load vs. displacement data of ReB$_2$/TaN multilayers at different t_{ReB2}:t_{TaN} and Λ.

3.4. Theoretical Model, Calculation, and Discussion

To further explain the hardness mechanism and the relationship between the mechanical properties and the multilayered structure, we use first-principles that are based on density functional theory (DFT) to simulate the interfacial structure for which we must choose the appropriate interfacial configuration models to calculate and compare the interfacial energy, adsorption energy, charge density, and density of states (DOS).

To avoid the interactions between repeated slabs, a uniform vacuum width of 15 Å is employed while performing the calculations for all of the surfaces. The slabs are fully relaxed until the system energy is minimized. As a result, a plane wave cutoff energy of 280 eV and 300 eV is employed for the ReB$_2$(001) and TaN(111) surfaces, which assures a total-energy convergence of 10^{-5} eV/atom. The Brillouin zone sampling is set with 6×6×1 and 5×5×1 Monkhorst-Pace k-point meshes for the ReB$_2$(001) and TaN(111) surfaces, respectively. For the interfaces, the cutoff energy of the plane wave is chosen as 330 eV. Integrations in the Brillouin zone are performed while using the special k-points that are generated with 5×5×1 mesh grids.

In the initial calculation, we first have to choose some possible interfacial models to calculate some of the preliminary results. We obtain the surface of TaN(111) through the cutting of the TaN bulk (space group P-6M2) and the surface of ReB$_2$ (001) through cutting the ReB$_2$ bulk (space group P63/MMc). For the ReB$_2$/TaN multilayers, six possible interfaces, including B-N, BB-N, Re-N, B-Ta, BB-Ta, and Re-Ta exist in the connection. The supercell, including 129 atoms of the BB-Ta interface and crystal structures of TaN and ReB$_2$, can be seen in Figure 7(a). There are some different symmetry sites in each type of interface connection in Figure 7(a1-a4). We use both the N- and Ta-terminated sites of the TaN to simulate the ReB$_2$(001)/TaN(111) interface and to choose three high-symmetry sites on which the interface atoms could bond. Using the B-N interface as an example, the B atoms could occupy the site on top of N atoms, which is called the B-N-Top configuration (Figure 7(b1)), or on the bridge site between two N atoms, which is called the B-N-bridge configuration (Figure 7(b2)). When there is a Ta atom in the second layer directly below the B atoms, we call it the hcp-hollow site, denoted as the B-N-hcp configuration (Figure 7(b3)). Eighteen different interface configurations of ReB$_2$/TaN are chosen in the initial calculation. To simulate the interfaces of the multilayers, TaN and ReB$_2$ slabs of thirteen layers are cut through using the CASTEP software. Each unit cell structure comprised of thirteen layers of TaN and thirteen layers of ReB$_2$ is separated with a 15-Å-thick vacuum layer.

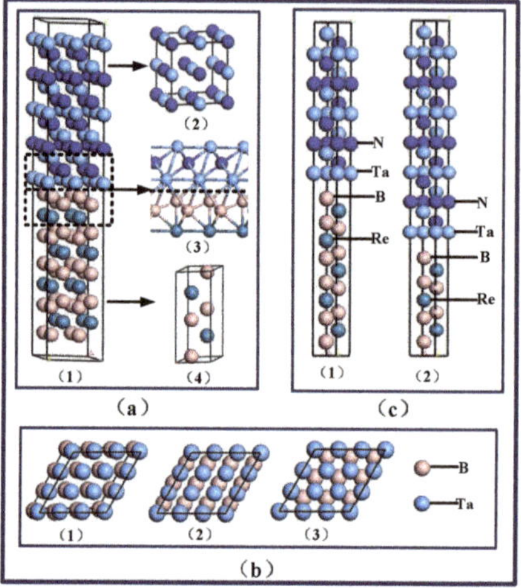

Figure 7. Supercell of BB-Ta interface (**a1**), crystal structures of TaN (**a2**) and ReB$_2$ (**a4**) and the interface bands (**a3**); Structures of BB-Ta-top configuration (**b1**), BB-Ta-bridge configuration (**b2**) and BB-Ta-hcp configuration (**b3**), respectively; Unit cell structures of 15ReB$_2$/15TaN interface (**c1**), and 9ReB$_2$/21TaN interface.(**c2**).

The adsorption energies (E_{ad}) corresponding to eighteen interfacial models with different interfacial atoms are calculated from the energy difference per unit area corresponding to the introduction of the interface in comparison to the two separate slabs [40,41]:

$$E_{ad} = \frac{E_{ReB_2(001)} + E_{TaN(111)} - E_{ReB_2(001)/TaN(111)}}{A} \qquad (4)$$

where $E_{ReB_2(001)/TaN(111)}$ is the total energy of the ReB$_2$/TaN, $E_{ReB_2(001)}$ and $E_{TaN(111)}$ are the total energies of the pure ReB$_2$ and TaN interlayers with the TaN and ReB$_2$ interlayers that are replaced by vacuum, respectively, in the same slab structure, and A is the area of the interface.

Table 2 lists the calculated E_{ad} of the ReB$_2$/TaN interface corresponding to eighteen different interfacial models. The general trend is that, the higher the adsorption energy, the stronger the chemical bonding at the interfaces. From Table 2, the obtained adsorption energy of the B-Ta interfaces is larger than that of other models, indicating that strength of this interfacial bonding is stronger than the others.

Table 2. The adsorption energy (E_{ad}) corresponding to eighteen interfacial models.

Stacking	E_{ad} (J/m^2)	Stacking	E_{ad} (J/m^2)	Stacking	E_{ad} (J/m^2)
B1-N hcp	−2.668	B1-N top	−3.244	B1-N bridge	−3.301
B2-N hcp	−2.668	B2-N top	−2.653	B2-N bridge	−2.665
Re-N hcp	0.145	Re-N top	0.063	Re-N bridge	1.110
B1-Ta hcp	6.341	B1-Ta top	6.028	B1-Ta bridge	7.046
B2-Ta hcp	6.316	B2-Ta top	6.296	B2-Ta bridge	7.213
Re-Ta hcp	4.814	Re-Ta top	4.090	Re-Ta bridge	5.142

To provide a detailed explanation of the relationship between the residual stress and interfacial energy, adsorption energy, and interfacial structure, the ReB$_2$(001)/TaN(111) interfacial configurations at five different thickness ratios. i.e., theoretical t_{ReB2}:t_{TaN}, are all built as B-Ta interfaces. When considering the ideal growth of the ReB$_2$ and TaN layers, B-terminated surfaces of ReB$_2$(001) and Ta-terminated surfaces of TaN(111) are built in the ReB$_2$(001)/TaN(111) configuration for the B-Ta interfaces because they exhibit higher adsorption energies than the others. To avoid the influence of the layer number on the total energy, each supercell contains thirty layers of $ReB_2(001)$ or $TaN(111)$. $ReB_2(001)/TaN(111)$ configurations at five different thickness ratios, namely, $3ReB_2/27TaN$, $9ReB_2/21TaN$, $15ReB_2/15TaN$, $21ReB_2/3TaN$, and $27ReB_2/3TaN$, are considered in the calculations. All the $ReB_2(001)/TaN(111)$ configurations at five different thickness ratios are B-Ta interfaces, each of which consists of a B-terminated $ReB_2(001)$ surface and a Ta-terminated $TaN(111)$ surface. Figure 7(c1) and (c2) display the supercells of the $15ReB_2/15TaN$ and $9ReB_2/21TaN$ configurations.

The interfacial energy (E_{inter}) shows how much weaker the interfacial bonding is, which is compared with the interlayer bonding in the bulk materials. The E_{inter} is calculated while using the following equation:

$$E_{inter} = \frac{E_{ReB_2(001)/TaN(111)} - \frac{N_{Re}}{n_{Re}}E_{Re}^{bulk} - \frac{N_B}{n_B}(E_{ReB_2}^{bulk} - E_{Re}^{bulk}) - N_{Ta}E_{Ta}^{bulk} - N_N(E_{TaN}^{bulk} - E_{Ta}^{bulk})}{A} - E_{sur}^{ReB_2(001)} - E_{sur}^{TaN(111)} \tag{5}$$

where $E_{ReB_2}^{bulk}$, E_{Re}^{bulk}, E_{TaN}^{bulk}, E_{Ta}^{bulk}, $E_{ReB_2(001)/TaN(111)}$, $E_{sur}^{ReB_2(001)}$, and $E_{sur}^{TaN(111)}$ are the total energy of the relaxed ReB_2 bulk material, Re bulk material, TaN bulk material, Ta bulk material, ReB_2/TaN interface, the energy of $ReB_2(001)$ surface, and the $TaN(111)$ surface, respectively; N_{Re}, N_B, N_{Ta}, and N_N denote the number of Re, B, Ta, and N atoms in the ReB_2/TaN interface, respectively; n_{Re} and n_B represent the number of Re atoms in the Re bulk material and that of B atoms in ReB_2 bulk material, respectively; and, A is the interfacial area. Formula (5) is attributed to the nonstoichiometric nature in the calculations to eliminate the effect of spurious dipole interactions that might bias the results. The E_{sur} that appears in Formula (5) is obtained from Formula (6),

$$E_{sur}^{XY} = \frac{E_{slab}^{total} - \frac{N_X}{n_X}E_X^{bulk} - \frac{N_Y}{n_Y}(E_{XY}^{bulk} - E_X^{bulk})}{2A} \tag{6}$$

where E_{slab}^{total}, E_{XY}^{bulk}, and E_X^{bulk} represent the total energies of a surface slab (ReB_2 or TaN), ReB_2 bulk (TaN bulk) and Re bulk (Ta bulk), respectively; N_X and N_Y denote the number of Re and B (Ta or N) atoms in the surface slabs, respectively; n_X and n_Y represent the number of Re atoms in Re bulk material (Ta atoms in Ta bulk material) and B atoms in ReB_2 bulk material (N atoms in TaN bulk material), respectively; and, A is the corresponding surface area. Due to the nonstoichiometric symmetric slabs in Formula (6), the effect of spurious dipole interactions can be eliminated to ensure the calculation is accurate [42].

The calculated results involving adsorption energy and interfacial energy are shown in Figure 8 as a function of the theoretical $t_{ReB2}:t_{TaN}$ for the $3ReB_2/27TaN$, $9ReB_2/21TaN$, $15ReB_2/15TaN$, $21ReB_2/3TaN$, and $27ReB_2/3TaN$ interfacial models. The higher the adsorption energy, the stronger the interfacial bonding and the lower the interfacial energy. Therefore, it can be clearly seen from Figure 8a,b that the change trend of adsorption energy is opposite to that of interface energy. The interfacial energy of a system measures the stability of the interface. The smaller the interfacial energy, the more stable the interfacial structure [43]. In particular, the interface will not form spontaneously when the interfacial energy of a system is higher than zero. According to Figure 8b, it is revealed that the $ReB_2(001)/TaN(111)$ interfacial structures at five different $t_{ReB2}:t_{TaN}$ that form spontaneously because the calculated interfacial energies of $ReB_2(001)/TaN(111)$ with five different models are negative. The interfacial energy of the $9ReB_2/21TaN$ structure (-1.297 J/m^2) is the lowest when compared with those of $27ReB_2/3TaN$, $21ReB_2/3TaN$, $15ReB_2/15TaN$, and $3ReB_2/27TaN$ by -0.920 J/m^2, -0.470 J/m^2, -0.424 J/m^2, and -0.505 J/m^2, respectively, indicating that $9ReB_2/21TaN$ is the most stable interfacial structure. This result is consistent with the adsorption energy. It is worth noting that the interfacial energy at different $t_{ReB2}:t_{TaN}$ follows the similar trend as that of the residual stress. A high residual stress easily causes some cracks in the coatings. Hence, the reduction of residual stress in the coatings is a key issue, which can improve their industrial applications. The interface with the $9ReB_2/21TaN$ structure, whose theoretical $t_{ReB2}:t_{TaN}$ is experimentally close to 1:3, shows the highest adsorption energy and the lowest interfacial energy, meaning that it is the most stable interfacial bond. It is clear from Figure 4 that the multilayer at a t_{ReB2}:t_{TaN} of 1:3 exhibits the lowest residual stress. We postulate that the low residual stress appears to be directly related to the stable interfacial bonding. In the theoretical calculation part, different atomic layer numbers are used to represent the different modulation ratios of nano-multilayers, which has been reflected in other papers [44] in which the TiAlN/ZrN nano-multilayer was investigated and the most stable interface was found at t_{TiAlN}:t_{ZrN} = 1:4. The difference for this work with the report of reference [44] in the modulation ratio of the most stable interface is due to the difference in the lattice structure and the interfacial bonding for different elements in the interface.

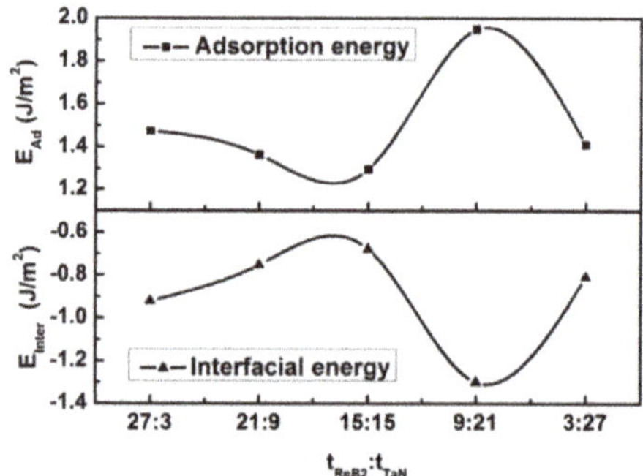

Figure 8. Adsorption energy (E_{ad}) and interfacial energy of ReB$_2$/TaN multilayers at different t_{ReB2}:t_{TaN} based on five B-Ta interfacial models.

To further clarify the behavior of the B-Ta interface with the 9ReB$_2$/21TaN structure, the charge densities, which express the spatial distribution of electrons in the system, and the charge density differences, which express the relative electron transfer for each atom during the construction of the interface system, are calculated. The charge density can be available directly, and the charge density difference is given in refs [45–48]:

$$\Delta\rho_{ReB_2/TaN} = \rho_{total} - \rho_{ReB_2} - \rho_{TaN} \tag{7}$$

where ρ_{total} is the total charge density of the ReB$_2$/TaN interface systems, and ρ_{ReB2} and ρ_{TaN} are the charge densities for the ReB$_2$ and TaN relaxed isolated slabs, respectively. The charge densities and the charge density differences in the interfaces are presented in Figure 9a,b, respectively. The short dashed line represents the interfaces because the selected plane can pass through the interfacial B and Ta atoms and directly provide the bonding interactions between them. From Figure 9a, due to the charge distributions in ReB$_2$ and TaN, the chemical bonding, which presents the covalent properties, can be found between the neighboring B and Ta atoms. Moreover, from Figure 9b, the charge transfer between the neighboring B and Ta atoms can be observed in which more blue parts means more electrons that the atoms loses, and more red parts means more electrons that the atoms receive. This indicates that some electrons transfer from Ta atoms to B atoms during the interface building, thereby forming ionic bonds between them. From the results, the chemical bonding between the neighboring B and Ta atoms in the interface system shows both covalency and iconicity.

For a more detailed understanding of the entire interfacial interactions, the density of states (DOS) for the entire B-Ta interface with the 9ReB$_2$/21TaN structure is calculated, as shown in Figure 10a. The DOS of the B-Ta interface shows that the wave overlaps the Fermi level, which demonstrates that the interface exhibits metallic characteristics. It is clear that the largest contribution to the total DOS is the d and p orbital electrons, while the s orbital electrons' contribution is relatively small. In Figure 10b, we calculate the density of states (DOS) of two interface linking atoms for the B-Ta interface in order to understand the interfacial interatomic influence. It is clear that the largest contribution to the interfacial DOS for the B1-Ta interface is the B-2p and Ta-5d orbitals. In addition, the covalency bonding also exists between the B and Ta atoms in the interface due to the electron orbitals.

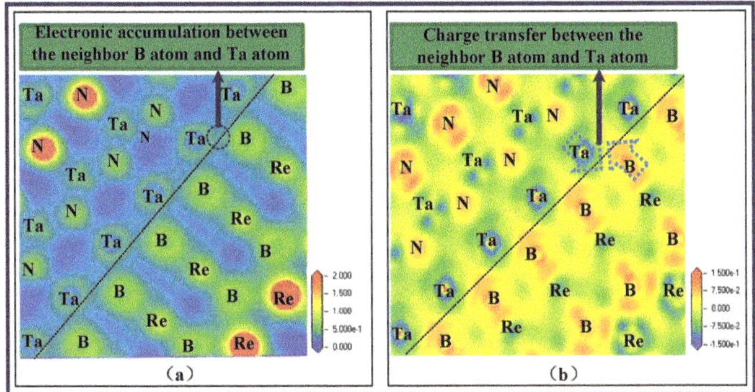

Figure 9. Charge density of $9ReB_2/21TaN$ interface (**a**) and charge density difference of $9ReB_2/21TaN$ interface (**b**), respectively, based on five B-Ta interfacial models.

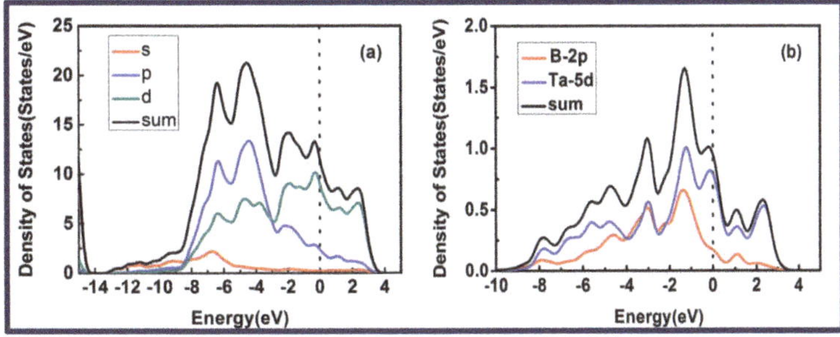

Figure 10. Density of states (DOS) of $9ReB_2/21TaN$ interface (**a**) and interface atoms of $9ReB_2/21TaN$ interface (**b**), respectively, based on five B-Ta interfacial models.

4. Conclusions

The ReB_2/TaN multilayers are well suited for fundamental and application studies of protective coatings, given their superior mechanical properties and their clear modulated structure. The interfacial models are set up and the hardness mechanism and the relationship between the mechanical properties and the multilayered structure are explained from the calculations of the interfacial energy, adsorption energy, charge density, and density of states (DOS) using first-principles based on density functional theory (DFT). The following conclusions are made:

(1) The microstructure evolutions in the ReB_2/TaN multilayers are carefully investigated by varying the modulation periods and modulation ratios. Clear coherent interfacial structures form between epitaxial layers at the optimal modulation period of 10 nm and the modulation ratio of 1:1. The fine nanocrystallites with small grain sizes are kept stable in individual layers at the optimal modulation condition.

(2) A maximum hardness of 38.7 GPa occurs at $\Lambda = 10$ nm and $t_{ReB2} : t_{TaN} = 1:1$. The highest multilayer also displays the highest fracture resistance and the highest resistance to plastic deformation.

(3) The shallow indentations show little difference in hardness between the monolithic coatings and the multilayers. However, variations in the load versus displacement curves are observed at

deeper indentation depths, indicating an enhancement of the hardness. One can deduce that the interface has a strong influence on the increase in the hardness.

(4)   Six possible multilayered interfaces, B-N, BB-N, Re-N, B-Ta, BB-Ta, and Re-Ta, including eighteen interface configurations of top, hcp, and bridge, are established. The highest adsorption energy, hence the best interface stability, occurs in the B-Ta interface configuration. The strengthening mechanisms of the multilayered structure are elucidated using the calculation results of the interfacial energies to advance the understanding of the relationship between the superior mechanical properties and the interfacial structure.

(5)   The $3ReB_2/27TaN$, $9ReB_2/21TaN$, $15ReB_2/15TaN$, $21ReB_2/3TaN$, and $27ReB_2/3TaN$ interfacial models are established to further explain the underlying mechanism for why the residual stress depends on the interfacial stability. The multilayers at a $t_{ReB2}{:}t_{TaN}$ of 1:3 exhibits the lowest residual stress, which agrees with the lowest interfacial energy and the highest adsorption energy of the $9ReB_2/21TaN$ interfacial structure. Therefore, the $9ReB_2/21TaN$ interfacial configuration is found to be the most stable interface, which is a main contribution to the residual stress release.

Author Contributions: Shangxiao Jin finished all experiments, measurements, and calculations. Dejun Li provide idea and design.

Funding: This work was supported by National Natural Science Foundation of China (51772209 and 51472180). This work was also supported by Program for Innovative Research in University of Tianjin (Grant No. TD13-5077).

Conflicts of Interest: The authors declare no conflict of interest.

References

1.   Gachon, J.; Rogachev, A.; Grigoryan, H.; Illarionova, E.; Kuntz, J.; Kovalev, D.; Nosyrev, A.; Sachova, N.; Tsygankov, P. On the mechanism of heterogeneous reaction and phase formation in Ti/Al multilayer nanofilms. *Acta Metall.* **2005**, *53*, 1225–1231. [CrossRef]

2.   Chung, H.; Weinberger, M.; Levine, J.; Kavner, A.; Yang, J.; Tolbert, S.; Kaner, R. Synthesis of Ultra-Incompressible Superhard Rhenium Diboride at Ambient Pressure. *Science* **2017**, *316*, 436. [CrossRef] [PubMed]

3.   Latini, A.; Rau, J.; Ferro, D.; Teghil, R.; Albertini, V.; Barinov, S. Superhard Rhenium Diboride Films: Preparation and Characterization. *Chem. Mater.* **2008**, *20*, 4507–4511. [CrossRef]

4.   Berger, M.; Karlsson, L.; Larsson, M.; Hogmark, S. Low stress TiB_2 coatings with improved tribological properties. *Thin Solid Films* **2001**, *401*, 179–186. [CrossRef]

5.   Losbichler, P.; Mitterer, C. Stoichiometry of unbalanced magnetron sputtered Al–Mg alloy coatings. *Surf. Coat. Technol.* **1997**, *97*, 567–573. [CrossRef]

6.   Kwon, K.; Ryu, C.; Sinclair, R.; Simon, S. Evidence of heteroepitaxial growth of copper on beta-tantalum. *Appl. Phys. Lett.* **1997**, *71*, 3069–3071. [CrossRef]

7.   Dupraz, M.; Poloni, R.; Ratter, K.; Rodney, D.; DeSantis, M.; Gilles, B.; Beutier, G.; Verdier, M. Wetting layer of copper on the tantalum (001) surface. *Phys. Rev. B* **2016**, *94*, 235427. [CrossRef]

8.   Lee, W.; Lin, J.; Lee, C. Characterization of tantalum nitride films deposited by reactive sputtering of Ta in N_2/Ar gas mixtures. *Mater. Chem. Phys.* **2001**, *68*, 266–271. [CrossRef]

9.   Yang, L.; Zhang, D.; Li, C.; Foo, P. Comparative study of Ta, TaN and Ta/TaN bi-layer barriers for Cu-ultra low-k porous polymer integration. *Thin Solid Films* **2004**, *462*, 176–181. [CrossRef]

10.   Muller, C.M.; Sologubenko, A.S.; Gerstl, S.; Suess, M.J.; Courty, D.; Spolenak, R.; Nanoscale, R. Cu/Ta multilayer deposition by co-sputtering on a rotating substrate. Empirical model and experiment. *Surf. Coat. Technol.* **2016**, *302*, 284–292. [CrossRef]

11.   Hecker, M.; Fischer, D.; Hoffmann, V.; Engelmann, H.; Voss, A.; Mattern, N.; Wenzel, C.; Vogt, C.; Zschech, E. Influence of N content on microstructure and thermal stability of Ta–N thin films for Cu interconnection. *Thin Solid Films* **2002**, *414*, 184–191. [CrossRef]

12.   Zhou, Y.M.; He, M.Z.; Xie, Z. Diffusion barrier performance of novel Ti/TaN double layers for Cu metallization. *Appl. Surf. Sci.* **2014**, *315*, 353–359. [CrossRef]

13. Yang, Y.H.; Wu, F.B. Microstructure evolution and protective properties of TaN multilayer coatings. *Surf. Coat. Technol.* **2016**, *308*, 108–114. [CrossRef]

14. Xu, J.; Li, G.; Gu, M. The microstructure and mechanical properties of TaN/TiN and TaWN/TiN superlattice films. *Thin Solid Films* **2000**, *370*, 45–49.

15. Baran, O.; Sukuroglu, E.; Efeoglu, I.; Totik, Y. The investigation of adhesion and fatigue properties of TiN/TaN multilayer coatings. *J. Adhes. Sci. Technol.* **2016**, *30*, 2188–2200. [CrossRef]

16. An, J.; Zhang, Q. Structure, Hardness and Tribological Properties of Nanolayered TiN/TaN Multilayer Coatings. *Mater. Charact.* **2007**, *58*, 439–446. [CrossRef]

17. Liu, G.; Zhang, S.; Liu, M.; Li, D. Effect of Modulation Period on Structure and Mechanical Properties of ReB_2/TaN Nanoscale Nanomultilayers. *J. Mater. Eng.* **2011**, *10*, 58–65.

18. Barnett, S. *Physics of Thin Films*; Academic Press: New York, NY, USA, 1993; Volume 1.

19. Anderson, P.; Li, C. Hall-Petch relations for multilayered materials. *Nanostruct. Mater.* **1955**, *5*, 349–362. [CrossRef]

20. Friedman, L.; Chrzan, D. Scaling Theory of the Hall-Petch Relation for Multilayers. *Phys. Rev. Lett.* **1998**, *81*, 2715–2718. [CrossRef]

21. Wang, P.; Wang, X.; Du, J.; Ren, F.; Zhang, Y.; Zhang, X.; Fu, E. The temperature and size effect on the electrical resistivity of Cu/V multilayer films. *Acta Mater.* **2017**, *126*, 294–301. [CrossRef]

22. Huang, H.; Spaepen, F. Tensile testing of free-standing Cu, Ag and Al thin films and Ag/Cu multilayers. *Acta Mater.* **2000**, *48*, 3261–3269. [CrossRef]

23. Misra, A.; Hirth, J.; Hoagland, R. Length-scale-dependent deformation mechanisms in incoherent metallic multilayered composites. *Acta Mater.* **2005**, *53*, 4817–4824. [CrossRef]

24. Philips, M.; Clemens, B.; Nix, W. A model for dislocation behavior during deformation of Al/Al3Sc (fcc/L12) metallic multilayers. *Acta Mater.* **2003**, *51*, 3157. [CrossRef]

25. Misra, A.; Hirth, J.; Kung, H. Single-dislocation-based strengthening mechanisms in nanoscale metallic multilayers. *Philos. Mag.* **2002**, *82*, 2935–2951. [CrossRef]

26. Torabinejad, V.; Aliofkhazraei, M.; Assareh, S.; Allahyarzadeh, M.; Rouhaghdam, A. Electrodeposition of Ni-Fe alloys, composites, and nano coatings-A review. *J. Alloys. Compd.* **2017**, *691*, 841–859. [CrossRef]

27. Jin, S.; Liu, N.; Zhang, S.; Li, D. The simulation of interface structure, energy and electronic properties of TaN/ReB_2 multilayers using first-principles. *Surf. Coat. Technol.* **2017**, *326*, 417–423. [CrossRef]

28. Chu, X.; Barnett, S. Model of superlattice yield stress and hardness enhancements. *J. Appl. Phys.* **1995**, *77*, 4403–4411. [CrossRef]

29. Wang, M.; Zhang, J.; Yang, J.; Li, D. Influence of Ar/N_2 flow ratio on structure and properties of nanoscale ZrN/WN multilayered coatings. *Surf. Coat. Technol.* **2007**, *201*, 5472–5476. [CrossRef]

30. Cullity, B. *Elements of X-ray Diffraction*; Addison Wesley: Reading, MA, USA, 1978; p. 512.

31. Lattemann, M.; Ulrich, S. Investigation of structure and mechanical properties of magnetron sputtered monolayer and multilayer coatings in the ternary system Si–B–C. *Surf. Coat. Technol.* **2007**, *201*, 5564–5569. [CrossRef]

32. Hamann, D.; Schluter, M.; Chiang, C. Norm-conserving pseudopotentials. *Phys. Rev. Lett.* **1979**, *43*, 1494–1497. [CrossRef]

33. Kresse, G.; Hafner, J. Ab initio molecular dynamics for liquid metals. *Phys. Rev. B* **1993**, *47*, 558–561. [CrossRef]

34. Perdew, J.; Wang, Y. Accurate and Simple Analytic Representation of the Electron-Gas Correlation Energy. *Phys. Rev. B* **1992**, *45*, 13244–13249. [CrossRef]

35. Vanderbilt, D. Soft self-consistent pseudopotentials in a generalized eigenvalue formalism. *Phys. Rev. B* **1990**, *41*, 7892–7895. [CrossRef]

36. Kong, M.; Shao, N.; Dong, Y.; Yue, J.; Li, G. Growth, microstructure and mechanical properties of (Ti, Al)N/VN nanomultilayers. *Mater. Lett.* **2006**, *60*, 874–877. [CrossRef]

37. Ohring, M. *The Materials Science of Thin Films*; Academic Press: New York, NY, USA, 1992; p. 416.

38. Shtanski, D.; Kulinich, S.; Levashov, E.; Moore, J. Structure and physical-mechanical properties of nanostructured thin films. *Phys. Solid. State* **2003**, *45*, 1177–1184. [CrossRef]

39. Chen, Y.; Polonsky, I.; Chung, Y.; Keer, L. Tribological properties and rolling-contact-fatigue lives of TiN/SiNx multilayer coatings. *Surf. Coat. Technol.* **2002**, *154*, 152–161. [CrossRef]

40. Finnis, M. The theory of metal-ceramic interfaces. *J. Phys. Condens. Matter.* **1996**, *8*, 5811–5836. [CrossRef]

41. Lin, Z.; Bristowe, Z. Microscopic characteristics of the Ag(111)/ZnO(0001) interface present in optical coatings. *Phys. Rev. B* **2007**, *75*, 205423. [CrossRef]

42. Yin, D.; Peng, X.; Qin, Y.; Wang, Z. Electronic property and bonding configuration at the TiN(111)/VN(111) interface. *J. Appl. Phys.* **2010**, *108*, 033714. [CrossRef]

43. Li, K.; Sun, Z.; Wang, F.; Zhou, N.; Hu, X. First-principles calculations on Mg/Al$_4$C$_3$ interfaces. *Appl. Surf. Sci.* **2013**, *270*, 584–589. [CrossRef]

44. Xu, Y.; Chen, L.; Pei, F.; Chang, K.; Du, Y. Effect of the Modulation Ratio on the Interface Structure of TiAlN/TiN and TiAlN/ZrN Multilayers: First-principles and Experimental Investigations. *Acta Mater.* **2017**, *130*, 281–288. [CrossRef]

45. Kohn, W.; Sham, L. Self-Consistent Equations Including Exchange and Correlation Effects. *Phys. Rev. A* **1965**, *140*, 1133–1138. [CrossRef]

46. Arya, A.; Carter, E. Structure, bonding, and adhesion at the TiC(100)/Fe(110) interface from first principles. *J. Chem. Phys.* **2003**, *118*, 8982–8996. [CrossRef]

47. Kim, H.; Yoo, S. Enhanced low field magnetoresistance in La$_{0.7}$Sr$_{0.3}$MnO$_3$-La$_2$O$_3$ composites. *J. Alloys Compd.* **2012**, *521*, 30–34. [CrossRef]

48. Wang, C.; Gao, H.; Dai, Y.; Ruan, X.; Shen, J.; Jun, W.; Sun, B. In-situ technique for synthesizing Fe–TiN composites. *J. Alloys Compd.* **2010**, *490*, 9–11. [CrossRef]

Article

Metal (Ag/Ti)-Containing Hydrogenated Amorphous Carbon Nanocomposite Films with Enhanced Nanoscratch Resistance: Hybrid PECVD/PVD System and Microstructural Characteristics

Marios Constantinou [1], Petros Nikolaou [1], Loukas Koutsokeras [1], Apostolos Avgeropoulos [2], Dimitrios Moschovas [2], Constantinos Varotsis [3], Panos Patsalas [4], Pantelis Kelires [1] and Georgios Constantinides [1,*]

[1] Research Unit for Nanostructured Materials Systems, Department of Mechanical Engineering and Materials Science and Engineering, Cyprus University of Technology, 3041 Lemesos, Cyprus; m.k.constantinou@cut.ac.cy (M.C.); petros.nikolaou@cut.ac.cy (P.N.); l.koutsokeras@cut.ac.cy (L.K.); pantelis.kelires@cut.ac.cy (P.K.)

[2] Department of Materials Science and Engineering, University of Ioannina, University Campus, 45110 Ioannina, Greece; aavger@cc.uoi.gr (A.A.); dmoschov@cc.uoi.gr (D.M.)

[3] Department of Environmental Science and Technology, Cyprus University of Technology, 3041 Lemesos, Cyprus; c.varotsis@cut.ac.cy

[4] Department of Physics, Aristotle University of Thessaloniki, 54124 Thessaloniki, Greece; ppats@physics.auth.gr

* Correspondence: g.constantinides@cut.ac.cy; Tel.: +357-25002626

Received: 27 February 2018; Accepted: 28 March 2018; Published: 30 March 2018

Abstract: This study aimed to develop hydrogenated amorphous carbon thin films with embedded metallic nanoparticles (a–C:H:Me) of controlled size and concentration. Towards this end, a novel hybrid deposition system is presented that uses a combination of Plasma Enhanced Chemical Vapor Deposition (PECVD) and Physical Vapor Deposition (PVD) technologies. The a–C:H matrix was deposited through the acceleration of carbon ions generated through a radio-frequency (RF) plasma source by cracking methane, whereas metallic nanoparticles were generated and deposited using terminated gas condensation (TGC) technology. The resulting material was a hydrogenated amorphous carbon film with controlled physical properties and evenly dispersed metallic nanoparticles (here Ag or Ti). The physical, chemical, morphological and mechanical characteristics of the films were investigated through X-ray reflectivity (XRR), Raman spectroscopy, Scanning Electron Microscopy (SEM), Atomic Force Microscopy (AFM), Transmission Electron Microscopy (TEM) and nanoscratch testing. The resulting amorphous carbon metal nanocomposite films (a–C:H:Ag and a–C:H:Ti) exhibited enhanced nanoscratch resistance (up to +50%) and low values of friction coefficient (<0.05), properties desirable for protective coatings and/or solid lubricant applications. The ability to form nanocomposite structures with tunable coating performance by potentially controlling the carbon bonding, hydrogen content, and the type/size/percent of metallic nanoparticles opens new avenues for a broad range of applications in which mechanical, physical, biological and/or combinatorial properties are required.

Keywords: hydrogenated amorphous carbon films; metallic nanoparticles; hybrid deposition system; nanoscratch; nanocomposites

---

1. Introduction

Hydrogenated amorphous carbon (a–C:H) thin films consist of carbon and hydrogen elements, where carbon atoms form dangling bonds with hydrogen or sp^1, sp^2, and sp^3 bonds with other two, three or four carbon atoms, respectively. The properties of such films are directly linked to the hydrogen content and the hybridization state of carbon bonds. In general, high sp^2 content promotes graphite-like properties, whereas high sp^3 percentages favor diamond-like characteristics. The sp^2 structure is relatively softer, more compliant and with lower coefficient of friction compared to the sp^3 structure that is stiff and hard but brittle. The introduction of hydrogen tends to soften the material, lower its coefficient of friction and increase its ductility [1]. Amorphous carbon films (hydrogenated or not) with a significant percentage of sp^3 hybridization, commonly referred to as diamond like carbon or DLC/DLCH in short, possess excellent properties like high stiffness and hardness, low coefficient of friction, high wear resistance, chemical inertness, low permeability, high melting point, optical transparency in infra-red (IR), high electrical resistivity and thermal conductivity [2–4]. These exceptional properties led to a series of industrial applications like protective coatings on microelectromechanical systems (MEMS), razor blades, hard disc drives and biomedical implants to gas barriers in polyethylene terephthalate (PET) bottles [5–7].

The main hindrance to further industrial exploitation of DLC films was poor scratch resistance which led to premature failure, primarily due to the high residual compressive stresses (in the GPa range) trapped during deposition and the ion subplantation process [8]. To circumvent this brittleness problem and allow for thicker and more stable films several recent studies proposed the doping of DLC with other elements [9–13]. Among the doping elements used were transition metals–typically termed metal containing hydrogenated amorphous carbon films, a–C:H:Me—or non-metal elements–typically termed as modified hydrogenated amorphous carbon films, a–C:H:X—such as silicon, oxygen, boron, nitrogen, and so forth. In essence, the idea lies in the formation of a particle reinforced nanostructured nanocomposite film for which the amorphous carbon matrix phase surrounds the soft metallic phase. For such composites, the toughening mechanism is based on continuum mechanics, where the cumulative composite response relates to the properties of the constituent phases, their relative fractions, the characteristic geometry of the dispersed phase (i.e., size, shape, dispersion and orientation) and finally the very nature of the bond between the matrix and the particulate nanoconstituents. Note that for thin films on substrate material the interfacial nature and residual stresses are also of important consideration. While there are other routes to improve the toughness of amorphous carbon films, like modifying the chemical structure, bonding characteristics or morphological details of the a–C matrix [7,14], we here concentrate on incorporating metallic nanoparticles into the hydrogenated amorphous carbon matrix.

The motivation of this study was the lack of (a) a concentrated study for the in-depth understanding of the structural and mechanical characteristics of a–C:H:Me and (b) deposition methodology in which the size of the reinforcing particles can be separately controlled and not be the result of the ion interaction with the substrate material. We aimed to develop a new generation of a–C:H:Ag and a–C:H:Ti films with improved combination of properties and performance. These characteristics have been achieved through the use of a hybrid deposition technology, a combination of Plasma Enhanced Chemical Vapor Deposition (PECVD) and Physical Vapor Deposition (PVD). The amorphous carbon matrix was generated by PECVD where carbon ions were generated by an RF plasma source by cracking methane, while the metallic nanoparticles (NPs) were generated through a nanoparticle source based on PVD technology. The motivation of using the above deposition system was associated with the flexibility of individually controlling the matrix quality and nanoparticle size/percent not currently achievable using conventional deposition techniques reported in the literature [15–19]. Such hybrid technologies have the ability to produce nanocomposites and laminar films for modern material needs, such as coatings for solar harvesting applications (i.e., by converting solar energy into thermal power), tribological coatings for energy reduction in vehicles (i.e., new tough coatings with low friction coefficient and density on engine components) and protective coatings in biomedical implants. The a–C:H:Ag and a–C:H:Ti nanocomposite films deposited herein were characterized using X-ray reflectivity (XRR), Raman spectroscopy, Atomic Force Microscopy (AFM) and Scanning/Transmission Electron Microscopy (SEM, TEM) for measuring density, thickness, chemical bonding, composition, roughness and nanoparticle size/shape. The nanoscratch response was quantified using an instrumented indentation platform. The transition metals of silver and titanium were selected as reinforcing nanoparticles due to their very good mechanical compatibility with a–C:H ($E_{Ag} = 70$ GPa, $H_{Ag} = 1$ GPa and $E_{Ti} = 115$ GPa, $H_{Ti} = 3\text{--}5$ GPa) and their tendency to reduce residual compressive stresses; an element with carbide-forming abilities, titanium, and an element with non-carbide forming tendency, silver, are investigated for comparison. Beyond their chemical and mechanical characteristics, Ag and Ti have been selected due to their optical characteristics—more precisely, due to the way they interact with light. For example nanoparticulate silver and titanium (in its nitride form), exhibit plasmon-resonance response when interacting with light [20], which translates into an enhanced optical absorption in the UV-visible-near IR range of wavelengths and can be opted in coatings for solar-harvesting applications [21,22]. In fact, their introduction into the amorphous carbon matrix generates a nanocomposite system with additional functionalities. Furthermore, beyond their optical characteristics, silver and titanium have been used in biomedical devices. While the therapeutic window where metallic nanoparticles are bactericidal but retain their biocompatibility characteristics with human cells is still a matter of debate and investigation [23–28], this type of nanocomposite systems (a–C:H:Me) could serve as candidate materials for such applications.

2. Materials and Methods

2.1. Deposition of Nanocomposite a–C:H:Me Films

2.1.1. Ion Beam Source

Metal-containing hydrogenated amorphous carbon films were deposited using a custom-made hybrid deposition system, which combines plasma enhanced chemical vapor deposition (PECVD) and physical vapor deposition (PVD) technologies. A schematic of the deposition chamber is presented in Figure 1a. PECVD was enabled through radio frequency (RF) ion-beam technology, a diagram of which is presented in Figure 1b. The ion beam source has an external RF antenna that spirals around the plasma tube in the form of a coil. The RF waves emitted by the antenna enter the transparent plasma tube to ionize the gas introduced therein to produce charged ions. The main chamber of the system was pumped down to 10^{-8} mbar (basic pressure) using a roughing and a turbo-molecular pump. The energetic carbon/hydrogen ions generated from this gas-cracking process (methane (CH_4) was used in this study) were accelerated towards the substrate by a voltage applied on a grid located between the plasma source and the substrate material. The voltage applied on the grid related to the kinetic energy of the ions. The transportation of ions from the source to the substrate occurs in line of sight conditions and a working pressure of approximately 10^{-3} mbar, the exact value of which depends on the total gas flow within the discharge tube. The accelerated ion species were deposited on the substrate material to grow hydrogenated amorphous carbon (a–C:H) films. The ion beam arrived at an incidence angle of $30°$ to the substrate which was located 22 cm away from the ion beam. In optimizing the density and deposition rate a parametric study on the effects of deposition conditions on the physical characteristics of a–C:H films preceded the deposition of the metal-containing nanocomposite films (a–C:H:Me).

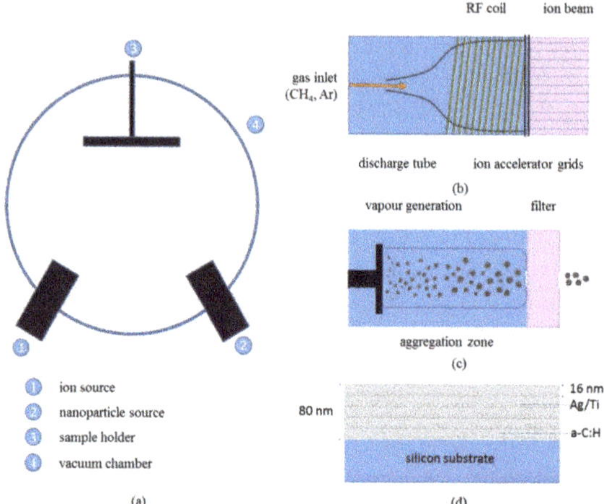

Figure 1. (**a**) Schematic of the hybrid PECVD/PVD system used within this study; Details of (**b**) the ion source and (**c**) the nanoparticle source; (**d**) Schematic of the metal containing hydrogenated amorphous carbon nanocomposite films deposited within this study (a–C:H:Ag and a–C:H:Ti).

2.1.2. Nanoparticle Source

Metal NPs were generated using NanoGen50 (Mantis Deposition Ltd., Thame, UK). The nanoparticle generator (Figure 1c) utilizes a variant of the PVD method, called Terminated Gas Condensation (TGC); this technology uses magnetron sputtering coupled with a condensation zone that is used to grow metallic NPs. More details on the operating mechanisms of TGC can be found in Ref. [29]. Here, silver or titanium were physically vapored by momentum transfer of argon ions onto the solid target (application of sputtering voltage/power promotes the creation and then acceleration of argon ions onto the solid target). The sputtered atoms of metal target nucleate and grow into larger clusters through collisions in the gas phase. The length of the condensation zone, which can be varied, affects the size distribution of the metallic clusters; longer times spend within the condensation zone result in higher collision in gas phase and thus larger clusters. The size of the metallic NPs have negative charge and can be consequently filtered by a quadrupole (complex of four rods) mass spectrometer MesoQ (Mantis Deposition Ltd., Thame, UK) which is located in series with the condensation zone (see Figure 1c). The quadrupole filter can be used to preselect NPs of specific size (mass) to pass through it and at the same time measure their current (which qualitatively relates to the number of NPs per unit area per time). In order to investigate the control on the particle size, Ag and Ti NPs were deposited on silicon substrates with sub-nanometer roughness of 0.4 nm (measured through AFM). Prior to deposition, the substrates were cleaned with compressed air to remove any possible dust particles or debris from their surfaces. Argon gas flow, sputtering current and condensation distance were set to 60 sccm, 60 mA and 8.5 cm, respectively; the working pressure in the main chamber was in the order of 10^{-3} mbar.

2.1.3. Hybrid Deposition of a–C:H:Me Nanocomposite Films

Nanocomposite films of a–C:H:Ag and a–C:H:Ti were deposited by sequential operation of the PECVD and PVD guns. A pattern of five layers of a–C:H films (each with ~16 nm thickness with a deposition rate of ~4 nm/min) and intermediate depositions of metallic NPs (selected nominal mean particle diameters of 4 nm for Ag and 10 nm for Ti) between each a–C:H layer was implemented. No energetic bias was applied on the substrate while the temperature remained at 25–35 °C. An RF power of 200 W was applied to a CH₄/Ar mixture of 6.0 sccm/0.5 sccm while the produced carbon ions were accelerated on the substrate material using 150 V grid voltage and a background pressure of 10^{-3} mbar to produce a–C:H layers with a density of 1.7 g/cm^3 (measured by X-ray reflectivity).

2.2. Characterization of Nanoparticles and Films

2.2.1. X-ray Reflectivity

An X-ray diffractometer (Rigaku Ultima IV, Tokyo, Japan) was used to measure the specular X-ray reflectivity of the various deposited films. The diffractometer was equipped with a Cu tube, operated at 40 kV accelerating voltage and 40 mA emission current. The incidence X-ray beam was collimated into a parallel beam with a 0.03 divergence and additionally monochromatized to Cu Ka (λ = 0.15419 nm) by a curved multilayer mirror. Density and thickness values of the thin films were extracted by fitting the respective experimental data to the theoretical reflectivity calculated using Parratt's formalism [13,30,31].

2.2.2. Raman Spectroscopy

The microstructural details of a–C:H:Me films were probed using Raman spectroscopy. This characterization method was employed in order to (a) access the bond characteristics of the deposited a–C:H films and indirectly link the information with sp^2/sp^3 configurations, and hydrogen content (i.e., through I_D/I_G and full-width at half-maximum of the G-peak, FWHM(G)) [32] and (b) trace the chemical modifications imparted on the nanocomposites, a–C:H:Ag and a–C:H:Ti, through the introduction of metal NPs. Raman data were collected by a confocal LabRAM (HORIBA Jobin Yvon, Kyoto, Japan) equipped with a CCD detector and 1800 grooves/mm grating. It is equipped with an Olympus BX41 microscope (10×, 15×, 40×, 50× and 100×). The 441.1 nm excitation laser beam was provided by a Helium-Cadmium laser (KIMMON KOHA, Fukushima, Japan). The laser power incident on the sample was 3–4 mW and the accumulation time was 15–20 min for each spectrum.

2.2.3. Atomic Force Microscopy

Atomic force microscopy (AFM) was used to quantify the size and number of metallic NPs and also to quantify the roughness of the resulting nanocomposite films. All measurements were performed in semi-contact mode using a scanning probe microscope (Ntegra Prima, NT-MDT, Moscow, Russia) equipped with an NT-MDT cantilever (NSG10) having a mean force constant of 11.8 N/m and a tip nominal radius of 6 nm. For NP size calibration AFM images were 500 nm × 500 nm in size collected with a tip scan rate of 1 Hz and a resolution 512 × 512 points in *x-y* direction. The surface topography of the deposited a–C:H and a–C:H:Me films was measured using contact mode and a CSG10 probe having a mean force constant of 0.11 N/m and a nominal tip radius of 6 nm. Images of 3 μm × 3 μm and 256 × 256 data density were collected and subsequently software-analyzed for quantifying the root mean square (RMS) roughness of the deposited a–C:H:Me surfaces.

2.2.4. Nanomechanical Testing

The nanotribological response of the deposited a–C:H:Me films was tested on an instrumented nanoindentation platform (Micro Materials Ltd, Wrexham, UK) using a friction bridge transducer and a conospherical diamond probe with a tip radius of 3.2 μm, as calibrated through elastic indentations on a material with known properties [33]. The three-pass experiment with 1 ramp-load scratch test between 2 topography passes was used for scratch testing all films with 6 repetitions, with a total scratch length of 300 μm, a load applied after 50 μm distance, a scan speed of 2 μm/s and a scratch loading rate of 1.6 mN/s to reach a maximum load of 200 mN. A 50 μm distance between each scratch was used. During all scratch tests friction data was collected that provided access to the friction coefficient.

2.2.5. Transmission/Scanning Electron Microscopy

Selected specimens were studied using TEM (JEM HR-2100, JEOL Ltd., Tokyo, Japan) operated at 200 kV in bright field mode. In studying the shape and size of the generated NPs, several depositions were made directly on formvar/carbon coated 300 mesh Cu TEM grids (Science Services GmbH, München, Germany). Residual scratches from nanomechanical measurements were also studied with SEM (Quanta 200, FEI, Hillsboro, OR, USA) at various magnifications in order to link microstructural characteristics with nanotribological metrics.

2.2.6. Residual Stress Measurements

Residual stresses generated within the film during deposition were estimated using Stoney's equation [34]:

$$\sigma_f = \frac{1}{6}\left(\frac{E_s}{1 - v_s}\right)\frac{t_s^2}{t_f}\left(\frac{1}{R}\right) \tag{1}$$

where $E_s = 170$ GPa and $v_s = 0.22$ are the silicon substrate modulus of elasticity and Poisson's ratio, and R is the radius of curvature calculated through height and cord measurements of cross-sectional data obtained from three dimensional non-contact optical profilometry surface profiles; height and cord data were recorded with 0.1 nm and 150 nm accuracy, respectively. The validity of Equation (1) was ensured by respecting that the fundamental assumptions on which it was derived persist [35,36]: (i) substrate and film thicknesses are significantly smaller than the plane dimensions, (ii) film thickness ($t_f \approx 80$ nm) is significantly smaller than the substrate thickness ($t_s = 375$ μm), (iii) substrate and film are homogeneous, and (iv) residual stresses induce equal bending in both x-y directions (i.e., spherical curvature and thus the biaxial stress is equal in the whole plate).

3. Results and Discussion

3.1. Ion Source and a–C:H Deposition Rate

Prior to the deposition of the nanocomposite films a series of pristine a–C:H films were deposited in order to study the effects of ion source characteristics on the deposition rate and physical properties of the films. The resulting thickness and density of the produced films were quantified through X-ray reflectivity measurements. The various depositions performed investigated the parameters that affect the ion beam characteristics and their link with microstructural characteristics of the films so to produce a–C:H with dense packing of carbon atoms and efficient deposition rates. In general, the ion beam process can be divided into three basic steps: (i) creation of ionized species, (ii) acceleration of species, and (iii) deposition and growth of the film; the controlling variables that affect those processes are the RF power, the gas flow, grid voltage, and substrate bias/temperature. For this study, substrate bias and temperature were kept at 0 V and 25 °C respectively.

In investigating the effect of gas flow rate and grid voltage a series of specimens were deposited while systematically varying these two parameters. Figure 2a shows that the deposition rate increases with grid voltage and the dependency is more pronounced for higher CH_4/Ar ratios. In the system presented herein the kinetic energy of the accelerated ions is controlled by the voltage applied to the plasma grid of the discharge quartz tube. It is apparent that the increase of grid voltage does not only affect the kinetic energy of the plasma generated ions but also the rate at which they pass through the grid, having as a result an increase in the amount of deposited species and subsequently film thickness. Furthermore, the deposition rate also increases when the CH_4 concentration in the tube increases. Figure 2b quantifies the effect of argon gas flow on the deposition rate of the a–C:H films. It is apparent that an increase in the relative volume fraction of argon flow relative to the total gas flow in the discharge tube (argon and methane) significantly reduces the resulting film thickness. It should be noted that the total gas flow rate was kept constant (6.5 sccm) such as the pressure within the chamber remained relatively unaffected. The highest thickness among the produced films is observed for Ar/CH$_4$ flow rates of 0.5 sccm/6.0 sccm (Ar volume fraction of ~8%) whereas for Ar/CH$_4$ flow rates of 2.5 sccm/4.0 sccm (Ar volume fraction of ~38%) the thickness of the deposited films is significantly reduced by −65%. This phenomenon is partly related to the fact that Ar ions generated within the plasma chamber and accelerated towards the substrate promote etching to the surface of the deposited material (it should be noted that argon atoms are significantly heavier than carbon atoms) thus decreasing deposition rate and thickness of a–C:H films. Furthermore, the increase of Ar volume fraction while maintaining a constant gas flow rate within the chamber leads to a gradual reduction in the CH$_4$ volume fraction which is the source of carbon ions, and its reduction inevitably leads to a reduction in the deposition rate and thickness of the deposited material. The density values of all films deposited herein vary within 1.52 g/cm^3–1.80 g/cm^3.

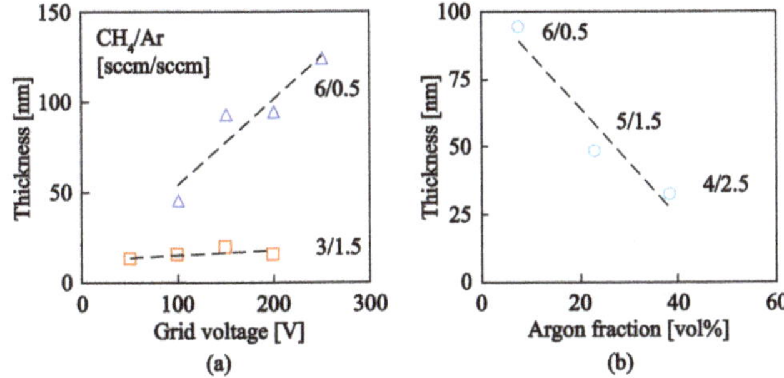

Figure 2. (a) Effect of grid voltage on film thickness for two different gas flow rate combinations: CH$_4$/Ar = 6.0 sccm/0.5 sccm = 12 and CH$_4$/Ar = 3.0 sccm/1.5 sccm = 2; results are for RF power of 200 W; (b) Effect of volume fraction on film thickness; results are for RF power of 200 W and grid voltage of 150 V and a constant total gas flow rate in the chamber (CH$_4$ + Ar = 6.5 sccm).

3.2. Ag and Ti Nanoparticles

A series of silver and titanium NP depositions were performed on silicon and TEM grid substrates in order to confirm the ability of the nanoparticle source to generate NPs with precise sizes and calibrate at the same time the deposition rates that were required for the controlled composition of the nanocomposite films. Table 1 shows details of the set of silver nanoparticle samples synthesized using NanoGen50 where the preselected nominal particle size has been varied while retaining a grounded substrate and a constant deposition flux (K), defined as the product of NP current (j) with deposition time (t):

$$K = j \times t \tag{2}$$

Table 1. Deposition details for a series of silver nanoparticle specimens with nominal diameters in the 4 to 12 nm range. Argon flow, magnetron position, and NP current where set at 60 sccm, 8.5 cm, and 60 mA respectively. The substrate was grounded and the size was selected using the MesoQ filter.

Samples	Working Pressure (mbar)	Nominal Size (nm)	Deposition Time (s)	J (nA)	K (nA·s)
Ag$_{12nm}$	4.9×10^{-3}	12	60	0.03	1.8
Ag$_{9nm}$	4.8×10^{-3}	9	18	0.10	1.8
Ag$_{6nm}$	4.8×10^{-3}	6	11	0.17	1.8
Ag$_{4nm}$	4.8×10^{-3}	4	45	0.04	1.8

Consequently, for a given ion current (which is experimentally tractable) the deposition flux can be easily controlled by tuning the duration of a given deposition.

Figure 3 shows TEM (Figure 3a,c) and AFM (Figure 3b,d) images of the generated Ag (Figure 3a,b) and Ti (Figure 3c,d) NPs. The images testify towards the ability of the nanoparticle source to generate a non-agglomerated group of almost spherical NPs with controlled diameters. In order to quantify the sizes of the Ag and Ti NPs the images were digitally analyzed. For all the samples, the size and number of NPs were counted by the threshold method [37,38] which makes the assumption that the NPs are spherical and rest on a substrate with minimal roughness. During the levelling process the particles were excluded from the fit to avoid local distortions of the data. Figure 4 shows the experimentally obtained diameters in comparison with their nominal values for the various Ag NPs synthesized herein. It is evident that the TEM results are in excellent agreement with the AFM data and confirm the ability of NanoGen to deposit NPs with controlled sizes and minimal size distributions. Some minor deviations between the experimental and nominal values can be attributed to the settings of the quadrupole filter which potentially could be optimized for even more refined correspondence between the two.

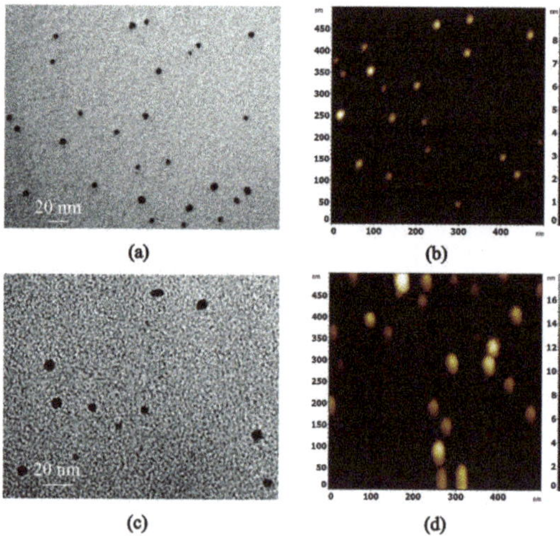

Figure 3. TEM (**a,c**) and AFM (**b,d**) images of Ag (**a,b**) and Ti (**c,d**) NPs synthesized using the nanoparticle source. Nominal diameters for Ag and Ti were 4 nm and 10 nm.

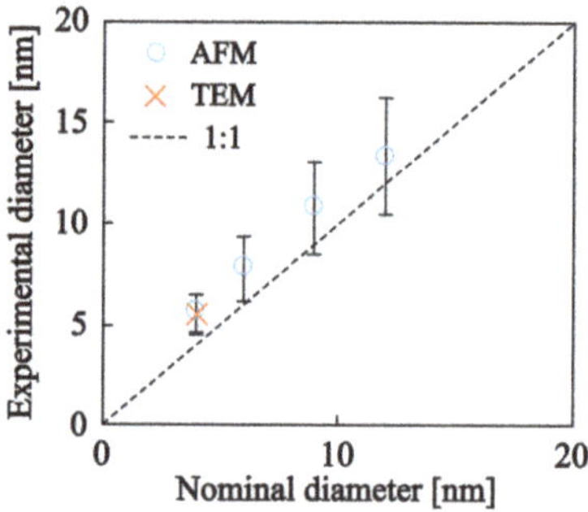

Figure 4. Experimental versus nominal NP diameters for a series of Ag nanoparticle depositions. Experimental diameters were obtained from AFM and TEM images after digital image analysis.

Figure 5a shows the particle density, calculated through AFM images in billions of NPs per square millimeter, as a function of the deposition time. The details of the samples synthesized towards this end are presented in Table 2 and the experiments were performed in order to quantify a measure of the deposition rate that allowed us to generate nanocomposite a–C:H:Me films with controlled compositions. The samples presented in Figure 5a had a nominal size of 4 nm and various deposition times preselected to achieve various coverages/densities. Figure 5b,c show AFM images of low and high NP coverage respectively. The AFM images were analyzed using the threshold routine and the values of NP density are plotted in Figure 5a. As expected, an increase in deposition time increases the number of NPs per mm^2 and this trend appears to be linear with a resulting deposition rate of 0.0722×10^9 mm^{-2} min^{-1}. The same experimental process has been followed for Ti NPs showing an almost identical deposition rate.

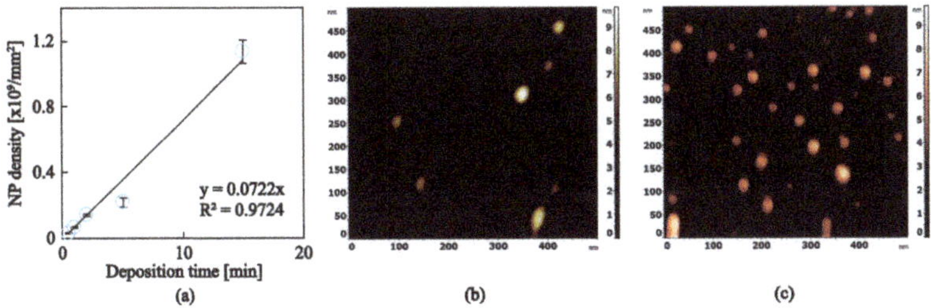

Figure 5. (**a**) Ag nanoparticle density for various deposition times; results relate to the samples shown in Table 2. AFM image of a (**b**) low deposition time and (**c**) high deposition time.

Table 2. Deposition conditions for 4 nm nominal diameter Ag NPs grown at various durations and 0 V substrate bias. Argon flow, magnetron position, and NP current were set at 60 sccm, 8.5 cm and 60 mA.

Samples	Working Pressure (mbar)	Nominal Size (nm)	Deposition Time (s)	*J* (nA)	*K* (nA·s)
Ag$_{30s}$	4.7×10^{-3}	4	30	0.3	9
Ag$_{60s}$	4.7×10^{-3}	4	60	0.3	18
Ag$_{120s}$	4.7×10^{-3}	4	120	0.3	36
Ag$_{300s}$	4.7×10^{-3}	4	300	0.3	90
Ag$_{900s}$	4.7×10^{-3}	4	900	0.3	270

3.3. a-C:H:Ag and a-C:H:Ti Nanocomposite Films

3.3.1. Microstructural Details and Bonding Characteristics

Details of the nanocomposite films prepared and tested within this study are shown in Table 3. The reported particle size relates to the experimentally obtained values (AFM and TEM analysis) whereas the metal contents were estimated through the calibration curves reported in Figures 4 and 5. The PECVD and PVD guns were operating in alternating turns (5 repetitions each) in order to generate multi-layer films of ~80 nm in thickness and intermediate nanoparticle depositions with concentrations as reported in Table 3. The hydrogenated amorphous carbon matrix used in all nanocomposite films was deposited using an RF power of 200 W, a gas mixture of CH$_4$/Ar = 6.0 sccm/0.5 sccm and grid voltage of 150 V, which resulted in a film density of 1.7 g/cm^3 as quantified through X-ray reflectivity measurements.

Table 3. Deposition details of the nanocomposite films grown in this study and the resulting surface roughness as quantified through AFM. PVD conditions: Argon flow, magnetron position, and sputtering current were set at 60 sccm, 8.5 cm and 60 mA. PECVD conditions: 6 sccm/0.5 sccm (CH_4/Ar), 200W RF power and 150 V grid current. The Me content is calculated through the calibration curve presented in Figure 5 and the measured NP size and a–C:H deposition rate.

Material	NP Size (nm)	Me Content (at.%)	RMS Roughness (nm)
a–C:H	-	0	0.4 ± 0.1
a–C:H:Ag$_{0.23at.\%}$	5.6	0.23	1.8 ± 1.0
a–C:H:Ag$_{0.33at.\%}$	5.6	0.33	2.0 ± 0.6
a–C:H:Ti$_{0.33at.\%}$	11.3	0.33	5.3 ± 0.9
a–C:H:Ti$_{0.56at.\%}$	11.3	0.59	4.8 ± 0.2

TEM images suggest that the metallic NPs are crystalline in nature (single crystals in most cases) and retain their original compositions, that is, neither Ag nor Ti NPs chemically react with the surrounding a–C:H matrix. Figure 6 shows a high resolution TEM image of a Ti NP embedded in an a–C:H matrix. A closer investigation at the Ti/a–C:H interphase coupled with a 2-dimensional fast Fourier transform of the cropped section suggests that (a) the nanoparticle exhibits d-spacings of 0.234 nm which correspond to the interplanar distance of Ti (002) planes, which translates to the fact that titanium carbide is not formed and (b) the aureole that is formed around the Ti NP shows characteristics of crystallinity with d-spacings on the order of 0.172 nm that can be related to the (004) planes of graphite, suggesting that the vicinity to the NPs surface graphitizes, the extend of which can be estimated up to 10 atomic planes (~3 nm). This is consistent with atomistic simulations [39] which reveal that the incorporation of transition metals into the amorphous carbon matrix leads to a process of C–C bond breaking and enables the formation of the lowest energy crystalline carbon state, that of graphite.

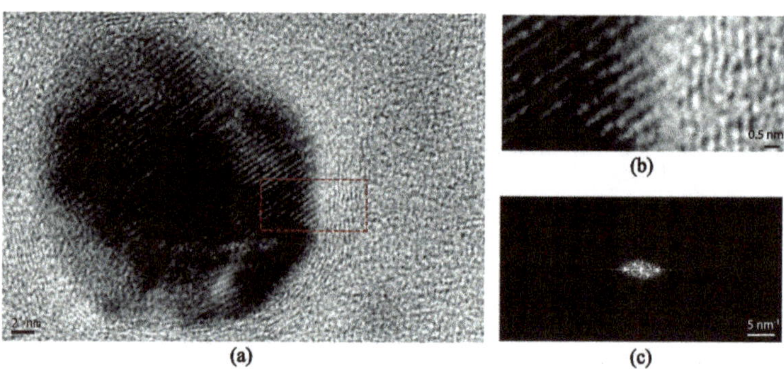

Figure 6. (a) High resolution TEM image of a Ti nanoparticle with the surrounding a–C:H matrix; (b) A magnified view of the a–C:H/Ti interface showing the transition in crystalline domains; (c) 2D FFT of image selection, identifying the two dominant d-spacings.

The Raman spectra for a–C:H and a–C:H:Me nanocomposites are presented in Figure 7a. All spectra exhibit the characteristic shape for amorphous carbon; the cumulative response was deconvoluted to the D-band (~1350 cm^{-1}) and G-band (~1550 cm^{-1}) contributions using Gaussian fits. Several important metrics, including the location of the G peak, the intensity ratio of D over G peaks (I_D/I_G) and the full width at half maximum of the G Peak (FWHM (G)) were extracted and the results are shown in Figure 7. In general, the D peak is due to the vibration of sp^2 rings and the G peak to the resonance of the sp^2 atoms organized in both rings and chains. Subsequently, the higher the I_D/I_G ratio the higher the sp^2 clustering within an a–C:H sample.

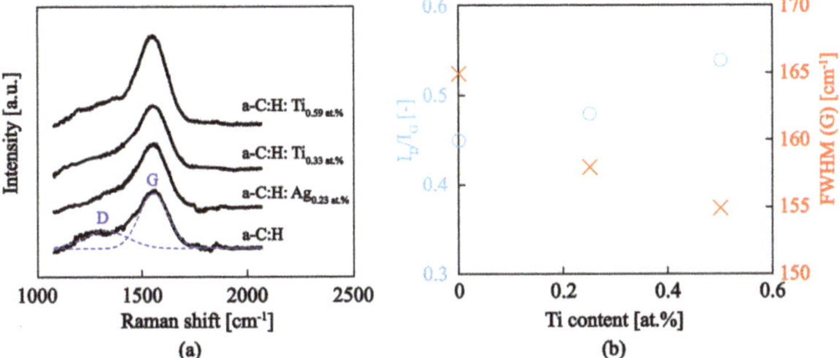

Figure 7. (**a**) Raman spectra for the a–C:H:Me nanocomposite films tested within this study showing also the deconvoluted Raman spectra for a–C:H, fitted by Gaussian curves at D and G carbon band resonant frequencies. (**b**) Extracted I_D/I_G and FWHM(G) as a function of titanium content for the a–C:H:Ti films synthesized herein.

The intensity ratio of a–C:H yields a value of $I_D/I_G = 0.45$. Empirical relations obtained from experimental data on a large collection of data on hydrogenated amorphous carbon films suggest that this intensity ratio is inversely related to both hydrogen content [32] and (indirectly through the Tauc gap) sp^3 hybridization state [40]. A comparison of this experimentally obtained value with literature data suggests that the pure a–C:H film synthesized within this study consists of a hydrogen content of 20–25 at.% and an sp^3 content of approximately 50 at.%. An a–C:H film with such characteristics is commonly referred to as diamond-like a–C:H (DLCH) with density values that vary between 1.5 g/cm^3 to 1.8 g/cm^3 and high sp^3 bonds (up to 70 at.%), a significant percentage of which are hydrogenated terminated [1,41]. Indeed, our a–C:H matrix with a density of 1.7 g/cm^3 (measured through XRR), hydrogen content of ~25 at.% and sp^3 content of ~50 at.% falls within the DLCH category [32]. It should be noted that the empirical relations utilized herein were obtained with an excitation wavelength of 514 nm, whereas the wavelength utilized in Ref. [32] was 441 nm. Nevertheless, experimental evidence suggest that the effect of excitation wavelength on the I_D/I_G is minimal and therefore the curves obtained with 514 nm excitation are not expected to significantly deviate from the 441 nm results [1,42].

The FWHM(G) probes the structural disorder of the sp^2 clustering in amorphous carbon material [40,43]. A lower FWHM(G) value denotes an a–C:H with less unstrained sp^2 clustering, whereas a higher FWHM(G) value suggests a material with an increased disordering in bond lengths and angles for the sp^2 clusters. The reduction of FWHM(G) with metallic doping (from 165 to 155, see Figure 7b) suggests that the introduction of Ti NPs tends to reduce the structural disorder which is consistent with the graphitization of the surrounding matrix as evidenced in TEM images (Figure 6) and the observed increase in the I_D/I_G ratios and the subsequent reduction of the residual stresses through the relief of strain energy introduced in distorted bond lengths and bond angles. Furthermore, the G peak position shifted slightly to higher wavenumbers, in particular from 1552 cm^{-1} to 1559 cm^{-1} with the incorporation of Ag or Ti metal within a–C:H matrix resulting to higher sp^2 clustering. This is consistent with the FWHM(G) reduction which also signifies an increase of sp^2 clustering as it falls following a reverse trajectory from amorphitization for composite samples. Our results are in very good agreement with literature data [17,44,45].

The surface roughness of a–C:H films deposited under controlled conditions onto a flat silicon substrate is measured at 0.4 ± 0.1 nm which is consistent with hydrogenated amorphous carbon values reported in the literature. The introduction of metallic NPs within the a–C:H matrix increases the surface roughness of the a–C:H:Ag and a–C:H:Ti nanocomposite films as presented in Table 3. The increase in roughness for nanocomposites can be attributed to nanoscale protrusions that are generated beyond the NPs as an overlay of a–C:H film is deposited. It is interesting to note that the increase in the roughness for the a–C:H:Ti is more pronounced which is probably related to the higher nanoparticle size used for these nanocomposites in comparison to a–C:H:Ag. Given that surface roughness is one of the parameters that control the interaction of cells with the substrate material they reside [46,47], the ability to manipulate roughness could be exploited for potential use of these materials in biomedical applications.

Residual compressive stresses as calculated through curvature measurements and the Stoney equation are shown in Figure 8. Consistent with literature data, a–C:H films deposited using PECVD contain a significant amount of stresses (~2.3 GPa) that are trapped within the material during the deposition process. The introduction of metallic NPs appears to have a positive effect on the nanocomposite response as a significant proportion of the residual stresses are relaxed: -26% for Ag and -33% for Ti. This is related to the graphitization of the matrix with the introduction of metallic NPs and the release of energy trapped within angular and linear bond-distortions. This observation is also in line with the Raman results presented above. This stress-reduction mechanism increases the stability of the film and enhances the required critical load for film delamination, as it is evidenced below.

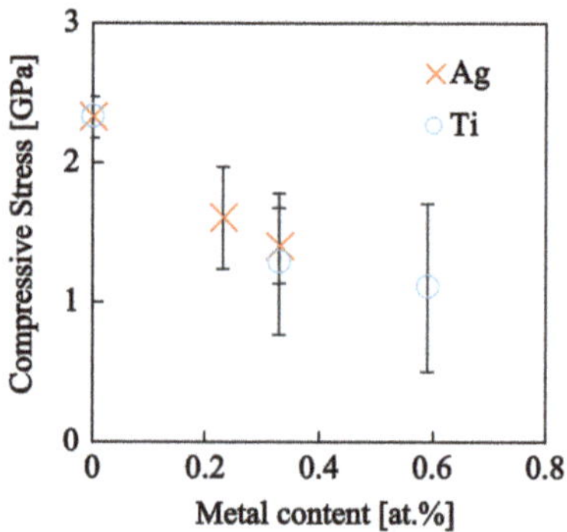

Figure 8. Residual stresses of the nanocomposite films as a function of the metal content.

3.3.2. Nanotribological Response: Scratch Resistance and Friction Coefficient

The tribological performance of all a–C:H and a–C:H:Me nanocomposites deposited within this study was investigated by the nanoscratch method [48–51] and the extracted critical loads for yielding, cracking, and delamination are reported in Table 4. Figure 9 shows typical data from a three-pass scratch test on a–C:H film, using a conospherical probe. A low load scan before and after the scratch test provided topography information that (a) were used to correct data for initial background tilt and topography and (b) detect film/substrate deformation responses due to applied contact pressures. The key features that are observed in scratch tests are: (a) the load related to the departure from elastic to plastic deformation (denoted as P_y), (b) the load to initiate film edge cracking (denoted as P_{CL1} and P_{CL2}), and (c) the load for full film fracture (denoted as P_{CL3}). The load required to initiate plastic deformations can be detected from the deviation of the residual scratch depth from initial topography that is an increase of the residual depth (topography scan during pass 3, Figure 9b). It should be noted that P_y is indicative to the substrate initial plastic deformation and not to the film properties itself as evidenced by calculations on the maximum stress location which is well within the substrate material [12,52]. An additional step on the residual scratch depth reveals film crack initiation while any large fluctuations of residual scratch depth relate to film fracture and delamination from the substrate. Furthermore, the friction force evolution with scratch distance (Figure 9c) is an alternative convenient means to detect/confirm P_{CL3} for delamination fracture. Plastic deformation, cracking and film failure changes were also confirmed microscopically from SEM images. For the pristine a–C:H the average critical loads for yielding, cracking and failure were P_y = 5.3 mN, P_{CL1} = 18.2 mN, P_{CL2} = 48.0 mN, and P_{CL3} = 114.8 mN. The extracted critical loads for all films tested herein are summarized in Table 4.

Table 4. Summary of tribomechanical metrics extracted from nanoscratch tests.

Samples	P_y (mN)	P_{CL1} (mN)	P_{CL2} (mN)	P_{CL3} (mN)	COF (-)
a–C:H	5.3 ± 0.6	18.2 ± 2.7	48.0 ± 7.7	114.8 ± 24.7	0.035 ± 0.002
a–C:H:Ag$_{0.23at.\%}$	4.5 ± 0.3	22.6 ± 3.7	54.2 ± 9.1	156.6 ± 7.5	0.019 ± 0.004
a–C:H:Ag$_{0.33at.\%}$	5.4 ± 0.7	23.5 ± 3.8	54.7 ± 9.3	162.8 ± 12.5	0.020 ± 0.003
a–C:H:Ti$_{0.33at.\%}$	4.7 ± 0.6	18.5 ± 2.9	49.8 ± 2.9	123.8 ± 31.5	0.041 ± 0.004
a–C:H:Ti$_{0.59at.\%}$	4.9 ± 0.4	18.8 ± 1.7	52.0 ± 5.0	133.0 ± 13.1	0.040 ± 0.006

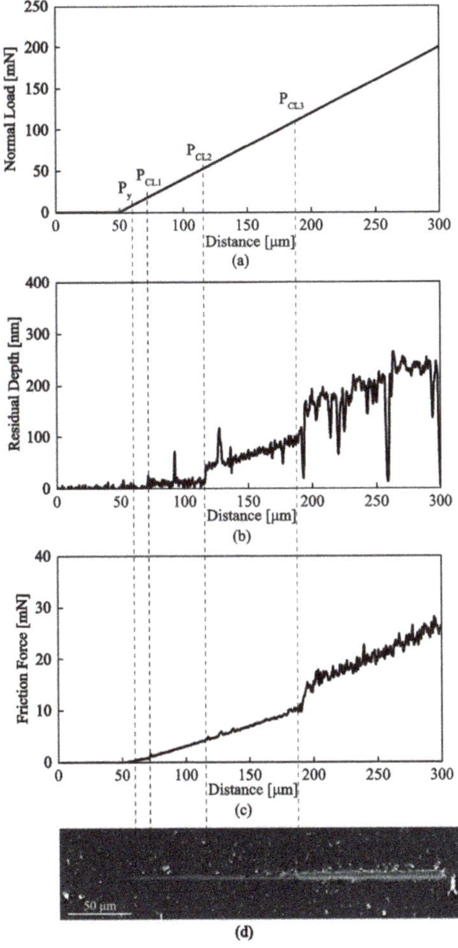

Figure 9. Typical results from a nanoscratch test on a C:H film. (a) Applied load, (b) residual depth and (c) resulting frictional force as a function of the scratch distance; (d) SEM image of the residual nanoscratched imprint.

The a–C:H, a–C:H:Ag, and a–C:H:Ti films exhibited similar critical loads for transitioning from elastic to plastic deformations. More precisely, the average P_y values for these films were found to be 5.3 mN, 4.5 mN, 5.4 mN, 4.7 mN and 4.9 mN respectively. Beyond that, the average values of the critical loads required for cracking initiation, P_{CL1}, were found to be 18.2 mN, 22.6 mN, 23.5 mN, 18.5 mN and 18.8 mN respectively; for all nanocomposite cases the average values were higher than the value for pristine a–C:H film. Furthermore, the delamination loads found for composite films in all other cases were higher compared to the neat a–C:H film; this can also be clearly observed by comparing the residual imprints. Indicative SEM images of residual imprints on pristine and metal containing nanocomposite films are shown in Figure 10, where it is evident that the initiation of delamination is shifted to higher distances/loads.

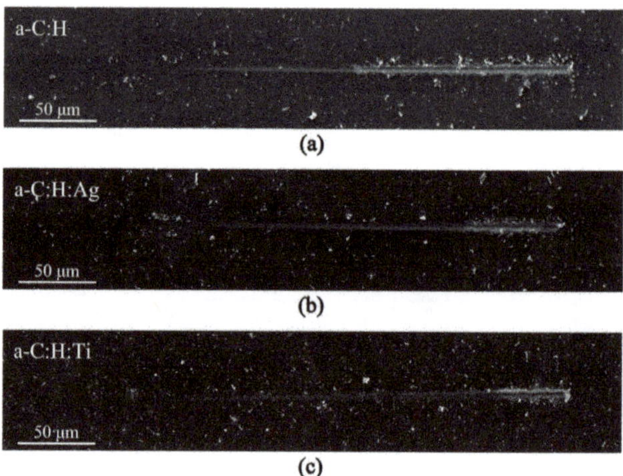

Figure 10. Characteristic residual scratches on (**a**) a–C:H, (**b**) a–C:H:Ag and (**c**) a–C:H:Ti films.

The enhancement in scratch resistance exhibited by a–C:H:Ag and a–C:H:Ti systems (Figure 11) is in line with bonding characteristics extracted from Raman measurements. The increase of (I_D/I_G) ratio and decrease of structural disorder through FWHM(G) value with metal content imply an increase of sp^2 clustering and deviation from amorphitization trajectory and hard materials. It is therefore clear that silver and titanium promote the graphite-like properties and both reduce hardness and increase ductility, toughness and abrasion resistance as well. These conclusions are well supported from experimental and theoretical studies in literature; concerning the increase of sp^2 clustering with the presence of amount and size of silver [39,45,53], while the decrease of hardness and increase of abrasion resistance is discussed in [12].

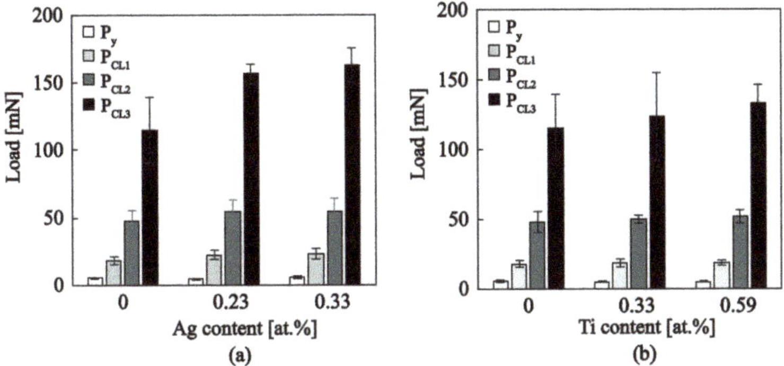

Figure 11. Critical loads for yield, cracking and delamination as quantified through scratch tests for (a) a–C:H:Ag and (b) a–C:H:Ti.

For such systems, residual stresses play a pivotal role in the delamination and early fracture during scratching. Residual stresses measured for a–C:H:Me films were found to decrease compared with a–C:H (see Figure 8). The reduction of residual stresses coupled with the chemical, physical and synthesis characteristics of metallic particles appears to be beneficial for the abrasion resistance of the nanocomposite films. For example, the low elastic modulus and hardness of Ag ($E \approx 70$ GPa, $H \approx 1$ GPa), coupled with its low wettability and inertness with carbon, and good dispersion and geometrical characteristics leads to a reduction of residual stresses, the development of a nanocomposite film which enhances ductility, toughness and scratch resistance to higher values in contrast to the pure a–C:H matrix. Titanium with an elastic modulus of $E \approx 115$ GPa and hardness of $H \approx -5$ GPa exhibits similar enhancements on the tribological response of the nanocomposite films. The lower improvements can be attributed to the higher roughness (probably caused by the higher nanoparticle size, 11.3 nm compared to 5.6 nm for silver) that amplifies the ploughing contribution of roughness and subsequently the frictional resistance (see Figures 12 and 13).

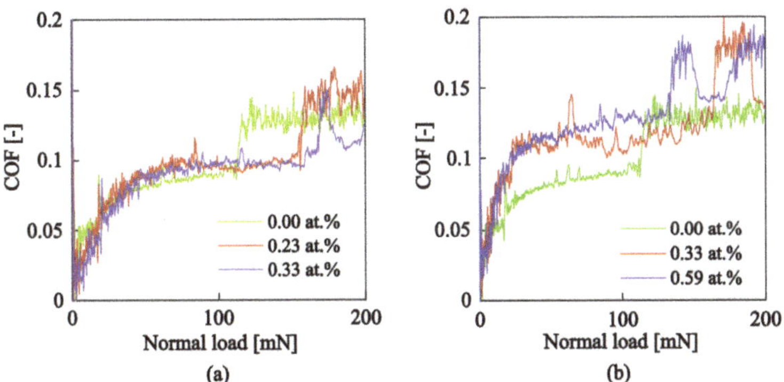

Figure 12. Coefficient of friction as a function of applied load for (a) a–C:H:Ag and (b) a–C:H:Ti nanocomposite films.

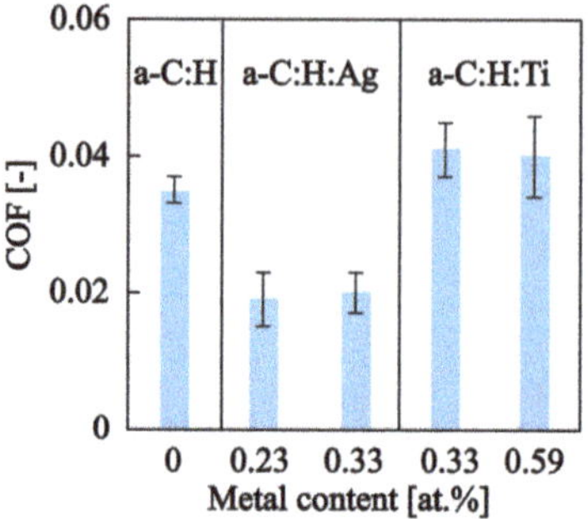

Figure 13. COF values for the a–C:H and a–C:H:Me films deposited in this study. The COF values reported correspond to the average values calculated within the elastic domain ($P < P_y$).

The coefficient of friction (COF) is another property of great significance which controls the tribological characteristics of these type of films. Figure 12a,b shows the evolution of COF with the applied load during the scratch test for a–C:H:Ag and a–C:H:Ti respectively. For low loads the contact pressure is below the yield limit of either the film or the substrate and the COF is very low (~0.02–0.04). As the contact pressure increases COF increases in a non-linear fashion with the applied load, subsequently reaching constant values until the film fails which can be noted with a sudden increase in the COF vs. applied load response. The increase of COF with the applied load relates to the plasticity and ploughing effect of diamond probe within the a–C:H film [54]. As COF increases the stress beneath the contacting probe becomes more significant with the possibility of failure to follow the probe [55]. Such a response is not directly detectable in the COF vs. applied load evolution but can be observed in the residual surface profiles through probe line scans or SEM images (see Figure 9).

The average COF values calculated within the elastic contact regime ($P < P_y$) for all films tested herein are shown in Figure 13. The COF value for pure a–C:H is 0.035 ± 0.002 which is in agreement with values reported in the literature for similar material systems and testing conditions [56]. The very low COF values have been attributed to the formation of a tribolayer between the probe and the film, which acts as a solid lubricant that suppresses both COF and wear. The introduction of silver into the system tends to lower the COF into even smaller values (~0.020). The enhancement of lubrication in the silver-doped amorphous carbon systems has been reported in [57,58] and can be attributed to the solid lubricant properties induced by the soft silver NPs [45,59]. The introduction of Ti NPs into a–C:H tends to slightly increase the COF (even though the increase is within the standard deviation of the experimental results) which might be attributed to the increased roughness generated by the larger size of Ti NPs and the increased elasticity/hardness of the Ti NPs which might delay the yielding of a–C:H and the generation of the transfer layer. An increased COF for surfaces with higher roughness has also been documented by Erdemir et al. [60] and has been attributed to the more extensive ploughing contribution that existed in such surfaces.

4. Conclusions

A novel hybrid (PECVD/PVD) deposition system is presented that can deliver nanocomposite a–C:H:Me films with dispersed NPs of controlled size and content. The matrix characteristics can be tailored by controlling the gas flow within the discharge tube, the RF power and the grid voltage. The nanoparticle source can deliver non-agglomerated spherical NPs whereas the quadruple filter (MesoQ) can narrow the particle size distribution and pre-select monodisperse particles with nanometer accuracy. AFM and TEM results testified towards the ability of particle size monitoring and were also used to calibrate the deposition rate for controlled nanocomposite film compositions. The PECVD system delivered an a–C:H matrix with 20–25% hydrogen content and sp^3 content of about 50% as estimated through Raman spectroscopy. Hydrogenated amorphous carbon films with embedded Ag or Ti NPs were deposited and exhibited very low coefficients of frictions (<0.05) and enhanced nanoscratch resistance (up to +50%). This improved nanotribological response can be traced to the reduced residual stresses, and the higher matrix ductility caused by the graphitization of the a–C:H through the release of strain energy and the nanocomposite morphology that increases the toughness of the material. These improved material systems retain their nanometer scale roughness and could be potentially exploited for biomedical or other protective applications.

Acknowledgments: The authors would like to acknowledge the financial support from the Strategic Infrastructure Project NEW INFRASTRUCTURE/STRATE/0308/04 of DESMI 2008, which is co-financed by the European Regional Development Fund, the European Social Fund, the Cohesion Fund, and the Research Promotion Foundation of the Republic of Cyprus. The continued support from Cyprus University of Technology towards RUNMS is gratefully acknowledged.

Author Contributions: P.K. and P.P. conceived the idea of this study; G.C. designed and coordinated the experiments, performed SEM measurements and analyzed nanomechanical data; M.C. and P.N. worked on the deposition of the nanocomposite films; A.A. and D.M. performed TEM measurements and image analysis; M.C. performed nanoscratch, residual stress measurements and AFM tests; L.K. performed XRR and SEM tests; C.V. performed Raman measurements and data analysis/interpretation; M.C. and G.C. wrote the paper; All authors discussed the results and critically reviewed/edited the manuscript.

Conflicts of Interest: The authors declare no conflict of interest. The funding sponsors had no role in the design of the study; in the collection, analyses, or interpretation of data; in the writing of the manuscript, and in the decision to publish the results.

References

1. Ferrari, A.C.; Robertson, J. Resonant Raman spectroscopy of disordered, amorphous, and diamondlike carbon. *Phys. Rev. B* **2001**, *64*, 75414. [CrossRef]
2. Robertson, J. Properties of diamond-like carbon. *Surf. Coat. Technol.* **1992**, *50*, 185–203. [CrossRef]
3. Robertson, J. Diamond-like amorphous carbon. *Mater. Sci. Eng. R Rep.* **2002**, *37*, 129–281. [CrossRef]
4. Lifshitz, Y. Diamond-like carbon—Present status. *Diam. Relat. Mater.* **1999**, *8*, 1659–1676. [CrossRef]
5. Robertson, J. Comparison of diamond-like carbon to diamond for applications. *Phys. Status Solidi Appl. Mater. Sci.* **2008**, *205*, 2233–2244. [CrossRef]
6. Shirakura, A.; Nakaya, M.; Koga, Y.; Kodama, H.; Hasebe, T.; Suzuki, T. Diamond-like carbon films for PET bottles and medical applications. *Thin Solid Films* **2006**, *494*, 84–91. [CrossRef]
7. Tsubone, D.; Hasebe, T.; Kamijo, A.; Hotta, A. Fracture mechanics of diamond-like carbon (DLC) films coated on flexible polymer substrates. *Surf. Coat. Technol.* **2007**, *201*, 6423–6430. [CrossRef]
8. Lifshitz, Y.; Kasi, S.R.; Rabalais, J.W.; Eckstein, W. Subplantation model for film growth from hyperthermal species. *Phys. Rev. B* **1990**, *41*, 10468–10480. [CrossRef]
9. Choi, H.W.; Choi, J.-H.; Lee, K.-R.; Ahn, J.-P.; Oh, K.H. Structure and mechanical properties of Ag-incorporated DLC films prepared by a hybrid ion beam deposition system. *Thin Solid Films* **2007**, *516*, 248–251. [CrossRef]
10. Cui, J.; Qiang, L.; Zhang, B.; Ling, X.; Yang, T.; Zhang, J. Mechanical and tribological properties of Ti-DLC films with different Ti content by magnetron sputtering technique. *Appl. Surf. Sci.* **2012**, *258*, 5025–5030. [CrossRef]
11. Tang, X.S.; Wang, H.J.; Feng, L.; Shao, L.X.; Zou, C.W. Mo doped DLC nanocomposite coatings with improved mechanical and blood compatibility properties. *Appl. Surf. Sci.* **2014**, *311*, 758–762. [CrossRef]
12. Constantinou, M.; Pervolaraki, M.; Nikolaou, P.; Prouskas, C.; Patsalas, P.; Kelires, P.; Giapintzakis, J.; Constantinides, G. Microstructure and nanomechanical properties of pulsed excimer laser deposited DLC:Ag films: Enhanced nanotribological response. *Surf. Coat. Technol.* **2017**, *309*, 320–330. [CrossRef]
13. Constantinou, M.; Pervolaraki, M.; Koutsokeras, L.; Prouskas, C.; Patsalas, P.; Kelires, P.; Giapintzakis, J.; Constantinides, G. Enhancing the nanoscratch resistance of pulsed laser deposited DLC films through molybdenum-doping. *Surf. Coat. Technol.* **2017**, *330*, 185–195. [CrossRef]
14. Yamamoto, G.; Okabe, T.; Ikenaga, N. Fracture behavior of diamondlike carbon films deposited on polymer substrates. *J. Vac. Sci. Technol. A Vac. Surf. Film* **2018**, *36*, 02C101. [CrossRef]
15. Baba, K.; Hatada, R.; Flege, S.; Ensinger, W.; Shibata, Y.; Nakashima, J.; Sawase, T.; Morimura, T. Preparation and antibacterial properties of Ag-containing diamond-like carbon films prepared by a combination of magnetron sputtering and plasma source ion implantation. *Vacuum* **2013**, *89*, 179–184. [CrossRef]

16. Venkatesh, M.; Taktak, S.; Meletis, E.I. Characterization of nanocomposite a-C:H/Ag thin films synthesized by a hybrid deposition process. *Phys. Met. Metallogr.* **2015**, *116*, 781–790. [CrossRef]
17. Wu, Y.; Chen, J.; Li, H.; Ji, L.; Ye, Y.; Zhou, H. Preparation and properties of Ag/DLC nanocomposite films fabricated by unbalanced magnetron sputtering. *Appl. Surf. Sci.* **2013**, *284*, 165–170. [CrossRef]
18. Delmdahl, R. The excimer laser: Precision engineering. *Nat. Photonics* **2010**, *4*, 286. [CrossRef]
19. Delmdahl, R.; Pätzel, R. Excimer laser technology trends. *J. Phys. D Appl. Phys.* **2014**, *47*, 34004. [CrossRef]
20. Guler, U.; Shalaev, V.M.; Boltasseva, A. Nanoparticle plasmonics: Going practical with transition metal nitrides. *Mater. Today* **2015**, *18*, 227–237. [CrossRef]
21. Zoubos, H.; Koutsokeras, L.E.; Anagnostopoulos, D.F.; Lidorikis, E.; Kalogirou, S.A.; Wildes, A.R.; Kelires, P.C.; Patsalas, P. Broadband optical absorption of amorphous carbon/Ag nanocomposite films and its potential for solar harvesting applications. *Sol. Energy Mater. Sol. Cells* **2013**, *117*, 350–356. [CrossRef]
22. Nikolaou, P.; Mina, C.; Constantinou, M.; Koutsokeras, L.E.; Constantinides, G.; Lidorikis, E.; Avgeropoulos, A.; Kelires, P.C.; Patsalas, P. Functionally graded poly(dimethylsiloxane)/silver nanocomposites with tailored broadband optical absorption. *Thin Solid Films* **2015**, *581*. [CrossRef]
23. Pauksch, L.; Hartmann, S.; Rohnke, M.; Szalay, G.; Alt, V.; Schnettler, R.; Lips, K.S. Biocompatibility of silver nanoparticles and silver ions in primary human mesenchymal stem cells and osteoblasts. *Acta Biomater.* **2014**, *10*, 439–449. [CrossRef] [PubMed]
24. Greulich, C.; Braun, D.; Peetsch, A.; Diendorf, J.; Siebers, B.; Epple, M.; Köller, M. The toxic effect of silver ions and silver nanoparticles towards bacteria and human cells occurs in the same concentration range. *RSC Adv.* **2012**, *2*, 6981. [CrossRef]
25. Castiglioni, S.; Cazzaniga, A.; Locatelli, L.; Maier, J.A.M. Silver Nanoparticles in Orthopedic Applications: New Insights on Their Effects on Osteogenic Cells. *Nanomaterials* **2017**, *7*, 124. [CrossRef] [PubMed]
26. Barbasz, A.; Oćwieja, M.; Roman, M. Toxicity of silver nanoparticles towards tumoral human cell lines U-937 and HL-60. *Colloids Surf. B Biointerfaces* **2017**, *156*, 397–404. [CrossRef] [PubMed]
27. Necula, B.S.; Van Leeuwen, J.P.T.M.; Fratila-Apachitei, L.E.; Zaat, S.A.J.; Apachitei, I.; Duszczyk, J. In vitro cytotoxicity evaluation of porous TiO_2-Ag antibacterial coatings for human fetal osteoblasts. *Acta Biomater.* **2012**, *8*, 4191–4197. [CrossRef] [PubMed]
28. Vrcek, I.V.; Zuntar, I.; Petlevski, R.; Pavicic, I.; Sikiric, M.D.; Curlin, M.; Goessler, W. Comparison of In Vitro Toxicity of Silver Ions and Silver Nanoparticles on Human Hepatoma Cells. *Environ. Toxicol.* **2009**, *24*, 296–303. [CrossRef]
29. Yang, Q.; Joyce, D.E.; Saranu, S.; Hughes, G.M.; Varambhia, A.; Moody, M.P.; Bagot, P.A.J. A combined approach for deposition and characterization of atomically engineered catalyst nanoparticles. *Catal. Struct. React.* **2015**, *1*, 125–131. [CrossRef]
30. Parratt, L.G. Surface studies of solids by total reflection of X-rays. *Phys. Rev.* **1954**, *95*, 359–369. [CrossRef]
31. Photiou, D.; Panagiotopoulos, N.T.; Koutsokeras, L.; Evangelakis, G.A.; Constantinides, G. Microstructure and nanomechanical properties of magnetron sputtered Ti−Nb films. *Surf. Coat. Technol.* **2016**, *302*, 310–319. [CrossRef]
32. Casiraghi, C.; Ferrari, A.C.; Robertson, J. Raman spectroscopy of hydrogenated amorphous carbons. *Phys. Rev. B* **2005**, *72*, 085401. [CrossRef]
33. Constantinides, G.; Silva, E.C.C.M.; Blackman, G.S.; Van Vliet, K.J. Dealing with imperfection: Quantifying potential length scale artefacts from nominally spherical indenter probes. *Nanotechnology* **2007**, *18*. [CrossRef]
34. Stoney, G.G. The tension of metallic films deposited by electrolysis. *Proc. R. Soc. A Math. Phys. Eng. Sci.* **1909**, *82*, 172–175. [CrossRef]
35. Guyot, N.; Harmand, Y.; Mezin, A. The role of the sample shape and size on the internal stress induced curvature of thin-film substrate systems. *Int. J. Solids Struct.* **2004**, *41*, 5143–5154. [CrossRef]
36. Mezin, A. Coating internal stress measurement through the curvature method: A geometry-based criterion delimiting the relevance of Stoney's formula. *Surf. Coat. Technol.* **2006**, *200*, 5259–5267. [CrossRef]
37. Gelinas, V.; Vidal, D. Determination of particle shape distribution of clay using an automated AFM image analysis method. *Powder Technol.* **2010**, *203*, 254–264. [CrossRef]
38. Gu, Y.; Xie, H.; Gao, J.; Liu, D.; Williams, C.T.; Murphy, C.J.; Ploehn, H.J. AFM characterization of dendrimer-stabilized platinum nanoparticles. *Langmuir* **2005**, *21*, 3122–3131. [CrossRef] [PubMed]
39. Tritsaris, G.A.; Mathioudakis, C.; Kelires, P.C.; Kaxiras, E. Optical and elastic properties of diamond-like carbon with metallic inclusions: A theoretical study. *J. Appl. Phys.* **2012**, *112*, 103503. [CrossRef]

40. Ferrari, A.C.; Robertson, J. Raman spectroscopy of amorphous, nanostructured, diamond-like carbon, and nanodiamond. *Philos. Trans. A Math. Phys. Eng. Sci.* **2004**, *362*, 2477–2512. [CrossRef] [PubMed]

41. Ferrari, A.C.; Libassi, A.; Tanner, B.K.; Stolojan, V.; Yuan, J.; Brown, L.M.; Rodil, S.E.; Kleinsorge, B. Density, sp 3 fraction, and cross-sectional structure of amorphous carbon films determined by X-ray reflectivity and electron energy-loss spectroscopy. *Phys. Rev. B* **2000**, *62*, 11089–11103. [CrossRef]

42. Meškinis, Š.; Vasiliauskas, A.; Šlapikas, K.; Niaura, G.; Juškenas, R.; Andrulevičius, M.; Tamulevičius, S. Structure of the silver containing diamond like carbon films: Study by multiwavelength Raman spectroscopy and XRD. *Diam. Relat. Mater.* **2013**, *40*, 32–37. [CrossRef]

43. Ferrari, A.C.; Robertson, J. Interpretation of Raman spectra of disordered and amorphous carbon. *Phys. Rev. B* **2000**, *61*, 14095–14107. [CrossRef]

44. Baba, K.; Hatada, R.; Flege, S.; Ensinger, W. Preparation and Properties of Ag-Containing Diamond-Like Carbon Films by Magnetron Plasma Source Ion Implantation. *Adv. Mater. Sci. Eng.* **2012**, *2012*, 536853. [CrossRef]

45. Qiang, L.; Zhang, B.; Zhou, Y.; Zhang, J. Improving the internal stress and wear resistance of DLC film by low content Ti doping. *Solid State Sci.* **2013**, *20*, 17–22. [CrossRef]

46. Huang, H.H.; Ho, C.T.; Lee, T.H.; Lee, T.L.; Liao, K.K.; Chen, F.L. Effect of surface roughness of ground titanium on initial cell adhesion. *Biomol. Eng.* **2004**, *21*, 93–97. [CrossRef] [PubMed]

47. Zareidoost, A.; Yousefpour, M.; Ghaseme, B.; Amanzadeh, A. The relationship of surface roughness and cell response of chemical surface modification of titanium. *J. Mater. Sci. Mater. Med.* **2013**, *23*, 1479–1488. [CrossRef] [PubMed]

48. Beake, B.D.; Harris, A.J.; Liskiewicz, T.W. Review of recent progress in nanoscratch testing. *Tribol. Mater. Surf. Interfaces* **2013**, *7*, 87–96. [CrossRef]

49. Beake, B.D.; Davies, M.I.; Liskiewicz, T.W.; Vishnyakov, V.M.; Goodes, S.R. Nano-scratch, nanoindentation and fretting tests of 5–80 nm ta-C films on Si(100). *Wear* **2013**, *301*, 575–582. [CrossRef]

50. Beake, B.D.; Vishnyakov, V.M.; Valizadeh, R.; Colligon, J.S. Influence of mechanical properties on the nanoscratch behaviour of hard nanocomposite TiN/Si 3 N 4 coatings on Si. *J. Phys. D Appl. Phys.* **2006**, *39*, 1392–1397. [CrossRef]

51. Shi, B.; Sullivan, J.L.; Beake, B.D. An investigation into which factors control the nanotribological behaviour of thin sputtered carbon films. *J. Phys. D Appl. Phys.* **2008**, *41*, 45303. [CrossRef]

52. Beake, B.D.; Goodes, S.R.; Shi, B. Nanomechanical and nanotribological testing of ultra-thin carbon-based and MoST films for increased MEMS durability. *J. Phys. D Appl. Phys.* **2009**, *42*, 65301. [CrossRef]

53. Matenoglou, G.; Evangelakis, G.A.; Kosmidis, C.; Foulias, S.; Papadimitriou, D.; Patsalas, P. Pulsed laser deposition of amorphous carbon/silver nanocomposites. *Appl. Surf. Sci.* **2007**, *253*, 8155–8159. [CrossRef]

54. Miyoshi, K. Adhesion, friction and micromechanical properties of ceramics. *Surf. Coat. Technol.* **1988**, *36*, 487–501. [CrossRef]

55. Hamilton, G.M. Explicit equations for the stresses beneath a sliding spherical contact. *Proc. Inst. Mech. Eng.* **1983**, *197C*, 53–59. [CrossRef]

56. Charitidis, C.A. Nanomechanical and nanotribological properties of carbon-based thin films: A review. *Int. J. Refract. Met. Hard Mater.* **2010**, *28*, 51–70. [CrossRef]

57. Pardo, A.; Gómez-Aleixandre, C. Friction and Wear Behavior of Plasma Assisted Chemical Vapor Deposited Nanocomposites Made of Metal Nanoparticles Embedded in a Hydrogenated Amorphous Carbon Matrix. *Surf. Coat. Technol.* **2012**, *206*, 3116–3124. [CrossRef]

58. Yu, X.; Qin, Y.; Wang, C.B.; Yang, Y.Q.; Ma, X.C. Effects of nanocrystalline silver incorporation on sliding tribological properties of Ag-containing diamond-like carbon films in multi-ion beam assisted deposition. *Vacuum* **2013**, *89*, 82–85. [CrossRef]

59. Hatada, R.; Flege, S.; Bobrich, A.; Ensinger, W.; Dietz, C.; Baba, K.; Sawase, T.; Watamoto, T.; Matsutani, T. Preparation of Ag-containing diamond-like carbon films on the interior surface of tubes by a combined method of plasma source ion implantation and DC sputtering. *Appl. Surf. Sci.* **2014**, *310*, 257–261. [CrossRef]

60. Erdemir, A.; Bindal, C.; Fenske, G.R.; Zuiker, C.; Wilbur, P. Characterization of transfer layers forming on surfaces sliding against diamond-like carbon. *Surf. Coat. Technol.* **1996**, *86*, 692–697. [CrossRef]

 *nanomaterials* 

Article

Enhancing the Microparticle Deposition Stability and Homogeneity on Planer for Synthesis of Self-Assembly Monolayer

An-Ci Shih, Chi-Jui Han, Tsung-Cheng Kuo and Yun-Chien Cheng *

Department of Mechanical Engineering, National Chiao Tung University, Hsinchu 300, Taiwan;
ancishih@gmail.com (A.-C.S.); dragon199357@gmail.com (C.-J.H.); lincecumepon@gmail.com (T.-C.K.)
* Correspondence: yccheng@nctu.edu.tw; Tel.: +886-3-571-2121 (ext. 55112)

Received: 26 January 2018; Accepted: 12 March 2018; Published: 14 March 2018

Abstract: The deposition stability and homogeneity of microparticles improved with mask, lengthened nozzle and flow rate adjustment. The microparticles can be used to encapsulate monomers, before the monomers in the microparticles can be deposited onto a substrate for nanoscale self-assembly. For the uniformity of the synthesized nanofilm, the homogeneity of the deposited microparticles becomes an important issue. Based on the ANSYS simulation results, the effects of secondary flow were minimized with a lengthened nozzle. The ANSYS simulation was also used to investigate the ring-vortex generation and why the ring vortex can be eliminated by adding a mask with an aperture between the nozzle and deposition substrate. The experimental results also showed that particle deposition with a lengthened nozzle was more stable, while adding the mask stabilized deposition and diminished the ring-vortex contamination. The effects of flow rate and pressure were also investigated. Hence, the deposition stability and homogeneity of microparticles was improved.

Keywords: microparticle deposition; self-assembly; homogeneity; monomer synthesis; mask

---

1. Introduction

Aerosol-based microparticle deposition technology is very commonly used for film fabrication process on a nanoscale [1–4], with several methods developed to deposit and pattern microparticles on substrates [5–11]. The advantages of aerosol-based microparticle deposition include cost effective equipment and the ability for microparticles to be made of various materials that are tailored to different applications. The applications of microparticle deposition include xerography printing [12], fuel-cell fabrication [13], electronic-element fabrication [14], film fabrication [15–18], peptide-array synthesis [19–21], etc. Among these applications, the particle-based peptide-array synthesis technique is a novel technique for microarray fabrication. It addresses solid amino-acid microparticles onto substrates and synthesizes high-density peptide arrays. Compared with other peptide array fabrication methods, the particle-based peptide array synthesis has higher spot density than SPOT synthesis [22] and is faster than photolithography synthesis [23,24].

The aerosol for particle-based peptide array synthesis was generated using compressed air, with the microparticles triboelectrically charged in turbulence. The charged microparticles will be attracted to the electrical field generated by microelectrodes, before being deposited on the microelectrode array in the desired pattern [19,25]. However, the same turbulence that generates aerosol and triboelectrically charges the microparticles also result in inhomogeneity in the microparticle deposition. Inhomogeneous microparticle deposition leads to unequal amounts of synthesized peptide, which decreases the reliability of array analysis. Hence, stabilizing the turbulence to create homogeneity in the microparticle deposition becomes an important issue. Although several studies have discussed how to pattern the microparticle deposition [5,7–10,26,27], stabilizing the aerosol turbulence for homogeneous microparticle deposition has not yet been fully investigated.

In this study, the turbulence was stabilized with lengthened aerosol nozzles and masks. A two-dimensional flow field simulation was used to investigate the ring vortex generated and to explain why the mask can eliminate ring vortex. Furthermore, we also enhanced the microparticle deposition stability and homogeneity by optimizing flow rates and pressures associated with lengthened nozzles and masks. The results in this study will help to improve the deposition homogeneity of particle-encapsulated monomers and hence, enhance the homogeneity of self-assembly synthesized peptide nanolayers.

2. Experimental and Simulation Setup

2.1. Aerosol Generation and Deposition System

In this study, the aerosol generation and deposition system was developed by referencing previous studies [25,28], which is shown in Figure 1. The system included an air compressor (JW-2525N, JUN WEI, Taichung, Taiwan), dehumidifying filter and pressure regulator (MATFR401, Mindman, Taipei, Taiwan), flow control valve (GRLSA-1/8-QS-6, Festo, Esslingen am Neckar, Germany), solenoid valve (MHJ10-S-0,35-QS-4-MF, Festo, Esslingen am Neckar, Germany), microparticle reservoir (Falcon™ 50 mL conical centrifuge tube, Corning, New York, NY, USA), micro sieve, flowmeter and our custom-made nozzle. The nozzles were divergent pipes to allow for stabilization of the flow [29–31]. Polytetrafluoroethylene (PTFE) tube was chosen to connect the aerosol system for the triboelectrical charge. Toner microparticles (Black, Xerox, Norwalk, CT, USA) were used for microparticle deposition experiments. For the particle size, 50% of the particles were smaller than 6.016 μm and 90% of particles were smaller than 8.309 μm [32]. Xeros Black particles were used in the study because the particle size is similar to the expensive bioparticles [33]. The result of this study will be applied to create homogeneity in bioparticle [11] deposition in the future. The deposited bioparticles can be addressed in different methods and used for monomer synthesis, such as electrical fields from microelectrodes [19,25] or laser [34–37], and can be used for several applications, such as antibody analysis [38] or surface functionalization [39]. The microparticles were dehumidified in the desiccator with a molecular sieve overnight before the experiment.

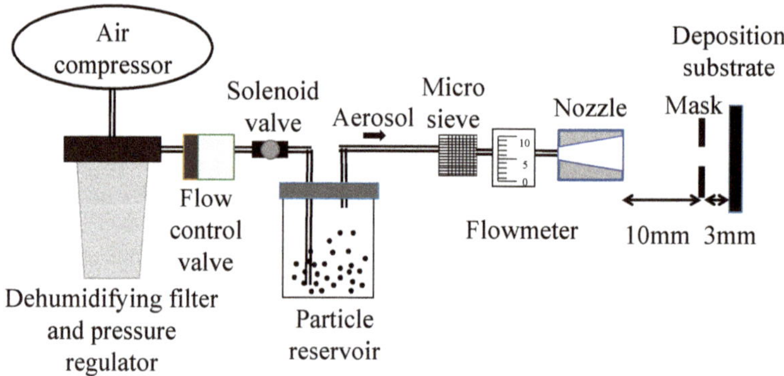

Figure 1. Aerosol generation and deposition system.

To generate the microparticle aerosol, the high pressure air was generated by the air compressor, which passed through the dehumidifying filter and pressure regulator. The air flow was controlled using a flow control valve and solenoid valve. Following this, the regulated air flow entered the microparticle reservoir and generated a microparticle aerosol with turbulence. After the agglomerated microparticles were broken up by passing through the micro-sieve, the microparticles were deposited on the substrate. In this study, the effect of including the mask between the nozzle and substrate was discussed.

2.2. Measurement and Analysis of Microparticle Deposition

To analyze the stability and homogeneity of the microparticle deposition, the area, symmetric value and thickness of microparticle deposition were measured. The standard deviations σ of these indexes were considered to be indexes of deposition stability in this study. In addition, if the deposition is stable and symmetrical, the deposition homogeneity can be easily improved by moving the deposition location. The area of microparticle deposition was calculated with Photoshop (Adobe Systems Incorporated, San Jose, CA, USA). The pixel number of black (microparticle) and blank (chip) area were calculated and converted to mm^2 according to Equation (1):

$$\text{Particle depopsition area} = \frac{n_{\text{Part}}}{n_{\text{Part}} + n_{\text{Chip}}} \times 400 \text{ mm}^2, \tag{1}$$

where n_{Part} represented the pixel number of microparticle area and n_{Chip} represented the pixel number of chip area.

The symmetric value was evaluated by dividing the longest axis ($L1$) of the microparticle deposition (Figure 2) by its shortest axis ($L2$), which is shown in Equation (2).

$$\text{Symmetric value} = L1/L2. \tag{2}$$

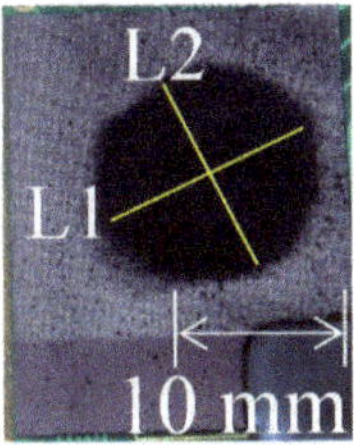

Figure 2. Symmetric value measurement of the deposited particle pattern where $L1$: longest axis and $L2$: shortest axis.

2.3. Simulation Method

The ANSYS FLUENT software (ANSYS, Canonsburg, PA, USA) was used to simulate the flow field in nozzles with different lengths and the flow field of deposition with or without mask. The aerosol flow field was considered as a single phase flow to simplify the simulation. The k-ε model was used to simulate the flow field. Equation (3) shows the continuity equation, which describes the mass that enters minus the mass that leaves the system. This is equal to the accumulation of mass in the system where ρ is fluid density; U is the flow velocity vector field; t is time; and x is the length. Equation (4) shows the momentum equation, which describes the relationship between acceleration of the gas and the force where S_M is the sum of body forces; p' is a modified pressure; and μ_{eff} is the effective viscosity accounting for turbulence. Equation (5) shows the Boussinesq buoyancy model in the k-ε model, which is used to predict the buoyancy effects on production and destruction of turbulence where P_{kb} is the buoyancy production term; μ_t is viscosity; ρ is fluid density; σ_p and β are constants; g_i is acceleration in x_i direction; and T is temperature.

$$\frac{\partial \rho}{\partial t} + \frac{\partial (\rho U_j)}{\partial x_j} = 0, \tag{3}$$

$$\frac{\partial \rho U_i}{\partial t} + \frac{\partial (\rho U_i U_j)}{\partial x_j} = -\frac{\partial p'}{\partial x_i} + \frac{\partial}{\partial x_j}\left[\mu_{\text{eff}}\left(\frac{\partial U_i}{\partial x_i} + \frac{\partial U_j}{\partial x_j}\right)\right] + S_{\text{M}},\tag{4}$$

$$P_{\text{kb}} = -\frac{\mu_{\text{t}}}{\rho\sigma_\rho}\rho\beta g_i\frac{\partial T}{\partial x_i}.\tag{5}$$

3. Results and Discussion

3.1. Nozzle Length and Flowrate Effects on the Stability of Deposited Microparticle Layer

In the simulation, the nozzles were divergent pipes according to the experimental setup. The nozzle length was set to 20-mm and 60-mm. The speed of air flow at nozzle inlet was 1 m/s and the pressure at the nozzle outlet was 0. The simulation results are shown in Figure 3. The flow direction was from left to right, so the speed values were shown as negative values.

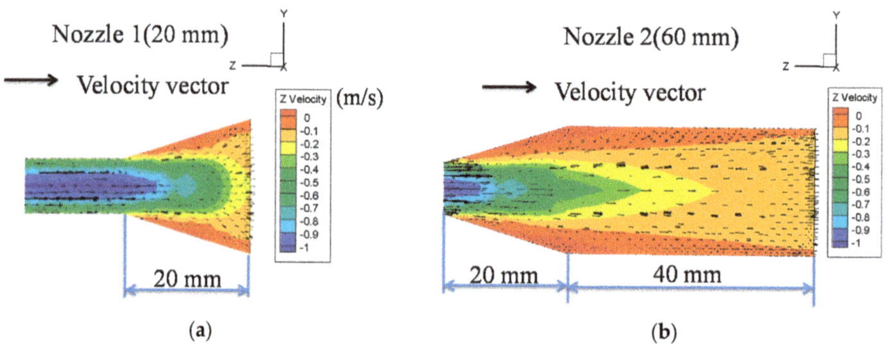

Figure 3. Fluid field simulation of different nozzle length, which was simulated by Fluent using: (**a**) 20 mm nozzle; and (**b**) 60 mm nozzle.

Figure 3a shows the simulation results of the 20-mm nozzle. When the air flow entered the divergent pipe, the flow slowed down. The boundary layer was formed along the inner wall of the divergent pipe due to flow viscosity. Flow along the inner wall influenced the flow closer to the nozzle center and formed a secondary flow. About halfway through the divergent pipe, the flow speed near the inner wall was close to zero, which might have occurred due to the formation of small vortexes. The simulation results also showed that the flow speed at the nozzle outlet was between −0.3 and −0.1 m/s.

Figure 3b shows the simulation results of the 60-mm nozzle. When the air flow entered the divergent pipe, the flow also slowed down. Since the nozzle was lengthened, the influence of boundary layer on inner flow increased. Therefore, the flow speed at nozzle outlet was between −0.2 and −0.1 m/s. The simulation shows that the flow-velocity distribution of 60-mm nozzle was more uniform than that of 20-mm nozzle. The flow direction of 60-mm nozzle was also more consistent than that of 20-mm nozzle. Hence, using 60-mm nozzle can result in more stable outlet flow compared to using 20-mm nozzle.

For the nozzle length experiment, the parameters used in the experiment were the same as those used in simulation. The air flow pressure, flow rate and distance between nozzle and substrate were 0.1 MPa, 5 L/min and 10 mm, respectively. Based on the flow rate and cross section of nozzle, the flow speed was calculated as 1 m/s.

The experimental results are shown in Figure 4. For the 20-mm nozzle, the average microparticle deposition area, standard deviation (σ) of deposition area and symmetric value were 137.683, 35.025 and 1.237 mm², respectively. For the 60-mm nozzle, the microparticle deposition area, σ of deposition area and symmetric value were 83.18, 15 and 1.152 mm², respectively. The standard deviation of the deposition area with the 60-mm nozzle was smaller than that with 20-mm nozzle, which indicated greater deposition stability of 60-mm nozzle over 20-mm nozzle. Furthermore, the symmetric value of the deposition area with 60-mm nozzle was smaller than that with 20-mm nozzle, which also means that the 60-mm-nozzle deposition was more symmetric and stable. This result supports the simulation result that the 60-mm nozzle can stabilize the air flow and thus, the deposition.

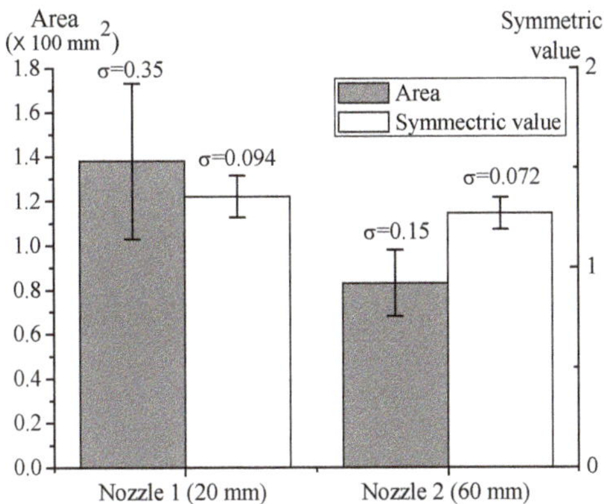

Figure 4. Effects of nozzle length on particle deposition. Data are shown with mean ± SD (*n* = 3).

Since the 60-mm nozzle resulted in more stable microparticle deposition, we investigated the pressure and flow rate effects on microparticle deposition with a 60-mm nozzle. The results are shown in Figure 5a. Both deposition area and symmetric value increased with the flow rate. The increase of flow rate resulted in turbulence, making the deposition unstable. Furthermore, the ring vortex (Figure 5b) appeared in every deposition. The pattern of ring vortex in deposition center was round, but there was a ring-shaped pattern outside. This is because the microparticles were blown away from the center. The ring vortex will contaminate the deposition area and decrease the homogeneity of the deposition.

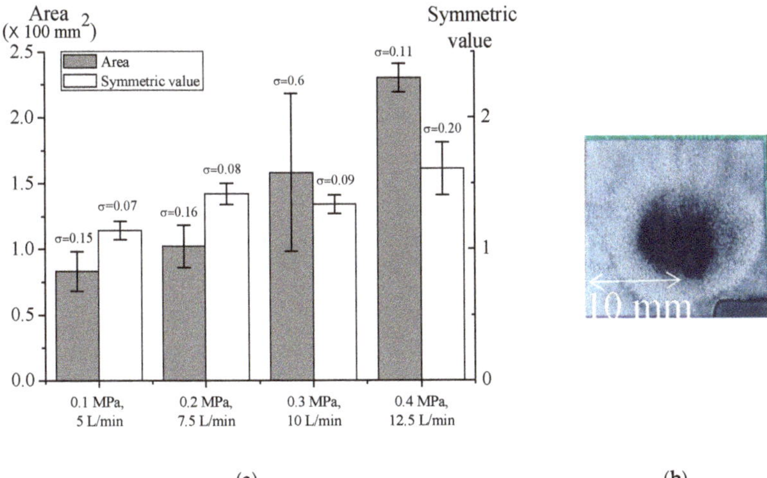

(a)　　　　　　　　　　(b)

Figure 5. (**a**) Effects of pressure and flow rate on particle deposition with 60-mm nozzle. Data are shown with mean ± SD (*n* = 3); and (**b**) Ring vortex.

3.2. Mask for Removing the Ring Vortex and Enhancing Deposition Stability

To solve the ring vortex problem occurring during deposition, a mask with fixed aperture was placed between nozzle and substrate, with the effects discussed with simulation and experiments.

The flow fields with and without mask using the 60-mm nozzle were simulated. The flow speed at the nozzle inlet was 4.145 m/s, which was calculated from the 12.5 L/min volume flow rate. The simulation results were shown in Figure 6. As the flow fields in our study are symmetric, we joined two simulations into one diagram for more intuitive side-by-side comparison. The left panel displays the simulation without mask, while the right displays the simulation with mask. The streamlines of flow field could be considered as the moving tracks of microparticles.

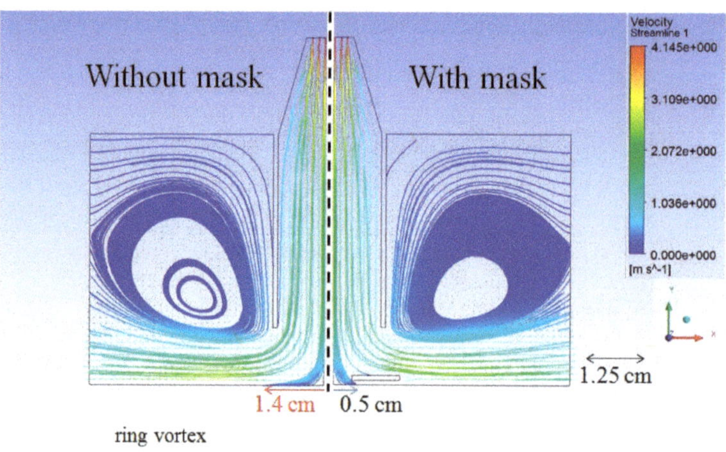

Figure 6. Simulation of flow field on the chip surface with and without mask.

Without using the mask, when the air flow reached the substrate, the inner flow stopped at the surface and the outer flow formed the ring vortex. The distribution area of flow field at the substrate was smaller with the mask than without due to partial blockage of the flow by the mask. The part of flow that generated the ring vortex was also blocked. Therefore, adding the mask resulted in smaller but more stable microparticle deposition. For the speed of flow field, the flow speed was faster in the periphery (2–3.1 m/s) than that close to the center. Therefore, the microparticles could be easily dispersed in the peripheral outer area. With the mask on the substrate, the dispersion of microparticles in the outer area was blocked and most of the deposited microparticles resulted from the stable flow in the inner area.

The effect of masking was also examined with actual experiments. A mask with a round aperture (radius of 4 mm) was placed between the nozzle and deposition substrate, which is shown in Figure 1. The air pressure was 0.3 MPa and the flow volume was 10 L/min. The deposition area, symmetric value and standard deviation (σ) are shown in Figure 7. With the mask, the average deposition area, σ of deposition and symmetric value were respectively 109.33, 18 and 1.044 mm^2. Without the mask, the respective values were 157.67, 60 and 1.337 mm^2. The smaller σ of deposition area indicates that the deposition with the mask was more stable. The symmetric value with mask was closer to 1, which also means that the deposition was more stable than the deposition without the mask. The difference between area with and area without the mask was not significant.

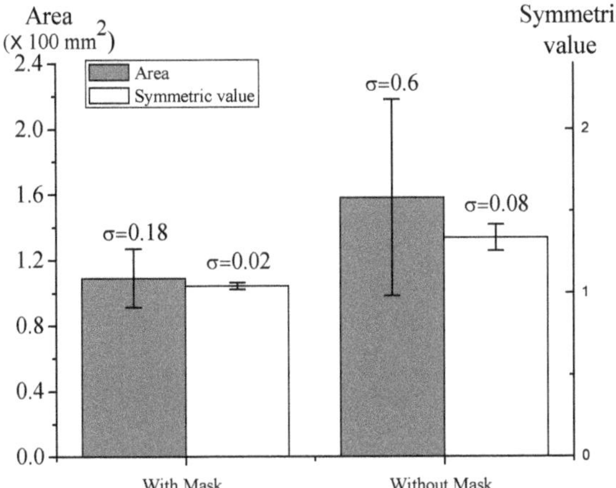

Figure 7. Particle deposition stability with and without mask. The pressure was 0.3 MPa; flow rate was 10 L/min; and distance was 10 mm. Data are shown with mean ± SD (*n* = 3).

3.3. Effects of Mask Size, Pressure, and Flow Rate on Microparticle Deposition Stability with Mask

Since the mask can stabilize the microparticle deposition, the effects of mask size, pressure and flow rate on microparticle deposition with the mask were investigated to further enhance the deposition stability.

3.3.1. Effects of Mask Size

The experimental results of microparticle deposition with different aperture sizes were shown in Figure 8. For the aperture radii of 3−5 mm, the average deposition area was 44.529, 80.099 and 120.105 mm², while the σ of deposition was 7.131, 8.116 and 30.045. The air flow through the mask still generated minor turbulence between the mask and substrate. The amount of air that passed through increased with mask aperture, with a corresponding increase in turbulence. The small mask aperture allowed less air to pass, with more stable and smaller deposition.

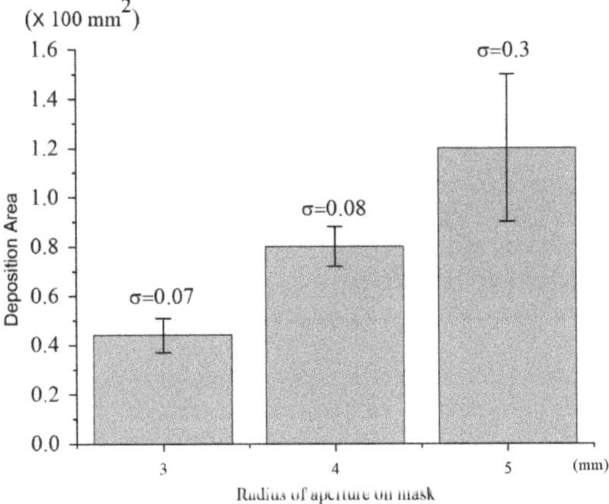

Figure 8. Effects of aperture size on particle deposition stability. Data are shown with mean ± SD (*n* = 3).

3.3.2. Effects of Pressure and Flow Rate

The aperture with a 3 mm radius resulted in most stable microparticle deposition. Hence, we investigated pressure and flow rate effects on deposition with a 3 mm aperture (Figure 9) and also compared the results against deposition without masks (Figure 5). The symmetric value of deposition with mask was closer to 1 than that without mask, so the mask did stabilize the microparticle deposition. On the other hand, the higher pressure and flow rate resulted in larger deposition area, since more microparticles were pumped out by higher pressure.

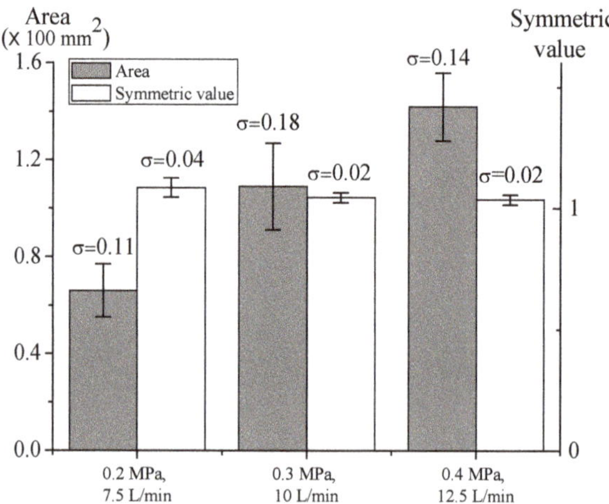

Figure 9. Effects of different pressures and flow rates on particle deposition with mask. Data are shown with mean ± SD (*n* = 3).

3.4. Particle Deposition in Initial Condition and after Optimization

Figure 10 shows the particle deposition under the initial conditions and after optimization. The deposition with a 6 mm aperture mask and low flow rate (Figure 10b) is more symmetric and stable than the deposition without the aperture and high flow rate.

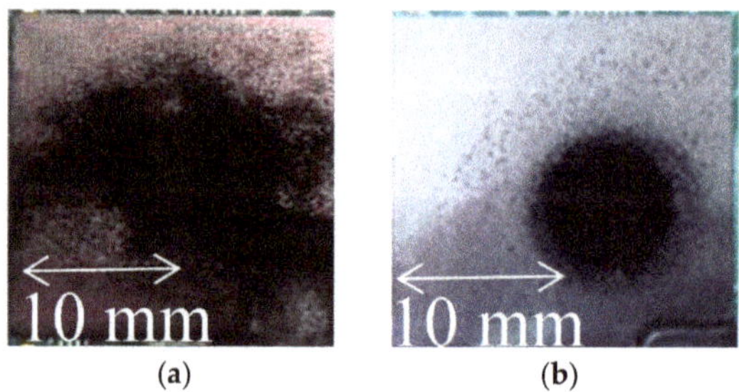

Figure 10. Particle deposition before and after optimization: (**a**) without mask at a flow rate of 12.5 L/min and (**b**) using mask with 6 mm aperture at a flow rate of 7.5 L/min.

4. Conclusions

The experiment and simulation results showed that a longer nozzle can reduce the effect of secondary flow and can make the microparticle deposition stable and thus, homogeneous. Increasing the pressure and flow rate generated turbulence and ring vortex, resulting in instability in the deposition. Using the mask can prevent the ring vortex from contaminating the deposition and stabilize the deposition. The mask with a 6 mm aperture had the most stable deposition, since there was more turbulence generated than when other masks were used. With a 6 mm aperture, decreasing the pressure or decreasing the flow rate increased the deposition stability. The experimental conditions that had the most stable deposition used a mask with a 6 mm aperture, 5 L/min flow rate and 0.15 MPa pressure. These results can facilitate the applications that need stable and homogeneous microparticle deposition, which will thus ensure the uniformity of microparticle-based nanolayer synthesis.

Acknowledgments: The authors acknowledge the financial support of Ministry of Science and Technology, Taiwan. The grant number is 103-2218-E-009-008 and 104-2221-E-009-099.

Author Contributions: Yun-Chien Cheng conceived and designed the experiments; An-Ci Shih and Chi-Jui Han performed the experiments; An-Ci Shih and Chi-Jui Han analyzed the data; Tsung-Cheng Kuo and Yun-Chien Cheng wrote the paper.

Conflicts of Interest: The authors declare no conflicts of interest.

References

1. Blandenet, G.; Court, M.; Lagarde, Y. Thin layers deposited by the pyrosol process. *Thin Solid Films* **1981**, *77*, 81–90. [CrossRef]
2. Mooney, J.B.; Radding, S.B. Spray pyrolysis processing. *Annu. Rev. Mater. Sci.* **1982**, *12*, 81–101. [CrossRef]
3. Su, B.; Choy, K. Microstructure and properties of the CdS thin films prepared by electrostatic spray assisted vapour deposition (ESAVD) method. *Thin Solid Films* **2000**, *359*, 160–164. [CrossRef]
4. Kashu, S.; Fuchita, E.; Manabe, T.; Hayashi, C. Deposition of ultra fine particles using a gas jet. *Jpn. J. Appl. Phys.* **1984**, *23*, L910. [CrossRef]
5. Adamczyk, Z.; Van De Ven, T.G. Deposition of particles under external forces in laminar flow through parallel-plate and cylindrical channels. *J. Colloid Interface Sci.* **1981**, *80*, 340–356. [CrossRef]
6. Park, J.; Jeong, J.; Kim, C.; Hwang, J. Deposition of charged aerosol particles on a substrate by collimating through an electric field assisted coaxial flow nozzle. *Aerosol Sci. Technol.* **2013**, *47*, 512–519. [CrossRef]
7. Kim, H.; Kim, J.; Yang, H.; Suh, J.; Kim, T.; Han, B.; Kim, S.; Kim, D.S.; Pikhitsa, P.V.; Choi, M. Parallel patterning of nanoparticles via electrodynamic focusing of charged aerosols. *Nat. Nanotechnol.* **2006**, *1*, 117–121. [CrossRef] [PubMed]
8. Qi, L.; McMurry, P.H.; Norris, D.J.; Girshick, S.L. Micropattern deposition of colloidal semiconductor nanocrystals by aerodynamic focusing. *Aerosol Sci. Technol.* **2010**, *44*, 55–60. [CrossRef]
9. Bowen, B.D.; Epstein, N. Fine particle deposition in smooth parallel-plate channels. *J. Colloid Interface Sci.* **1979**, *72*, 81–97. [CrossRef]
10. Lee, H.; You, S.; Woo, C.G.; Lim, K.; Jun, K.; Choi, M. Focused patterning of nanoparticles by controlling electric field induced particle motion. *Appl. Phys. Lett.* **2009**, *94*, 53104. [CrossRef]
11. Stadler, V.; Felgenhauer, T.; Beyer, M.; Fernandez, S.; Leibe, K.; Güttler, S.; Gröning, M.; König, K.; Torralba, G.; Hausmann, M.; et al. Combinatorial synthesis of peptide arrays with a laser printer. *Angew. Chem. Int. Ed.* **2008**, *47*, 7132–7135. [CrossRef] [PubMed]
12. Pai, D.M.; Springett, B.E. Physics of electrophotography. *Rev. Mod. Phys.* **1993**, *65*, 163. [CrossRef]
13. Li, W.; Liang, C.; Zhou, W.; Qiu, J.; Li, H.; Sun, G.; Xin, Q. Homogeneous and controllable Pt particles deposited on multi-wall carbon nanotubes as cathode catalyst for direct methanol fuel cells. *Carbon* **2004**, *42*, 436–439. [CrossRef]
14. Imanaka, Y.; Amada, H.; Kumasaka, F. Dielectric and insulating properties of embedded capacitor for flexible electronics prepared by aerosol-type nanoparticle deposition. *Jpn. J. Appl. Phys.* **2013**, *52*, 5DA2. [CrossRef]
15. Huang, C.; Becker, M.F.; Keto, J.W.; Kovar, D. Annealing of nanostructured silver films produced by supersonic deposition of nanoparticles. *J. Appl. Phys.* **2007**, *102*, 54308. [CrossRef]
16. Lebedev, M.; Akedo, J.; Akiyama, Y. Actuation properties of lead zirconate titanate thick films structured on Si membrane by the aerosol deposition method. *Jpn. J. Appl. Phys.* **2000**, *39*, 5600. [CrossRef]

17. Tsukamoto, M.; Fujihara, T.; Abe, N.; Miyake, S.; Katto, M.; Nakayama, T.; Akedo, J. Hydroxyapatite coating on titanium plate with an ultrafine particle beam. *Jpn. J. Appl. Phys.* **2003**, *42*, L120. [CrossRef]

18. Murakami, H.; Nishino, J.; Yaegashi, S.; Shiohara, Y.; Tanaka, S. Preparation of $YBa_2CuZ_3O_{7-x}$ and $YBa_2Cu_4O_8$ Thick Films by Gas Deposition Using Fine Powders. *Jpn. J. Appl. Phys.* **1991**, *30*, L185. [CrossRef]

19. Beyer, M.; Nesterov, A.; Block, I.; König, K.; Felgenhauer, T.; Fernandez, S.; Leibe, K.; Torralba, G.; Hausmann, M.; Trunk, U. Combinatorial synthesis of peptide arrays onto a microchip. *Science* **2007**, *318*, 1888. [CrossRef] [PubMed]

20. Breitling, F.; Felgenhauer, T.; Nesterov, A.; Lindenstruth, V.; Stadler, V.; Bischoff, F.R. Particle-based synthesis of peptide arrays. *ChemBioChem* **2009**, *10*, 803–808. [CrossRef] [PubMed]

21. Maerkle, F.; Loeffler, F.F.; Schillo, S.; Foertsch, T.; Muenster, B.; Striffler, J.; Schirwitz, C.; Bischoff, F.R.; Breitling, F.; Nesterov-Mueller, A. High-density peptide arrays with combinatorial laser fusing. *Adv. Mater.* **2014**, *26*, 3730–3734. [CrossRef] [PubMed]

22. Frank, R. Spot-synthesis: An easy technique for the positionally addressable, parallel chemical synthesis on a membrane support. *Tetrahedron* **1992**, *48*, 9217–9232. [CrossRef]

23. Fodor, S.P.; Read, J.L.; Pirrung, M.C.; Stryer, L.; Lu, A.T.; Solas, D. Light-directed, spatially addressable parallel chemical synthesis. *Science* **1991**, *251*, 767–773. [CrossRef] [PubMed]

24. Breitling, F.; Nesterov, A.; Stadler, V.; Felgenhauer, T.; Bischoff, F.R. High-density peptide arrays. *Mol. Biosyst.* **2009**, *5*, 224–234. [CrossRef] [PubMed]

25. Löffler, F.; Wagner, J.; König, K.; Märkle, F.; Fernandez, S.; Schirwitz, C.; Torralba, G.; Hausmann, M.; Lindenstruth, V.; Bischoff, F. High-precision combinatorial deposition of micro particle patterns on a microelectronic chip. *Aerosol Sci. Technol.* **2011**, *45*, 65–74. [CrossRef]

26. Huh, S.H.; Riu, D.H.; Naono, Y.; Taguchi, Y.; Kawabata, S.; Nakajima, A. Generation of high-quality lines and arrays using nanoparticle controlling processes. *Appl. Phys. Lett.* **2007**, *91*, 93118. [CrossRef]

27. Loeffler, F.; Schirwitz, C.; Wagner, J.; Koenig, K.; Maerkle, F.; Torralba, G.; Hausmann, M.; Bischoff, F.R.; Nesterov-Mueller, A.; Breitling, F. Biomolecule arrays using functional combinatorial particle patterning on microchips. *Adv. Funct. Mater.* **2012**, *22*, 2503–2508. [CrossRef]

28. Tang, P.; Fletcher, D.; Chan, H.-K.; Raper, J.A. Simple and cost-effective powder disperser for aerosol particle size measurement. *Powder Technol.* **2008**, *187*, 27–36. [CrossRef]

29. Sahu, K.C.; Govindarajan, R. Stability of flow through a slowly diverging pipe. *J. Fluid Mech.* **2005**, *531*, 325–334. [CrossRef]

30. Peixinho, J.; Besnard, H. Transition to turbulence in slowly divergent pipe flow. *Phys. Fluids* **2013**, *25*, 111702. [CrossRef]

31. Peixinho, J. Flow in a Slowly Divergent Pipe Section. In *Seventh IUTAM Symposium on Laminar-Turbulent Transition*; Springer: Dordrecht, the Netherlands, 2010; Available online: https://www.amazon.com/Seventh-IUTAM-Symposium-Laminar-Turbulent-Transition/dp/9048137225 (accessed on 1 March 2018).

32. Cheng, Y.C. CMOS-Chip Based Printing System for Combinatorial Synthesis. Ph.D. Thesis, Technischen Universität Darmstadt, Darmstadt, Germany, 2012.

33. Loeffler, F.; Cheng, Y.; Muenster, B.; Striffler, J.; Liu, F.; Bischoff, R.F.; Doersam, E.; Breitling, F.; Nesterov-Mueller, A. Printing peptide arrays with a complementary metal oxide semiconductor chip. *Adv. Biochem. Eng. Biotechnol.* **2013**, *137*, 1–23. [PubMed]

34. Loeffler, F.; Foertsch, T.; Popov, R.; Mattes, D.; Schlageter, M.; Sedlmayr, M.; Ridder, B.; Dang, F.; Bojničić-Kninski, F.; Weber, L.; et al. High-flexibility combinatorial peptide synthesis with laser-based transfer of monomers in solid matrix material. *Nat. Commun.* **2016**, *7*, 11844. [CrossRef] [PubMed]

35. Ridder, B.; Foertsch, T.; Wellec, A.; Mattes, D.; Bojnicic-Kninski, C.; Loeffler, F.; Nesterov-Mueller, A.; Meier, M.; Breitling, F. Development of a poly(dimethylacrylamide) based matrix material for solid phase high density peptide array synthesis employing a laser based material transfer. *Appl. Surf. Sci.* **2016**, *389*, 942–951. [CrossRef]

36. Bojnicic-Kninski, C.; Bykovskaya, V.; Maerkle, F.; Popov, R.; Palermoa, A.; Mattes, D.; Weber, L.; Ridder, B.; Foertsch, T.; Loeffler, F.; et al. Microcavity functionalization: Selective functionalization of microstructured surfaces by laser-assisted particle transfer. *Adv. Funct. Mater.* **2016**, *26*, 7067–7073. [CrossRef]

37. Bojnicic-Kninski, C.; Popov, R.; Dörsam, E.; Loeffler, F.; Breitling, F.; Nesterov-Mueller, A. Combinatorial particle patterning. *Adv. Funct. Mater.* **2017**, *27*, 1703511. [CrossRef]

38. Webera, L.; Palermoa, A.; Kügler, J.; Nesterov-Mueller, A.; Breitling, F.; Loeffler, F. Single amino acid fingerprinting of the human antibody repertoire with high density peptide arrays. *J. Immunol. Methods* **2017**, *443*, 45–54. [CrossRef] [PubMed]
39. Ridder, B.; Mattes, D.; Nesterov-Mueller, A.; Breitling, F.; Meier, M. Peptide array functionalization via the Ugi four-component reaction. *Chem. Commun.* **2017**, *53*, 5553–5556. [CrossRef] [PubMed]

 *nanomaterials*

Article

Passive Sampling of Gaseous Elemental Mercury Based on a Composite TiO$_2$NP/AuNP Layer

Antonella Macagnano [1,2,*], Paolo Papa [1], Joshua Avossa [1], Viviana Perri [1], Marcello Marelli [3], Francesca Sprovieri [4], Emiliano Zampetti [1], Fabrizio De Cesare [1,2], Andrea Bearzotti [1] and Nicola Pirrone [4]

1   Institute of Atmospheric Pollution Research—National Research Council (IIA-CNR), Research Area of Rome 1, Via Salaria km 29,300, 00016 Monterotondo, Italy; p.papa@iia.cnr.it (P.P.); joshua.avossa@iia.cnr.it (J.A.); v.perri@iia.cnr.it (V.P.); e.zampetti@iia.cnr.it (E.Z.); decesare@unitus.it (F.D.C.); a.bearzotti@iia.cnr.it (A.B.)
2   Department of Innovation in Biological Systems, Food and Forestry University of Tuscia (DIBAF), Via S. Camillo de Lellis, 00100 Viterbo, Italy
3   Institute of Molecular Science and Technologies—National Research Council (ISTM-CNR), Via G. Fantoli 16/15, 20138 Milano, Italy; m.marelli@istm.cnr.it
4   Institute of Atmospheric Pollution Research—National Research Council (IIA-CNR), Division of Rende, c/o UNICAL-Polifunzionale, 87036 Arcavacata di Rende (CS), Italy; sprovieri@iia.cnr.it (F.S.); pirrone@iia.cnr.it (N.P.)
*   Correspondence: antonella.macagnano@cnr.it or a.macagnano@iia.cnr.it; Tel.: +39 06 90672395-2401

Received: 6 September 2018; Accepted: 5 October 2018; Published: 7 October 2018

Abstract: Passive sampling systems (PASs) are a low cost strategy to quantify Hg levels in air over both different environmental locations and time periods of few hours to weeks/months. For this reason, novel nanostructured materials have been designed and developed. They consist of an adsorbent layer made of titania nanoparticles (TiO$_2$NPs, ≤25 nm diameter) finely decorated with gold nanoparticles. The TiO$_2$NPs functionalization occurred for the photocatalytic properties of titania-anatase when UV-irradiated in an aqueous solution containing HAuCl$_4$. The resulting nanostructured suspension was deposited by drop-casting on a thin quartz slices, dried and then incorporated into a common axial sampler to be investigated as a potential PAS device. The morphological characteristics of the sample were studied by High-Resolution Transmission Electron Microscopy, Atomic Force Microscopy, and Optical Microscopy. UV-Vis spectra showed a blue shift of the membrane when exposed to Hg^0 vapors. The adsorbed mercury was thermally desorbed for a few minutes, and then quantified by a mercury vapor analyzer. Such a sampling system reported an efficiency of adsorption that was equal to ≈95%. Temperature and relative humidity only mildly affected the membrane performances. These structures seem to be promising candidates for mercury samplers, due to both the strong affinity of gold with Hg, and the wide adsorbing surface.

Keywords: TiO$_2$NPs; AuNPs; photocatalysis; mercury vapors adsorbing layer; PAS device

---

1. Introduction

A thin film is commonly thought of as a layer with a thickness ranging from fractions of one nanometer to several micrometers; however, it is difficult to draw a border line between thick and thin films on the basis of their thickness, overall when the film is a nanocomposite or hybrid structure [1,2]. When a layer is designed to selectively adsorb gas or VOCs (volatile organic compounds), it has to be both chemically (to favor more specific interaction forces between adsorbent and adsorbate) and physically (to increase the number of the adsorbing sites, e.g., by acting on roughness and porosity) treated. The surface layer in contact with the environment is the main responsibility of the adsorbing process, and the relationship between the gas phase concentration and the adsorbed phase concentration at a constant temperature is reported as an adsorption isotherm, whereas the shape depicts the affinity relationships between the adsorbate and the absorbent [3]. Therefore, most of the chemical sensors and samplers for monitoring mercury in the atmosphere have been designed with these features taken into account. Mercury is a toxic pollutant, and it is considered by WHO (World Health Organization) as one of the top 10 chemicals of major public health concern [4]. It is continuously traveling through water, soil, and atmosphere, in various forms, to different parts of the world, and it is commonly emitted from both natural sources, as volcanoes, wildfire, and soil, and human activities, as fossil fuel burning, waste incineration, power plants, and artisanal mining [5,6]. There is a huge and quickly growing body of scientific literature on the distribution of mercury in several ecosystems: atmosphere is the principal transport pathway of Hg emissions, whereas soil and water play a significant role in the reallocation of Hg in several ecosystems [7]. Mercury in the atmosphere can be carried as gaseous elemental mercury (GEM) and gaseous oxidized mercury (GOM), which together are named as total gaseous mercury (TGM, C_{mean}: $\approx 1.4 \pm 0.15$ ng m^{-3}) and particle-bound mercury (PBM). Among them, GEM holds the most long atmospheric residence time (≈ 1 year) due to the relatively high vapor pressure and inertness to atmospheric oxidation [5,8]; conversely, GOM and PBM have much shorter atmospheric stays, being put down closer to their source locations [9,10]. Adsorbents based on sulfur-treated carbons, alumina, and zeolites, have been among the most commonly investigated materials that are able to capture mercury from the environment [11]. More recently, ZnS nanoparticles (NPs) have been developed and variously implemented as alternatives to remove Hg^0 from polluted environments [12]. The strong affinity between heavy metals (e.g., mercury) and noble metals, such as gold and silver, has been also investigated in literature as a suitable strategy to capture and reveal such a pollutant in air. Therefore, a series of filters, and detecting systems exploiting such features, have been designed to remove and detect, respectively, these pollutants within the environment. Specifically, when a removal process is required, as well as the detection of a very low concentration, cartridge-like structures with a high surface-volume ratio are preferred [13,14], and large amounts of volumes containing the pollutant are fluxed throughout the cartridge/filter, entrapping it inside. Conversely, when real-time detection or diffusion processes are investigated, thin film structures are commonly preferred. Both sensors and adsorbing devices based on metal thin films, and more recently, porous or nanostructured, have been reported in literature [15–17]. Nanostructured thin film layers, made with nanoparticles, are preferentially assembled in ordered structures, conforming to the surface with precise control over chemical and physical properties, in reproducible scaffolding. Some examples of deposition techniques that are commonly used include the electrodeposition of metal oxide and metal nanoparticles [18], the deposition of nanoparticle monolayers via the Langmuir Blodgett technique [19], sol–gel chemistry-based deposition of nanoparticles [20], layer-by-layer dip coating [21], in situ synthesis of nanoparticles using polymeric thin films as templates [22,23] self-assembling [24], and electrospray [25]. Further, depending on the functionalization or charges on the nanoparticle (NP) shells, ordered thin layer or 3D structures can also be designed by drop-casting, which is one of the simplest and cheapest deposition techniques [26,27], even if it is rarely able to assemble homogeneous layers, especially on large surfaces, mainly due to differences in evaporation rates through the substrate, or concentration fluctuations that can lead to variations in internal structure and film thickness. When the substrate is a porous matrix, the dropped nanocomponents, depending on their affinity to the surface, can penetrate and decorate the pores' surfaces, thus assuming a peculiar 3D-structure that is more or less conformal to the substrate scaffold. Such a porous layer, comprising interconnected void volumes and high surface-area-to-volume ratio, facilitates gas and VOC diffusion through its bulk, as well as gas/VOC adsorption onto its binding sites.

In the present study, highly pure quartz (SiO$_2$) microfibrous filters have been used as suitable substrates decorated with a nanocomposite material by dropping, capable of adsorbing, and then quantifying vapors of mercury from the atmosphere, as a promising thin structure for passive air sampler (PAS). The latter are generally designed to be cheap, simple to operate, and to work without electricity. For mercury analysis, the most basic requirement is that a PAS is able to sorb a sufficient amount of mercury for accurate and precise quantification. The peculiarity of the passive samplers relies on unassisted molecular diffusion of gaseous agents (i.e., volatile vapors of elemental mercury) through a diffusive surface onto an adsorbent material. Unlike active (pumped) sampling, passive samplers require no electricity (expensive pumps), have no moving parts, and are simple to use (no pump operation or calibration). After sampling, the adsorbed mercury should be desorbed off the adsorbent by solvent (chemical procedure) or thermal desorption (physical procedure). Passive samplers have to be commonly compact, portable, unobtrusive, and inexpensive. They are able to give information on the average pollution levels over time periods of a few hours, to weeks/months. They do not require supervision, and can be used in hazardous environments. Passive samplers have been designed, using a variety of synthetic materials (like sulfur-impregnated carbon (SIC), chlorine-impregnated carbon (CIC), bromine-impregnated carbon (BIC), gold-coated sorbents (GCS), etc.) and housings for Hg collection [28,29]. Activated carbon is suggested to be the most suitable sorbent material for PASs, since it is low cost and provides a large surface area [30]. On the other hand, the amalgamation between Au and Hg is also considered as an effective alternative strategy to design mercury samplers, even if it is not very popular since more expensive [28]. However, a large variety of materials based on nanotechnology have already been applied for this purpose, even if the state of art in those nanomaterial-based passive samplers is still in the early stages [31–37]. Sampling rate and adsorption capacity are the two key factors to evaluate the performance of a passive sampler. The PAS sampling rate depends on the shape of the sampler, and it is affected by meteorological factors [38–40]; meanwhile, the adsorption capacity depends on the affinity between the adsorbate and adsorbent, as well as the adsorbing layer structure (i.e., specific surface area, pore size distribution and number of binding site), and it is affected by temperature and potential chemical interferents in the air. All of these samplers work on the basis of diffusion. Most commercially available passive/diffusive samplers are planar or axial in shape [31]. Commonly, the adsorbing matrix is disk-shaped structured, with different thicknesses and porosities. Alternatively, they can have a radial shape, consisting of a columnar sorbent surrounded by a cylindrical diffusive barrier, with the purpose of increasing the sampling rate by maximizing the surface area across which diffusion occurs (Radiello® [41]). Other passive samplers for mercury vapor collection on the basis of diffusion have been constructed using a variety of synthetic materials (i.e., gold and silver surfaces, and sulfate-impregnated carbon) and housings [31,39,42].

In this work, an alternative approach adopted in the place of conventional ones has been described, designing onto a SiO_2 micro-fibrous filter, a coating (partially penetrated inside the filter) made of an aggregation of densely packed nanoparticles of TiO_2 finely decorated with smaller nanoparticles of gold AuNPs. Such a layer was achieved by exploiting the capability of TiO_2 (anatase) to photoreduce aqueous $HAuCl_4$ into elemental gold when irradiated with UV-light [43]. Furthermore, when a polymer (e.g., polyvinylpyrrolidone, PVP) was added to the $HAuCl_4$ aqueous solution, globose gold nanoparticles could be grown onto the metal oxide particles, obtaining a nanocomposite material with promising properties in adsorbing mercury from the atmosphere. These novel nanocomposite structures have here been considered as being very attractive adsorbing layers for passive sampling, due to both the strong affinity between mercury and gold, wide adsorbing surface due to the nanosize of the materials (expected high efficiency and lifetime), the robustness of the materials and the chance to use them for many times, thus avoiding the spread of waste materials into the environment.

2. Materials and Methods

All chemicals were acquired from Sigma Aldrich (Milan, Italy), and used without further purification: polyvinylpyrrolidone (PVP, Mn = 1,300,000 g/mol), titanium (IV) oxide (anatase, $\leq$25 nm diameter, Sigma Aldrich, Milan, Italy) and gold (III) chloride hydrate ($HAuCl_4$, 99.999%). Ultrapure water (5.5 10^{-8} S cm^{-1}) was produced by MilliQ-EMD Millipore. Quartz slice filters (Whatman™, Little Chalfont, UK) were 400 µm thick, and 2 cm wide with $\approx$2 µm pore size and $\leq$3 µm fiber diameter.

Titanium (IV) oxide (anatase) were suspended in an aqueous solution of PVP/ HAuCl$_4$ for preliminary investigations (600 mg TiO$_2$/ 60 mL PVP$_{aq}$/ 0.03 mg HAuCl$_4$). Such a suspension was UV-irradiated for 1 hr (365 nm, Helios Italquartz, Italy), thus changing the color from yellow to blue-violet, and subsequently centrifuged and rinsed with water at least three times to remove PVP excess (Thermo Scientific SL16R 230V, Langenselbold, Germany; T: 4 °C; t: 15 min ($\times$3); RCF: 5000 g). The resulting precipitated was vortexed, diluted with a few cc of ultrapure water, and deposited on the quartz slices by drop casting using a customized mask of Teflon$^{®}$ (d: 11.25 mm, $\approx$10 mg), then heated to 80 °C, and finally to 450 °C under a clean air flow, in order to remove both possible traces of polymer and mercury absorbed during preparation.

The nanostructured layers were analyzed by UV-Vis spectrophotometry (Spectrophotometer UV-2600, Shimadzu, integrating sphere ISR-2600Plus, Duisburg, Germany) before and after gold nano-functionalization, and by AFM (Nanosurf Flex-AFM, Liestal, Switzerland), which captured the layer surface images in tapping mode using 190Al-G tips, 190 kHz, 48 N/m.

Powder samples for transmission electron microscopy (TEM) measurements were gently grounded in an agate mortar, dispersed in isopropyl alcohol, sonicated for 10 minutes, and dropped onto a holey-carbon coated copper TEM grid. TEM and scanning transmission electron microscopy (STEM) analysis were performed by a ZEISS Libra 200FE microscope (Oberkochen, Germany) after complete solvent evaporation overnight in air. The size distribution were manually calculated counting more than 400 NPs by iTEM software (Olympus SIS, Muenster, Germany).

Optical micrographs were provided by Zeiss Axiophot Stereomicroscope equipped with a color videocamera (Axio Cam MRC, Wexford, Ireland) using a computer assisted image analysis system (AxioVision, Wexford, Ireland). The side-view of the quartz slice coated with AuNPs/TiO$_2$NPs film was provided by a Portable USB Digital Microscope 1$\times$–5000$\times$ Magnification Mini Microscope Camera (1x-5000X, Bangweier, Guangdong, China) by placing the sample between a microscope slide and a base support: the remaining floating sample was 45° tilted and displayed in the same picture.

A prototype of thermal desorption system was also planned in CNR-IIA and developed (Spaziani Rolando, Italy) in order to be connected to the most commons analytical systems of mercury. The prototype was manufactured in quartz and housed in a heater system (De Marco Forneria, Italy) to allow the fast desorption of the Hg^0 adsorbed on the thin layer of the nanostructured material, flowing pure (Air 5.0, Praxair-Rivoira, Italy) throughout the desorption chamber.

Mercury Vapor Analyzer Tekran 2537A (Tekran Instruments C., Toronto, ON Canada) was used to quantify the mercury desorbed from the nanocomposite film. The exposure of the adsorbing layers to injected volumes (μL) of mercury vapors (Tekran 2505, Mercury Calibration Unit) were carried out within customized sealed Duran glass samplers (V: 8.5 mL) to study the efficiency of the membrane. Conversely, the AuNPs-TiO$_2$NPs adsorbing discs were placed into customized glass samplers with diffusive caps (nylon membrane) for sampling rate calculation, and deployed with concentrations of mercury in relative humidity (%RH)- and temperature (T)-controlled measuring rooms. Such measuring rooms were monitored by the mercury vapor analyzer and a humidity-temperature transmitter (Relative Humidity and Temperature Probe HMP230, Vaisala Corporation, Helsinki, Finland).

3. Results and Discussion

Exploiting the photocatalytic properties of TiO$_2$, gold nanoparticles were selectively grown under UV light irradiation on titania nanoparticles through the photoreduction of HAuCl$_4$ in the presence of an organic capping reagent (PVP). The light yellow-colored aqueous suspension of TiO$_2$NPs containing HAuCl$_4$ and PVP when exposed to UV-light irradiation for 1 hr under magnetic stirring assumed a blue-purple color, due to the formation of gold nanoparticles (Figure 1). Subsequently to the centrifugation and then the resuspension of the pellet by ultrapure water, known amounts of the heterogeneous mixture, under stirring, were picked up and deposited onto several substrates to be characterized. Before each morphological and physical-chemical measurements, the samples were heated at 450 °C per 1 hr to eliminate the PVP traces and the potential mercury adsorbed.

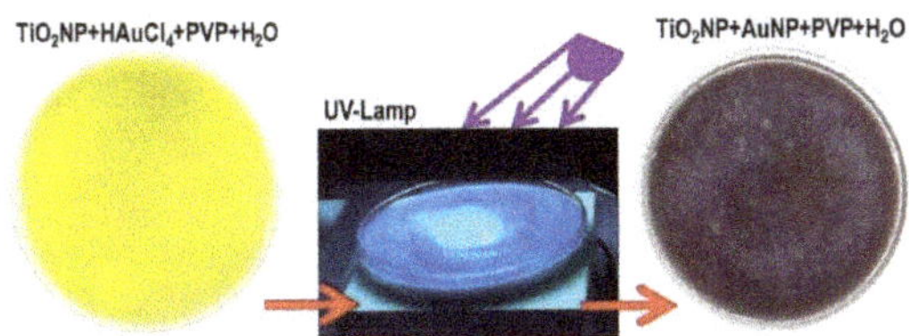

Figure 1. Functionalization of TiO$_2$ nanoparticles by UV-irradiation in a 0.1 M PVP (polyvinylpyrrolidone) aqueous suspension.

Due to the inhomogeneity of the film, all the spectra here depicted were carried out only over a selected area of the Au/TiO$_2$NPs film coating a flat substrate (SiO$_2$ wafer). The sample reported a reflectance minimum (about 15%) at 550 nm wavelength, thus confirming the growth of metal gold on the TiO$_2$ nanoparticles, and another stronger signal, with an onset at 380 nm, related to the charge transfer from the valence band to the conduction band of the titania nanoparticles [44]. The UV-Vis diffuse reflectance spectrum is depicted in Figure 2. The broad band (550 nm) was attributable to the characteristic localized surface plasmon resonance (LSPR) band of AuNPs, ranging between 500 and 600 nm [45,46]. This visible band was presumed to arise from the combined oscillations of the valence electrons confined in a cage of nanometer dimensions [47]: the position and shape of the surface plasmon band is affected by many parameters, including the dielectric constant of the medium, the particle size and shape and the coulomb charge of the nanoparticle. When AuNPs are joined to metal oxides, such as TiO$_2$, the material appears to be purple-brown colored, due to the characteristic surface plasmon band of gold.

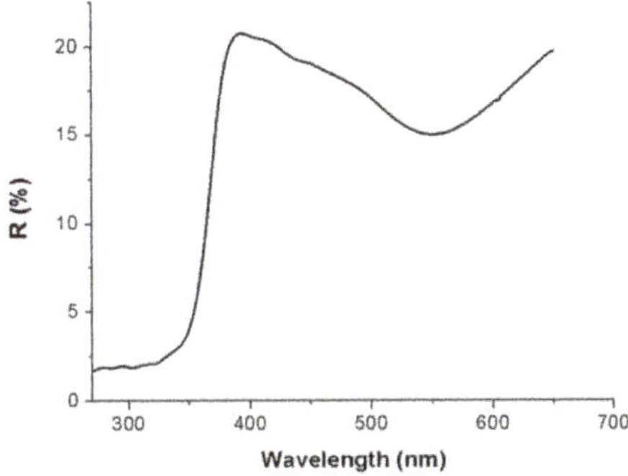

Figure 2. UV-Vis diffuse reflectance spectrum of AuNPs/TiO$_2$ layer in 270–670 nm wavelength range.

However, the minimum value of the reflectance band could be blue- or red-shifted, depending on the value of the average size of particles (i.e., the peak absorbance wavelength increases with particle diameter), as their aggregation (which is enhanced by a red-shift in the spectrum, as well as the broadening of adsorption peaks, and a decrease in peak intensities), functionalization, and inter-particles distance [48]. In literature, it was observed that the surface plasmon oscillation of gold nanoparticles in a suspension red shifted from ~520 to 530 nm as the particle diameter increased from 5 to 40 nm [49].

Both AuNP shape and size should be mainly related to the PVP concentration (the capping agent) and UV-light intensity, respectively [43,50] over the photocatalytic process. Specifically, the average particle size decreases as the intensity of the light increases. This effect of light intensity on the gold particle size could be very general, and it could be used to tune the average particle size to the optimum value when preparing Au/TiO$_2$ using this route. Additionally, both the size and the number (or density) of the nanoparticles can increase directly with the duration of irradiation.

The gold nanoparticle size distribution of the sample (Figure 3b) was centered around the mean value of 32.6 nm (70% of NPs ranging between 5 to 40 nm). A STEM micrograph (Figure 3a) showed a good particle dispersion onto the support (Figure 3b), whereas the HRTEM (High-Resolution Transmission Electron Microscopy) ones (Figure 3c,d) revealed well-shaped and highly crystalline gold nanoparticles that were in intimate contact with the anatase crystalline support (AuNPs appear slightly darker with respect to the anatase support). Interestingly, the gold NPs shared similar sizes with the support grains, assembling together into homogeneous aggregates.

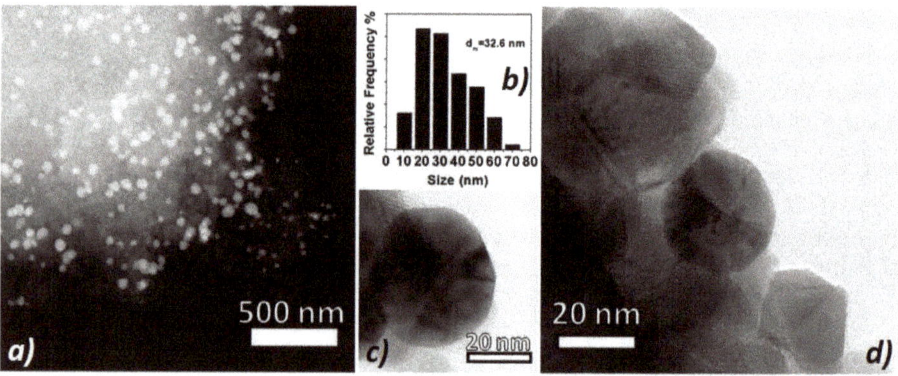

Figure 3. (**a**) A representative STEM micrograph of an Au/TiO$_2$ anatase sample at low magnification—AuNPs (gold nanoparticles) appear brighter over the greyish support; (**b**) AuNPs size distribution and representative HRTEM (High-Resolution Transmission Electron Microscopy) micrographs of (**c**) a gold NP and (**d**) a AuNP/TiO$_2$NPs cluster where a AuNP appears rounder and slightly darker than anatase ones.

Exploiting the properties of gold nanostructures due to the mercury adsorption, miniaturized sensing devices were demonstrated to be able to detect picograms of mercury in the air, like gold-microcantilevers [16] by changing their resonant frequency in real time. Au-TiO$_2$NPs deposited onto gold electrodes have been investigated as electrochemical sensors to detect Hg (II) in water [51]. Nanocomposite sensors made of titania nanofibers decorated with gold nanoparticles showed a limit of detection of 6 pptv (parts per trillion by volume) and 2 pptv, respectively, depending on the strategy of sampling [52]. Furthermore, devices based on resistivity changes in very thin gold films are also commercially available (Jerome® J405 mercury vapor analyzer) with a 0.01 µg m^{-3} resolution and a 750 ± 50 cc min^{-1} flow rate [53]. In the case of passive samplers, gold nanostructures have been used, for instance, by deploying very thin gold electrodes (50 nm) to Hg^0 for 100 min, and then measuring the changes in resistance (Limit of Detection, LOD: 1 µg/m^3): in this case, the analysis was provided by the same device [54]. Mercury also induces changes in the optical properties of gold films [55]. Exploiting this feature, in literature, porous glass discs were coated with AuNPs, showing a linear range within 0.1–15 ng Hg^0 and a LOD of 0.4 ng [56]: the exposure to Hg^0 caused a color change from red to violet-purple.

Here, the AuNP–TiO$_2$NP layer was exposed to a known concentration of elemental mercury vapors (14 mg/m^3) at increasing time (room temperature, 35% RH).

The diffuse reflectance spectra (Figure 4A) depicted an apparent wavelength blue shift of $\approx$ 2.7 nm after 15 min of exposure, up to $\approx$ 4.6 nm, and $\approx$ 6.6 nm after 60 and 120 min, respectively, according to a non-linear curve (Figure 4B), suggesting a quicker process initially and slower afterwards, also reported in literature [57]. A possible explanation has been provided by Mie theory [58]: since mercury is expected to be adsorbed strongly onto gold surface and Au is more electronegative than Hg (2.5 and 2.0, respectively) [59], mercury may be able to donate the electron density to gold NPs [60] causing the surface plasmon mode to blue-shift [58]. The related changes (0.1%) in reflectance, which seem to increase with dependence on the exposure time (up to 60 min), may be due to the change of the refractive index for mercury entrapment inside the nanocomposite layer [49].

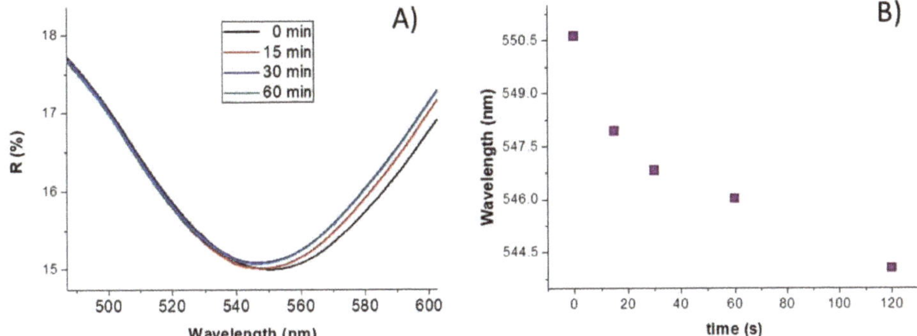

Figure 4. Comparison of UV-Vis diffuse reflectance spectra of AuNPs/TiO_2 layer in a 450–650 nm wavelength range at different exposure times to Hg^0 vapor pressure at 20 °C and 35%RH (**A**); a curve plot depicting the wavelength values of the AuNPs/TiO_2 layer at increasing exposure time to Hg^0 (**B**).

Similarly, a coating dispersed on a fibrous quartz surface of about 0.99×10^2 mm^2 was achieved by slowly dropping hundreds microliters of the aqueous suspension through a suitable mask, defining the area to be covered, and allowing water in excess to flow away through the disk filter. After deposition, the layer looked quite compact and it conformed to the quartz disc surface (Figure 5). The coating penetrated inside the quartz filter for about 25% of its thick layer, confirmed by the color change of cross filter section in Figure 5, right. Optical micrographs show that the composite film was conformal to the microfibrous surface of the supporting scaffold (Figure 5, inset), keeping its roughness characterized by valleys and overhangs.

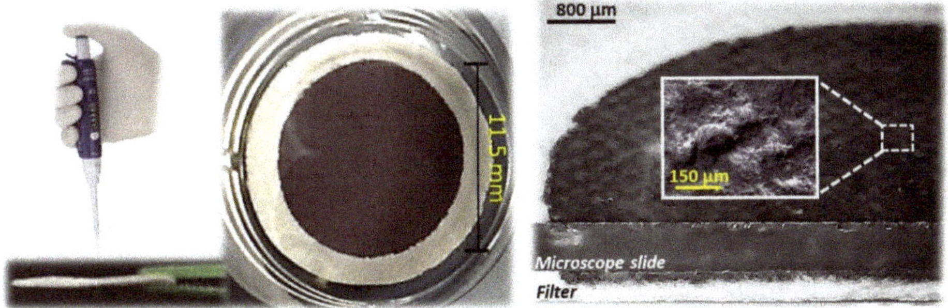

Figure 5. (**Left**) Drop deposition of TiO_2/AuNPs aqueous suspension; top view (**middle**) and side view (**right**) of the resulting disc. The inset depicts an optical magnification of the film surface.

Such a layer, since partially withheld by quartz microfibers, was stable and easy to be handled without suffering apparent damages or detachment. Figure 6 presents Atomic Force Microscope (AFM) surface topography images of the functionalized quartz filters. The coating surface showed a rough material provided of different sized grains (Ra: 5.4 ± 1.4 nm, average roughness) with ridges and valleys conforming to the fibrous substrate (Figure 6a). At higher magnification (Figure 6b,c), grains

(Ra: 140 ± 85 nm) appeared densely packed with a series of void spaces (darker areas), due to their uneven boundaries.

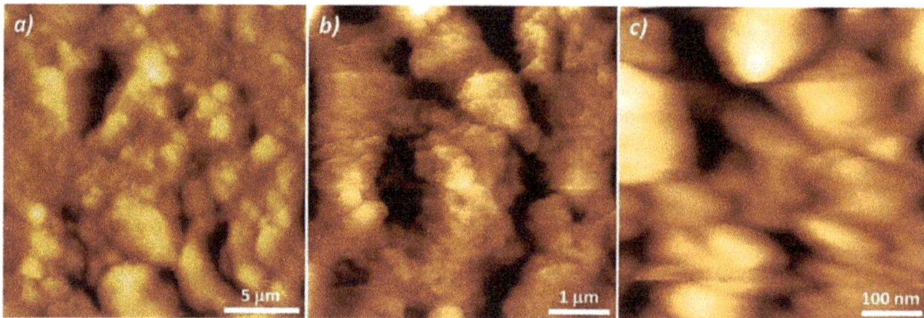

Figure 6. Atomic Force Microscope topography images of the quartz filter surface coated with a TiO_2/AuNP layer, at different magnifications.

In order to calibrate the adsorbing membrane to mercury vapors, the adsorbing disc was placed on the bottom of a suitable sealed glass chamber that was 2.5 cm height, and with a volume of 8.5 mL, where increasing amounts of Hg^0 vapors (μL) were injected by a gas-tight syringe at ~20 °C (under dry air). The injected volumes were selected to lower the experimental error as much as possible. Further, each amount was first theoretically estimated and then experimentally measured by injection into the analytical instrument. Finally, the disc was subjected to thermal desorption under an air flow that collected the desorbed mercury and delivered it to the mercury analyzer. In Figure 7, the adsorbed mass of Hg^0 versus the exposure time is reported. It was noted that 15 min of film exposure appeared to be sufficient to adsorb 90% of the injected mass. However, all of the estimated injected mass values were not completely adsorbed, even after an hour, probably due to the partial nonspecific sorption of mercury to the glass container.

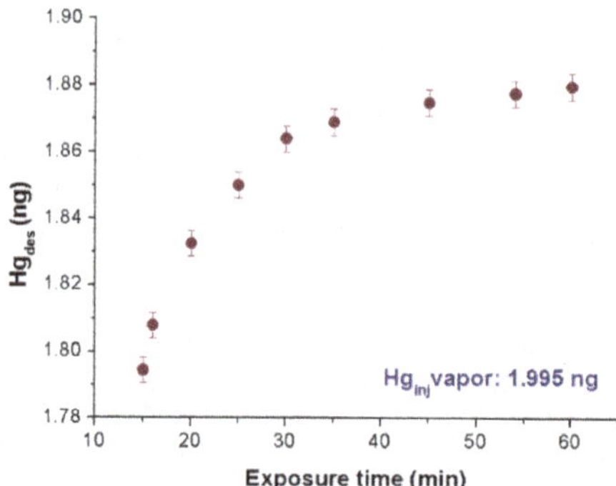

Figure 7. Adsorption curve of a known amount of injected Hg^0 vapors (C: 0.235 ng/mL) versus exposure times ranging between 15 and 60 min.

Therefore, to investigate the effects of humidity and temperature, further measurements of adsorption at different Hg^0 vapor concentrations were carried out by exposing the same membrane to the pollutant just for 15 min. The relative humidity changes were controlled and generated by a mass flow controller flowing dry air, and increasing the concentrations of the water vapors throughout the measuring chamber. Temperature values were provided by dipping the measuring chamber into a thermal bath. In order to obtain the membrane performances at −20 °C, the measuring chamber was put into a fridge at −20 °C. After exposure to Hg^0 by injection (15 min) the sample was resumed and desorbed for quantification. Experimental results reported a slight increase of the adsorbed mass when the relative humidity increased up to 70% RH (relative humidity) within the measurement chamber. Specifically, when 645 pg of Hg vapor were injected into the differently humidified measuring chamber, the nanostructured material was slightly affected by the water molecules, improving entrapment by an additional amount of 0.4 $\pm$ 0.01 pg of the analyte per %RH unit (Figure 8, right).

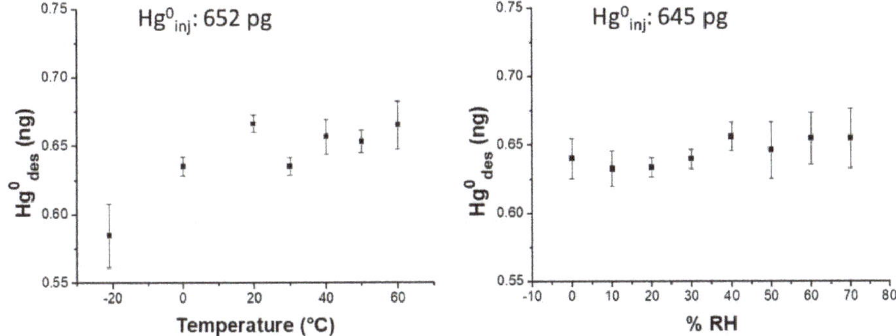

Figure 8. Plots depicting the desorbed mercury from a AuNP/TiO$_2$NP film upon exposure, within a sealed glass vessel that is similar to the passive sampler (PAS) container, to: (**left**) a known mass of Hg^0 vapor (about 652 $\pm$ 0.01 pg) when temperature of the sampler changed (ranging between -20 and 60 °C; dry air); (**right**) a known mass of Hg^0 vapor (about 645 $\pm$ 0.01 pg) when the %RH in the sampler changed (ranging between 5% and 70% relative humidity; room temperature). Each point in both the graphs is the resulting mean value from 5-times repeated-measures (error bars depict standard deviations).

Notably, when %RH was ranging between 50–70%, the desorbed values oscillated between 0.632 ng and 0.656 ng, i.e. they increased the values spread, and then the error. Similarly, temperature also slightly affected the analyte adsorption onto the PAS membrane, since the curve slope increased by 0.62 pg per Celsius degree over a thermal range of -20–60 °C when approximately 645 pg Hg^0 mass was injected (Figure 8, left).

Injected mercury vapor mass values were compared to the amount of Hg^0 that was actually adsorbed onto the exposed layer, in order to value its efficiency. Such a parameter was measured by adding a few microliters containing increasing masses of Hg^0 vapor into the measuring chamber at room temperature and in dry air. Five replicate measurements were provided for each injection. Upon a 15 min-deployment time, the adsorbing disc was thermally desorbed, and the collected vapors were delivered to the analytical instrument and then measured. The linear relationships between the mass injected and the mass adsorbed until 10 ng were depicted in the plot of Figure 9, suggesting a high absorbance of the nanostructured material. The affinity between mercury and the nanostructured material was confirmed also by the slope value of the linear fitting (S: 0.950 $\pm$ 0.005, R^2: 0.998) calculated on 15 min of sampling per each step. The mercury adsorbed mass when the disc was exposed to saturated mercury vapors for 18 h (T: 20 °C) was estimated to be more than 15 μg, confirming that such a thin layer was a very highly sorbent device for mercury.

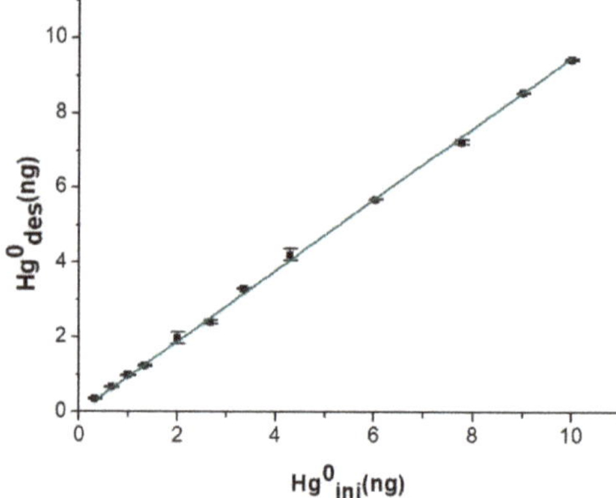

Figure 9. Adsorption curve of mercury vapors (measured as PAS-desorbed mercury) at increasing injected amounts of mercury vapors (ng).

Samples were completely restored after dozens of cycles of measurements, confirming the potential to use the same sample for many exposures. The functioning of the diffusive samplers is based on the movement of the contaminant molecules across a concentration gradient. In the collecting device (the case of a passive sampler, Figure 10) the contaminants diffuse from an area of higher concentration towards an area of lower concentration. According to the first Fick's Law, the rate at which the chemicals diffuse is represented by the following formula:

$$Q = D\left(\frac{A}{L}\right)C\,t, \tag{1}$$

where Q is the amount of the sample collected (ng), D is the diffusion coefficient (cm^2/min), A is the cross-sectional area of the diffusion path (cm^2), L is the diffusive path length (cm), C is the airborne concentration (mg/m^3) and t is the sampling time (min). Each contaminant has its own diffusion coefficient that is determined by its unique chemical and physical properties. The A/L parameter is determined by the sampler's geometry; the product of D (A/L) is the theoretical sampling rate of a diffusive sampler for a specific compound (e.g., elementary mercury).

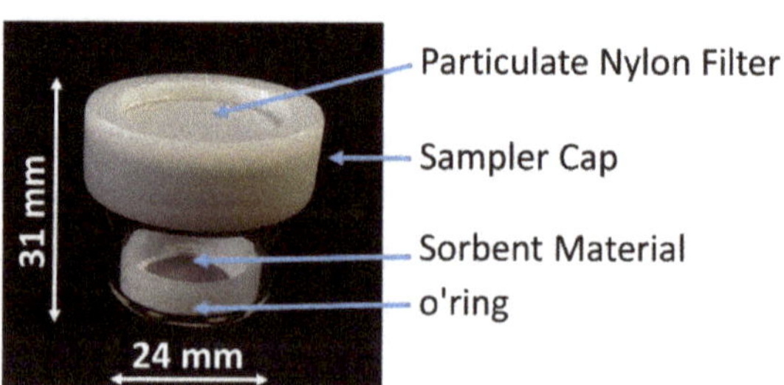

Figure 10. Prototype of mercury passive sampler comprising a diffusive membrane (particulate filter), the fibrous disc coated with AuNP/TiO$_2$NP (sorbent material), and a borosilicate vessel allowing the axial diffusion of Hg^0 from the cap to the bottom. The O-ring stops the disk on the tube bottom.

In order to experimentally evaluate the sampling rate (*SR*) of the proposed passive sampler, a useful method is given by the use of the empirical equation (2) [28]:

$$SR = Q/(Ct) \tag{2}$$

where Q is the amount of the adsorbed mercury (ng), C is the exposition concentration (ng m^{-3}), and t is the deployed time (days) of the sample, named as PS. In our case, several measurements have been performed, exposing three passive samplers (PS1, PS2, PS3) at three different vapor mercury concentrations for 3, 7, and 15 days of sampling time. All of the measurements were performed in three measuring chambers, where three different ambient vapor mercury concentrations, of 1.2, 3.5, and 4.5 ng m^{-3}, were kept constant. As previously mentioned, the PAS device here described works exploiting the unassisted axial diffusion process of the mercury vapor through the diffusive membrane, along the glass vessel (diffusion path), up to the adsorbing film placed on the vessel bottom. This PAS comprises a see-through borosilicate vessel, a cap made of a nylon membrane for gas diffusion and particulate stopping, a locking ring to keep the adsorbing membrane to the vessel bottom, and finally the adsorbing membrane (the violet discs in Figures 10 and 11). The PAS fabrication was easy and quite reproducible, since all of the membranes that were decorated by a given volume of the suspension, reported the same weight (10.00 ± 0.25 mg), obviously with the uncertainty (2.5%) generated by the deposition technique and by the irregularity of the hosting substrate. The resulting nanostructures looked very stable, since they were partially entrapped inside the filter and did not appear to be decolored or scratched, even after 1 year of measurements. The pictures in Figure 11 shows a batch of the fibrous quartz discs decorated with the AuNP–TiO$_2$NP layers, and their placement into the device in order to be characterized as potential PAS for mercury. Each adsorbing disc was mercury thermally desorbed before (to have a clean substrate) and after (to measure the concentration in air) each exposure, and the desorbed vapors were delivered to the analytical instrument. Commonly, 10 min of heating was sufficient to both restore the adsorbing disc, and to measure the amount of Hg^0 that was adsorbed throughout the deployment time. For each measure, quartz discs were heated under clean and dry air flow until the Tekran Analyzer displayed values of Hg^0 that were close to zero.

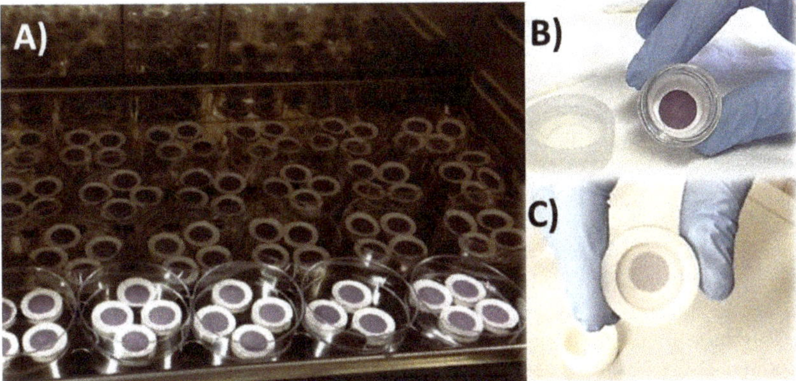

Figure 11. A fabrication step of the PAS devices: many of AuNPs-TiO$_2$/quartz discs drop-deposited and placed in an oven to facilitate the solvent evaporation (**A**); the placement of the adsorbing disc on the borosilicate vessel bottom (**B**); the PAS device with the particulate filter mounted (**C**).

After each exposure, using Equation (2), the sampling rate (*SR*) of each passive was calculated, and the relative results were reported in Table 1.

Table 1. Passive devices sampling rate values.

Sampler	$^{(1)}$Hg^0 Vapor Concentration (ng/m^3)	Hg^0 Adsorbed Mass (ng)	Exposure Time (days)	Sampling Rate (m^3/days)
PS1	3.5	0.140	3	0.013
PS2	3.5	0.143	3	0.014
PS3	3.5	0.144	3	0.014
PS1	1.2	0.124	7	0.015
PS2	1.2	0.124	7	0.015
PS3	1.2	0.116	7	0.014
PS1	3.5	0.357	7	0.015
PS2	3.5	0.355	7	0.014
PS3	3.5	0.359	7	0.015
PS1	4.5	0.463	7	0.015
PS2	4.5	0.471	7	0.015
PS3	4.5	0.452	7	0.014
PS1	1.2	0.279	15	0.016
PS2	1.2	0.272	15	0.015
PS3	1.2	0.269	15	0.015
PS1	3.5	0.727	15	0.014
PS2	3.5	0.736	15	0.014
PS3	3.5	0.720	15	0.014
PS1	4.5	0.978	15	0.014
PS2	4.5	0.962	15	0.014
PS3	4.5	0.951	15	0.014

^{1} measured by TEKRAN 2537A unit, the temperature was 24 °C (with a fluctuation of 1 °C), and the RH% was 40% (with a fluctuation of 2%) over the whole experiment.

The mean value of sampling rate was estimated to be 0.014 ± 0.0007 (m^3/day). Figure 12 depicts the *SR* values related to the exposure to increasing concentrations of mercury vapors. The reported error bars are referred to the standard deviation (*SD*) of uncertainty, showing a low dispersion of the overall values at increasing concentrations. In literature, generally depending on the range of environmental mercury concentration to be monitored, several PASs have been designed with different *SRs* [28], ranging from 0.00031 m^3/day based on gold-coated silica placed in an axial tube without diffusive membrane [61], to 0.13 m^3/day based on sulfur-impregnated activated carbon (axial tube, no diffusive membrane) [38,42], and to 6.6 m^3/day based on gold-coated quartz fiber filters [62]. Furthermore, since *SRs* are commonly affected by environmental conditions (strong wind, pressure, proximity to the coast or to desert and sandy areas, rain), protective shells have also been also used, improving the reliability of the measurements (*SR*: 0.121 ± 0.005 m^3/day) [63]. Physically, *SR* quantifies the volume of air that is effectively stripped of the pollutant per unit of time. As previously described, it depends on the diffusion coefficient of the compound in air, but also on other parameters as the diffusive path length of the PAS device: by changing parameters as the length or the quality of the diffusive barriers, *SR* can be modulated. Thus, higher *SR* values are commonly preferred when very low-polluted environments may be measured with a certain accuracy and for a short time (e.g., wearable devices that are commonly suitable for 1 or 2 deployment days). Conversely, PASs with lower *SR* values are desired for longer times of monitoring (up to one year) to ensure the presence of free adsorbing sites on the membrane. On the other hand, *SR* values that are too low could be responsible for a low resolution of the PAS devices when exposed in poorly polluted sites, making them unattractive as accurate measurement tools, since, accordingly, highly sensitive analytical techniques are required.

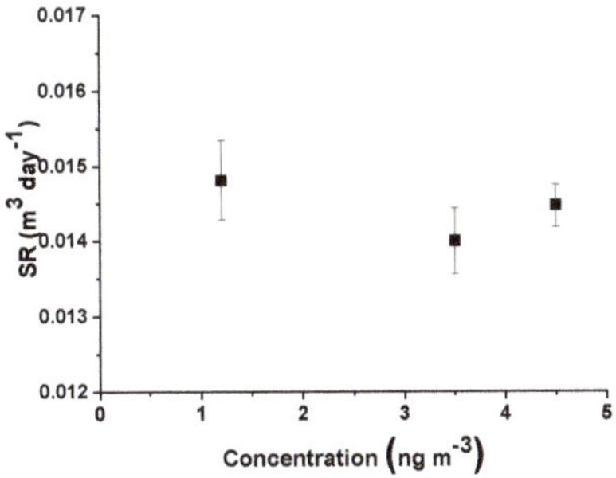

Figure 12. Estimated sampling rate vs concentration (ranging from 1.2 to 4.5 ngm^{-3}).

A comparison between the estimated concentrations, calculated using the experimental sampling rate (*SR*) and the measured values by the vapor mercury analyzer, has been reported in Figure 13.

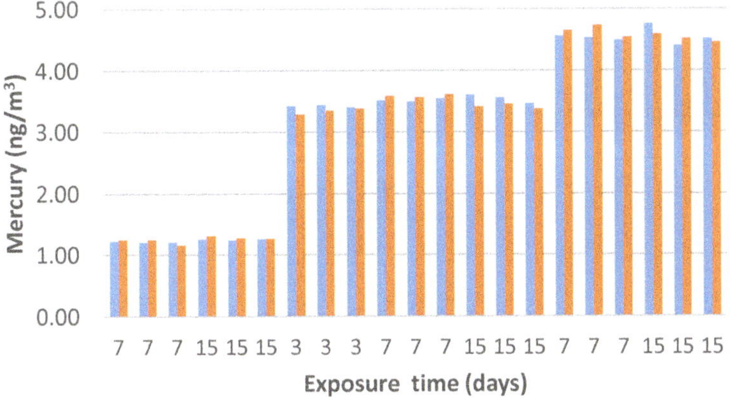

Figure 13. Comparison between the PAS estimated concentration using the experimental sampling rate (red bar) and the measured values by the vapor mercury analyzer (blue bar).

The resulting PAS values were similar to those that were reported by the analytical device over exposure times ranging between three and 15 days and to different concentrations of Hg^0 vapors, ranging between 1.2 and 4.6 ng/m^3. Specifically, when the mercury analyzer measured average concentration values of 1.23, 3.49, and 4.59 ng/m^3, PAS values were reported to be 1.25, 3.44, and 4.57 ng/m^3, respectively, with an average deviation of ~1.2%.

4. Conclusions

Since thin film structures are preferred for the development of passive samplers, a nanostructured mercury vapor-adsorbing layer made of TiO$_2$NPs and photo-decorated with AuNPs was designed and assembled by drop-casting onto a microporous filter of SiO$_2$. The decorated disc looked quite compact and it conformed to the SiO$_2$ microfibrous surface, keeping its roughness made of valleys and overhangs, and penetrating inside the quartz filter for less than a quarter of its thickness. Such a resulting layer was stable and easy to be handled without suffering apparent damage or detachment.

The gold nanoparticles, grown on TiO$_2$NPs, shared similar size with the support grains, assembling together into homogeneous aggregates. However if the AuNP shapes were regular (spherical, highly crystalline), their size distribution became heterogeneous, ranging between 5 and 40 nm.

Exploiting both the high surface/volume ratio and the strong affinity between gold and mercury, the membranes, investigated as potential passive samplers for gaseous elemental mercury, showed a high absorbency (up to 15 µg), together with a 95% efficiency of absorption, with only slight effects due to temperature (+0.1% per Celsius degree, in a thermal range between −20 and 60 °C) and relative humidity (+0.06% per %RH unit, between a dry and a 70% humid environment).

Samples could be restored after dozens of cycles of measurements by thermal desorption, confirming their potential to use the same sample for many exposures.

When the adsorbing discs were placed inside axial passive samplers, a sampling rate of 0.014 m^3/day was estimated when they were tested in controlled environments. Their features were compared to those of the analytical measuring instrument, reporting an average deviation of ~1.2%. Such a value suggested the chance to be applied for both short and longer monitoring campaigns. Therefore, due to their ease of preparation, their high sensitivity to gaseous elemental mercury due to the strong affinity between mercury and gold, high efficiency and a long lifetime, the AuNP/TiO$_2$NP-based devices are expected to be promising candidates for passive sampling strategies.

On the other hand, further studies are needed to evaluate the effect of thickness on the efficiency and adhesiveness of the quartz support. Similarly, the AuNP size and shape, as well as its density will be investigated too, in order to assess their effects onto the sensitivity and compactness of the aggregated hybrid nanoparticles inside the film. Finally, further investigations are in progress in environments that are polluted with potential interferents, such as chlorides and sulfides, and in extreme environmental conditions, in order to be considered for monitoring campaigns over the globe.

Author Contributions: Conceptualization and Design, A.M.; Methodology, A.M, E.Z., A.B., F.D.C, F.S., J.A., V.P. and N.P.; Validation, P.P., E.Z. and V.P.; Material Characterization, M.M., A.M., F.D.C. and J.A.; Experimental Investigation, P.P., M.M., V.P. and J.A.; Resources, A.M., F.S., N.P.; Project administration, N.P.; Data Curation, E.Z., P.P., F.D.C., A.B. and A.M.; Writing—Original Draft Preparation, A.M.; Writing—Review & Editing, All the Authors; Supervision, A.M., N.P.

Funding: This research was funded by UNEP-GEF Project named "Development of a Plan for Global Monitoring of Human Exposure to and Environmental Concentrations of Mercury", grant number 4E59 of the Global Environment Facility (GEF) to UNEP, and with the co-finance of WHO Europe, CNR-IIA and UNEP.

Acknowledgments: Authors gratefully thank Capocecera A. and Esposito G. for their technical support in providing Hg^0 vapor measurements and assembling PAS devices, respectively.

Conflicts of Interest: The authors declare that the research was conducted in the absence of any commercial or financial relationships that could be construed as a potential conflict of interest.

References

1. Ohring, M. *Material Science of Thin Film*, 2nd ed.; Academic Press: Cambridge, MA, USA, 2001; ISBN 9780125249751.

2. Macagnano, A. New Applications of Nanoheterogeneous Systems. In *Nanostructured Polymer Blends*; Thomas, S., Shanks, R., Chandrasekharakurup, S., Eds.; Elsevier Science: Exeter, UK, 2013; pp. 449–493. ISBN 9781455731596.

3. Huang, H.; Haghighat, F.; Blondeau, P. Volatile organic compound (VOC) adsorption on material: Influence of gas phase concentration, relative humidity and VOC type. *Indoor Air* **2006**, *16*, 236–247. [CrossRef] [PubMed]

4. Mercury and Health. Available online: http://www.who.int/news-room/fact-sheets/detail/mercury-and-health (accessed on 4 September 2018).

5. Pirrone, N.; Cinnirella, S.; Feng, X.; Finkelman, R.B.; Friedli, H.R.; Leaner, J.; Mason, R.; Mukherjee, A.B.; Stracher, G.B.; Streets, D.G.; et al. Global mercury emissions to the atmosphere from anthropogenic and natural sources. *Atmos. Chem. Phys.* **2010**, *10*, 5951–5964. [CrossRef]

6. Pacyna, E.G.; Pacyna, J.M.; Sundseth, K.; Munthe, J.; Kindbom, K.; Wilson, S.; Steenhuisen, F.; Maxson, P. Global emission of mercury to the atmosphere from anthropogenic sources in 2005 and projections to 2020. *Atmos. Environ.* **2010**, *44*, 2487–2499. [CrossRef]

7. Kim, M.K.; Zoh, K.D. Fate and transport of mercury in environmental media and human exposure. *J. Prev. Med. Public Heal.* **2012**, *45*, 335–343. [CrossRef] [PubMed]

8. Skov, H.; Christensen, J.H.; Goodsite, M.E.; Heidam, N.Z.; Jensen, B.; Wåhlin, P.; Geernaert, G. Fate of elemental mercury in the arctic during atmospheric mercury depletion episodes and the load of atmospheric mercury to the arctic. *Environ. Sci. Technol.* **2004**, *38*, 2373–2382. [CrossRef] [PubMed]

9. Pirrone, N.; Mason, R. *Mercury Fate and Transport in the Global Atmosphere: Emissions, Measurements and Models*, 1st ed.; Springer Science+Business Media: New York, NY, USA, 2009; ISBN 978-0-387-93958-2.

10. United Nations Environment Programme (UNEP). *Global Mercury Assessment 2013: Sources, Emissions, Releases and Environmental Transport*; United Nations: New York, NY, USA, 2015; ISBN 978-9211587272.

11. Otani, Y.; Emi, H.; Kanaoka, C.; Uchijima, I.; Nishino, H. Removal of mercury vapor from air with sulfur-impregnated adsorbents. *Environ. Sci. Technol.* **1986**, *22*, 708–711. [CrossRef]

12. Li, H.; Feng, S.; Qu, W.; Yang, J.; Liu, S.; Liu, Y. Adsorption and oxidation of elemental mercury on chlorinated ZnS surface. *Energy Fuels* **2018**, *32*, 7745–7751. [CrossRef]

13. Macagnano, A.; Zampetti, E.; Bearzotti, A.; De Cesare, F. Electrospinning for Air Pollution Control. In *Electrospinning for Advanced Energy and Environmental Applications*; Cavaliere, S., Ed.; CRC Press: Boca Raton, FL, USA, 2017; ISBN 9781138749085.

14. Alagappan, P.N.; Heimann, J.; Morrow, L.; Andreoli, E.; Barron, A.R. Easily regenerated readily deployable absorbent for heavy metal removal from contaminated water. *Sci. Rep.* **2017**, *7*, 6682. [CrossRef] [PubMed]

15. Mcnerney, J.J.; Buseck, P.R.; Hanson, R.C. Mercury detection by means of thin gold films. *Science* **1972**, *178*, 611–612. [CrossRef] [PubMed]

16. Thundat, T.; Wachter, E.A.; Sharp, S.L.; Warmack, R.J. Detection of mercury vapor using resonating microcantilevers. *Appl. Phys. Lett.* **1995**, *66*, 1695–1697. [CrossRef]

17. Rex, M.; Hernandez, F.E.; Campiglia, A.D. Pushing the limits of mercury sensors with gold nanorods. *Anal. Chem.* **2006**, *78*, 445–451. [CrossRef] [PubMed]

18. Zhao, L.Y.; Eldridge, K.R.; Sukhija, K.; Jalili, H.; Heinig, N.F.; Leung, K.T. Electrodeposition of iron core-shell nanoparticles on a H-terminated Si(100) surface. *Appl. Phys. Lett.* **2006**, *88*, 033111. [CrossRef]

19. Paul, S.; Pearson, C.; Molloy, A.; Cousins, M.A.; Green, M.; Kolliopoulou, S.; Dimitrakis, P.; Normand, P.; Tsoukalas, D.; Petty, M.C. Langmuir-Blodgett film deposition of metallic nanoparticles and their application to electronic memory structures. *Nano Lett.* **2003**, *3*, 533–536. [CrossRef]

20. Fan, H.; Wright, A.; Gabaldon, J.; Rodriguez, A.; Brinker, C.J.; Jiang, Y.B. Three-dimensionally ordered gold nanocrystal/silica superlattice thin films synthesized via sol-gel self-assembly. *Adv. Funct. Mater.* **2006**, *16*, 891–895. [CrossRef]

21. Lee, D.; Rubner, M.F.; Cohen, R.E. All-nanoparticle thin-film coatings. *Nano Lett.* **2006**, *6*, 2305–2312. [CrossRef] [PubMed]

22. Boontongkong, Y.; Cohen, R.E. Cavitated block copolymer micellar thin films: Lateral arrays of open nanoreactors. *Macromolecules* **2002**, *35*, 3647–3652. [CrossRef]

23. Segalman, R.A. Patterning with block copolymer thin films. *Mater. Sci. Eng. R Rep.* **2005**, *48*, 191–226. [CrossRef]

24. Kinge, S.; Crego-Calama, M.; Reinhoudt, D.N. Self-assembling nanoparticles at surfaces and interfaces. *ChemPhysChem* **2008**, *9*, 20–42. [CrossRef] [PubMed]

25. Suh, J.; Han, B.; Okuyama, K.; Choi, M. Highly charging of nanoparticles through electrospray of nanoparticle suspension. *J. Colloid Interface Sci.* **2005**, *287*, 135–140. [CrossRef] [PubMed]

26. Fontana, L.; Fratoddi, I.; Venditti, I.; Ksenzov, D.; Russo, M.V.; Grigorian, S. Structural studies on drop-cast film based on functionalized gold nanoparticles network: The effect of thermal treatment. *Appl. Surf. Sci.* **2016**, *369*, 115–119. [CrossRef]

27. Bearzotti, A.; Papa, P.; Macagnano, A.; Zampetti, E.; Venditti, I.; Fioravanti, R.; Fontana, L.; Matassa, R.; Familiari, G.; Fratoddi, I. Environmental Hg vapours adsorption and detection by using functionalized gold nanoparticles network. *J. Environ. Chem. Eng.* **2018**, *6*, 4706–4713. [CrossRef]

28. McLagan, D.S.; Mazur, M.E.E.; Mitchell, C.P.J.; Wania, F. Passive air sampling of gaseous elemental mercury: a critical review. *Atmos. Chem. Phys.* **2016**, *16*, 3061–3076. [CrossRef]

29. McLagan, D.S.; Mitchell, C.P.J.; Steffen, A.; Hung, H.; Shin, C.; Stupple, G.W.; Olson, M.L.; Luke, W.T.; Kelley, P.; Howard, D.; et al. Global evaluation and calibration of a passive air sampler for gaseous mercury. *Atmos. Chem. Phys.* **2018**, *18*, 5905–5919. [CrossRef]

30. Lee, S.H.; Rhim, Y.J.; Cho, S.P.; Baek, J.I. Carbon-based novel sorbent for removing gas-phase mercury. *Fuel* **2006**, *85*, 219–226. [CrossRef]

31. Huang, J.; Lyman, S.N.; Hartman, J.S.; Gustin, M.S. A review of passive sampling systems for ambient air mercury measurements. *Environ. Sci. Process. Impacts* **2014**, *16*, 374–392. [CrossRef] [PubMed]

32. James, J.Z.; Lucas, D.; Koshland, C.P. Gold nanoparticle films as sensitive and reusable elemental mercury sensors. *Environ. Sci. Technol.* **2012**, *46*, 9557–9562. [CrossRef] [PubMed]

33. Sedghi, R.; Heidari, B.; Behbahani, M. Synthesis, characterization and application of poly(acrylamide-*co*-methylenbisacrylamide) nanocomposite as a colorimetric chemosensor for visual detection of trace levels of Hg and Pb ions. *J. Hazard. Mater.* **2015**, *285*, 109–116. [CrossRef] [PubMed]

34. Zhang, Y.; Zeng, G.M.; Tang, L.; Li, Y.P.; Chen, Z.M.; Huang, G.H. Quantitative detection of trace mercury in environmental media using a three-dimensional electrochemical sensor with an anionic intercalator. *RSC Adv.* **2014**, *4*, 18485–18492. [CrossRef]

35. Kang, T.; Yoo, S.M.; Kang, M.; Lee, H.; Kim, H.; Lee, S.Y.; Kim, B. Single-step multiplex detection of toxic metal ions by Au nanowires-on-chip sensor using reporter elimination. *Lab Chip* **2012**, *12*, 3077–3081. [CrossRef] [PubMed]

36. Botasini, S.; Heijo, G.; Méndez, E. Toward decentralized analysis of mercury (II) in real samples. A critical review on nanotechnology-based methodologies. *Anal. Chim. Acta* **2013**, *800*, 1–11. [CrossRef] [PubMed]

37. Santos, E.B.; Ferlin, S.; Fostier, A.H.; Mazali, I.O. Using gold nanoparticles as passive sampler for indoor monitoring of gaseous elemental mercury. *J. Braz. Chem. Soc.* **2017**, *28*, 1274–1280. [CrossRef]

38. Guo, H.; Lin, H.; Zhang, W.; Deng, C.; Wang, H.; Zhang, Q.; Shen, Y.; Wang, X. Influence of meteorological factors on the atmospheric mercury measurement by a novel passive sampler. *Atmos. Environ.* **2014**, *97*, 310–315. [CrossRef]

39. Gustin, M.S.; Lyman, S.N.; Kilner, P.; Prestbo, E. Development of a passive sampler for gaseous mercury. *Atmos. Environ.* **2011**, *45*, 5805–5812. [CrossRef]

40. Tuduri, L.; Harner, T.; Hung, H. Polyurethane foam (PUF) disks passive air samplers: Wind effect on sampling rates. *Environ. Pollut.* **2006**, *144*, 377–383. [CrossRef] [PubMed]

41. Król, S.; Zabiegała, B.; Namieśnik, J. Monitoring VOCs in atmospheric air II. Sample collection and preparation. *TrAC-Trends Anal. Chem.* **2010**, *29*, 1101–1112. [CrossRef]

42. Zhang, W.; Tong, Y.; Hu, D.; Ou, L.; Wang, X. Characterization of atmospheric mercury concentrations along an urban-rural gradient using a newly developed passive sampler. *Atmos. Environ.* **2012**, *47*, 26–32. [CrossRef]

43. Li, D.; McCann, J.T.; Gratt, M.; Xia, Y. Photocatalytic deposition of gold nanoparticles on electrospun nanofibers of titania. *Chem. Phys. Lett.* **2004**, *394*, 387–391. [CrossRef]

44. Rahulan, K.M.; Ganesan, S.; Aruna, P. Synthesis and optical limiting studies of Au-doped TiO$_2$ nanoparticles. *Adv. Nat. Sci. Nanosci. Nanotechnol.* **2011**, *2*. [CrossRef]

45. Jain, P.K.; Lee, K.S.; El-Sayed, I.H.; El-Sayed, M.A. Calculated absorption and scattering properties of gold nanoparticles of different size, shape, and composition: Applications in biological imaging and biomedicine. *J. Phys. Chem. B* **2006**, *110*, 7238–7248. [CrossRef] [PubMed]

46. Njoki, P.N.; Lim, I.-I.S.; Mott, D.; Park, H.-Y.; Khan, B.; Mishra, S.; Sujakumar, R.; Luo, J.; Zhong, C.-J. Size correlation of optical and spectroscopic properties for gold nanoparticles. *J. Phys. Chem. C* **2007**, *111*, 14664–14669. [CrossRef]

47. Primo, A.; Corma, A.; García, H. Titania supported gold nanoparticles as photocatalyst. *Phys. Chem. Chem. Phys.* **2011**, *13*, 886–910. [CrossRef] [PubMed]

48. Amendola, V.; Meneghetti, M.; Amendola, V.; Meneghetti, M. Size evaluation of gold nanoparticles by UV-VISs spectroscopy size evaluation of gold nanoparticles by UV-VIS spectroscopy. *J. Phys. Chem. C* **2009**, *113*, 4277–4285. [CrossRef]

49. Morris, T.; Copeland, H.; McLinden, E.; Wilson, S.; Szulczewski, G. The effects of mercury adsorption on the optical response of size-selected gold and silver nanoparticles. *Langmuir* **2002**, *18*, 7261–7264. [CrossRef]

50. McGilvray, K.L.; Decan, M.R.; Wang, D.; Scaiano, J.C. Facile photochemical synthesis of unprotected aqueous gold nanoparticles. *J. Am. Chem. Soc.* **2006**, *128*, 15980–15981. [CrossRef] [PubMed]

51. Zhou, L.; Xiong, W.; Liu, S. Preparation of a gold electrode modified with Au–TiO$_2$ nanoparticles as an electrochemical sensor for the detection of mercury(II) ions. *J. Mater. Sci.* **2015**, *50*, 769–776. [CrossRef]

52. Macagnano, A.; Perri, V.; Zampetti, E.; Ferretti, A.M.; Sprovieri, F.; Pirrone, N.; Bearzotti, A.; Esposito, G.; De Cesare, F. Elemental mercury vapor chemoresistors employing TIO$_2$ nanofibers photocatalytically decorated with Au-nanoparticles. *Sens. Actuators B Chem.* **2017**, *247*, 957–967. [CrossRef]

53. Gold Film Mercury Vapor Analyzer. Available online: https://www.azic.com/jerome/j405/ (accessed on 4 September 2018).

54. Mattoli, V.; Mazzolai, B.; Raffa, V.; Mondini, A.; Dario, P. Design of a new real-time dosimeter to monitor personal exposure to elemental gaseous mercury. *Sens. Actuators B Chem.* **2007**, *123*, 158–167. [CrossRef]

55. Morris, T.; Kloepper, K.; Wilson, S.; Szulczewski, G. A spectroscopic study of mercury vapor adsorption on gold nanoparticle films. *J. Colloid Interface Sci.* **2002**, *254*, 49–55. [CrossRef] [PubMed]

56. De Barros Santos, E.; Moher, P.; Ferlin, S.; Fostier, A.H.; Mazali, I.O.; Telmer, K.; Brolo, A.G. Proof of concept for a passive sampler for monitoring of gaseous elemental mercury in artisanal gold mining. *Sci. Rep.* **2017**, *7*, 16513. [CrossRef] [PubMed]

57. Ravindranath, R.; Periasamy, A.P.; Roy, P.; Chen, Y.W.; Chang, H.T. Smart app-based on-field colorimetric quantification of mercury via analyte-induced enhancement of the photocatalytic activity of TiO2–Au nanospheres. *Anal. Bioanal. Chem.* **2018**, *410*, 4555–4564. [CrossRef] [PubMed]

58. Henglein, A. Physicochemical properties of small metal particles in solution: "Microelectrode" reactions, chemisorption, composite metal particles, and the atom-to-metal transition. *J. Phys. Chem.* **1993**, *97*, 5457–5471. [CrossRef]

59. Pauling, L. *Nature of the Chemical Bond and the Structure of Molecules and Crystals: An Introduction to Modern Structural Chemistry*, 3rd ed.; Cornell University Press: Nwe York, NY, USA, 1960; ISBN 9780801403330.

60. Bajpai, R.P.; Kita, H.; Azuma, K. Adsorption of water on au and mercury on Au, Ag, Mo and Re: Measurements of the changes in work function. *Jpn. J. Appl. Phys.* **1976**, *15*, 2083–2086. [CrossRef]

61. Brown, R.J.C.; Burdon, M.K.; Brown, A.S.; Kim, K.H. Assessment of pumped mercury vapour adsorption tubes as passive samplers using a micro-exposure chamber. *J. Environ. Monit.* **2012**, *14*, 2456–2463. [CrossRef] [PubMed]

62. Huang, J.; Choi, H.-D.; Landis, M.S.; Holsen, T.M. An application of passive samplers to understand atmospheric mercury concentration and dry deposition spatial distributions. *J. Environ. Monit.* **2012**, *14*, 2976. [CrossRef] [PubMed]

63. McLagan, D.S.; Mitchell, C.P.J.; Huang, H.; Lei, Y.D.; Cole, A.S.; Steffen, A.; Hung, H.; Wania, F. A High-Precision Passive Air Sampler for Gaseous Mercury. *Environ. Sci. Technol. Lett.* **2016**, *3*, 24–29. [CrossRef]

Article

Strongly Iridescent Hybrid Photonic Sensors Based on Self-Assembled Nanoparticles for Hazardous Solvent Detection

Ayaka Sato, Yuya Ikeda, Koichi Yamaguchi and Varun Vohra *

Department of Engineering Science, the University of Electro-communications, 1-5-1 Chofugaoka, Chofu, Tokyo 182-8585, Japan; s1413079@edu.cc.uec.ac.jp (A.S.); b1366569@crystal.ee.uec.ac.jp (Y.I.); kyama@ee.uec.ac.jp (K.Y.)
* Correspondence: varun.vohra@uec.ac.jp; Tel.: +81-42-443-5359

Received: 15 February 2018; Accepted: 14 March 2018; Published: 16 March 2018

Abstract: Facile detection and the identification of hazardous organic solvents are essential for ensuring global safety and avoiding harm to the environment caused by industrial wastes. Here, we present a simple method for the fabrication of silver-coated monodisperse polystyrene nanoparticle photonic structures that are embedded into a polydimethylsiloxane (PDMS) matrix. These hybrid materials exhibit a strong green iridescence with a reflectance peak at 550 nm that originates from the close-packed arrangement of the nanoparticles. This reflectance peak measured under Wulff-Bragg conditions displays a 20 to 50 nm red shift when the photonic sensors are exposed to five commonly employed and highly hazardous organic solvents. These red-shifts correlate well with PDMS swelling ratios using the various solvents, which suggests that the observable color variations result from an increase in the photonic crystal lattice parameter with a similar mechanism to the color modulation of the chameleon skin. Dynamic reflectance measurements enable the possibility of clearly identifying each of the tested solvents. Furthermore, as small amounts of hazardous solvents such as tetrahydrofuran can be detected even when mixed with water, the nanostructured solvent sensors we introduce here could have a major impact on global safety measures as innovative photonic technology for easily visualizing and identifying the presence of contaminants in water.

Keywords: hazardous organic solvents; photonic nanostructures; self-assembly; polymer nanoparticles; biomimetic solvent sensors; iridescence

---

1. Introduction

Organic solvents are widely used in multiple industries and for daily chores. For instance, they are employed for industrial printing, coatings, adhesives, painting or cleaning, and in numerous types of advanced technological research and production such as the organic electronics field [1–4]. A large number of these organic solvents are considered as hazardous and can lead to either environmental or health issues [1–8]. Chlorinated solvents such as chloroform (CF) or chlorobenzene (CB), as well as heterocyclic ethers such as tetrahydrofuran (THF), can damage numerous organs including liver, kidneys, and the central nervous system. Although strict rules on hazardous solvent waste disposal are in place in developed countries, this may not be the case on a global scale. Consequently, developing simple and low-cost technologies to visually detect the presence of contaminants (hazardous organic solvents and their vapors) in air or in water is of major importance for reducing the risks that result from organic solvent usage.

Photonic crystals found in nature have well-defined nano- or micro-structures [9–15] and have inspired strategies for the development of biomimetic sensing technologies based on structural coloration [13–17]. In particular, some animals provide perfect examples of visible color-tunable photonic materials [10–12]. The panther chameleons (*Furcifer Pardalis*) can, for instance, change their skin color by modifying the lattice parameters of their iridophores between the excited and the relaxed states [10]. Various methods for fabricating bio-inspired materials with similar structural coloration properties have been explored over the past two decades, which are either based on lithographic techniques (e.g., electron-beam lithography, imprint lithography, holographic lithography, or two-photon lithography) [18–21] or on self-assembled materials [22–24]. Although lithographic techniques can generate well-defined architectures, colloidal self-assembly has been drawing increasing interest in the field, as it provides a low-cost alternative to fabricating materials with structural coloration properties. This was first achieved with silicon-based monodisperse spherical particles with submicrometer diameters [25,26], but these hard particles were quickly replaced with soft polymer nanoparticles, as they can be produced at a lower cost and their properties can be easily adapted through simple chemistry (e.g., to generate core-shell structures with different refractive indices) [16,23,27].

Several studies have demonstrated that polymer nanoparticles as fused films or embedded in elastomer matrices can be applied to water or organic solvent sensors fabrication [16,27,28]. Similarly to the chameleon's camouflage properties, the color variation observed from these films can be correlated with changes in the lattice parameters of the photonic crystals and/or changes in refractive indices upon diffusion of the organic molecules inside the matrix. The spectral changes are generally described by the Bragg diffraction equation that is derived from Bragg and Snell's laws (Equation (1)), which relates the maximum peak wavelength (λ) with the incident angle (θ) and the effective refractive index (n_{eff}):

$$m\lambda = 2\mathrm{d}_{111}\sqrt{n_{\text{eff}}{}^2 - \sin^2\theta} \tag{1}$$

in which d_{111} corresponds to the distance between adjacent nanoparticle centers in the (111) plane and m is the order of diffraction. d_{111} and n_{eff} can be calculated following Equations (2) and (3).

$$d_{111} = \sqrt{2/3}\,D_{\text{particle}} \tag{2}$$

$$n_{\text{eff}}{}^2 = \sqrt{0.74\,n_{\text{particle}}^2 + 0.26\,n_{\text{fill}}^2} \tag{3}$$

in which D_{particle} and n_{particle} correspond to the diameter and refractive index of the nanoparticle, respectively, and n_{fill} is the refractive index of the filling matrix. One of the most commonly used matrices for this application is a cross-linked silicon-based elastomer, polydimethylsiloxane (PDMS). Depositing uniform three dimensional nanoparticle structures on hyrdophobic substrates has been one of the challenges to produce strongly iridescent materials. This was successfully achieved by using techniques such as dip-coating [16,29] or Langmuir-Blodgett deposition [30,31]. However, the commonly employed thin film deposition technique—spin-coating—is usually avoided, as it has a tendency to create non-uniform films due to the fast drying kinetics resulting from the quick rotation of the substrate [24]. Uniform nanoparticle photonic crystals from aqueous dispersions could be deposited on PDMS substrates by solving the wetting issue through a time-consuming chemical surface functionalization of the substrate [32].

Here, we introduce an innovative hybrid material based on spin-coated monodisperse polystyrene (PS) nanoparticles coated with a thin metallic layer and embedded in a PDMS matrix. By increasing the wettability of PDMS using a surface plasma treatment, we could form spin-coated, crack-free, close-packed, three-dimensional photonic structures on PDMS which, once coated with silver (Ag), exhibit strong iridescent colors. After depositing a second layer of PDMS on top of the Ag-coated photonic crystals, we produce a mechanically resistant material that can repeatedly be used as photonic chemical sensor for numerous organic solvents. Unlike previous studies on photonic sensors fabricated using self-assembled PS nanoparticles [16,28,33], we demonstrate that this approach is not limited to water and alcohols but can be applied to the detection and identification of hazardous solvents such as CF, CB, and THF, which can contaminate water and represent a real danger to human health and the environment. Using a set of five test organic solvents, we observe the dynamic reflectance peak shifts that result from variations in d_{111} as a result of the PDMS swelling. Furthermore, we verified that the prepared materials can be employed to detect small amounts of hazardous solvents mixed in water and, consequently, our study opens the path to low-cost photonic sensors for water contamination detection.

2. Materials and Methods

A schematic representation of the fabrication procedure for the PDMS-based photonic sensors is presented in Figure 1. 2.5 × 2.5 cm^2 PDMS substrates (Dow Corning, Sylgard$^®$ 184, Midland, MI, USA) were fabricated by pouring a 10:1 mixture of the base and curing agent after intense mixing of the two components and depositing it in a square-shaped container before curing it at 80 °C for 2 h. The PDMS thickness (1.5 mm) was controlled by the deposited volume of uncured mixture. After the curing step, the PDMS substrates were exposed to oxygen plasma for 30 min using a pressure of 500 mTorr and a light intensity of 10 W at 10 MHz. The contact angle between the 600 nm diameter PS nanoparticle aqueous dispersion (Thermo Fisher 5060A, Waltham, MA, USA, size uniformity ≤ 3%, $n_{particle}$ = 1.59 at 589 nm), and the PDMS substrates was accurately measured using a plug-in for the ImageJ 1.47v software, which follows a computational method that is described elsewhere [34]. The monodisperse PS nanoparticle dispersion diluted to 5 wt % was spin-coated on top of the modified surface PDMS substrates at 600 rpm for 1 min.

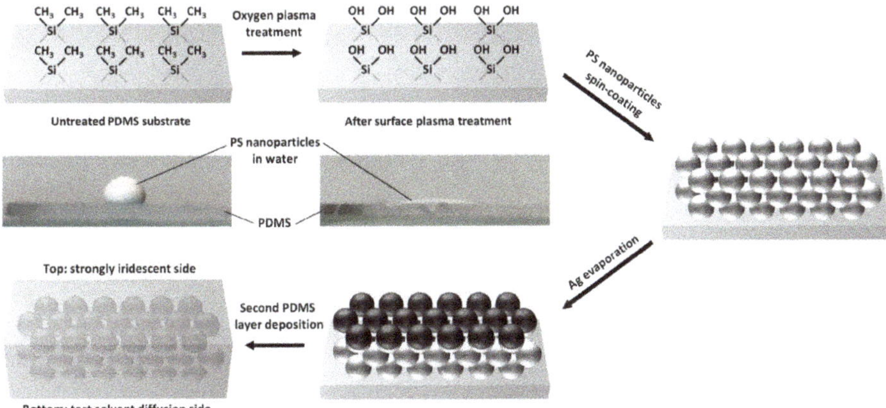

Figure 1. Schematic description of the multistep fabrication process for strongly iridescent hybrid photonic materials. The photographs correspond to PS nanoparticle aqueous dispersions deposited on PDMS substrates before and after oxygen plasma treatment.

The PS-coated PDMS substrates used for metallization were placed in an evaporation chamber until a vacuum level of 10^{-6} Torr was reached and then an 70 nm-thick Ag layer was deposited on the substrates at an evaporation rate of 0.3 nm·s^{-1}. The Ag-coated substrates employed for PDMS embedded photonic sensor fabrication were then once again placed in a square-shaped box and covered with the same volume of PDMS mixture that was cured in the same conditions as the substrate fabrication step. The resulting photonic sensors have a thickness of approximately 3 mm, which ensures that no buckling will occur when they are exposed to the organic solvents.

SEM images of the metalized PS nanoparticles that are deposited on PDMS substrates were collected using a Field Emission SEM (Hitachi, SU3500, Tokyo, Japan) at 10 kV and with magnifications of 5000× and 10,000×. The reflectance measurements were carried out by placing the samples on a horizontal stage with an incident light oriented 52.3° with respect to the vertical axis (Figure 2a). The light collected at an angle of 18.3° with respect to the vertical axis was analyzed using a

BlueWave-VIS Spectrometer (StellarNet, Inc., Tampa, FL, USA) from 450 to 750 nm. These angles correspond to the Wulf-Bragg conditions ($\theta = 17°$) for the close-packed nanoparticles in the (111) plane. The reflectance spectra were collected using a standard halogen lamp (StellarNet, Inc., Tampa, FL, USA, SL1 Tungsten Halogen Light Source) and normalized using the reflection spectra of Ag-covered glass substrates. All measurements were performed in the experimental room's atmospheric conditions with temperatures of 25 °C and a relative humidity of 55%.

The PDMS swelling ratios were calculated by immersing the hybrid photonic materials (3 mm-thick, area of 2.5×2.5 cm^2) into the various test solvents. The difference in weight before and after immersion for 1 h at room temperature and the molecular weight of the molecules was employed to calculate the amount of solvent that diffused into the hybrid material. Testing of the photonic materials as hazardous organic solvent sensors was performed using the same geometry as described above, and reflectance spectra were measured at regular time intervals after dropping 100 µL of each test solvent on the hybrid material surface (test solvent diffusion side in Figure 1). The sensitivity measurements were performed by gradually increasing the amount of THF placed on the photonic sensor surface. Similarly, 5 µL of THF were mixed into 95 µL of water, and the solvent mixture was deposited on top of the PDMS-based material to observe the changes in reflectance peak.

3. Results and Discussion

3.1. Strongly Iridescent Hybrid Photonic Films

After spin-coating the PS nanoparticles on a PDMS substrate, a 70 nm-thick Ag layer is deposited on top of these films, which are then covered with a second PDMS layer to ensure that the photonic structure is not damaged by repeated solvent diffusion. As demonstrated by the photographs in Figure 1, surface plasma treatment of the PDMS substrate is an essential step for ensuring that PS nanoparticles can be deposited from aqueous dispersions onto these high surface energy elastomeric substrates. Surface plasma treatment is a well-known technique that is often applied to improving the wetting properties of aqueous solutions onto hydrophobic layers such as poly(3,4-ethylenedioxythiophene) polystyrene sulfonate or PDMS [16,35,36].

The oxygen plasma treatment performed on PDMS modifies the functional groups present at its surface. In fact, the surface of pristine PDMS substrates is mainly composed of methyl groups. During plasma treatment, oxidation of these methyl groups takes place, which results in the formation of silanols that enable wetting from aqueous solutions and dispersions [37]. In the case of the PS nanoparticles dispersion we employed for our photonic material fabrication, average contact angles with the PDMS substrate of 107.5° and 24.9° were obtained before and after plasma treatment, respectively (Figure 1). This major change in contact angle and surface wetting properties enabled the spin-coating of wet layers of PS nanoparticle aqueous suspensions on PDMS substrates. Although the nanoparticles are dispersed in water, PS is intrinsically hydrophobic. The interactions between PS nanoparticles and the hydrophobic substrate (PDMS) are consequently likely to form uniform close-packed colloid photonic crystals. The scanning electron microscope (SEM) images in Figure 2b clearly demonstrate that a honeycomb close-packed arrangement of the PS nanoparticles on plasma-treated PDMS was successfully achieved. The photonic materials exhibits crack-free iridescence over areas as large as 5 cm^2, confirming that, unlike previous attempts to fabricate PDMS-PS nanoparticle composites based on drop-casting, uniform arrangements were obtained [28]. Furthermore, coating the nanoparticles with a thin Ag layer considerably increased the reflectance of the PDMS-PS nanoparticle composites with values approximately 2.4 times higher than materials based on bare PS nanoparticles embedded in the silicon elastomer matrix (Figure 2c).

The face-centered cubic arrangements exhibit a (111) diffraction plane that is oriented at an angle of 35.3° with respect to the sample (Figure 2b). Consequently, for θ values, we consider the angle with respect to the [111] normal direction rather than the direction that is normal to the sample plane. To facilitate the visual detection of hazardous solvent traces, we selected the geometry in which θ has a value of 17°, as it results in a green structural coloration with a reflectance peak maximum around 550 nm (Figure 2c). Similarly to previous studies on photonic opal structures, we could observe a small spatial distribution of the diffraction peak, which may be a consequence of multiple diffractions from different Bragg planes that interact with each other (Figure 2d) [38]. This may also explain why the observed reflectance peak at 550 nm in Bragg conditions does not perfectly match the results from calculations using Equation (1), which predict a reflectance peak around 497 nm. The presence of packing defects (Figure 2b) and the dispersity in nanoparticle dimensions may also contribute to the observed differences between theoretical and experimental data. As the material present on the sample dropping side of these strongly iridescent multilayer films is PDMS, they may be employed to detect the presence of hazardous organic solvents, which can diffuse inside the silicon elastomer matrix [39].

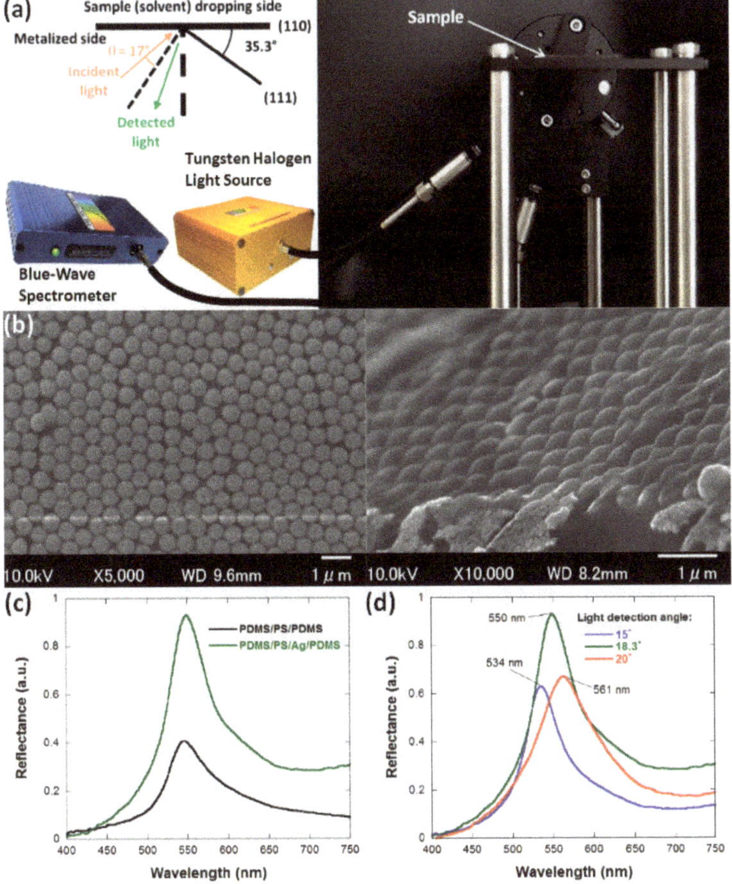

Figure 2. (**a**) Schematic representation and experimental set-up for the reflectance measurements; (**b**) top-view SEM image of the surface (left) and tilted-view SEM image of the cross-section (right) of Ag-coated PS nanoparticles deposited on plasma-treated PDMS; (**c**) reflectance spectrum of bare and Ag-coated PS nanoparticles embedded in PDMS for the third order of Bragg diffraction from the (111) plane observed at 17° with respect to the [111] direction; and (**d**) spatial and spectral distribution of the reflectance peak with incident light oriented 53.3° with respect to the vertical axis.

3.2. Hazardous Solvent Detection in Biomimetic Photonic Sensors

To test the hybrid photonic materials we developed for hazardous organic solvent sensing applications, we selected five test solvents with various PDMS swelling ratios (Table 1). In addition to CF, CB, and THF, we used dichloromethane (DCM) and dimethoxyethane (DME), which are suspected of causing cancer or may damage fertility and unborn children.

Table 1. Summarized data on tested solvent and observed reflectance peak shift when 100µL of solvent are deposited on the hybrid sensor.

Solvent	THF	DME	CF	CB	DCM
Refractive index ($n_{solvent}$)	1.40	1.38	1.45	1.52	1.42
PDMS swelling ratio (mmol/g of PDMS)	17.8	11.6	17.7	7.7	7.9
Reflectance peak (nm)	601	592	597	569	575

The five tested solvents exhibit dynamic shifts of the reflectance peak that were initially observed at 550 nm. Taking into account Equations (1) and (3), the measured red shifts (Figure 3) of the reflectance peak either originate from an increase in the distance between neighboring centers in the (111) plane or larger n_{eff} values. The refractive indices of the test solvents ($n_{solvent}$) are summarized in Table 1. Interestingly, although THF and PDMS are relatively similar from an optical point-of-view, with n_{THF} and n_{PDMS} having values of 1.40 and 1.41, respectively, a 50 nm red shift of the reflectance peak can be observed, which suggests that the key parameter for the response from the photonic sensor is the PDMS swelling ratio. In fact, CF, which has a refractive index of 1.45 but almost the same swelling ratio as THF, yields similar values of reflectance peak shift.

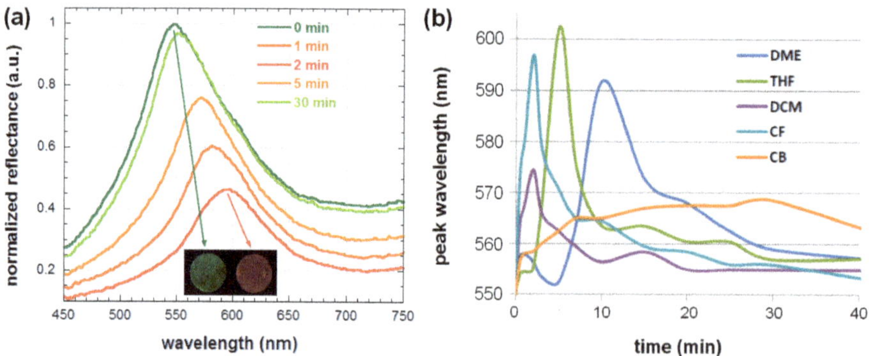

Figure 3. (a) Time-resolved reflectance spectra of 100 µL CF dropped from the non-metal side onto the hybrid photonic sensor; (b) time-dependent evolution of the reflectance peak wavelength of various solvents dropped on the hybrid photonic sensor. The inset in (a) corresponds to the observed sample with an area of 5 cm^2 before deposition of CF and 2 min after the deposition.

Using the solvent with a high swelling ratio (CF), we verified that the changes in reflected color from green to red (597 nm) can be visually recognized within a few minutes (inset of Figure 3a). In fact, the five test solvents can be categorized in three classes of swelling ratios with values close to 18 (CF and THF), 12 (DME), and 8 (DCM and CB) mmol/g of PDMS, respectively. These three classes of solvents yield shifts of approximately 50 nm, 40 nm, and 20 nm, respectively. Consequently, it is safe to assume that the working mechanism of these biomimetic photonic sensors is very similar to the color-tuning properties of the chameleon skin and principally relies on a deformation of the photonic crystal lattice upon swelling of PDMS by the solvents (increase in d_{111}) [10]. Note that the standard deviation for 10 measurements using the same solvent is less than 1 nm, which implies that even small differences in reflectance peak wavelength (e.g., 601 and 597 nm for THF and CF, respectively) can be employed to identify the nature of the solvent that diffuses into the photonic sensor. Nonetheless, to ensure that each solvent can be precisely identified, we used a dynamic approach based on the solvent diffusion rate into the PDMS matrix. As displayed in Figure 3b, CF and DCM reach their maximum reflectance shifts within 2 min, whereas THF, DME, and CB take longer times of approximately 5, 10, and 30 min, respectively.

The time-dependent measurements not only allow for a better recognition of the solvents but also provide important information for understanding the differences observed in reflectance shifts from solvents with similar PDMS swelling ratios. For instance, CF and THF have PDMS swelling ratios of 17.7 and 17.8 mmol/g, respectively. Nonetheless, as CF has a higher refractive index compared to THF, a larger red shift is expected. Repeated measurements confirm that the reflectance measured from the sensors in the presence of THF is red-shifted by approximately 5 nm as compared to the results from CF. Our hypothesis for this contradictive result is that the solvent diffusion kinetics, which determine the amount of solvent present in the vicinity of the PS nanoparticles at a given time, will affect the increase in d_{111} as represented in Figure 4.

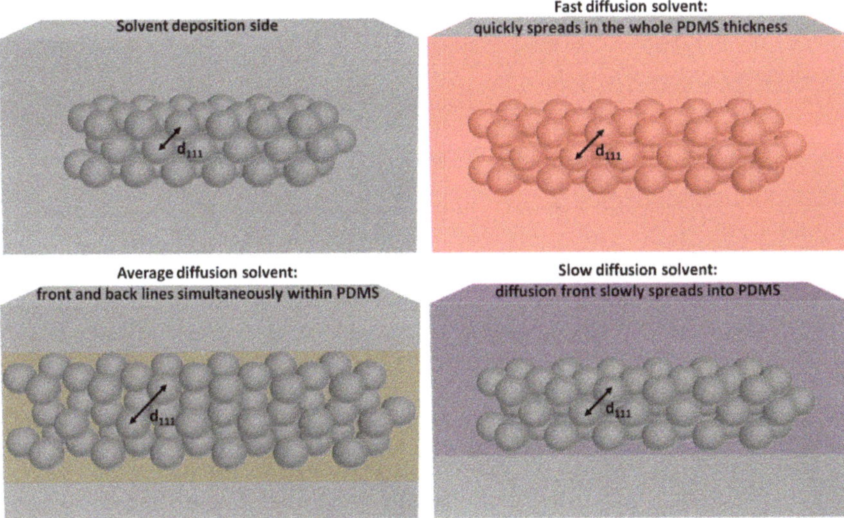

Figure 4. Schematic representation of the increased distance between adjacent PS nanoparticles embedded in the PDMS matrix when solvents with fast, average, and slow diffusion speeds are deposited on the surface of the photonic sensors.

As the reflectance peak shifts back to the initial position (550 nm) after a certain time, we neglected the effect of PS solubility on the various test solvents and considered that the solvents preferentially diffuse into PDMS. When CF is deposited on top of the photonic sensor, it quickly diffuses inside PDMS, which results in an almost homogenous distribution of the solvent molecules inside the whole elastomeric matrix thickness (approximately 3 mm). On the other hand, in the case of THF, the diffusion is slower and, consequently, the solvent swollen area gradually moves in the vertical direction within PDMS. The schematic representations in Figure 4 are exaggerated for clarity of presentation. In the case of fast diffusing solvents (e.g., CF), a smaller amount of solvent molecules remain in the direct proximity of the PS nanoparticles at the time when the maximum reflectance peak shift is recorded. This also explains why a smaller shift is observed for CF with respect to THF, which has slower diffusion kinetics even though the two solvents have similar swelling ratios (measured by immersing the photonic sensors into the various solvents). A slower diffusion rate consequently seems to be beneficial for precise measurement of the maximum peak shift due to swelling of the PDMS matrix. Nonetheless, if the diffusion of the solvent inside PDMS is too slow, the distance between the diffusion front and back will increase, which will also result in less solvent molecules in the vicinity of the PS nanoparticles at a given time. Consequently, a larger red-shift can be observed when comparing DCM to CB despite the fact that they have similar swelling ratios. The hypothesis of time-dependent swelling of PDMS suggests that the reflectance shifts should also depend on the volume of solvent that is in contact with the sensor.

We verified the detection threshold (sensitivity) of the photonic sensors we developed here by varying the amounts of solvents used as test samples (Table 2). All the reflectance wavelengths in Table 2 correspond to the maximum shift observed during each test. The time at which these maximum shifts were observed varies depending on the test solvent. Additionally, considering the experimental error on the reflectance peak measurements, we can suppose that the minimum volume of solvent that can be detected by the photonic sensor is 5 µL (shifts larger than 5 nm). To verify whether this detection threshold can be applied to, for example, THF mixed in water, we prepared a low concentration (5 vol %) mixture of THF in water. The mixed solvent was deposited on the photonic sensors, which resulted in a reflectance peak wavelength of 569 nm collected 5 min after deposition. As this value is very close to the one measured for 5 µL of THF deposited on the sensor, we can speculate that all the THF molecules mixed in water diffuse from the mixed solvent into the PDMS based photonic sensors. These sensors consequently have a great potential as innovative technology for precisely and rapidly detecting the presence of hazardous, solvent-contaminated water.

Table 2. Photonic sensor reflectance peak wavelength with respect to deposited amount of solvent.

Solvent Amount (µL)		0	2	5	10	20	50	100
	THF	550	552	567	574	599	602	601
	DME	550	555	562	571	578	591	592
Reflectance Peak (nm)	**CF**	550	551	558	569	589	599	597
	CB	550	552	557	567	571	570	569
	DCM	550	554	561	569	574	574	575

4. Conclusions

In summary, we have successfully fabricated strongly iridescent hybrid photonic sensors based on self-assembled spin-coated PS nanoparticles on plasma-treated PDMS substrates. The addition of the reflective Ag layer resulted in strong structural coloration properties that can be easily observed by naked eye. The formation of close-packed structures was confirmed by SEM measurements, and we employed the green Bragg reflection peak (λ = 550 nm) to probe the diffusion of hazardous organic solvents into the photonic sensor.

Using a set of five organic solvents, we verified that these multilayer nanostructured materials can be used for the detection and identification of hazardous water contaminants. Direct visualization of the changes in reflectance peak wavelengths for a given amount of solvent can be highly beneficial for rapid analysis of the water purity with respect to these organic solvents, which can be harmful to the human health and the environment. After closely studying the behavior of the various solvents and correlating the observed reflectance peak shifts with the parameters present in the Bragg diffraction equation, we concluded that, similarly to the chameleon skin, the changes in visible color of the sensors strongly depend on the PDMS swelling by the test solvent. As each solvent has a different diffusion rate into PDMS, reliable contaminant identification can be achieved by analyzing the data in terms of time-dependent reflectance peak shift.

Lastly, we verified that hazardous solvent amounts as low as 5 µL can be detected using the hybrid photonic sensors, even when they are tested as low concentration contaminants in water. This opens the path for facile and rapid recognition of a variety of hazardous solvents that can either affect some organs such as the liver or the lungs but also result in infertility or increase the probability of harm to unborn children. As this technology can be easily employed as simple water purity test in developing countries in which sometimes no strict rules on industrial solvent disposal exist, the sensors we produced for our study could find major applications in global safety measure implementation.

Acknowledgments: The authors would like to acknowledge the University of Electro-Communications for providing financial support through the Research Support for Young Faculty Members Program. The authors would also like to express their gratitude to Prof. Takashi Hirano for kindly accepting to share his solvents for the purpose of this research.

Author Contributions: V.V. and K.Y. conceived and designed the experiments; A.S. and Y.I. performed the experiments and analyzed the data; V.V. wrote the paper.

Conflicts of Interest: The authors declare no conflict of interest.

References

1. Kristensen, P.; Irgens, L.M.; Kjersti Daltveit, A.; Andersen, A. Perinatal Outcome among Children of Men Exposed to Lead and Organic Solvents in the Printing Industry. *Am. J. Epidemiol.* **1993**, *137*, 134–144. [CrossRef] [PubMed]

2. Olson, B.A. Effects of organic solvents on behavioral performance of workers in the paint industry. *Neurobehav. Toxicol. Teratol.* **1982**, *4*, 703–708.

3. Agnesi, R.; Valentini, F.; Mastrangelo, G. Risk of spontaneous abortion and maternal exposure to organic solvents in the shoe industry. *Int. Arch. Occup. Environ. Health* **1997**, *69*, 311–316. [CrossRef] [PubMed]

4. Vohra, V.; Mróz, W.; Inaba, S.; Porzio, W.; Giovanella, U.; Galeotti, F. Low-Cost and Green Fabrication of Polymer Electronic Devices by Push-Coating of the Polymer Active Layers. *ACS Appl. Mater. Interfaces* **2017**, *9*, 25434–25444. [CrossRef] [PubMed]

5. Hardell, L.; Eriksson, M.; Lenner, P.; Lundgren, E. Malignant lymphoma and exposure to chemicals, especially organic solvents, chlorophenols and phenoxy acids case-control study. *Br. J. Cancer* **1981**, *43*, 169–176. [CrossRef] [PubMed]

6. Sallmén, M.; Lindbohm, M.-L.; Kyyrönen, P.; Nykyri, E.; Anttila, A.; Taskinen, H.; Hemminki, K. Reduced fertility among women exposed to organic solvents. *Am. J. Ind. Med.* **1995**, *27*, 699–713. [CrossRef] [PubMed]

7. Sherman, J.; Chin, B.; Huibers, P.D.; Garcia-Valls, R.; Hatton, T.A. Solvent replacement for green processing. *Environ. Health Perspect.* **1998**, *106*, 253–271. [CrossRef] [PubMed]

8. Prat, D.; Hayler, J.; Wells, A. A survey of solvent selection guides. *Green Chem.* **2014**, *16*, 4546–4551. [CrossRef]

9. Sun, J.; Bhushan, B.; Tong, J. Structural coloration in nature. *RSC Adv.* **2013**, *3*, 14862–14889. [CrossRef]

10. Teyssier, J.; Saenko, S.V.; van der Marel, D.; Milinkovitch, M.C. Photonic crystals cause active colour change in chameleons. *Nat. Commun.* **2015**, *6*, 6368. [CrossRef] [PubMed]

11. Mäthger, L.M.; Land, M.F.; Siebeck, U.E.; Marshall, N.J. Rapid colour changes in multilayer reflecting stripes in the paradise whiptail, *Pentapodus paradiseus. J. Exp. Biol.* **2003**, *206*, 3607–3613. [CrossRef] [PubMed]

12. Yoshioka, S.; Matsuhana, B.; Tanaka, S.; Inouye, Y.; Oshima, N.; Kinoshita, S. Mechanism of variable structural colour in the neon tetra: Quantitative evaluation of the Venetian blind model. *J. R. Soc. Interface* **2011**, *8*, 56–66. [CrossRef] [PubMed]

13. Fenzl, C.; Hirsch, T.; Wolfbeis, O.S. Photonic Crystals for Chemical Sensing and Biosensing. *Angew. Chem. Int. Ed.* **2014**, *53*, 3318–3335. [CrossRef] [PubMed]

14. Yi, W.; Xiong, D.B.; Zhang, D. Biomimetic and Bioinspired Photonic Structures. *Nano Adv.* **2016**, *1*, 62–70. [CrossRef]

15. Li, Q.; Zeng, Q.; Shi, L.; Zhang, X.; Zhang, K.-Q. Bio-inspired Sensors Based on Photonic Structures of Morpho Butterfly Wings: A Review. *J. Mater. Chem. C* **2016**, *4*, 1752–1763. [CrossRef]

16. Kuo, W.-K.; Weng, H.-P.; Hsu, J.-J.; Yu, H.H. A bioinspired color-changing polystyrene microarray as a rapid qualitative sensor for methanol and ethanol. *Mater. Chem. Phys.* **2016**, *173*, 285–290. [CrossRef]

17. Burgess, B.; Loncar, M.; Aizenberg, J. Structural colour in colourimetric sensors and indicators. *J. Mater. Chem. C* **2013**, *1*, 6075–6086. [CrossRef]

18. Zhang, S.; Chen, Y. Nanofabrication and coloration study of artificial Morpho butterfly wings with aligned lamellae layers. *Sci. Rep.* **2015**, *5*, 16637. [CrossRef] [PubMed]

19. Wu, C.-S.; Lin, C.-F.; Lin, H.-Y.; Lee, C.-L.; Chen, C.-D. Polymer-based photonic crystals fabricated with single-step electron-beam lithography. *Adv. Mater.* **2007**, *19*, 3052–3056. [CrossRef]

20. Aryal, M.; Ko, D.-H.; Tumbleston, J.R.; Gadisa, A.; Samulski, E.T.; Lopez, R. Large area nanofabrication of butterfly wing's three dimensional ultrastructures. *J. Vac. Sci. Technol. B* **2012**, *30*, 061802. [CrossRef]

21. Siddique, R.H.; Hünig, R.; Faisal, A.; Lemmer, U.; Hölscher, H. Fabrication of hierarchical photonic nanostructures inspired by Morpho butterflies utilizing laser interference lithography. *Opt. Mater. Express* **2015**, *5*, 996–1005. [CrossRef]

22. Vohra, V.; Galeotti, F.; Giovanella, U.; Anzai, T.; Kozma, E.; Botta, C. Investigating phase separation and structural coloration of self-assembled ternary polymer thin films. *Appl. Phys. Lett.* **2016**, *109*, 103702. [CrossRef]

23. Zulian, L.; Emilitri, E.; Scavia, G.; Botta, C.; Colombo, M.; Destri, S. Structural Iridescent Tuned Colors from Self-Assembled Polymer Opal Surfaces. *ACS Appl. Mater. Interfaces* **2012**, *4*, 6071–6079. [CrossRef] [PubMed]

24. Zheng, H.; Ravaine, S. Bottom-Up Assembly and Applications of Photonic Materials. *Crystals* **2016**, *6*, 54. [CrossRef]

25. Míguez, H.; López, C.; Meseguer, F.; Blanco, A.; Vázquez, L.; Mayoral, R.; Ocaña, M.; Fornés, V.; Mifsud, A. Photonic crystal properties of packed submicrometric SiO_2 spheres. *Appl. Phys. Lett.* **1997**, *71*, 1148–1150. [CrossRef]

26. Blanco, A.; Chomski, E.; Grabtchak, S.; Ibisate, M.; John, S.; Leonard, S.W.; Lopez, C.; Meseguer, F.; Miguez, H.; Mondia, J.P.; et al. Large-scale synthesis of a silicon photonic crystal with a complete three-dimensional bandgap near 1.5 micrometres. *Nature* **2000**, *405*, 437–440. [CrossRef] [PubMed]

27. Duan, L.; You, B.; Zhou, S.; Wu, L. Self-assembly of polymer colloids and their solvatochromic-responsive properties. *J. Mater. Chem.* **2011**, *21*, 687–692. [CrossRef]

28. Fenzl, C.; Hirsch, T.; Wolfbeis, O.S. Photonic Crystal Based Sensor for Organic Solvents and for Solvent-Water Mixtures. *Sensors* **2012**, *12*, 16954–16963. [CrossRef] [PubMed]

29. Shieh, J.-Y.; Kuo, J.-Y.; Weng, H.-P.; Yu, H.H. Preparation and Evaluation of the Bioinspired PS/PDMS Photochromic Films by the Self-Assembly Dip−Drawing Method. *Langmuir* **2013**, *29*, 667–672. [CrossRef] [PubMed]

30. Reculusa, S.; Ravaine, S. Synthesis of Colloidal Crystals of Controllable Thickness through the Langmuir–Blodgett Technique. *Chem. Mater.* **2003**, *15*, 598–605. [CrossRef]

31. Kuo, W.-K.; Hsu, J.-J.; Nien, C.-K.; Yu, H.H. Moth-Eye-Inspired Biophotonic Surfaces with Antireflective and Hydrophobic Characteristics. *ACS Appl. Mater. Interfaces* **2016**, *8*, 32021–32030. [CrossRef] [PubMed]

32. Ko, Y.G.; Shin, D.H.; Lee, G.S.; Choi, U.S. Fabrication of colloidal crystals on hydrophilic/hydrophobic surface by spin-coating. *Colloids Surf. A* **2011**, *385*, 188–194. [CrossRef]

33. Kuo, W.-K.; Weng, H.-P.; Hsu, J.-J.; Yu, H.H. Photonic Crystal-Based Sensors for Detecting Alcohol Concentration. *Appl. Sci.* **2016**, *6*, 67. [CrossRef]

34. Stalder, A.F.; Kulik, G.; Sage, D.; Barbieri, L.; Hoffmann, P. A Snake-Based Approach to Accurate Determination of Both Contact Points and Contact Angles. *Colloids Surf. A* **2006**, *286*, 92–103. [CrossRef]

35. Vohra, V.; Anzai, T.; Inaba, S.; Porzio, W.; Barba, L. Transfer-printing of active layers to achieve high quality interfaces in sequentially deposited multilayer inverted polymer solar cells fabricated in air. *Sci. Technol. Adv. Mater.* **2016**, *17*, 530–540. [CrossRef] [PubMed]

36. Zappia, S.; Scavia, G.; Ferretti, A.M.; Giovanella, U.; Vohra, V.; Destri, S. Water-Processable Amphiphilic Low Band Gap Block Copolymer: Fullerene Blend Nanoparticles as Alternative Sustainable Approach for Organic Solar Cells. *Adv. Sustain. Syst.* **2018**, *2*, 1700155. [CrossRef]

37. Bhattacharya, S.; Datta, A.; Berg, J.M.; Gangopadhyay, S. Studies on surface wettability of poly(dimethyl) siloxane (PDMS) and glass under oxygen-plasma treatment and correlation with bond strength. *J. Microelectromech. Syst.* **2005**, *14*, 590–597. [CrossRef]

38. Shishkin, I.; Rybin, M.V.; Samusev, K.B.; Golubev, V.G.; Limonov, M.F. Multiple Bragg diffraction in opal-based photonic crystals: Spectral and spatial dispersion. *Phys. Rev. B* **2014**, *89*, 035124. [CrossRef]

39. Ng Lee, J.; Park, C.; Whitesides, G.M. Solvent Compatibility of Poly(dimethylsiloxane)-Based Microfluidic Devices. *Anal. Chem.* **2003**, *75*, 6544–6554. [CrossRef]

 *nanomaterials*

Article

Protective Properties of a Microstructure Composed of Barrier Nanostructured Organics and SiO$_x$ Layers Deposited on a Polymer Matrix

Radek Prikryl [1], Pavel Otrisal [2,*], Vladimir Obsel [3], Lubomír Svorc [4], Radovan Karkalic [5] and Jan Buk [6]

[1]   Institute of Materials Science, Faculty of Chemistry, Brno University of Technology, Purkynova 464/118, 612 00 Brno, Czech Republic; prikryl@fch.vut.cz
[2]   Nuclear, Biological and Chemical Defence Institute of the University of Defence, Sidliste Vita Nejedleho, 68201 Vyškov, Czech Republic
[3]   DEZA, Hochmanova 1, 628 01 Brno, Czech Republic; vobsel@seznam.cz
[4]   Institute of Analytical Chemistry, Faculty of Chemical and Food Technology, Slovak University of Technology in Bratislava, Radlinskeho 9, 812 37 Bratislava, Slovakia; lubomir.svorc@stuba.sk
[5]   Department of Military Chemical Engineering, Military Academy, GeneralaPavlaJurisicaSturma 33, 11000 Belgrade, Serbia; radovan.karkalic@va.mod.gov.rs
[6]   P A R D A M, Ltd., Zizkova 2494, 413 01 Roudnicenad Labem, Czech Republic; jan.buk@pardam.cz
*   Correspondence: pavel.otrisal@unob.cz; Tel.: +420-973-452-335

Received: 25 June 2018; Accepted: 29 August 2018; Published: 31 August 2018

Abstract: The SiO$_x$ barrier nanocoatings have been prepared on selected polymer matrices to increase their resistance against permeation of toxic substances. The aim has been to find out whether the method of vacuum plasma deposition of SiO$_x$ barrier nanocoatings on a polyethylene terephthalate (PET) foil used by Aluminium Company of Canada (ALCAN) company (ALCAN Packaging Kreuzlingen AG (SA/Ltd., Kreuzlingen, Switzerland) within the production of CERAMIS® packaging materials with barrier properties can also be used to increase the resistance of foils from other polymers against the permeation of organic solvents and other toxic liquids. The scanning electron microscopy (SEM) microstructure of SiO$_x$ nanocoatings prepared by thermal deposition from SiO in vacuum by the Plasma Assisted Physical Vapour Deposition (PA-PVD) method or vacuum deposition of hexamethyldisiloxane (HMDSO) by the Plasma-enhanced chemical vapour deposition (PECVD) method have been studied. The microstructure and behavior of samples when exposed to a liquid test substance in relation to the barrier properties is described.

Keywords: barrier material; nanocoating of SiO$_x$; polymeric matrix; plasma deposition; PVD; PA-PVD; PECVD; permeation; CERAMIS®; SorpTest

1. Introduction

Multilayer structures generally increase the protective properties of barrier materials [1–4]. The effect of a combination of the location of the barrier layers on the Protection Factor (BF) is shown in Figure 1.

Within the production of packaging materials for sensitive commodities and in chemical protection, this principle has been used for many years [5,6]. Multilayer barrier materials usually have better performance properties than the individual components from which they are made [7,8]. They are mostly cheaper and better processed [9–11]. This principle is also used by Aluminium Company of Canada (ALCAN) company in the production of SiO$_x$-based CERAMIS® [12] packaging materials, which form a barrier layer between polymer foils (usually polyethylene terephthalate

(PET) and polypropylene (PP)) during their lamination [13–16]. The principle of this technology of reactivating steaming the SiO_x barrier layer from the SiO substrate in vacuum is illustrated in Figure S1.

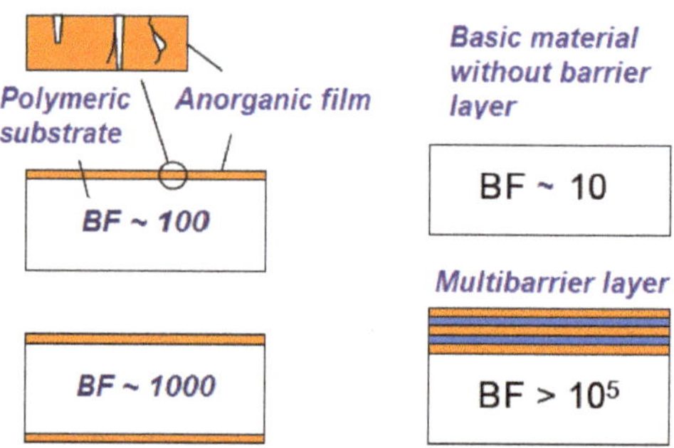

Figure 1. Influence of the barrier nanolayers arrangement on the protective properties of polymeric barrier materials.

In the Aluminium Company of Canada (ALCAN) brochure, this packaging material states that it creates an excellent barrier against gaseous and liquid harmful substances, independent of temperature and humidity. It is transparent for microwaves, suitable for heat sterilization and pasteurization, printable by common printing processes and it allows detection of metals. Due to the patented SiO_x barrier coating technology used by ALCAN was not available, one of the obtained samples was evaluated for protective properties against sulphur mustard (HD) and 1,6-dichlorohexane (DCH). For a CERAMIS® CO 20 sample weighing 18.2 g/m² (PET 12 μm, SiO_x 60 nm, PP 20 μm), the protective properties against the permeation of HD and DCH have been found to be greater than 70 h.

Although this material exhibits high resistance to the permeation of selected toxic compounds, it has one major disadvantage from the point of view of its possible utility in anti-gas protection. Its considerable stiffness and imprecision does not allow it to be used for the construction of protective clothing. It was therefore the effort to apply the SiO_x barrier layer to more suitable polymeric matrices.

Vacuum plasma deposition technology of barrier layers within normal temperatures is particularly suitable for the preparation of advanced surfaces with specific properties on flexible polymer and textile substrates [17–20]. A synoptical example of this advanced technology is the advanced novoFlex®600 equipment for coating polymeric materials in the footage developed by the Fraunhofer Institute for Organic Electronics, Electron Beam and Plasma Technology FEP (Dresden, Germany).

Another way to prepare SiO_x barrier nanolayers on polymeric or textile materials is a method of plasma deposition from a mixture of siloxane substrates and oxygen. As the suitable substrates hexamethyldisiloxane (HMDSO) and tetraethylorthosilicate (TEOS), which most closely approximate to the structure of SiO_x are considered [21].

The barrier layers could be prepared as flexible based on polymeric materials. One of the materials is Parylene. It is a commercial name for the group of poly-*p*-xylylene polymers. Dimmer ofpoly-*p*-xylylene is thermally evaporated (sublimated) from a powder within a low pressure of tens of Pascals. Steams of this dimmer joint a pyrolitic chamber. Within the temperature of approximately 680 °C comes to the split of the dimmer on monomer units. These molecules are reactive and devaporate on the surface of coated materials (in the same way as on walls of the reactive chamber). They polymerize in radical mechanism and form a linear polymeric polymer that is deposited as a very thin barrier layer. This layer is very homogenous and inert. The deposition of parylene was first observed by Szwarc in 1947, and Gorham later described a more efficient, general synthetic way to prepare poly-*p*-xylylene through the vacuum vapour-phase pyrolysis of paracyclophane at has been used till nowadays [22,23]. This polymerisation is self-initiated and non-terminated, requires no solvent or catalyst, produces no by-products except for the unreacted precursor. The polymer thin film is deposited spontaneously at the room temperature thus no thermal stress is induced in a coated object. Recently, parylene is one of the most well-known chemical vapour deposited (CVD) polymers and number of its applications has grown dramatically over the years.

The excellent resistance of polymeric packaging materials with the SiO_x barrier layer against the permeation of oxygen and humidity has conducted to the conception that the plasma deposition of SiO_x nanolayer and/or parylene film on the suitable polymeric matrix could be used for preparation of even constructive materials for protective garments resistant against toxic liquids and gases. This phenomenon has become a subject of our research. The aim of the study was to test these structures on the various polymers be used to increase the resistance against the permeation of organic solvents and other toxic liquids. The swelling of the polymeric substrate and material of nanostructured films is the main problem of barrier films functionality because of some destructions.

2. Materials and Methods

2.1. Source Materials, Substrates and Polymeric Matrices

SiO powder provided by Merck KGaA (Darmstadt, Germany) was used as source material for Plasma Assisted Physical Vapour Deposition (PA-PVD) process, hexamethyldi-siloxane (HMDSO) CAS 107-46-0 (Sigma-Aldrich, Prague, Czech Republic) was utilized as source materials for PE-CVD method. For PVD process of organics film the 1,3,5-triazine-2,4,6-triamine (melamine), CAS 108-78-1 (Sigma-Aldrich, Prague, Czech Republic) was used. Dimer of chlorinated *p*-xylylene (parylene C) (KunShanZhangShengNaNo Tech Company, Kunshan, China), CAS 28804-46-8 was utilized as source material for CVD method. Bis(2-chloroethyl) sulphide (sulphur mustard, HD) CAS 505-60-2 (Military Technical Institute of Protection, Brno, Czech Republic) and 1,6-Dichlorohexane CAS 218-491-7 (Sigma-Aldrich, Prague, Czech Republic) were used as test chemicals for verification of the quality of new developed barrier materials with nanolayers. Polyvinylidenchlorid (PVDC), CAS 9002-85-1 foil of 20 μm was used for lamination of PEVA foil. Thin films were prepared on polymeric foils from polyethylenevinyl acetate (PEVA), CAS 24937-78-8, foil of 200 μm thickness, polypropylene (PP) foil of 30 um thickness and polyethylentereftalate (PET) foil of 8 μm thickness. Si wafer WFP 6005 provided by ON Semiconductor (ON Semiconductor Czech Republic, Rožnov pod Radhoštěm, Czech Republic) was used as substrate for thin film characterization by energy-dispersive X-ray spectroscopy—EDX (ON Semiconductor Czech Republic, Ltd., Rožnov pod Radhoštěm, Czech Republic). EDX is a part of Scanning Electron Microscope ZEISS EVO LS 10 (Oberkochen, Germany). This spectroscopy provides elemental information about the composition of the structure of the surface of a sample. Performed in conjunction with SEM. Elements with atomic numbers down to carbon can be viewed with Energy Dispersive Spectroscopy (EDS).

2.2. Apparatuses for the Samples Preparation

The equipment for thermal lamination of plastic films without the use of binders was used as method for special multilayer structure fabrication. The original vacuum apparatus (Faculty of Chemistry, Brno University of Technology) for the preparation of barrier nanocoatings base on SiO_x and melamine on a polymeric matrix by the PVD, PA-PVD and PECVD method was used. The construction is evident from Figure S2.

Plasma surface activation was carried out before each first layer preparation under pressure of 10 Pa and oxygen flow of 10 sccm. The plasma discharge RF power of 13.56 MHz was set to 50 W for 10 min.

PVD process of SiO and Melamine based layers was carried out in following steps. The vacuum system was evacuated to base pressure of 1×10^{-5} Pa, then process started at pressure 1×10^{-4} Pa of source material vapour. Sample thickness varied from 50 to 150 nm controlled by deposition time.

PA-PVD process of SiO_x based films preparation was carried out under process pressure of 3×10^{-2} Pa and oxygen mass flow of 5 sccm. The RF power of 13.56 MHz plasma discharge was 200 W. The sample thickness varied from 50 nm to 150 nm and was controlled by adjusting the deposition time.

PE-CVD process of SiO_x based films deposition from HMDSO vapour was used under following conditions. The process pressure of 8–10 Pa was controlled by pump speed and monomer HMDSO flow, argon mass flow was kept as constant of 2 sccm and oxygen mass flow of 5 sccm. RF power of 13.56 MHz was used in range of 10–100 W, typical Self Bias Voltage was around 100 V. Sample thickness varied from 50–150 nm and was controlled by tuning the deposition time.

For the CVD process of parylene film preparation the original vacuum system developed on Brno University of Technology (Brno, Czech Republic) was used. It works on the principle of Gorham method described above. The conditions used for the process were as follows: pressure of monomer vapour was controlled to 8–10 Pa by temperature of dimer evaporator, the temperature of pyrolytic chamber was kept to 680 °C. The sample thickness was controlled by initial weight of dimer load.

2.3. Measuring Instruments

The high-resolution JEOL-6700F scanning electron microscope (JEOL, Tokyo, Japan) was used for the evaluation of nanocoatings surface microstructure at Institute of Scientific Instruments of the Czech Academy of Sciences, Brno, Czech Republic. This microscope is suitable for observing fine structures, such as multi-layer coatings and nanoparticles produced by nanotechnologies. Thanks to very slow electrons, it is also suitable for observing non-conducting samples.

The original SorpTest device was utilized for the study and quantitative testing of the barrier materials resistance against the permeation of volatile toxic substances, including chemical warfare agents [24]. The SorpTest devise is a semiautomatic measuring system using an innovated PVDF permeation cell enabling continual monitoring dynamic and static permeation rate of gases of volatile toxic compound through barrier materials with the employment of the reversible quartz crystal microbalance (QCM) sensor with the polymeric detection layer. With the help of this device it is possible to determine the period during which the limit of the toxic substance penetrates through the barrier material. The rate of permeation of vapours of the tested chemical through the barrier material within its contamination with the liquid phase is monitored. Schema of the SorpTest system is presented Figure S3.

The original equipment developed at the Faculty of Chemistry of the Brno University of Technology was used to determine oxygen transmission rate (OTR) under static conditions. The principle of measurement is similar to principle given in American Society for Testing Material (ASTM) D3985-17. OTR is the steady state rate at which oxygen gas permeates through a film at specified conditions of temperature and relative humidity. Values are expressed in $cc/m^2/24$ h units.

3. Results and Discussion

3.1. Protective Properties

As a carrier material for barrier layers, PEVA has been used as a representative of an easily swellable polymer and PP as a representative of a low swellable polymer. Barrier layers of melamine, parylene, PVDC and SiO_x have been applied to them occasionally combined with each other [25]. The prepared samples have been evaluated from the point of view of protective properties, microstructure and behaviour when exposed to the liquid phase of the test chemical. The PET foil is not suitable as the carrier material; otherwise testing the protective properties would be unbearably long.

Combinations of PEVA material with layers of parylene (by CVD), PVDC (by lamination) and SiO_x (PE-CVD) have been prepared and evaluated. Due to the low protection properties of PEVA foil itself against liquid toxic substances, in our case, sulphur mustard (HD) and 1,6-dichlorohexane (DCH), the protective properties of these materials have been verified within one and both-sides coating with a thin layer of PVDC and SiO_x. For one-sided coating, the PVDC has a breakthrough time value (BT_Y)

of 55 min and 129 min from the site of PEVA. This difference has been caused due to the presence of the directional effect of permeation caused by the different swell ability of both materials. The PEVA foil itself had the RD_Y only 27 min. For both-sided PVDC lamination, BT_Y was found to be 188 min on one side and the second side. For DCH, these values are slightly lower and approximately equal to 51 min on PEVA side and 78 min on both sides covering from both sides (Figure 2).

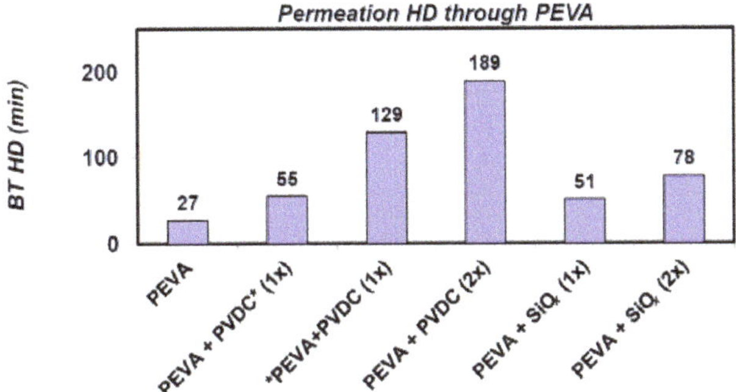

Figure 2. Breakthrough times of modifiedpolyethylenevinyl acetate (PEVA) foil determinate by the MikroTest method.

Further combinations of PEVA foil with barrier layers of SiO_x (by PE-CVD or PA-PVD), melamine (by PVD) and PVDC (by lamination) have been prepared [26,27]. These have been subsequently evaluated for resistance against the permeation of DCH. Interestingly, the resulting BT values for DCH in most combinations have been approximately the same and ranged from 23 to 33 min. The exception was only a combination with PVDC, with a single-sided coverage of BT 70 min from the PEVA side and 166 min with double-sided coverage. These results indicate that the exposure to DCH results in the rapid destruction of most barrier layers as a result of swelling of the carrier polymer so that the resulting protective properties of the starting material do not increase too much (Figure 3). The change in microstructure of the barrier layers following the DCH exposure described below also corresponds to these considerations.

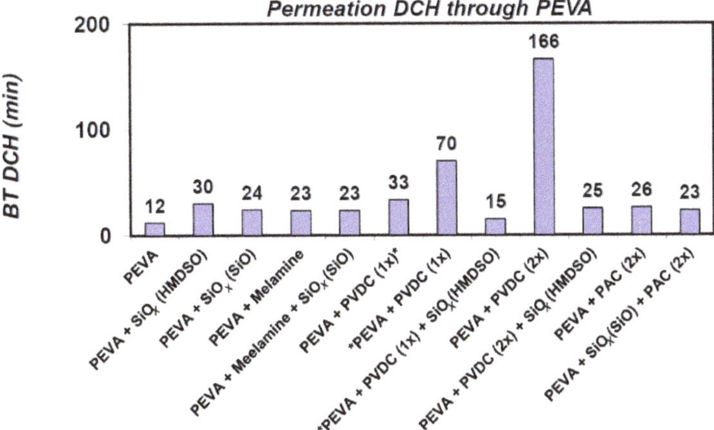

Figure 3. The resistance of the PEVA foil modified with different barrier layers against 1,6-dichlorohexane (DCH).

In order to remove the impact of the swelling on the protective properties of the studied barrier layers, polypropylene has been used as the carrier polymer. The results shown in Figure 4 confirm previous considerations of the negative effect of the swelling of the carrier polymer on the protective properties of the applied barrier nanolayers.

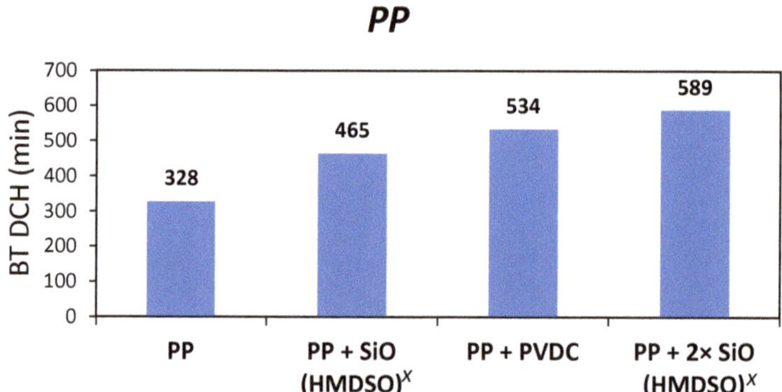

Figure 4. Resistance of the PP foil modified with different barrier layers against DCH permeation.

A similar situation occurs when assessing the protective properties with the help of oxygen permeation (OTR), which also does not come to swelling of the carrier polymer (Figure 5).

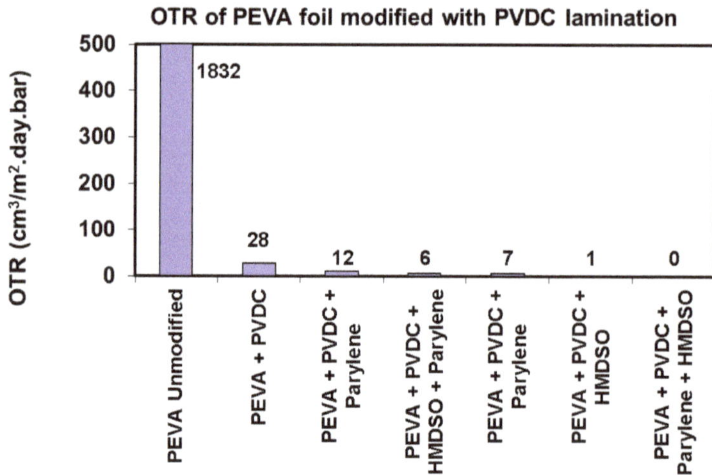

Figure 5. Oxygen permeation values detected for PEVA foil with barrier layers of PVDC, plasma polymerized HMDSO and parylene.

3.2. Microstructure

The analysis is mainly focused on microstructure assessment of inorganic nanocoatings SiO_x, which, when exposed to liquid test chemicals, unlike organic nanocoatings, resulted in significant destruction due to swelling of the polymer matrix. From a large number of SEM micrographs have been selected primarily those that were typical for the studied material and regularly repeated on different samples. Within large magnification, most researched SiO_x layers were found to have characteristic granular substructures of tens of nanometres dimensions, which are the basic building blocks of these nanocoatings.

It is necessary to realize that within the plasma deposition of nanostructured SiO_x barrier layers is a key parameter their elasticity and ductility, which usually is not more than 3% [27,28]. With gradual deformation of the substrate, the effect of ductility on protective properties can also be evaluated by OTR measurements, as there is no swelling of the carrier polymer which causes cracking and thus deterioration of the protective properties. When exposed to a liquid test chemical, the evaluation of the quality of the barrier layers is strongly dependent on the properties of the carrier substrate [29,30]. If a polymer resistant to swelling is used, the results of the protective properties evaluation are similar to those of the OTR. When using an easily swellable polymer, the cracks and defects in the barrier layer polymer are rapidly contaminated by the liquid test chemical and the subsequent swelling causes further destruction of the barrier layer and thereby deterioration of the protective properties.

However, the thickness of the barrier layer is also a crucial parameter. Basically, all publications dealing with the effect of the SiO_x nanocoating thickness on barrier properties indicate an optimum where OTR values are best. The principle of thickness dependence on OTR is explained by the fact that in the nucleation area the discrete centres are first formed on the polymer substrate. The best OTR values are then achieved when they are interconnected. However, if a uniform nanolayer is rapidly formed in another region, its thickness is preferably increased and the growth of discontinuous centres is slowed down. However, a faster increase of the thickness of the barrier nanolayer at these sites above the optimum value, however, results in microcracks resulting from internal stress in the layer material which getting the OTR value worse.

As a basic comparison layer, a nanostructured SiO_x layer on the silicon substrate has been prepared by PECVD by plasma deposition. It is apparent from Figure 6 and Figure S6 that the prepared nanocoating is homogeneous with a uniform microstructure that does not change even after prolonged exposure to DCH.

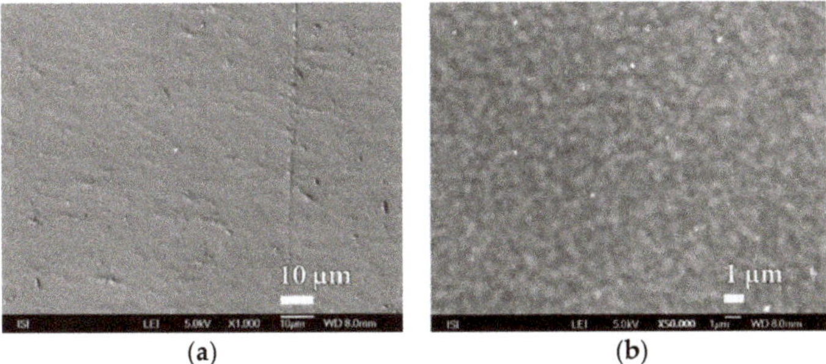

(a)   (b)

Figure 6. Scanning electron microscopy (SEM) micrographs of the surface of the nanolayer of SiO_x prepared by HMDSO method on the silicon substrate at various magnifications before exposure to DCH (**a**) 1000× and (**b**) 50,000×.

When assessing the quality of the SiO_x nanocoatings, the surface quality of the carrier polymer foil should be taken into account in order to determine if found defects do not copy the underlying material. Figure 7 and Figure S7 shows the typical non-homogeneity of the surface of the PEVA foil to be taken into account when assessing the microstructure of SiO_x nanocoatings.

For this reason, efforts were made to optimize plasma deposition conditions in order to improve the surface homogeneity of the polymer film used. One from the options was to modify the polymer matrix for less swelling. After verifying a number of PEVA surface treatment options, it was finally chosen to laminate it (at 110 °C without the binder) with a thin PVDC foil that showed good adhesion to the material. Another option was to replace the PEVA foil with a less-swelling PP film with a better surface. The use of PET foil that would be most suitable for plasma barrier coating was problematic because we could not measure too high BT values changes in protective properties in real time after applying different barrier layers. Polymeric matrices thus prepared were subjected to SiO_x barrier nanocoatings with plasma deposition. Prepared samples have been evaluated not only from the point of permeation resistance and for microstructure of barrier nanocoatings surface before and after exposure of the DCH test chemical with high resolution SEM. It is evident from Figure 8 and Figure S8 that the quality of the PVDC surface is significantly better than the original PEVA foil. It has been shown that this material is not porous even at high magnification (80,000×) and therefore it can be advantageously used for the deposition of other barrier layers based on melamine, parylene and mainly SiO_x plasma technology.

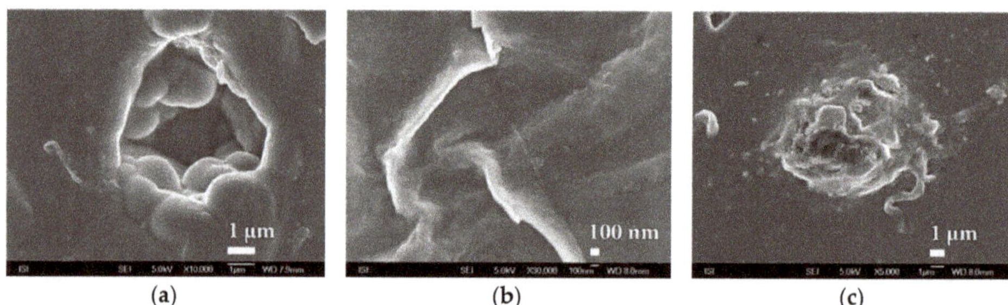

Figure 7. The exhibitions of different inhomogeneity on the surface of the used polymer foil PEVA at various magnifications of (**a**) 10,000×; (**b**) 30,000×;(**c**) 5000×.

Figure 8. SEM micrographs of PVDC foil surface laminated at 110 °C to PEVA foil at magnification of (**a**) 1000× and (**b**) 80,000×.

However, it is interesting to note that even with short-term oxygen plasma treatment, the surface of the PVDC foil is significantly wrinkled, while its homogeneity remains unchanged (Figure 9 and Figure S9).

A homogeneous layer of SiO_x with a well-developed microstructure which is probably very thin is formed with plasma deposition by PVD method on the surface of the PVDC foil (Figure 10 and Figure S10).

Similarly, within plasma deposition with the PVD method, at this time on the surface of PEVA foil, behaves even melamine and parylene C. The characteristic microstructure of the surface of these barrier nanocoatings is shown in Figures 11 and 12. In both cases a compact and fairly homogeneous barrier nanocoating with a well-developed microstructure copying the PEVA foil surface is formed.

A comparison of the microstructures in Figures 10–12 and Figure S10–S12 show that melamine produces a coarser microstructure than SiO_x but thicker than parylene C. Further SiO_x nanocoatings on PEVA film have been prepared by plasma deposition with the HMDSO method. Within this method of preparation, it is possible, in accordance with conditions during exposure, to form a SiO_x nanocoating of the different character than within plasma deposition by PVD method.

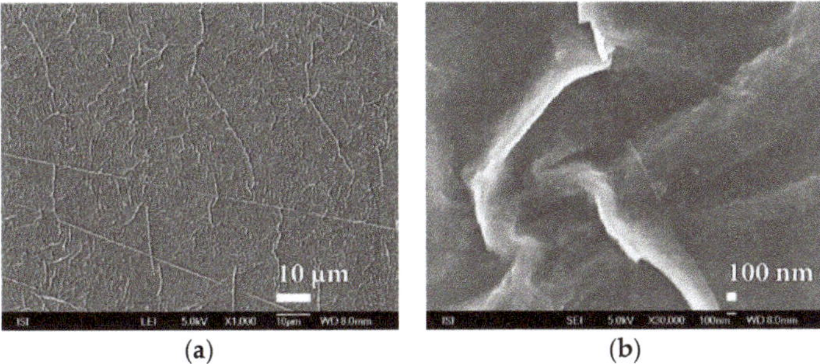

Figure 9. SEM micrographs of PVDC foil surface laminated to PEVA foil after short exposure to oxygen plasma treatment at magnification of (**a**) 1000× (**b**) 30,000×.

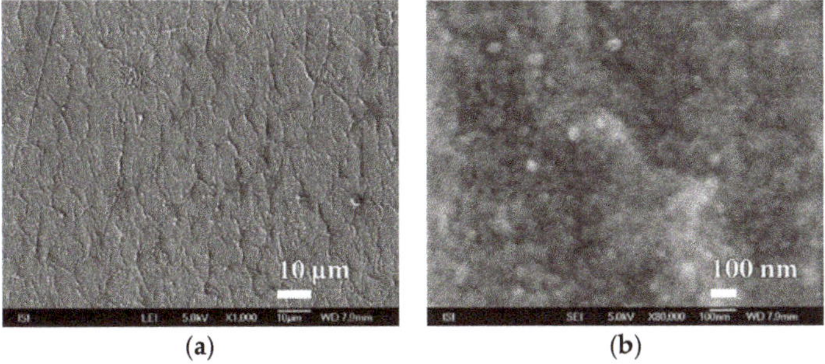

Figure 10. SEM micrographs of the surface SiO_x barrier nanocoating prepared by plasma deposition with the PVD method on PVDC foil laminated on the PEVA foil, at magnification of (**a**) 1000× and (**b**) 80,000×.

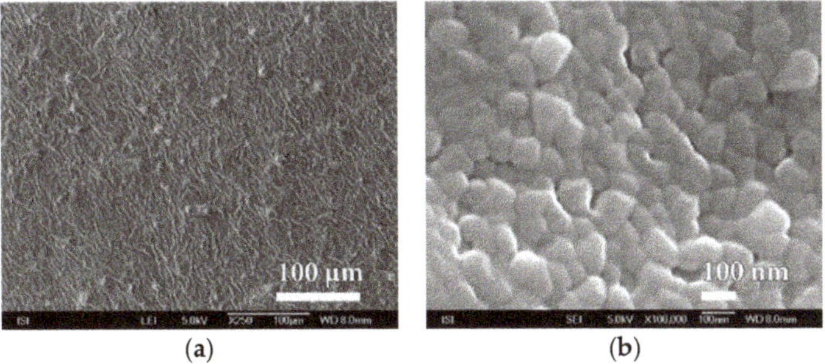

Figure 11. SEM micrographs of melamine nanocoating surface on PEVA SiO_x foil prepared by PVD plasma deposition method at magnification of (**a**) 250× and (**b**) 100,000×.

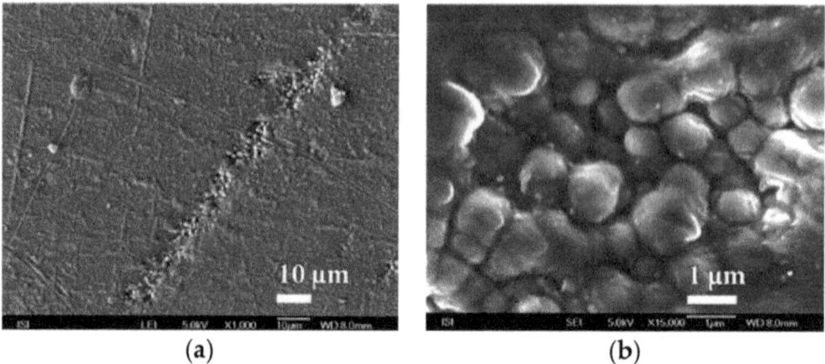

(a) (b)

Figure 12. SEM micrographs of parylene nanocoating surfaces on PEVA SiO_x foil prepared by PVD plasma deposition method at magnification of (**a**) 1000× and (**b**) 15,000×.

The different appearance of SiO_x nanocoatings prepared under various experimental conditions, documented in Figure 13 and Figure S13, is probably related to the nanocoating thickness and properties of the carrier substrate.

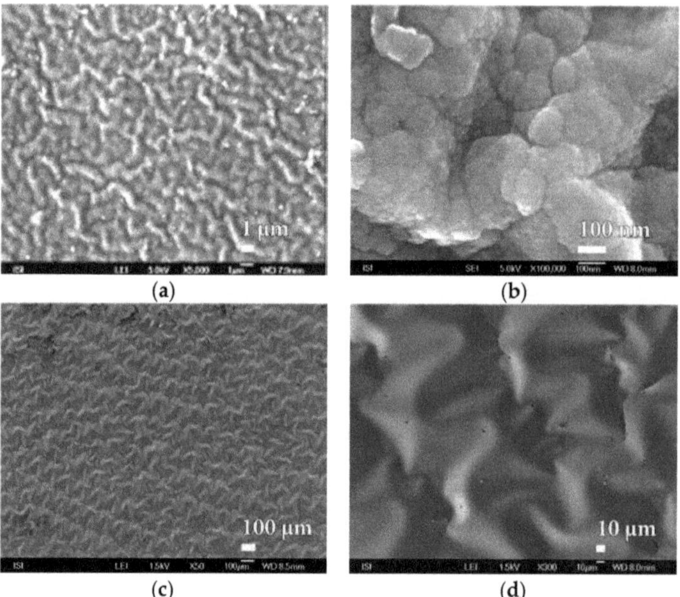

(a) (b)

(c) (d)

Figure 13. SEM micrographs of SiO_x barrier nanocoating surface prepared by the plasma deposition by HMDSO method on PEVA foil at magnification of (**a**) 5000× and (**b**) 100,000× and on PEVA foil stained with aluminium at magnification of (**c**) 50× and (**d**) 300×.

This assumption confirms the deposition of SiO_x by the HMDSO method on the silicon wafer surface, documented in Figure 14 and Figure S14.

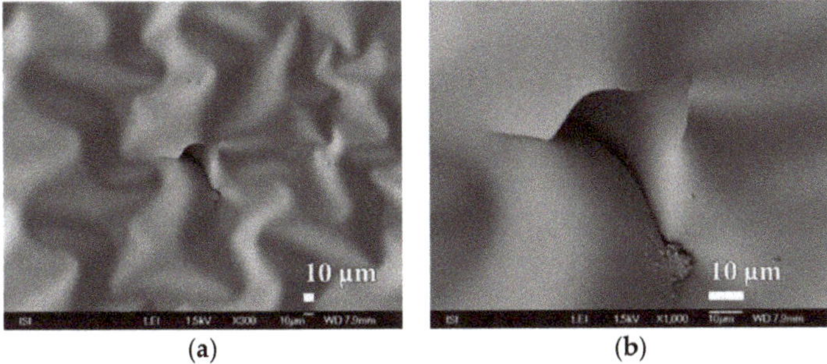

(a) (b)

Figure 14. SEM micrographs of surface SiO_x barrier nanocoating prepared by plasma deposition by HMDSO on the silicon wafer at magnification of (**a**) 300× and (**b**) 1000×.

The SiO_x nanocoating has been deposited on the surface of the Viton elastomer by the HMDSO method for comparison, documented in Figure 15 and Figure S15. Interestingly, in this case a homogeneous barrier layer has been formed, however with a porous microstructure.

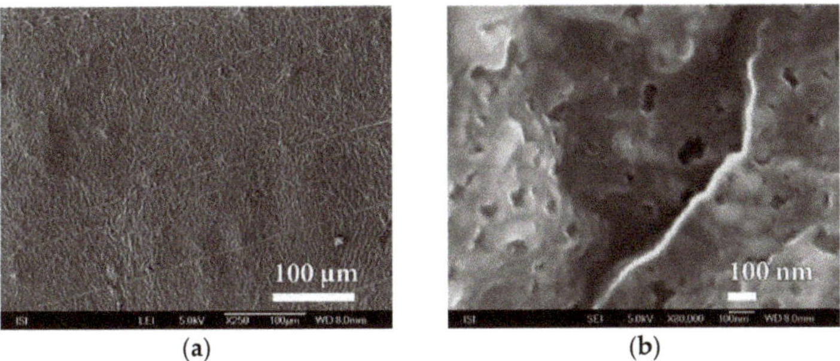

(a) (b)

Figure 15. SEM micrographs of surface SiO_x barrier nanocoating prepared by plasma deposition by HMDSO method on the Viton fluorelastomer at magnification of (**a**) 250× and (**b**) 80,000×.

The nanocoating SiO_x has been deposited on the surface of both PP and PET foils by the HMDSO method. The formed nanocoating has been both nonporous and homogeneous with a well-developed microstructure copying the surface of the PP foil, as shown in Figure 16.

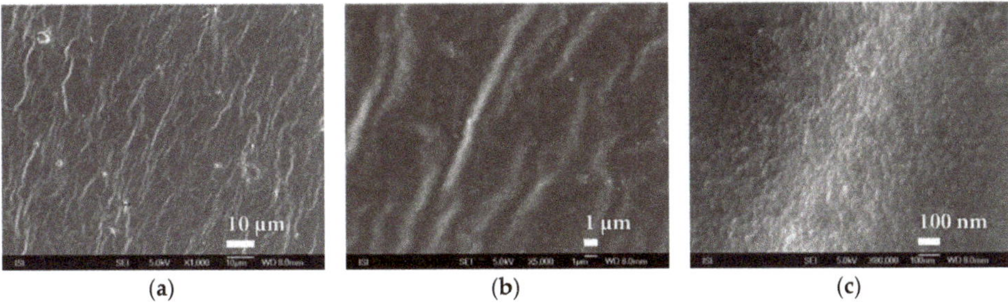

(a) (b) (c)

Figure 16. SEM micrographs of SiO_x barrier nanocoating surface prepared by plasma deposition by HMDSO method on the PP foil at magnification of (**a**) 1000×, (**b**) 5000× and (**c**) 80,000×.

If samples of polymeric materials with a SiO_x barrier layer are exposed to liquid DCH, an easily swellable PEVA polymer matrix will dramatically change the appearance of this nanocoating. Liquid DCH permeating through micro cracks or other leaks to the polymeric matrix apparently will not prevent its rapid swelling which causes destruction of the SiO_x nanopowder [30–32]. This effect will not appear at all or only to a small extent in the scope of PP foils or PET foils. It is conformable with the discussion above related to the protective properties of these materials. The destruction of the SiO_x barrier layer due to the polymer matrix embossment can be demonstrated by many of the examples shown in Figure S4. However, visual changes due to the polymer matrix swelling from visual changes due to mechanical damage to the SiO_x barrier layer, as documented in Figure S5, are required.

During the study of SiO_x nanocoatings after destruction, sites with such great damage have been found. It was possible to precisely determine the thickness of deposited barrier layer moving around 100 nm (Figure 17 and Figure S16). It even appears that in some cases the SiO_x nanocoating has a multilayer character of approximately 100 nm (Figure 17a).

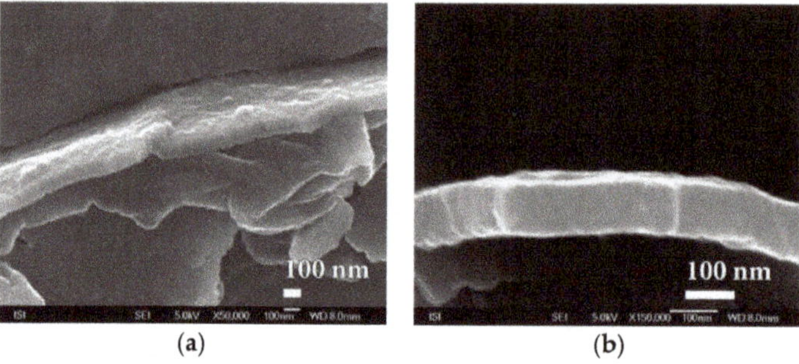

(a)   (b)

Figure 17. SEM photographs of SiO_x barrier nanocoating fracture prepared by plasma deposition by HMDSO method on PEVA foil at magnification of (**a**) 50,000× and (**b**) 150,000×.

Another interesting finding is the typical spherical substructure characteristic of most studied SiO_x nanocoatings. This substructure, clearly recognizable only with sufficiently large magnification, is also found inside deep material defects and even creates interesting structures on the surface and on the smooth surface of the PEVA foil, as can be seen from Figure 18.

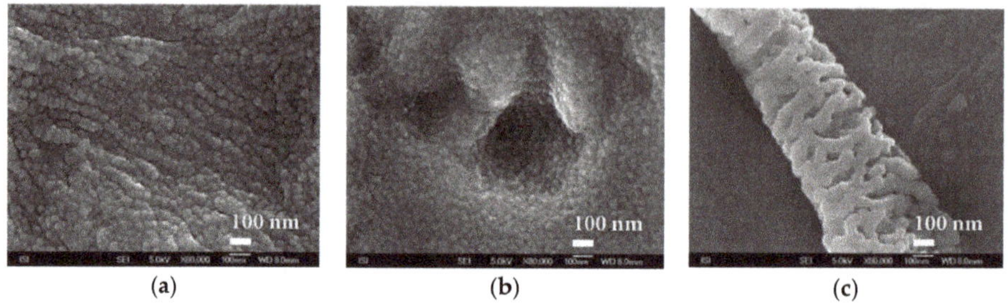

(a)   (b)   (c)

Figure 18. SEM micrographs of SiO_x nanocoating substructure prepared by plasma deposition by HMDSO method on the PEVA foil at magnification of 80,000× of three different areas. (**a**) Typical surface morphology; (**b**) layer defect and (**c**) surface abnormality.

4. Conclusions

It has been shown that the plasma deposition in vacuum can form barrier nanopowders from SiO_x and other suitable substrates on the polymer matrix. It is clear from the results that the HMDSO method, which makes it possible to influence the SiO_x nanocoating thickness and quality better, is more appropriate. Within forming practically applicable barrier materials with nanocoating of SiO_x it is necessary to be respect several essential requirements. The nanocoating SiO_x must be deposited primarily on a low swellable polymeric foil and, due to its brittleness, it must immediately be laminated with another foil due to prevention of its mechanical damaged. This is in a harmony with the production process of ALCAN Company within the production of packaging with barrier properties of CERAMIS® technology. The aim of further research in this field should be to improve the fixation of barrier nanocoatings on the polymer matrix and their resistance to mechanical stress. If it is real to prepare reactive coatings or deposition of suitable metallic oxides with catalytic properties it would enable this process to be used for the preparation of nanocoatings and nanomaterials with self-decontaminating or photocatalytic properties during further research.

Supplementary Materials: The following are available online at http://www.mdpi.com/2079-4991/8/9/679/s1, Figure S1: The principle of CERAMIS® reactive steam technology and examples of SiO_x® coated materials, Figure S2: The scheme of the PE-CVD/PA-PVD apparatus for the preparation of SiO_xnanocoatingsSiO$_x$ with steaming form powder SiO in vacuum, Figure S3: Diagram and real appearance of the SorpTest device for testing the resistance of barrier materials against the permeation of gaseous and liquid toxic substances, Figure S4: Scanning electron microscopy (SEM) micrographs of examples of visual changes of the SiO_x barrier layer caused by polymer matrix swelling with visible different kinds of cracks, Figure S5: Scanning electron microscopy (SEM) micrographs of examples of visual changes of the SiO_x barrier layer caused by mechanical damage, Figure S6: Scanning electron microscopy (SEM) micrographs of the surface of the nanolayer of SiO_x prepared by HMDSO method on the silicon substrate at various magnifications of $20,000\times$ before exposure to DCH, Figure S7: The exhibitions of different inhomogeneity on the surface of the used polymer foil PEVA at various magnification of (a) $30,000\times$, (b) $15,000\times$, (c) $80,000\times$, Figure S8: SEM micrographs of PVDC foil surface laminated at $110\,°C$ to PEVA foil at magnification of $30,000\times$, Figure S9: SEM micrographs of PVDC foil surface laminated to PEVA foil after short exposure to oxygen plasma treatment at magnification of 5000, Figure S10: SEM micrographs of the surface SiO_x barrier nanocoating prepared by plasma deposition with the PVD method on PVDC foil laminated on the PEVA foil, at magnification of $5000\times$, Figure S11: SEM micrographs of melamine nanocoating surface on PEVA SiO_x foil prepared by PVD plasma deposition method at magnification of $30,000\times$, Figure S12: SEM micrographs of parylenenanocoating surface on PEVA SiO_x foil prepared by PVD plasma deposition method at magnification of $5000\times$, Figure S13: SEM micrographs of SiO_x barrier nanocoating surface prepared by the plasma deposition by HMDSO method on PEVA foil at magnification of (a) $1000\times$ and on PEVA foil stained with aluminium at magnification (b) $1000\times$, Figure S14: SEM micrographs of surface SiO_x barrier nanocoating prepared by plasma deposition by HMDSO on the silicon wafer at magnification of $1000\times$, Figure S15: SEM micrographs of surface SiO_x barrier nanocoating prepared by plasma deposition by HMDSO method on the Viton fluorelastomer at magnification of $5000\times$, Figure S16: SEM photographs of SiO_x barrier nanocoating fracture prepared by plasma deposition by HMDSO method on PEVA foil at magnification of $100,000\times$.

Author Contributions: Conceptualization, P.O. and V.O.; Methodology, P.O. and V.O.; Validation, R.P., P.O., V.O. and R.K.; Formal Analysis, R.P., P.O., V.O and L.S.; Investigation, P.P., V.O., L.S., R.K. and J.B.; Resources, R.P., V.O., R.K. and J.B.; Data Curation, R.P., P.O., V.O and L.S.; Writing-Original Draft Preparation, P.O. and V.O.; Writing-Review & Editing, P.O. and R.P.; Supervision, P.O. and R.P; Project Administration, P.O. and V.O.; Funding Acquisition, R.P.

Funding: This work was supported by Czech Ministry of Education, Youth and Sports, Project number LO1211, Material Research Centre at Faculty of Chemistry, Brno University of Technology.

Acknowledgments: We would like to thank all subjects who volunteered in the study.

Conflicts of Interest: The authors declare no conflict of interest.

References

1. Kogelschatz, U.; Eliasson, B.; Egli, W. Dielectric-barrier discharges principle and applications. *J. Phys. Arch.* **1997**, *7*, C4-47–C4-66. [CrossRef]

2. Otrisal, P.; Florus, S.; Svorc, L.; Barsan, G.; Mosteanu, D. A new colorimetric assay for determination of selected toxic vapours and liquids permeation through barrier materials using the minitest device. *Rev. Plast.* **2017**, *54*, 748–751.

3. Kmochova, A.; Tichy, A.; Zarybnicka, L.; Sinkorova, Z.; Vavrova, J.; Rehacek, V.; Durisova, K.; Kubelkova, K.; Pejchal, J.; Kuca, K. Modulation of ionizing radiation-induced effects by NU7441, KU55933 and VE821 in peripheral blood lymphocytes. *J. Appl. Biomed.* **2016**, *14*, 19–24. [CrossRef]

4.   Krumpolec, R.; Zahoranova, A.; Cernak, M.; Kovacik, D. Chemical and physical evaluation of hydrophobic pp-HMDSO layers deposited by plasma polymerization at atmospheric pressure. *Chem. Listy* **2012**, *106*, 1450–1454.

5.   Radovan, K.M.; Negovan, I.D.; Dalibor, J.B.; Smiljana, M.M.; Dejan, I.R.; Marija, M.D.; Branko, K.V. Dynamic adsorption characteristics of thin layered activated charcoal materials used in chemical protective overgarments. *Indian J. Fibre Text. Res.* **2016**, *41*, 402–410.

6.   Mosteanu, D.; Barsan, G.; Otrisal, P.; Giurgiu, L.; Oancea, R. Obtaining the volatile oils from wormwood and tarragon plants by a new microwave hydrodistillation method. *Rev. Chim.* **2017**, *68*, 2499–2502.

7.   Sun, Y.J.; Fu, Y.B.; Chen, Q.; Zhang, C.M.; Sang, L.J.; Zhang, Y.F. Silicon dioxide coating deposited by PDPs on PET films and influence on oxygen transmission rate. *Chin. Phys. Lett.* **2008**, *25*, 1753–1756.

8.   Felts, J.T. Transparent barrier coatings update: Flexible substrates. *J. Plast. Film. Sheeting* **1993**, *9*, 139–158. [CrossRef]

9.   Macák, J.M.; Hromádko, L.; Bulánek, R.; Koudelková, E.; Buk, J.; Tejkl, M. Process for Preparing Submicron Fibers of Amorphous Silica and Submicron Fibers of Amorphous Silica Prepared in Such a Manner. Available online: https://1url.cz/qtyDc (accessed on 12 June 2018).

10. Madocks, J.; Marcus, P.; Morse, P. Silicon dioxide coatings on plastic substrates by large area plasma enhanced chemical vapour deposition process. In Proceedings of the 49th Annual Technical Conference, Washington, DC, USA, 22–27 April 2006; pp. 569–573.

11. Hoskova-Mayerova, S.; Maturo, A. Algebraic hyperstructures and social relations. *Italian J. Pure. Appl. Math.* **2018**, *39*, 701–709.

12. Anonymous. CERAMIS® Barrier Films: Transparent SiOx Barrier Films for Ultimate Product Protection. Available online: https://1url.cz/Pt734 (accessed on 23 June 2018).

13. Roberts, A.P.; Henry, B.M.; Sutton, A.P.; Grovenor, C.R.M.; Briggs, G.A.D.; Miyamoto, T.; Kano, M.; Tsukahara, Y.; Yanaka, M. Gas permeation in silicon-oxide/polymer (SiO_x/PET) barrier films: Role of the oxide lattice, nano-defects and macro-defects. *J. Membr. Sci.* **2002**, *208*, 75–88. [CrossRef]

14. Erlat, A.G.; Spontak, R.J.; Clarke, R.P.; Robinson, T.C.; Haaland, P.D.; Tropsha, Y.; Harvey, N.G.; Vogler, E.A. SiO_x gas barrier coatings on polymer substrates: Morphology and gas transport considerations. *J. Phys. Chem. B* **1999**, *103*, 6047–6055. [CrossRef]

15. Felts, J.; Finson, E.J. Transparent SiO_2 barrier coatings: Conversion and production status. In Proceedings of the 37th Annual Technical Conference Proceedings, Boston, MA, USA, 8–13 May1994; pp. 139–143.

16. Foest, R.; Schafer, J.; Quade, A.; Ohl, A.; Weltmann, K.D. Miniaturized atmospheric pressure plasma jet (APPJ) for deposition of SiO_x films with different Silicon-organics compounds—A comparative study. In Proceedings of the 35th EPS Conference on Plasma Phys, Hersonissos, Greece, 9–13 June 2008.

17. Gaur, S.; Vergason, G. Plasma polymerization: For industrial production. In Proceedings of the 44th Annual Technical Conference Proceedings, Philadelphia, PA, USA, 21–26 April 2001.

18. Huang, X. Fabrication and Properties of Carbon Fibers. *Materials* **2009**, *1*, 938–947. [CrossRef]

19. Horova, I.; Kolacek, J.; Vopatova, K. Full bandwidth matrix selectors for gradient kernel density estimate. *Comput. Stat. Data Anal.* **2013**, *57*, 364–376. [CrossRef]

20. Raynaud, P.; Clergereaux, R.; Segui, Y.; Gherardi, N.; Despax, B. Plasma Nanolayering, Nanoclustering, Nanotexturing: Challenges for Future Processes for Mass Produstion. Available online: https://1url.cz/Yt73j (accessed on 23 June 2018).

21. Starostin, S.A.; Aldea, E.; Vries, H.W.; Creatore, M.; van de Sanden, M.C.M. Application of atmospheric pressure glow discharge (APGD) for deposition of thin silica like films on polymeric webs. In Proceedings of the 28th International Conference on Phenomena in Ionized Gases (ICPIG), Prague, Czech Republic, 15–20 July 2007.

22. Fortin, J.B.; Lu, T.-M. *Chemical Vapor Deposition Polymerization: The Growth and Properties of Parylene Thin Films*; Springer Science & Business Media: New York, NY, USA, 2003. [CrossRef]

23. Gorham, W.F. A new, general synthetic method for the preparation of linear poly-*p*-xylylenes. *J. Polym. Sci. Part A* **1966**, *4*, 3027–3039. [CrossRef]

24. Míša, M.; Obšel, V. System for Measuring Permeation of Gases and Vapors Through Barrier Membranes. Available online: https://1url.cz/Bty4h (accessed on 11 June 2018).

25. Yasuda, H. *Plasma Polymerization*; 1st Issue; Academic Press: Orlando, FL, USA, 1985.

26. Smith, A.W.; Copeland, N. Clear barrier coatings by plasma CVD for packaging applications. In Proceedings of the 43th Annual Technical Conference Proceedings, Washington, DC, USA, 22–27 April 2006; pp. 636–641.
27. Izu, M.; Dotter, B.; Ovshinsky, S.R. Roll-to-roll microwave PECVD machine for high-barrier film coatings. *J. Photopolym. Sci. Technol.* **1995**, *8*, 195–204. [CrossRef]
28. Hones, P.; Levy, F.; Randall, N.X. Influence of deposition parameters on mechanical properties of sputter-deposited Cr_2O_3 thin films. *J. Mater. Res.* **1999**, *14*, 3623–3629. [CrossRef]
29. Jahromi, S.; Moosheimer, S.U. Oxygen barrier coatings based on supramolecular assembly of melamine. *Macromolecules* **2002**, *33*, 7582–7587. [CrossRef]
30. Howells, D.G.; Henry, B.M.; Leterrier, Y.; Manson, J.A.E.; Madocks, J.; Assender, H.E. Mechanical properties of SiO_x gas barrier coatings on polyester films. *Surf. Coat. Technol.* **2008**, *202*, 3529–3537. [CrossRef]
31. Ozeki, K.; Nagashima, Y.; Ohgoe, Y.; Hirakuri, K.K.; Mukaibayashi, H.; Masuzawa, T. Gas barrier properties of diamond-like carbon films coated on PTFE. *Appl. Surf. Sci.* **2009**, *25*, 7286–7290. [CrossRef]
32. Stodola, P. Using metaheuristics on the multi-depot vehicle routing problem with modified optimization criterion. *Algorithms* **2018**, *11*, 74. [CrossRef]

Article

Electrospinning of Polystyrene/Polyhydroxybutyrate Nanofibers Doped with Porphyrin and Graphene for Chemiresistor Gas Sensors

Joshua Avossa [1], Roberto Paolesse [1,2], Corrado Di Natale [1,3], Emiliano Zampetti [1], Giovanni Bertoni [4], Fabrizio De Cesare [1,5], Giuseppe Scarascia-Mugnozza [5] and Antonella Macagnano [1,5,*]

[1] Institute of Atmospheric Pollution Research–National Research Council (IIA-CNR), Research Area of Rome 1, Via Salaria km 29.300, 00016 Monterotondo, Italy; joshua.avossa@iia.cnr.it (J.A.); paolesse@uniroma2.it (R.P.); dinatale@uniroma2.it (C.D.N.); e.zampetti@iia.cnr.it (E.Z.); decesare@unitus.it (F.D.C.)

[2] Department of Chemical Science and Technology, University of Tor Vergata, Via della Ricerca Scientifica 00133 Rome, Italy

[3] Department of Electronic Engineering, University of Tor Vergata, Via del Politecnico 1, 00133 Rome, Italy

[4] Institute of Materials for Electronics and Magnetism–National Research Council (IMEM-CNR), Parco Area delle Scienze 37/A, 43124 Parma, Italy; giovanni.bertoni@imem.cnr.it

[5] Department for Innovation in Biological, Agro-food and Forest Systems (DIBAF), Via S. Camillo de Lellis, 00100 Viterbo, Italy; gscaras@unitus.it

* Correspondence: antonella.macagnano@cnr.it or a.macagnano@iia.cnr.it; Tel.: +39-069-067-2395-2401

Received: 17 January 2019; Accepted: 12 February 2019; Published: 17 February 2019

Abstract: Structural and functional properties of polymer composites based on carbon nanomaterials are so attractive that they have become a big challenge in chemical sensors investigation. In the present study, a thin nanofibrous layer, comprising two insulating polymers (polystyrene (PS) and polyhydroxibutyrate (PHB)), a known percentage of nanofillers of mesoporous graphitized carbon (MGC) and a free-base tetraphenylporphyrin, was deposited onto an Interdigitated Electrode (IDE) by electrospinning technology. The potentials of the working temperature to drive both the sensitivity and the selectivity of the chemical sensor were studied and described. The effects of the porphyrin combination with the composite graphene–polymer system appeared evident when nanofibrous layers, with and without porphyrin, were compared for their morphology and electrical and sensing parameters. Porphyrin fibers appeared smoother and thinner and were more resistive at lower temperature, but became much more conductive when temperature increased to 60–70 °C. Both adsorption and diffusion of chemicals seemed ruled by porphyrin according its combination inside the composite fiber, since the response rates dramatically increased (toluene and acetic acid). Finally, the opposite effect of the working temperature on the sensitivity of the porphyrin-doped fibers (i.e., increasing) and the porphyrin-free fibers (i.e., decreasing) seemed further confirmation of the key role of such a macromolecule in the VOC (volatile organic compound) adsorption.

Keywords: electrospinning deposition; chemosensor; nanocomposite conductive polymers; polyhydroxibutyrate; polystyrene; H$_2$TPP; VOCs selectivity; mesoporous graphene

---

1. Introduction

Chemical sensors are usually conceived as electronic devices comprising a sensing material, in charge of interacting with the target analyte, and a transducer, to transform such an interaction into an electric/optical signal [1]. Among the main drivers for the design of advanced chemical sensors, the key characteristics include sensitivity, selectivity, and rapid detection of target molecules. Over the last decade, the combination of nanostructured materials with many transducers has boosted the advances in this area, leading to significant enhancements in their sensing performance [2,3]. Consequently, together with a plethora of complex nanostructures, polymer nanocomposites have been designed and investigated as promising candidates for developing advanced materials for sensors. These joint materials benefit from the synergy between filler nanoparticles and polymer chains: they are both on similar length scales and with a very large interfacial surface area when compared to the volume of the material [4]. On the other hand, polymers are one of the most extensively exploited classes of materials due to the great variety of available chemical moieties with their relatively low cost, easy processing, and potential for designing and fabricating recycled and sustainable materials for sensors of the last generation [5]. More specifically, the development of polymeric composites based on carbon nanomaterials, such as carbon nanotubes (CNTs) and graphene (G), has been given a great deal of attention as a path to achieve new sensing materials with new structural (e.g., mechanical stability) and functional properties, likely better performing than pure components. Among the remarkable features of these two carbon allotropes to be used to design a chemical sensor, there are high electrical conductivity and large surface area [6]. Both nanomaterials have the same honeycomb lattice with sp^2-hybridized carbon atoms, but with a two-dimensional sheet in graphene and its rolling up in one (SWNT, single-wall nanotubes) or more concentric tubes (MWNT, multi-wall nanotubes). Chemiresistors based on carbon-polymer combinations comprise features such as great stability, improvement of lifetime, tunable selectivity, ability to work at room temperature, good reversibility and reproducibility, low power consumption and cost effectiveness [7]. Generally, in these systems, current passes through continuous pathways of the conductive carbon particles between the parallel electrodes of the transducers. The sorption of a chemical vapor can cause softening/swelling of the polymer film, breaking some of the continuous pathways and increasing the resistance of the composite. Therefore, polymers can be designed to be more or less selective to different classes of VOCs, taking into account their solvation parameters according to vapor solubility (linear solvation energy relationships (LSER)) [8]. Obviously, much attention has been paid to properly modifying both polymers (functionalization of polymer chains) and polymer films (layer structure) to be simultaneously more selective to defined VOCs and optimized to host carbon nanofillers [9–11]. Therefore, chemiresistors based on Nafion–CNT [12] and poly(2,5-dimethylaniline)–CNT [13] have revealed intriguing performances in measuring air humidity and acid vapors, respectively. More recently, many nanocomposite polymer–graphene materials have been investigated and used successfully as sensors for industrial chemical reagents, drugs and explosives [14,15], thus highlighting comparable [16], and in some studies even better performances [17], than those of CNT-based sensors. This aspect sounds attractive since graphene is synthesized according to lower cost procedures and it is a material with the highest electrical conductivity known at room temperature (6000 Sc m^{-1}) [18], a huge surface area (2.63 $\times$ 10^3 m^2 g^{-1}) and a complete impermeability to any gases [19]. The efficiency of graphene–polymer composite seems to be related to the molecular-level dispersion of graphene [20] that commonly occurs by using a proper surfactant and/or selecting a polymer matrix that, through $\pi-\pi$ stacking or hydrophobic (van der Waals) interactions, preserve the intrinsic electronic properties of graphene and allow a nanofiller homogeneous distribution [21]. Regarding a graphene hosting layer framework, thin and porous films, facilitating gas/VOCs diffusion, are preferred to get fast sensor responses and to avoid layer poisoning (hysteresis effects). A very porous layer with a controlled distribution of the nanofillers can be developed by electrospinning (ES) deposition. This is a technology able to produce advanced multifunctional polymer nanocomposites (2D and 3D micro- or nanofibrous layers) and has been conceived as one of the most promising strategies to design and fabricate highly sensitive nanocomposite films at low cost and with high production rates [22]. The process uses a high voltage to provide sufficient charges in the polymer solution such that a jet is ejected from the tip of a spinneret toward a grounded collector: the solvent evaporates on the path and the polymer nanofibers can be grown according to various arrangements and different morphologies. Thus, graphene nanoparticles can be added in the polymer solution for electrospinning, and nanofibers can be investigated as potential conductive material for chemical sensors. In a recent study, Avossa et al. [23] used the temperature to modulate the sensitivity of nanocomposite polymer nanofibers (polystyrene, polyhydroxibutyrate, mesoporous graphitized carbon (PS-PHB-MGC)) to gas and VOCs, making the sensor more selective to NO_2 (LOD: 2 ppb, limit of detection) when operated at 80 °C, whereas VOCs adsorption decreases.

A polymer doping agent is an alternative to the functionalization or substitution of polymer matrix for tuning the sensor selectivity. Different porphyrin species have been used to dope organic polymer. For instance, porphyrins combined with poly(2-phenyl-1,4-xylylene) have led to a huge variation in responses depending on the selected analytes (toluene, ethylacetate, ethanol, and propanone) [24]. Porphyrins have excellent sensing properties (their framework, peripheral substituents and the core that could be practically occupied by all metals of the periodic table) that make them an effective object of study and sensor applications of the last thirty years [25]. In the literature, there are several cases of porphyrins subjected to electrospun deposition in combination with electrospinnable polymers, so that their dispersion [26] and arrangement inside fibers as well as their photocatalytic [27] and sensing features [28,29] have been investigated extensively [30]. On the other hand, the best performances seem to be achieved when porphyrin occupies the outer part of the fiber [31] or when polymer fibers are very porous. Electrospun polymer fibers with a ternary composite combination (porphyrin, graphene oxide and nylon) are also manufactured [32] to be investigated towards more advanced applications.

In this paper, we present the design and the creation of a nanofibrous conductive chemical sensor based on a quaternary combination of two insulating polymers (PS and PHB, named as PsB) doped with 5,10,15,20-tetraphenylporphyrin (H_2TPP) and mesoporous graphene nanopowder. Porphyrin, due to its molecular structure, is expected to give a significant contribution to the sensing properties of the polymer composite fibers [23], by coordinating the planar surfaces of MGC and the phenyl rings of PS (effects on fiber morphology and structure) and "capturing" selectively volatile organic compounds (VOCs) (effects on sensor sensitivity and selectivity). The polymers were selected because they are versatile (widely used in many consumer products), eco-compatible, biodegradable (PHB), recyclable (PS) [33,34] and resistant to thermal excursions (thermoplastics). Additionally, they are both soluble in chloroform and insoluble in H_2O, meaning that they could be deposited by a unique electrospun mixture with a single needle, and were expected to be stable to changes in environmental humidity. In this more complex fibrous matrix, a preliminary study on the dependence of the sensor selectivity on the temperature was also carried out and is described below.

2. Materials and Methods

Mesoporous graphitized carbon nanopowder (MGC) (<500 nm; available surface area: 50–100 m^2/g; average pore diameter: 137 Å), hexadecyltrimethylammonium bromide (CTAB) (~99%), polystyrene (PS) (Mw = 192,000 g/mol), chloroform (≥99%), toluene (≥99.8%), acetic acid (≥99%), and poly[(R)-3-hydroxybutyric] acid (PHB) (natural origin) were purchased from Sigma-Aldrich (Merck KGaA, Darmstadt, Germany). Ethanol (≥99.8%) was obtained from Fluka (Buchs, Switzerland). All chemicals were used without further purification. A 5,10,15,20-tetraphenylporphyrin (H_2TPP) was prepared following literature protocol [35]. Standardized pure air (5.0) was purchased from Praxair-RIVOIRA (Rome, Italy) and stored in cylinders.

Interdigitated Electrodes (IDEs), provided by Micrux Technologies (Oviedo, Spain), were fabricated on borosilicate substrate (IDE sizes: 10 mm × 6 mm × 0.75 mm; Pt/Ti electrodes, 120 pairs, 10 μm wide × 5 mm long × 150 nm thick, with 10 μm gap) and rinsed with soap and a "base piranha" mixture at 60 °C for ~15 min, (3:1, v/v, ammonia water and hydrogen peroxide water solution) and finally with Milli-Q water (~18 MΩ cm) before any use.

The electrospun dispersion was prepared by first solubilizing 450 mg of PS pellets into 9 mL of chloroform under magnetic stirring. After complete dissolution, 60 mg of PHB were added into the solution and mixed at 50 °C for 2 h. Then, 150 mg of CTAB and 1 mL of ethanol were poured into the system and mixed overnight at 50 °C under magnetic stirring. Suspensions composed of 1.3 mg of MGC with and without 15 mg of H_2TPP were poured into 2 mL of the PS/PHB/CTAB solution and sonicated for at least 1 h.

The resulting polymer dispersions were loaded into glass syringes (1 cm long stainless steel and blunt tips) and connected to a syringe pump (Model KDS 200, KD Scientific). The fibers depositions were carried out in a home-made (IIA-CNR, Monterotondo, Rome, Italy) and ventilated clean box equipped at ambient condition. The electrospinning apparatus consisted of a high power AC-DC converter, a high voltage oscillator (100 V) driving a high voltage (ranging from 1 to 50 kV), a syringe pump and a rotating conductive pipe with a 45 mm diameter grounded collector. The fibrous layers were fabricated by applying ~6 kV DC voltage between the syringe tip and the collector (8 cm of distance), at a pump feeding rate of 700 µL h^{-1}. Deposition time was fixed at 2 min to obtain a thin and adhering coverage of the surface (IDEs and SiO_2 wafers and High Precision Quartz slices). UV-Vis spectrophotometer (UV-2600 Shimadzu, Kyoto, Japan) was used to collect UV spectra of the fibrous layer at the solid state.

Optical micrographs were captured by a Leitz-Wetzlar (Metallux 708082, Wetzlar, Germany) microscope, for the evaluation of the quality coverage of the fibers deposited onto the IDE.

Fibers morphological analyses were carried out by means of micrographs from Scanning Electron Microscopy (SEM) and Transmission Electron Microscopy (TEM). The electrospun nanofibrous fabrics deposited on thin SiO_2 wafers and sputter-coated with gold in a Balzers MED 010 unit were analyzed for SEM by a JEOL JSM 6010LA electron microscope (High Equipment Centre, University of Tuscia, Viterbo, Italy). Scanning transmission electron microscopy (STEM) images were acquired in annular dark field mode (ADF) on a JEOL JEM-2200FS microscope operated at 200 kV and with a spot size of 2 nm (IMEM-CNR, Parma, Italy). The H_2TPP/PS-PHB-MGC fibers were deposited on a lacey carbon coated copper grid, to reduce electrostatic charging of the fibers under the electron beam. Chemical mapping of carbon and oxygen were obtained from energy dispersive X-ray spectroscopy (EDXS) using a Si-Li detector (JEOL JED-2400, Akishima, Tokyo, Japan).

The resulting chemiresistors (IDEs + NFs, where NFs means nanofibers) were sealed in a measurement glass chamber (~100 mL volume) and connected to an electrometer (Keithley 6517, Solon, Ohio, USA) capable of measuring their electrical parameters and sending data to a PC (LabVIEW 2014 Software, National Instruments, Austin, TX, USA). The current, provided under dry and clean air, was recorded by applying potential values from −4.0 to 4.0 V in steps of 0.4 V at different temperatures (25, 40, 50, 60, and 70 °C). Current versus applied voltage values were used to calculate the resistance of the fibrous coated IDE and its correlation to the temperature, as well as a potential hysteresis of the material when it was electrically stressed. All batches of the chemiresistors fabricated on different dates but keeping the identical deposition parameters reported the same electrical features, confirming the reproducibility of the deposition technique.

Dynamic sensor measurements were carried out at different working temperature (T_w: 50, 60, and 70 °C, generated by a micro-heater placed below each sensing area of the IDE) using: (i) 4-channel MKS 247 managing up to four MKS mass flow controllers (MFC), set in the range 0–200 sccm (standard cubic centimeter per minutes); and (ii) Environics S4000 (Environics, Inc., Tolland, CT, USA) flow controller, containing three MFCs supplying three flow rates (up to 500, 250 and 2.5 sccm, respectively), managed by its own software. Pure air was used as the gas carrier and it was blended with increasing concentrations of vapors of water (H_2O), acetic acid (AcAc) and toluene (Tol), respectively, obtained through air bubbling in customized borosilicate bubblers (Rolando Spaziani S.r.l., Nettuno, Rome, Italy). A total gas flow of 300 sccm passed through the measurement chamber, housing the IDEs. Each measurement was carried out after the complete recovery of the starting current (the baseline) under dry and clean air flow. IDE responses were calculated as $\Delta I/I_0$, where ΔI is the current variation and I_0 is the current when the air flowed.

3. Results and Discussion

Electrospinning technology was used to create nanocomposite nanofibrous layers in a single step using a single needle. The depositions were easily carried out onto several substrates, specifically silicon dioxide thin slices (for fibers morphological, chemical and optical characterization) and customized borosilicate IDE transducers (for measuring electrical and sensing features of the thin nanofibrous coating). Each substrate, fixed onto the grounded rotating cylinder facing the needle tip, was able to collect the ejected fibers within the deposition cone (Figure 1). Fibers did not look aligned over the electrode, but arranged to form a porous network placed on the IDE surface. Despite the heterogeneity of the polymer suspensions, the electrospun jet streams occurred without discontinuity, so that fibers were collected for a few minutes.

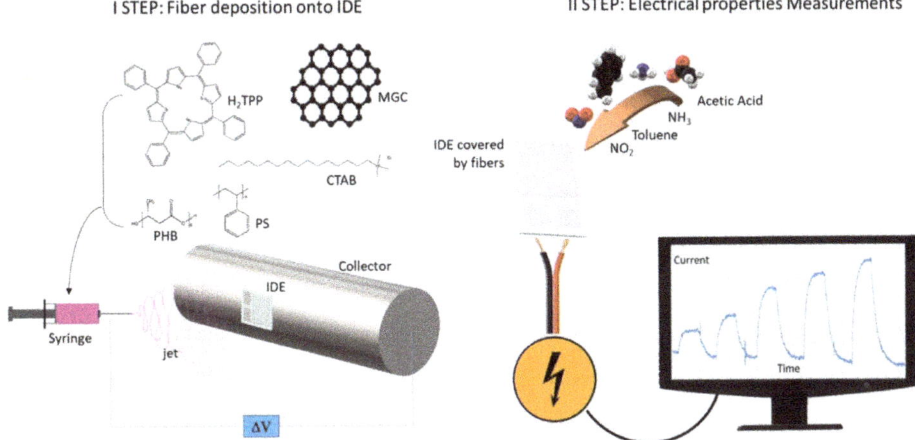

Figure 1. Sketch of electrospinning technique able to coat an IDE (interdigitated electrode) with composite nanofibers (**left**); and current measurements upon interaction of the chemosensor with gaseous molecules (**right**).

Shortly, the deposition process was generated by the application of a definite electrical field between the polymer suspension droplet at the metal nozzle and the grounded substrate placed at a distance. Due to the application of the electrical field, the polymer drop first changed shape (from spherical to conical) and then was elongated until the electrostatic forces exceeded the surface tension of the polymer suspension and forced the ejection of the liquid jet. Finally, dry and fine fibers were collected following the jet bending and stretching processes (by forces with opposing effects) [36], solvent evaporation and then splaying. The resulting fabrics appeared soft and cotton candy-like, and resulted easy to peel when electrospun processes were carried out for a longer time. Thus, to improve the fibers adhesion, it was necessary to thoroughly clean all the surfaces (base-piranha solution) and, following the deposition, incubate all the substrate at 60 °C under slight vacuum. The nanocomposite fabrics were pink (Figure 2c), turning to orange when the thickness increased but white/light gray (Figure 2d) if porphyrin-free. Nanocomposite fibers without porphyrin (PsB-MGC) previously investigated [23] appeared extremely rough on the surface and decorated with brighter islands, but fairly uniform in term of shape (cylindrical) and size (d: 550 ± 170 nm) (Figure 2b). The fibers combined with H₂TPP (H₂TPP-PsB-MGC) kept the same circular cross-sectional shape but appeared much smaller in size (d: 174 ± 50 nm) (Figure 2a). A possible reason for this phenomenon could be attributed to the polymer percentage decreasing in the final electrospun mixture [37,38] when porphyrin molecules were added. Furthermore, a higher voltage applied to the porphyrin mixture necessary to engage the electrospun process ($V_{H2TPP\text{-}PsB\text{-}MGC} = \approx 6$ kV; $V_{PsB\text{-}MGC} = 2.9$ kV) could also be responsible for the fiber diameter reduction when combined to a lower feed rate (i.e., 700 µL h^{-1} and 900 µL h^{-1} used for H$_2$TPP- PsB-MGC and PsB-MGC suspensions, respectively) [39,40]. Porphyrin fibers looked smoother and had small spherical/elliptical bumps protruding from the whole the surface of the fibers (Figure 2a, inset). Therefore, the addition of the porphyrin to the composite system seemed to substantially change the morphology of the resulting fibers: in addition to dispersion forces, H$_2$TPP could interact with PS and graphene by $\pi-\pi$ interactions and with PHB by hydrogen bond. However, long, continuous and unbeaded fibers proved an appropriate combination of electrospinning set of parameters.

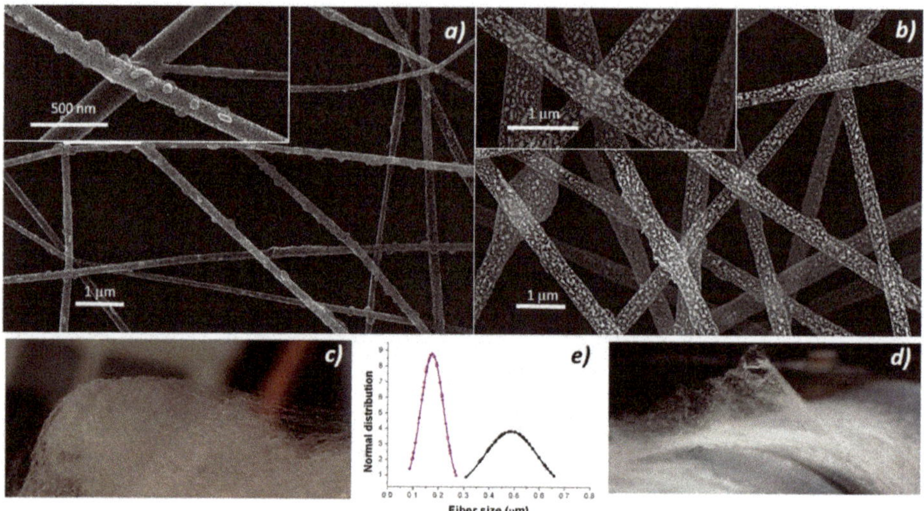

Figure 2. SEM micrographs of H$_2$TPP-PsB-MGC (**a**) and PsB-MGC (**b**) [23] and their respective pictures placed under (**c,d**). Diameter distribution graph (**e**) of H$_2$TPP-PsB-MGC (purple) (**a**) and PsB-MGC fibers (black) (**b**).

Figure 3 shows a magnified TEM micrograph focusing on a single H$_2$TPP- PsB-MGC fiber. The shape is regular but the heterogeneity of the fiber seems to be confirmed by areas with different contrast among the diverse nanoaggregates. For ADF-STEM imaging, the contrast (brightness/darkness) is approximately proportional to the square of the averaged atomic number projected in beam direction z and it depends linearly on the thickness [41] (in particular, the bigger is the atomic number, the brighter is the image). The bright regions inside the polymer/porphyrin fiber could be due to Br$^-$, counterion to CTA$^+$ (Cetyltrimethylammonium) in the surfactant. The cationic surfactant was used to decrease the aggregation of MGC particles and improve their solubility/stability in the polymer matrix. Thus, the brightest area (a higher scattering, i.e., higher intensity in the image) distributed along the inner part of the fiber is supposed to be consequently and indirectly related also to graphene dispersion. As concerns the distribution of both polymers (PS and PHB), the electrospun nanofibers were expected to result in a bulk matrix mainly composed of one of the two polymers hosting an approximately inhomogeneous dispersion of the second polymer, as a consequence of the poor polymer–polymer miscibility. According to the literature, when two polymers are soluble in the same solvent but incompatible with each other, they solidify in different domains [42] during fiber formation [43]. In the inset of Figure 3, the EDXS chemical map is shown, as obtained from C-K (blue) and O-K peaks (green). Oxygen looks to be more concentrated at the surface of the fiber, leading us to suppose a higher presence of PHB at the surface (carbonyl, hydroxyl and ether groups). On the other hand, the distribution of the porphyrin can hardly be highlighted by this technique since H$_2$TPP is also substantially constituted by C atoms with the exception of the four N atoms in the core of the macrocycle (but having close atomic number); thus, the contrast is too low to recognize the porphyrin. A high affinity between PS and H_2TPP ($\pi-\pi$ stacking of the aromatic rings) is reported in literature where the homogeneous dispersion of the porphyrin in the surface and inside PS electrospun fibers is described using fluorescence microscopy (homogeneously red colored fibers) [28]. A weak $\pi-\pi$ stacking could also occurs between graphene flakes surface and porphyrin molecules [44]. Therefore, porphyrin could be fairly dispersed among PS chains and MGC nanofillers in fiber inner part, and PHB arranged to the outermost part.

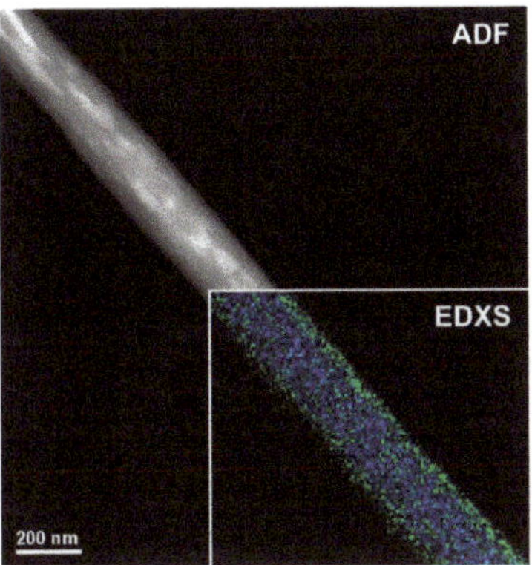

Figure 3. Annular dark field mode-scanning - transmission electron microscopy (ADFM-STEM) image of a porphyrin doped fiber. The inset shows the corresponding energy dispersive X-ray spectroscopy (EDXR) chemical map from carbon (blue) and oxygen (green).

The UV-Vis diffuse reflectance (R%) spectrum of a H_2TPP-PsB-MGC fibrous layer (Figure 4) showed the characteristic features of the H_2TPP chromophore, with the Soret (reflectance minimum about 2.5% at 415 nm) and Q bands well defined (VI: 516 nm, R: 13%; III: 550 nm, R: 19%; II: 591 nm, R: 23%; I: 648, R: 22%). Although the Soret band showed the expected broadening in the solid state, the absence of wavelength shifts seemed to indicate that the porphyrin could be well dispersed in the polymeric matrix. This hypothesis is also supported by the narrow Q bands.

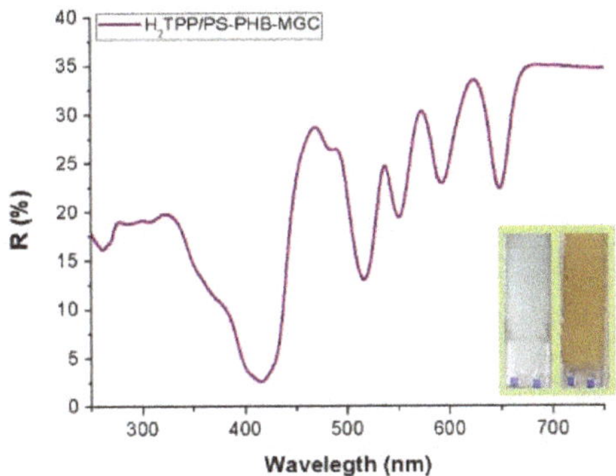

Figure 4. Diffuse reflectance ultraviolet-visible (DR-UV-Vis) spectrum of a H_2TPP PsB-MGC thick fibrous layer (the orange one in inset). Inset shows also a porhyrin-free fibrous coating (the white-grey one).

The nanocomposite fibrous layer, comprising many interfaces between each component. was expected to be an intriguing system for the development of chemical sensors, due to both the wide adsorption surface and the surface energy potentials involved [45]. Optical microscope pictures (Figure 5b) depicted a Pt-Ti microtransducer coated with 2 min-deposited fibers, which appeared optically transparent and with some small black MGC aggregates spread inside fibers, suggesting that the graphene distribution was not completely homogeneous but enclosed within the polymer wires. Increasing the deposition time, a thicker layer was obtained, but the adhesion resulted inhomogeneous and it was more easily peelable from the transducer. To measure the electrical parameters of the resulting chemiresistor, both at room and increasing temperatures up to 70 °C, each IDE, during the current vs. voltage measurements, was positioned onto a customized micro-heater fabricated on alumina substrate. The supplied voltage ranged between −4 V and +4 V. Current–voltage curves displayed a quasilinear relationship between the current changes and the imposed increasing voltage values. Indeed, both PS and PHB, being thermoplastics, can be heated to their melting point (T_{M_PS}: 240 °C; T_{M_PHB}: 175 °C), cooled, and reheated again without substantial degradation. The experimental melting point of porphyrin is also high enough (T_{M_H2TPP}: ≥300 °C [46]) to allow the sensor to work properly in the established range. However, further heating involved an initial increase in current followed by a slow and irreversible increase in resistance, probably due to an irreversible arrangement of MGC inside fibers. For this reason, all experiments were carried out up to 70 °C. To investigate the contribution of the porphyrin to the chemiresistor electrical features, the same amount of graphene was used [23] to produce the fibers with and without porphyrin: indeed, upon the addition of porphyrin, MGC final mass percentage resulted about 1%. However, at room temperature (25 °C), porphyrin chemiresistor was more resistive (about 1.2×10^8 MΩ) than the porphyrin-free one (about 3.4×10^6 MΩ). The current inside fibers is supposed to occur by tunneling of electrons among MGC particles through a small insulating barrier (percolation theory). The electrical properties of such fibrous layers depend on both the quantity and the quality of the "texture" covering the electrodes. Therefore, the resulting measured electrical resistance is related to the individual fiber resistance (due to its dimension and shape), the fiber density (number of fibers per unit of surface area) and the electrode coverage. In porphyrin-fibers, nanofillers appeared distributed within the inner part of fibers, whereas the MGC aggregation inside PsB-MGC was supposed outer, then closer to the IDE metal-electrodes and with a smaller polymer barrier. A reason of this effect could be due to the arrangement of MGC inside fibers due to the porphyrin addition. Furthermore, it is known that conductivity can increase more than one order of magnitude when fiber diameter increases [47]. Porphyrin fibers were estimated to be much thinner than PsB-MGC one. On the other hand, thinner fibers are commonly preferred for sensor applications, since smaller-diameter wires are expected to have a faster response associated with a quicker diffusion of gas molecules through the fiber. Additionally, the collected porphyrin fibrous coatings showed a lower density of the PsB-MGC nanofibers over the electrodes (i.e., layer more porous, Figure 5b), although it increased with the same deposition time. This result could be a further reason to explain a lower conductivity of H_2TPP fibrous chemiresistor. The electrical signals at 25 and 40 °C were noisy, thus the porphyrin chemiresistor did not seem to work properly. Conversely, increasing the electrode working temperature, current increased considerably, especially when temperature value was set at 70 °C. Consequently, the signal to noise ratio increased too, and the baseline looked more stable.

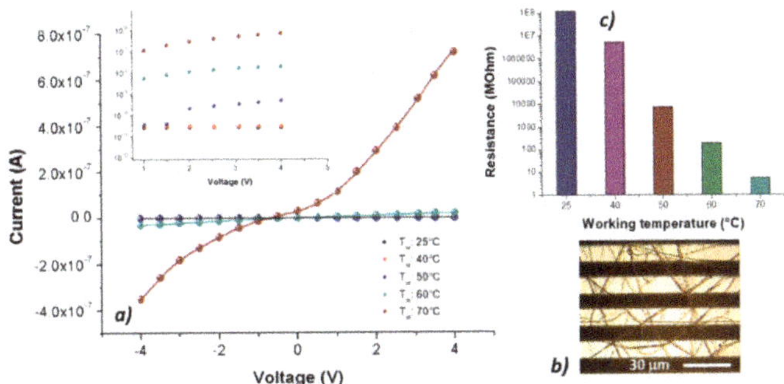

Figure 5. Current vs. voltage diagram at 25, 40, 50, 60 and 70 °C and 25, 40, 50 and 60 °C in the inset (**a**); optical image of the IDE covered by H_2TPP-PsB-MGC fiber (**b**); and resistance values diagram at 25, 40, 50, 60 and 70 °C (**c**).

When the sensor was heated from 25 to 40 °C, the electrical resistance changed by two orders of magnitude (from 10^8 to 10^6 MΩ) (Figure 5c and inset). The exponential decreasing of the resistance values to the increasing 10 °C steps has been reported in the semi-logarithm bar-plot of Figure 5c, where resistance changed from $\approx 7 \times 10^3$ to ≈ 5 MΩ, going from 50 to 70 °C. Such a non-linear dependence of conductivity on the heating apparently confirms the prevalence of the tunneling current (depending on the small dielectric barriers (insulating polymer) between the particles [48]) in comparison with the contact one [49,50], which usually should dominate in highly filled composites in contact with each other. The tunnel contribution is described by Equation (1):

$$\rho_T = e^{\left(\frac{\pi w}{2}\sqrt{\frac{2mV_0}{\left(\frac{h}{2\pi}\right)^2}}\right)}$$

(1)

where ρ_T is the tunnel resistivity, m is the electron rest mass, h is Planck's constant, V_0 is the height of the barrier, and w is its width [51]. Commonly, in a polymer nanocomposite matrix, the gap among nanofillers tends to increase with temperature due to the polymer phase volume expansion (i.e., polymer crystalline phase melting [52]) or the amorphous phase softening [53] during melting, resulting in a resistivity rise of several decades. In the fibrous chemiresistors here described, with and without porphyrin [23], the different results could be explained by the phenyl group rotation in polystyrene (polymer backbone chain reorientation [54]), which could favor strong π–π interactions existing between aromatic organic molecules and the basal plane of MGC. Such a rearrangement could be responsible for the connectivity improvement of the conductive network of the nanofillers, inducing the enhancement of the electrical conductivity [55]. The interfacial force between graphene and PS could be enhanced by surface modification, which reduced the interfacial thermal resistance and dispersed graphene more uniformly [56]. A significant contribution to the MGC rearrangement inside fibers could also be generated by the aromatic planes of H_2TPP facing graphene surfaces. Indeed, comparing both the fibrous layers resistance values, H_2TPP-PsB-MGC became less resistive than PsB-MGC, when the working temperature reached and went over 60 °C (Figure 6), notwithstanding the disadvantaged parameters listed above of the porphyrin layers to be conductive. More specifically, its resistance value reached ≈ 5.7 MΩ versus ≈ 81.6 MΩ of PsB-MGC measured at the same temperature (T_w: 70 °C) in dry and clean air. Furthermore, the steep slope of H_2TPP-PsB-MGC curve (Figure 6) could also be affected by the temperature effects on charge transport among porphyrin that is strongly temperature dependent [57]. It means that at lower temperature H_2TPP could work as barrier while at higher temperature it should promote conduction.

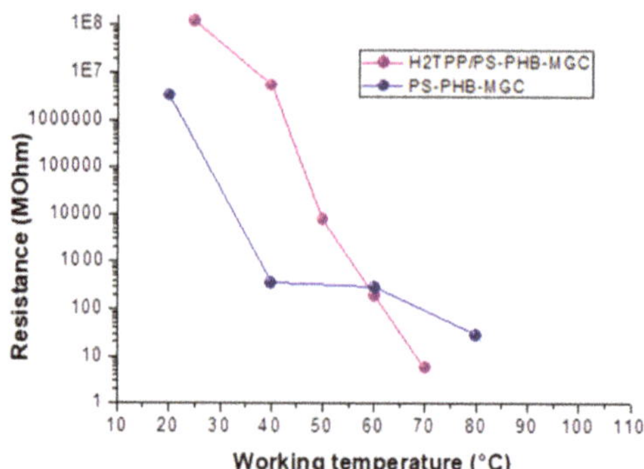

Figure 6. Resistance vs. working temperature for H_2TPP-PsB-MGC (purple) and PsB-MGC (blue) fibers.

Vapor measurements were carried out according to a dynamic mode: the sensor was exposed to a gaseous stream of molecules, the content of which was ruled by blending a stream of pure air with a second stream saturated with the vapor to be analyzed. Thus, the fibrous chemosensor was firstly deployed to increasing percentages of water vapors, ranging 0–50% with increments of 10%, and with working temperature values set at 40, 50 and 60 °C. Both transient responses and response curves are depicted in Figure 7a,b. The current, reported as I/I_0 (I: the current value; I_0: starting current value under clean air) linearly increased when humidity percentage increased, too. A reason for this positive trend was the presence of the cationic surfactant inside the fibers that facilitated the dispersion of the carbon nanostructures [58]. Another reason could be related to the structure of the nanofillers, having here a mesoporous configuration, capable of easily entrapping water molecules and then interacting by the oxygen atoms making part of the framework of each MGC sheet [23]. When the chemosensor (T_w: 50 °C) was exposed to 50% relative humidity, it became seven times more conductive than the dried one. An inverse relationship occurred when working temperature increased: under 50% RH (relative humidity), the sensor became six (60 °C) and three times (70 °C) more conductive than under dry air, therefore generally the sensor responses to water vapors decreased with temperature. The related sensitivity values, defined as the change of measured signal per analyte concentration unit, i.e., the slope of the sensor responses graph [59] (Figure 7c), showed a decrease of 64% and 78% when the sensor worked, respectively, at 60 and 70 °C. Sensitivity is a key parameter in sensor design, because it represents an index related to the sensor ability to "capture" an analyte. In fact, such a fibrous layer was designed to be scarcely affine to water molecules, having two hydrophobic polymers (PS and PHB) and planar structures of H_2TPP and MGC preferring π–π interactions (despite H-bonds due to MGC structural defects and a cationic surfactant water soluble). Further, they were insoluble in H_2O, meaning that the resulting fibers could be exposed to a wide range of relative humidity percentages without undergoing structural changes. The lowering of the affinity index to vapor due to the heating should be presumably caused by the decrease of the vapor molecules diffusion mainly caused by the backbone polymer chains motion [60,61]. In addition, in agreement with the kinetic theory of matter, the sorbed molecules (H-bonds and Van der Waals forces), gaining kinetic energy when heated, were able to "fly" out of the binding site. Indeed, furnishing the sufficient kinetic energy to desorb the "captured" material is the usual strategy to restore/regenerate an adsorbent [62]. A very similar behavior is reported for PsB-MGC nanosensors [23].

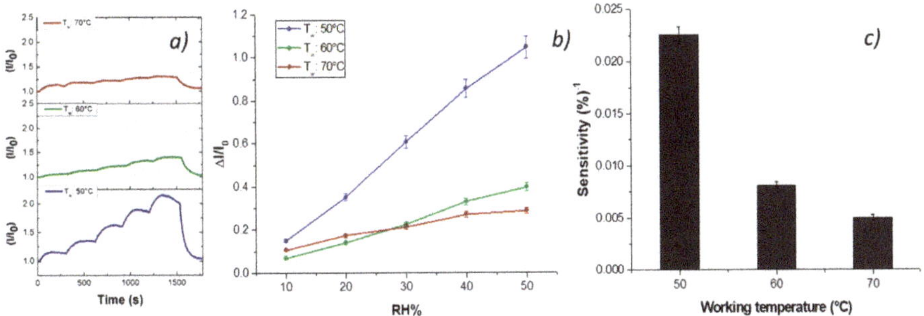

Figure 7. Normalized current (I/I_0) versus time during water vapor measurement (10%, 20%, 30%, 40% and 50% RH) at 70, 60 and 50 °C (**a**); $\Delta I/I_0$–RH percentage diagram (response curves) (**b**); and sensitivity of the H_2TPP-PsB-MGC electrode at the different working temperatures (**c**).

To investigate a potential role of the porphyrin as a selective sensing agent, the fibrous chemosensor was deployed to toluene and acetic acid vapors, the two chemicals that, among the VOC chemical classes previously tested, reported the lowest and the highest, respectively, affinity to PsB-MGC fibers. Additionally, when working temperature increased, the PsB-MGC sensitivity values decreased greatly, making that sensor suitable to work at room (or close to room) temperature for VOC detection. Furthermore, the interest in testing both the chemicals is reinforced by the fact they have proved to be common organic indoor air pollutants [63] and extremely toxic by inhalation [64,65]. Finally, both acetic acid and toluene, being solvents of PHB and PS, respectively, were expected to be able to easily penetrate inside fibers. However, the analyzed vapors flowed in low concentrations (acetic acid and toluene up to 1000 and 1200 ppm, respectively) to avoid the poisoning of the sensing layers. Such a missing effect (i.e., the poisoning) was supported by no change in the baseline of the transient sensor responses, confirming that no chemical interaction had resulted in a permanent variation of the polymer structure. Thus, known concentrations of toluene vapors were generated and flowed throughout the sensor measuring chamber, and the related electrical changes are depicted in Figure 8. The shape of the transient responses (Figure 8a) pointed out quick responses to toluene at each temperature, ranging between 50 and 70 °C, suggesting a Langmuir-like kinetics and reaching the plateau in a few minutes. However, unexpectedly, the sensor response (current values) increased by heating (Figure 8), reporting the highest sensitivity at 70 °C (S_{Tol70}: 1.76×10^{-3} ppm^{-1}; S_{Tol50}: 9.63×10^{-5} ppm^{-1}). The exposure to acetic acid vapors also induced fast responses, linearly related to increasing concentrations of the sample (Figure 9). At higher temperature, sensitivity to acetic acid increased too (S_{AcAc70}: 14×10^{-3} ppm^{-1}; S_{AcAc50}: 4.4×10^{-4} ppm^{-1}) according to an exponential rate (Figure 9d,e). Further, a better signal/noise ratio was gathered as the temperature increased.

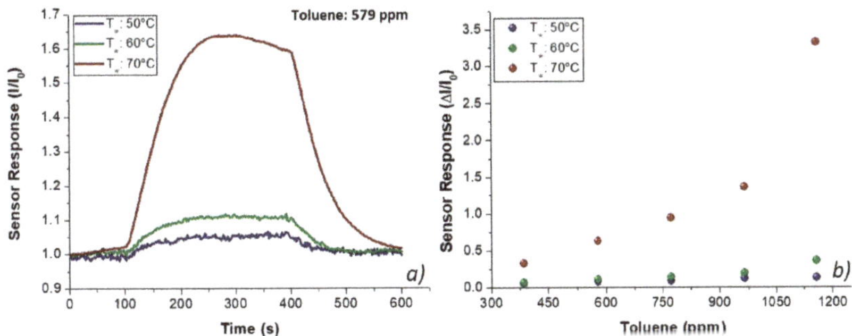

Figure 8. Transient responses (I/I_0 versus time) upon injection of 579 ppm of toluene in dry air (**a**); and response curves ($\Delta I/I_0$) to different toluene concentrations at increasing temperature values (50, 60 and 70 °C) (**b**).

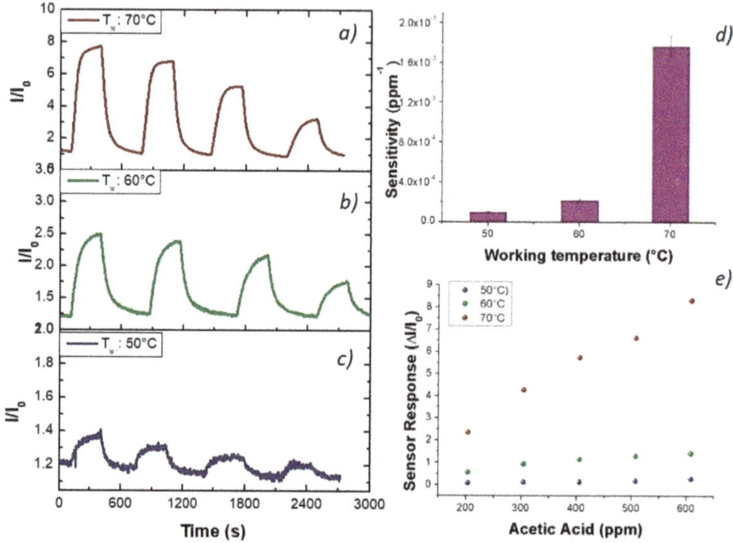

Figure 9. Transient responses (I/I_0 versus time) upon injection of different concentration of acetic acid in dry air at: 70 °C (**a**); 60 °C (**b**); and 50 °C (**c**); sensitivity dependence on the working temperature (**d**); and response curves ($\Delta I/I_0$) to increasing acetic acid concentrations and increasing temperature (**e**).

Figure 10 depicts, in semi-logarithm scale, a comparison among the sensitivities to the selected chemicals, enhancing the different affinity of the material to each analyte, the same positive trend for the VOCs and the divergent trend (i.e., sensitivity decreasing) for water vapors. Therefore, going from 50 to 70 °C, the sensitivity to water vapors became five times lower; conversely, to acetic acid and toluene, it was about 32 and 18 times greater, respectively. Another significant sensor feature is the limit of detection (LOD) defined as the lowest concentration of the analyte that can be detected by the sensor under given conditions, particularly at a given temperature. Thus, LOD_{AcAc70} and LOD_{Tol70} (three standard deviations of the blank) were lowered up to ~1 and ~3 ppm, respectively.

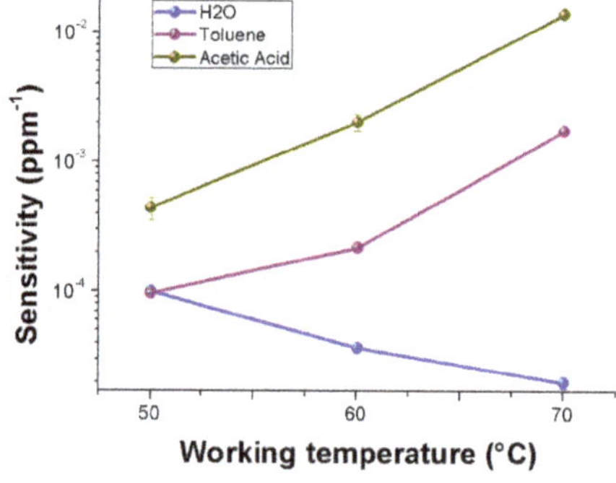

Figure 10. Sensitivity values changes to water vapors, toluene and acid acetic, respectively, depending on the sensor working temperature.

Porphyrin contribution to the sensor features was highlighted by the shape of the normalized response rate, which described the current variation per ppm in time during the exposure to the VOCs flow (Figure 11a). When toluene molecules kept in touch with the fibrous surface within the first 60 s (60 °C), a seven times higher response rate to toluene was measured for H_2TPP-PsB-MGC sensor than PsB-MGC (specifically from $6.75 \times 10^{-4} \pm 1.95018 \times 10^{-5}$ ppm^{-1} s^{-1} to $4.90 \times 10^{-3} \pm 6.27 \times 10^{-5}$ ppm^{-1} s^{-1}). This temperature value was chosen because the current values of both sensors were comparable (Figure 11). The kinetic adsorption profile followed a classical Langmuir profile in both sensors but with different magnitude and adsorption rate. Since both sensors were operating at the same temperature and same toluene concentration, the main parameters involved in the sensor response were expected to be related to the number of the available binding sites and the adsorption energy (features defining the affinity between analyte and surface) [66]. Additionally, toluene could efficiently mediate the electron transfer between porphyrin and MGC. On the other hands, this organic compound could provide conformational changes of both macromolecules and hosting polymer chains (PS), thus contributing to the redistribution of the graphene network, which is responsible for the charge flow. When the sensors were exposed to acetic acid, the H_2TPP inside fibers apparently was responsible for the changes in both response magnitude and adsorption curve shape, Langmuir-like and Henry-like to with and without porphyrin sensor, respectively (Figure 11), indicating a higher affinity of porphyrin fibers to the analyte. Both transient response and calibration curves are the results of the ad/absorbing processes that depend on the chemical affinity of the VOCs to the material. The nanofibrous thin film can be considered as a complex and heterogeneous system where MGCs, and their arrangement through the fibers, are responsible for the resulting electrical features. Thus, the analytes have to be able to diffuse through the polymer–porphyrin matrix to be adsorbed onto the mesopores (or structural defects) and/or the planar surfaces of graphene, determining the changes in the charge density. The mesoporous structure could work as nucleation center for entrapping and growing molecules, such as AcAc, with multiple functional groups. Simultaneously, specific hydrogen bond interaction with the four nitrogen atoms, arranged in the core of the macrocycle structure, could occur [67]. Since all VOCs induced a rise in current, the effect on network distribution inside fibers could be considered the dominant one. The active contribution of porphyrin to the VOCs adsorption and to the electrical mechanisms is visualized in Figure 11b,d, whereas the changes of sensitivities in temperature for both sensors enhanced the opposite curve trend: sensitivity decreasing by heating in PsB-MGC sensor and conversely increasing in H_2TPP-PsB-MGC.

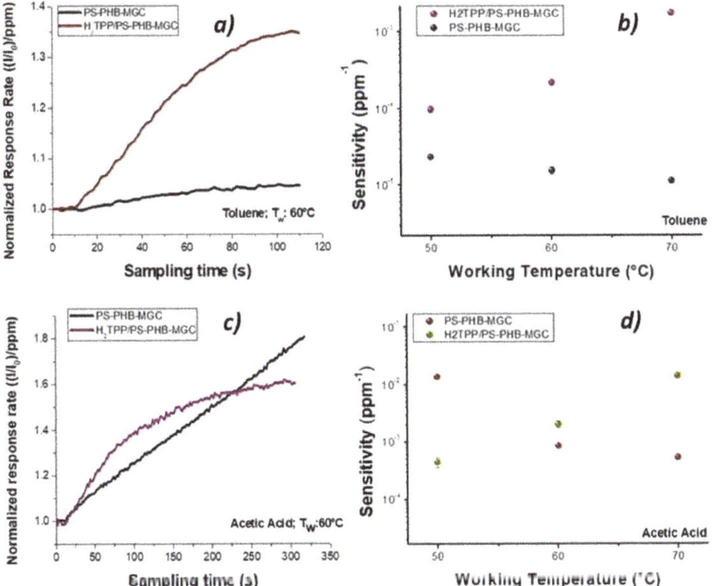

Figure 11. Comparison of the normalized response rate (**a,c**) and sensitivities (**b,d**) of H_2TPP-PsB-MGC and PsB-MGC to toluene (**a,b**) and acetic acid (**c,d**), respectively.

4. Conclusions

The present study reported the development of a conductive nanofibrous and nanocomposite polymer sensor combined with a free-base tetraphenylporphyrin, having the role of driving the selectivity and sensitivity of the polymer layer. Electrospinning technology allowed, in a single step and for 2 min, the fabrication of a pink-colored and highly porous layer adhering to the surface of the electrode. The sensor was able to work in a stable and reproducible way between 50 and 70 °C without any significant degradation, and revealing non-linear relationships between the conductivity and the temperature. The electrical conductivity increased when temperature increased, presumably due to the improving of the connectivity of the MGC networks. The effects of the porphyrin appeared significantly in the morphology of fibers (which were smoother and thinner than the porphyrin-free fibers) and in the electrical features. In fact, H_2TPP-Ps-MGC resulted more resistive at lower temperature, but became much more conductive than PsB-MGC when the chemiresistor worked at 60–70 °C, due to the rearrangement of MGC through the polymer fiber and presumably favored by the aromatic planes of H_2TPP facing graphene surfaces and phenyl groups of PS. It means that, at a higher temperature, H_2TPP tends to promote the fibrous layer conductivity. Furthermore, porphyrin not only increased the sensor sensitivity to toluene vapor (i.e., adsorption and diffusion favored), which was not revealed by PsB-MGC, but also increased with increasing temperature, differently from what occurred in PsB-MGC, whereas the sensitivity to VOCs decreased with heating. Further studies are needed to understand the whole mechanism of ad/absorption occurring between MGC–polymer–porphyrin and the VOCs/gas as well as the role of each polymer inside the fibers when the working temperature changed. However, this preliminary study suggests that this complex and nanostructured polymer matrix is expected as a challenging tool, where sensitivity and selectivity, now driven by temperature and a free-base porphyrin, would be designed and ruled taking into account a series of new combinations to create polymer composite sensors able to work alone or in array, at low cost, with fast responses, easy to be produced in large-scale and to be applied for multifaceted environments.

Author Contributions: Conceptualization, R.P., C.D.N. and A.M; Data curation, A.M.; Formal analysis, J.A. and C.D.N.; Investigation, J.A., G.B., F.D.C., G.S.-M. and A.M.; Methodology, J.A., E.Z. and A.M.; Resources, R.P., E.Z., G.B., F.D.C. and G.S.-M; Supervision, A.M.; Validation, J.A.; Writing—original draft, A.M.; and Writing—review and editing, all authors.

Funding: This research was funded by a 2-Year National Project, BRIC ID.12 2016—National Institute for Insurance against Accidents at Work (INAIL), titled: "Design and development of a sensory system for the measurement of volatile compounds and the identification of job-related microorganisms".

Acknowledgments: The authors gratefully thank G. Tranfo of National Institute for Insurance against Accidents at Work for her crucial support in the BRIC ID.12 2016 Project and A.R. Taddei of High Equipment Centre, Electron Microscopy Section, University of Tuscia (VT), Italy, for providing SEM micrographs. Finally, the authors thank A. Capocecera for his technical support in developing customized software for laboratory measurements.

Conflicts of Interest: The authors declare no conflict of interest.

References

1. Bhattacharya, S.; Agarwal, A.K.; Chanda, N.; Pandey, A.; Sen, A.K. (Eds.) *Environmental, Chemical and Medical Sensors*; Springer: Singapore, 2018; ISBN 978-981-10-7751-7.
2. Tuantranont, A. Applications of Nanomaterials in Sensors and Diagnostics. In *Nanomaterials for Sensing Applications: Introduction and Perspective*; Springer: Berlin/Heidelberg, Germany, 2013; ISBN 978-3-642-36025-1.
3. Banica, F.G. Nanomaterials applications in chemical sensors. In *Chemical Sensors and Biosensors: Fundamentals and Applications*; John Wiley & Sons Ltd.: Chichester, UK, 2012; p. 135, ISBN 9780470710661.
4. Salvagione, H.J.; Diez-Pascual, A.M.; Lazaro, E.; Vera, S.; Gomez-Fatou, M.A. Chemical sensors based on polymer composites with carbon nanotubes and graphene: the role of the polymer. *J. Mater. Chem. A* **2014**, *2*, 14289–14328. [CrossRef]
5. Macagnano, A.; Perri, V.; Zampetti, E.; Bearzotti, A.; De Cesare, F. Humidity effects on a novel eco-friendly chemosensor based on electrospun PANi/PHB nanofibres. *Sens. Actuators B Chem.* **2016**, *232*, 16–27. [CrossRef]

6. Morris, J.E.; Iniewski, K. *Graphene, Carbon Nanotubes and Nanostructures (Techniques and Applications)*; CRC Press: Boca Raton, FL, USA, 2013; ISBN 1466560568.

7. Quercia, L.; Loffredo, F.; Francia, G.D. Influence of filler dispersion on thin film composites sensing properties. *Sens. Actuators B Chem.* **2005**, *109*, 153–158. [CrossRef]

8. Grate, J.W.; Abraham, M.H.; Wise, B.M. *Design and Information Content of Arrays of Sorption-Based Vapor Sensors Using Solubility Interactions and Linear Solvation Energy Relationships*; Springer: New York, NY, USA, 2009.

9. Feller, J.F.; Castro, M.; Kumar, B. Polymer–carbon nanotube conductive nanocomposites for sensing. In *Polymer–Carbon Nanotube Composites Preparation, Properties and Applications*; McNally, T., Pötschke, P., Eds.; Woodhead Publishing: Cambridge, UK, 2011; pp. 760–803. ISBN 978-1-84569-761-7.

10. Lonergan, M.C.; Severin, E.J.; Doleman, B.J.; Beaber, S.A.; Grubbs, R.H.; Lewis, N.S. Array-Based Vapor Sensing Using Chemically Sensitive, Carbon Black—Polymer Resistors. *Chem. Mater.* **1996**, *4756*, 2298–2312. [CrossRef]

11. Briglin, S.M.; Lewis, N.S. Characterization of the Temporal Response Profile of Carbon Black—Polymer Composite Detectors to Volatile Organic Vapors. *J. Phys. Chem. B.* **2003**, 11031–11042. [CrossRef]

12. Chang, C.P.; Yuan, C.L. The fabrication of a MWNTs—Polymer composite chemoresistive sensor array to discriminate between chemical toxic agents. *J. Mater. Sci.* **2009**, 5485–5493. [CrossRef]

13. Bavastrello, V.; Stura, E.; Carrara, S.; Erokhin, V.; Nicolini, C. Poly (2,5-dimethylaniline)–MWNTs nanocomposite: A new material for conductometric acid vapours sensor. *Sens. Actuators B Chem.* **2004**, *98*, 247–253. [CrossRef]

14. Hierlemann, A.; Gutierrez-osuna, R. Higher-Order Chemical Sensing. *Chem. Rev.* **2008**, *108*, 563–613. [CrossRef]

15. Qureshi, A.; Kang, W.P.; Davidson, J.L.; Gurbuz, Y. Diamond & Related Materials Review on carbon-derived, solid-state, micro and nano sensors for electrochemical sensing applications. *Diam. Relat. Mater.* **2009**, *18*, 1401–1420. [CrossRef]

16. Huang, X.; Qi, X.; Boeya, F.; Zhang, H. Graphene-based composites. *Chem. Soc. Rev.* **2012**, *41*, 525–944. [CrossRef]

17. Zheng, D.; Vashist, S.K.; Dykas, M.M.; Saha, S.; Al-rubeaan, K.; Lam, E.; Luong, J.H.T.; Sheu, F. Graphene versus Multi-Walled Carbon Nanotubes for Electrochemical Glucose Biosensing. *Materials* **2013**, *6*, 1011–1027. [CrossRef] [PubMed]

18. Berger, C.; Song, Z.M.; Li, T.B.; Li, X.B.; Ogbazghi, A.Y.; Feng, R.; Dai, Z.N.; Marchenkov, A.N.; Conrad, E.H.; First, P.N.; et al. Ultrathin Epitaxial Graphite: 2D Electron Gas Properties and a Route toward. *J. Phys. Chem. B* **2004**, 19912–19916. [CrossRef]

19. Bunch, J.S.; Verbridge, S.S.; Alden, J.S.; van der Zande, A.M.; Parpia, J.M.; Craighead, H.G.; McEuen, P.L. Impermeable Atomic Membranes from Graphene Sheets. *Nanoletters* **2008**, *8*, 2458–2462. [CrossRef] [PubMed]

20. Kuilla, T.; Bhadra, S.; Yao, D.; Kim, N.H.; Bose, S.; Lee, J.H. Progress in Polymer Science Recent advances in graphene based polymer composites. *Prog. Polym. Sci.* **2010**, *35*, 1350–1375. [CrossRef]

21. Sinha, R.S. (Ed.) *Processing of Polymer-Based Nanocomposites*; Springer International Publishing: Cham, Switzerland, 2018; ISBN 978-3-319-97779-9.

22. Macagnano, A.; Zampetti, E.; Kny, E. (Eds.) *Electrospinning for High Performance Sensors*; Springer International Publishing: Cham, Switzerland, 2015; ISBN 978-3-319-14405-4.

23. Avossa, J.; Zampetti, E.; De Cesare, F.; Bearzotti, A.; Scarascia-Mugnozza, G.; Vitiello, G.; Zussman, E. Thermally Driven Selective Nanocomposite PS-PHB/MGC Nanofibrous Conductive Sensor for Air Pollutant Detection. *Front. Chem.* **2018**, *6*, 1–14. [CrossRef]

24. Esteves, C.H.A.; Iglesias, B.A.; Li, R.W.C.; Ogawa, T.; Araki, K.; Gruber, J. New composite porphyrin-conductive polymer gas sensors for application in electronic noses. *Sens. Actuators B. Chem.* **2014**, *193*, 136–141. [CrossRef]

25. Paolesse, R.; Nardis, S.; Monti, D.; Stefanelli, M.; Di Natale, C. Porphyrinoids for Chemical Sensor Applications. *Chem. Rev.* **2017**, 2517–2583. [CrossRef]

26. Olkhov, A.A.; Karpova, S.G.; Tuhayeva, P.M.; Lobanov, A.V.; Kurnosov, A.S.; Mastalygina, E.E.; Iordanskii, A.L. Supramolecular structure of electrospun ultrathin fibers based on poly-(3-hydroxibutirate) with zinc-tetraphenylporphyrin complex. *AIP Conf. Proc.* **2018**, *2051*. [CrossRef]

27. Shao, L.; Hu, B.; Dong, P.; Ji, W.; Qi, C. Electrospinning Fe(III)porphyrin/TiO2/poly(styrene) mixture: Formation of a novel nanofiber photocatalyst for the photodegradation of methyl orange. *J. Porphyr. Phthalocyanines* **2010**, *14*, 993–999. [CrossRef]

28. Hu, M.; Kang, W.; Zhao, Y.; Shi, J.; Cheng, B. A fluorescent and colorimetric sensor based on a porphyrin doped polystyrene nanoporous fiber membrane for HCl gas detection. *RSC Adv.* **2017**. [CrossRef]

29. Yang, Y.; Wang, H.; Su, K.; Long, Y.; Peng, Z.; Li, N.; Liu, F. A facile and sensitive fluorescent sensor using electrospun nanofibrous film for nitroaromatic explosive detection. *J. Mater. Chem.* **2011**, 11895–11900. [CrossRef]

30. Arai, T.; Tanaka, M.; Kawakami, H. Porphyrin-Containing Electrospun Nano fi bers: Positional Control of Porphyrin Molecules in Nano fibers and Their Catalytic Application. *Appl. Mater. Interfaces* **2012**. [CrossRef] [PubMed]

31. Jang, K.; Baek, I.W.; Back, S.Y.; Ahn, H. Electrospinning of porphyrin/polyvinyl alcohol (PVA) nanofibers and their acid vapor sensing capability. *J. Nanosci. Nanotechnol.* **2011**, *11*, 6102–6108. [CrossRef] [PubMed]

32. García-Pérez, C.; Menchaca-Campos, C.; García-Sánchez, M.A.; Pereyra-Laguna, E.; Rodríguez-Pérez, O.; Uruchurtu-Chavarín, J. Nylon/Porphyrin/Graphene Oxide Fiber Ternary Composite, Synthesis and Characterization. *Open J. Compos. Mater.* **2017**, *7*, 146–165. [CrossRef]

33. Uyar, T.; Besenbacher, F. Electrospinning of uniform polystyrene fibers: The effect of solvent conductivity. *Polymer* **2008**, *49*, 5336–5343. [CrossRef]

34. Bychuk, M.A.; Kil'deeva, N.R.; Kurinova, M.A.; Bogdanov, N.V.; Kalinin, M.V.; Novikov, A.V.; Vikhoreva, G.A. Electrospinning of Biodegradable Polymer Scaffolds. *Fibre Chem.* **2015**, *46*, 345–348. [CrossRef]

35. Fuhrhop, J.H.; Smith, K.M. Laboratory methods. In *Porphyrins and Metalloporphyrins*; Smith, K.M., Ed.; Elsevier Science: Amsterdam, The Netherlands, 1975; pp. 757–869.

36. Garg, K.; Bowlin, G.L. Electrospinning jets and nanofibrous structures. *Biomicrofluidics* **2011**, *5*, 013403. [CrossRef]

37. Zhu, G.; Zhao, L.Y.; Zhu, L.T.; Deng, X.Y.; Chen, W.L. Effect of Experimental Parameters on Nanofiber Diameter from Electrospinning with Wire Electrodes. *IOP Conf. Ser. Mater. Sci. Eng.* **2017**, *230*. [CrossRef]

38. Fong, H.; Chun, I.; Reneker, D.H. Beaded nanofibers formed during electrospinning. *Polymer* **1999**, *40*, 4585–4592. [CrossRef]

39. Lee, J.S.; Choi, K.H.; Ghim, H.D.; Kim, S.S.; Chun, D.H.; Kim, H.Y.; Lyoo, W.S. Role of molecular weight of atactic poly(vinyl alcohol) (PVA) in the structure and properties of PVA nanofabric prepared by electrospinning. *J. Appl. Polym. Sci.* **2004**, *93*, 1638–1646. [CrossRef]

40. Wang, X.; Cao, J.; Hu, Z.; Pan, W.; Liu, Z. Jet shaping nanofibers and the collection of nanofiber mats in electrospinning. *J. Mater. Sci. Technol.* **2006**, *22*, 536–540.

41. Su, D. Advanced electron microscopy characterization of nanomaterials for catalysis. *Green Energy Environ.* **2017**, *2*, 70–83. [CrossRef]

42. Zhong, G.; Wang, K.; Zhang, L.; Li, Z.M.; Fong, H.; Zhu, L. Nanodroplet formation and exclusive homogenously nucleated crystallization in confined electrospun immiscible polymer blend fibers of polystyrene and poly(ethylene oxide). *Polymer* **2011**, *52*, 5397–5402. [CrossRef]

43. Bognitzki, M.; Czado, W.; Frese, T.; Schaper, A.; Hellwig, M.; Steinhart, M.; Greiner, A.; Wendorff, J.H. Nanostructured fibers via electrospinning. *Adv. Mater.* **2001**, *13*, 70–72. [CrossRef]

44. Kim, S.J.; Song, W.; Kim, S.; Kang, M.-A.; Myung, S.; Lee, S.S.; Lim, J.; An, K.-S. Tunable functionalization of graphene nanosheets for graphene-organic hybrid photodetectors. *Nanotechnology* **2016**, *27*. [CrossRef] [PubMed]

45. Gardner, D.J.; Blumentritt, M.; Kiziltas, A.; Kiziltas, E.E.; Peng, Y.; Yildirim, N. Polymer Nanocomposites from the Surface Energy Perspective. *Rev. Adhes. Adhes.* **2013**, *2*, 175–215. [CrossRef]

46. ChemSpider. Search and Share Chemistry. Available online: http://www.chemspider.com/Chemical-Structure.10291672.html (accessed on 16 January 2019).

47. MacDiarmid, A.G.; Jones, W.E., Jr.; Norris, I.D.; Gao, J.; Johnson, A.T., Jr.; Pinto, N.J.; Hone, J.; Han, B.; Ko, F.K.; Okuzaki, H.; et al. Electrostatically-generated nanofibers of electronic polymers. *Synth. Met.* **2001**, 27–30. [CrossRef]

48. Sheng, P. Fluctuation-induced tunneling conduction in disordered materials. *Phys. Rev. B* **1980**, *21*, 2180–2195. [CrossRef]

49. Syurik, J.; Ageev, O.A.; Cherednichenko, D.I.; Konoplev, B.G.; Alexeev, A. Non-linear conductivity dependence on temperature in graphene-based polymer nanocomposite. *Carbon* **2013**, *63*, 317–323. [CrossRef]

50. Gao, C.; Liu, P.; Ding, Y.; Li, T.; Wang, F.; Chen, J.; Zhang, S.; Li, Z.; Yang, M. Non-contact percolation of unstable graphene networks in poly(styrene-co-acrylonitrile) nanocomposites: Electrical and rheological properties. *Compos. Sci. Technol.* **2018**, *155*, 41–49. [CrossRef]

51. Celzard, A.; McRae, E.; Marêché, J.F.; Furdin, G.; Sundqvist, B. Conduction mechanisms in some graphite–polymer composites: Effects of temperature and hydrostatic pressure. *J. Appl. Phys.* **1998**, *83*, 1410–1419. [CrossRef]

52. Feller, J.F.; Linossier, I.; Grohens, Y. Conductive polymer composites: comparative study of poly(ester)—Short carbon fibres and poly(epoxy)—Short carbon fibres mechanical and electrical properties. *Polym. Adv. Technol.* **2002**, *57*, 64–71. [CrossRef]

53. Pillin, I.; Feller, J.F.; Pimbert, S.; Levesque, G. Conductive poly(ester)/poly(olefin)/carbon black composites: Influence of glass transition temperature and crystallinity of the poly(ester) matrix on the electrical properties. *Plast. Rubbers Compos.* **2002**, *31*, 300–306. [CrossRef]

54. Tonelli, A.E. Phenyl Group Rotation in Polystyrene. *Macromolecules* **1973**, *6*, 682–683. [CrossRef]

55. Cao, Q.; Song, Y.; Tan, Y.; Zheng, Q. Thermal-induced percolation in high-density polyethylene/carbon black composites. *Polymer* **2009**, *50*, 6350–6356. [CrossRef]

56. Li, A.; Zhang, C.; Zhang, Y.-F. Thermal Conductivity of Graphene-Polymer Composites: Mechanisms, Properties, and Applications. *Polymers* **2017**, *9*, 437. [CrossRef]

57. Keun, B.; Aratani, N.; Kuk, J.; Kim, D.; Osuka, A.; Yoo, K. Length and temperature dependence of electrical conduction through dithiolated porphyrin arrays. *Chem. Phys. Lett.* **2005**, *412*, 303–306. [CrossRef]

58. Koutsioukis, A.; Georgakilas, V.; Belessi, V.; Zboril, R. Highly Conductive Water-Based Polymer/Graphene Nanocomposites for Printed Electronics. *Chem. Eur. J.* **2017**, 8268–8274. [CrossRef]

59. Kalantar-Zadeh, K. *Sensors: An Introductory Course*; Springer Science & Business Media: New York, NY, USA, 2013; ISBN 978-1-4614-5052-8.

60. Haslam, R.T.; Hershey, R.I.; Kean, R.H. Effect of gas velocity and temperature on rate of absorption. *Ind. Eng. Chem.* **1924**, *16*, 1224–1230. [CrossRef]

61. Ramesh, N.; Duda, J.L. Diffusion in polymers below the glass transition temperature: Comparison of two approaches based on free volume concepts. *Korean J. Chem. Eng.* **2000**, *17*, 310–317. [CrossRef]

62. Technical Bulletin: Choosing an Adsorption System for VOC: Carbon, Zeolite, or Polymers? Available online: https://www3.epa.gov/ttn/catc/dir1/fadsorb.pdf (accessed on 16 January 2019).

63. Salthammer, T.; Uhde, E. (Eds.) *Organic Indoor Air Pollutants: Occurrence, Measurement, Evaluation*, 2nd ed.; WILEY-VCH: Weinheim, Germany, 2009; ISBN 978-3-527-31267-2.

64. Toluene. Available online: https://www.epa.gov/sites/production/files/2016-09/documents/toluene.pdf (accessed on 5 February 2019).

65. TOXNET. Toxicology Data Network. Available online: https://toxnet.nlm.nih.gov/cgi-bin/sis/search2/r?dbs+hsdb:@term+@DOCNO+40 (accessed on 5 February 2019).

66. Ndiaye, A.; Bonnet, P.; Pauly, A.; Dubois, M.; Brunet, J.; Varenne, C.; Guerin, K.; Lauron, B. Noncovalent Functionalization of Single-Wall Carbon Nanotubes for the Elaboration of Gas Sensor Dedicated to BTX Type Gases: The Case of Toluene. *J. Phys. Chem. C* **2013**, *117*, 20217–20228. [CrossRef]

67. Nardis, S.; Pomarico, G.; Tortora, L.; Capuano, R.; D'Amico, A.; Di Natale, C.; Paolesse, R. Sensing mechanisms of supramolecular porphyrin aggregates: a teamwork task for the detection of gaseous analytes. *J. Mater. Chem.* **2011**, *21*, 18638. [CrossRef]

 *nanomaterials*

Article

Nanostructured Hydrogels by Blend Electrospinning of Polycaprolactone/Gelatin Nanofibers

Lode Daelemans [1,†], **Iline Steyaert** [1,2,†], **Ella Schoolaert** [1], **Camille Goudenhooft** [1], **Hubert Rahier** [2] and **Karen De Clerck** [1,*]

1   Department of Materials, Textiles and Chemical Engineering (MaTCh), Ghent University, Technologiepark 907, 9052 Ghent, Belgium; lode.daelemans@ugent.be (L.D.); iline.steyaert@ugent.be (I.S.); ella.schoolaert@ugent.be (E.S.); camille.goudenhooft@univ-ubs.fr (C.G.)
2   Research Unit of Physical Chemistry and Polymer Science, Department of Materials and Chemistry, Vrije Universiteit Brussel, Pleinlaan 2, 1050 Brussels, Belgium; hrahier@vub.ac.be
*   Correspondence: karen.declerck@ugent.be; Tel.: +32-9-2645-740
†   These authors contributed equally to this work.

Received: 25 June 2018; Accepted: 17 July 2018; Published: 20 July 2018

Abstract: Nanofibrous membranes based on polycaprolactone (PCL) have a large potential for use in biomedical applications but are limited by the hydrophobicity of PCL. Blend electrospinning of PCL with other biomedical suited materials, such as gelatin (Gt) allows for the design of better and new materials. This study investigates the possibility of blend electrospinning PCL/Gt nanofibrous membranes which can be used to design a range of novel materials better suited for biomedical applications. The electrospinnability and stability of PCL/Gt blend nanofibers from a non-toxic acid solvent system are investigated. The solvent system developed in this work allows good electrospinnable emulsions for the whole PCL/Gt composition range. Uniform bead-free nanofibers can easily be produced, and the resulting fiber diameter can be tuned by altering the total polymer concentration. Addition of small amounts of water stabilizes the electrospinning emulsions, allowing the electrospinning of large and homogeneous nanofibrous structures over a prolonged period. The resulting blend nanofibrous membranes are analyzed for their composition, morphology, and homogeneity. Cold-gelling experiments on these novel membranes show the possibility of obtaining water-stable PCL/Gt nanofibrous membranes, as well as nanostructured hydrogels reinforced with nanofibers. Both material classes provide a high potential for designing new material applications.

Keywords: biomaterial; biomedical; nanofibers; scaffolds; reinforced; hybrid material; thermal analysis; nanofibrous membranes

---

1. Introduction

Electrospun nanofiber membranes have gained a lot of attention the past few years. Their striking resemblance to the morphology of the extracellular matrix (ECM) made them especially interesting for biomedical applications. Nanofibrous membranes can, unlike structures with porosity on a larger scale, provide an environment closely mimicking the natural ECM by providing appropriate cell binding sites. Cell behavior and functionality can be controlled by man-made nanofibrous scaffolds, which are therefore increasingly being used in tissue engineering and wound-healing applications [1,2]. Solvent electrospinning provides a relatively simple way to produce nanofibers from polymer solutions and is therefore currently the most commonly applied processing technique. To fully exploit the potential of electrospun nanofibrous materials, further research toward the use of less toxic, more economical solvents and a stable electrospinning process suitable for upscaling, is essential.

Polycaprolactone (PCL) is commonly used in biomedical products, ranging from sutures or staples for wound closure to contraceptive devices. This is mainly due to its tunable degradation profile, in combination with low cost, easy production and availability of medical-grade material [3,4]. It is thus not surprising that PCL nanofibers for biomedical applications have also been developed [5–10]. The practical application of PCL nanofibers in the biomedical field, however, is severely limited due to the hydrophobic properties adversely affecting cell affinity. Although the nanofiber morphology closely resembles the ECM, its complexity in terms of composition, biological functionality and biophysical properties makes it difficult to reproduce using single polymer systems. Blending synthetic and natural polymers offers the opportunity to tune physical properties and bioactivity while minimizing the disadvantages of both polymer components, thus more closely resembling the native ECM [2,11–13]. Compared to the use of copolymers or modified nanofibers, blending offers an alternative approach that is simple and economical [14].

A wide variety of natural polymers are available for biomedical applications, including several polypeptides and polysaccharides. They provide the desired biological properties in a blend and are biodegradable. Gelatin (Gt) is a commercial polypeptide widely available in medical-grade material and very suitable for biomimetic applications since it displays many integrin binding sites for cell adhesion and differentiation [15–19]. Although it is possible to produce pure Gt nanofibers [20], blending with PCL can give rise to superior mechanical properties [21–23], increasing its application potential.

Our previous work showed that Gt nanofibers are cold-water-soluble due to their high surface-to-volume ratio which promotes water penetration and dissolution. They can be used as an instant cold gel. This presents opportunities for PCL/Gt blended nanofibrous membranes to design new hybrid materials for biomedical applications. In this article, we develop a non-toxic and economical solvent system for the production of PCL/Gt blend nanofibers by electrospinning. An acetic acid (AA)/formic acid (FA) solvent system is selected based on our previous work in which pure PCL and pure Gt nanofibers were produced [5,20]. Several PCL/Gt blend nanofibrous membranes with varying compositions are electrospun and their morphology is characterized. Cold-gelling experiments are performed on nanofibrous membranes with varying PCL/Gt ratio to study its effect on the gel formation of the Gt component and the resulting structure of these novel materials.

2. Materials and Methods

2.1. Materials

Polycaprolactone (average M_n of 80 kg·mol^{-1}) was purchased from Sigma-Aldrich (Overijse, Belgium). Commercial gelatin isolated from pigskin by the acidic process (type A pharma-grade, 300 Bloom) was kindly supplied by Rousselot, Ghent, Belgium. Both polymers were used as-received for the electrospinning process. AA (99.8 vol%) and FA (98.0 vol%) were supplied by Sigma-Aldrich (Overijse, Belgium).

Efforts in optimization also included testing of a lower molecular weight PCL (average M_w 14 kg·mol^{-1}) purchased from Sigma-Aldrich (Overijse, Belgium) and a gelatin isolated from bovine bones by the alkaline process (type B pharma-grade, 260 Bloom) kindly supplied by Rousselot, Ghent, Belgium.

2.2. Electrospinning

The electrospinning process allows for easy production of nanofibers using polymer solutions, and is more thoroughly described for PCL and gelatin in our previous papers [5,20]. Blend nanofibers are obtained when the solution used for electrospinning contains two (or more) polymer components.

The PCL/Gt electrospinning solutions were obtained by simultaneous dissolution of PCL and gelatin, allowing for variations in solvent ratio, total polymer concentration and polymer ratio. The electrospinning solutions were characterized prior to use by their viscosity and conductivity. Viscosity was measured using a Brookfield viscometer LVDV-II (standard deviation was on average

8%, Brookfield, Middleboro, MA, USA) and conductivity was determined by a CDM210 conductivity meter of Radiometer Analytical (standard deviation was on average 11%, Düsseldorf, Germany).

All electrospinning trials for process optimization were performed using a mono-nozzle setup, leading to a scalable electrospinning process. Nanofibrous membranes for subsequent analysis were electrospun using a multinozzle or rotating drum setup at low speed [24,25], already providing upscaled samples. The latter samples were electrospun using the AA/FA solvent system (solvent ratio as specified in the text), a polymer concentration of 13 wt% (PCL/Gt ratio as specified in the text), a flow rate of 1 mL·h^{-1}, a tip-to-collector distance of 12.5 cm and the applied voltage adapted for stable electrospinning.

The electrospun samples were examined by scanning electron microscopy (FEI Quanta 200 F, Thermo Fisher Scientific, Eindhoven, The Netherlands) at an accelerating voltage of 20 kV. Sample preparation was done using a sputter coater (Emitech SC7620, coating with Au, Rotterdam, The Netherlands). The nanofiber diameters were measured using ImageJ version 1.48, https://imagej.net. The average fiber diameters and their standard deviations are based on 50 measurements per sample.

2.3. Water Treatment of PCL/Gt Blends

Water treatment of PCL/Gt blend nanofibrous membranes was done by washing the samples in a large quantity of demineralized water at 35 °C (maximal temperature without affecting PCL nanofibrous structure) for 30 s. Four repetitions were done before further drying at ambient temperature for over 24 h.

2.4. Materials Characterization

Solution cast films (prepared by solution casting in Teflon evaporation dishes of the electrospinning solution) and nanofibrous membranes were characterized using a Fourier Transform Infrared (FTIR, Thermo Fisher Scientific, Eindhoven, The Netherlands) spectrometer with Attenuated Total Reflectance (ATR) accessory (diamond crystal) from Thermo Scientific. The spectra were recorded in the range 4000–400 cm^{-1} with a resolution of 4 cm^{-1}. 32 scans were averaged for each spectrum. The films were prepared by sampling the electrospinning solution and solution casting onto Teflon evaporation dishes. All nanofibrous membranes were measured as-spun. Additionally, PCL/Gt nanofibrous membranes were also measured after (partial) removal of gelatin by water treatment as detailed above.

Differential scanning calorimetry (DSC) measurements were performed using a TA Instruments (Asse, Belgium) Q2000 equipped with a refrigerated cooling system and using nitrogen as purge gas (50 mL·min^{-1}). The instrument was calibrated using Tzero technology, including a temperature calibration with indium. DSC measurements were performed on samples of 3.00 ± 0.30 mg conditioned at 65% RH for 24 h, enclosed in hermetic Tzero aluminum crucibles and at a heating rate of 2.5 K·min^{-1}. The standard deviation of the melting enthalpy of PCL nanofibers was 5%.

Dynamic Mechanical Analysis (DMA) measurements were done using a TA Instruments Q800 (Asse, Belgium), equipped with liquid nitrogen cooling (LNCS) and film tension clamps. Calibration was done according to a manufacture-defined procedure. Temperature calibration was performed by means of the melting transition of gallium. Films and nanofibrous membranes were measured in film tension mode, using a Poisson's ratio of 0.44, a frequency of 1 Hz, a strain of 0.5%, a static force of 0.01 N and a heating rate of 2.5 K·min^{-1}. The measurements were performed on as-spun nanofibrous membranes at room humidity (40 ± 10% RH). The samples were not heated above 30 °C to avoid melting of the PCL component.

3. Results and Discussion

3.1. Electrospinning of PCL/Gt Blend Nnanofibers Using AA/FA

Our previous investigation of the pure components illustrated that both PCL and gelatin are electrospinnable when dissolved in AA/FA, and this for several concentrations and solvent ratios [5,20].

There is a significant overlap of solution and processing parameters for the stable electrospinning of the pure polymers. Based hereon, a total polymer concentration of 13 wt% and a 70/30 AA/FA solvent system were chosen as a starting point for the blend electrospinning (tip-to-collector distance (TCD)) of 12.5 mm, flow rate of 1 mL·h^{-1}. This solvent choice minimizes the amount of FA, which minimizes degradation of PCL [5], while guaranteeing electrospinning of fine bead-free PCL and gelatin nanofibers.

As illustrated in Figure 1, the full range of PCL/Gt blend ratios is well electrospinnable using the chosen parameters. Indeed, uniform bead-free PCL/Gt nanofibers could be produced in a stable manner for all compositions. Although the conductivity of the electrospinning solution increases significantly, and the viscosity markedly drops with increasing gelatin content (Figure S1), no remarkable differences in nanofiber diameter were observed. Variations in total polymer concentration with a fixed polymer ratio, on the other hand, do affect fiber diameter. Electrospinning of a 50/50 PCL/Gt blend solution results in a stable process and bead-free nanofibers for polymer concentrations between 9 wt% and 17 wt%. Within this electrospinnable window, the nanofiber diameter could be tuned between 140 nm and 550 nm respectively (Figure 2). Blend electrospinning of PCL/Gt using the AA/FA solvent system is thus highly flexible, with both the PCL/Gt ratio and the fiber diameter adjustable to the end application.

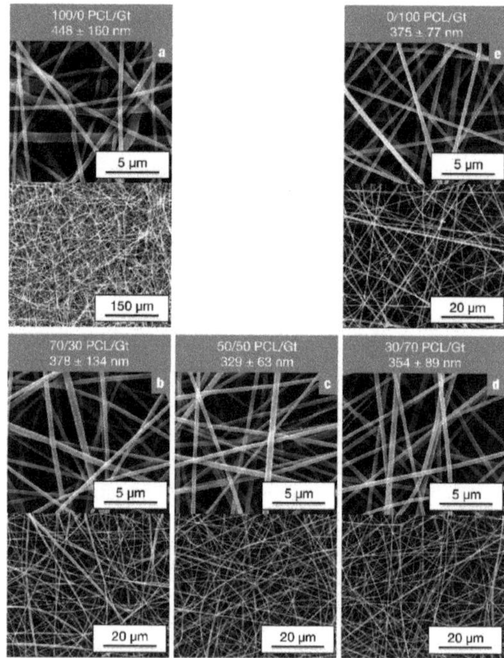

Figure 1. Scanning electron microscopy (SEM) images of PCL/Gt blend nanofibers electrospun using a total polymer concentration of 13 wt% and 70/30 AA/FA (TCD of 12.5 cm, flow rate of 1 mL·h^{-1} and voltage adjusted in the range of 15–20 kV for stable electrospinning): (**a**) 100/0 PCL/Gt, (**b**) 70/30 PCL/Gt, (**c**) 50/50 PCL/Gt, (**d**) 30/70 PCL/Gt and (**e**) 0/100 PCL/Gt. All PCL/Gt blend ratios were well electrospinnable.

Although this clearly did not impede electrospinnability, all the PCL/Gt blend solutions were opaque due to a phase separation resulting in emulsions. The homogeneous emulsions gradually evolved toward two completely separated clear phases over the course of a few hours after complete dissolution. Within this time frame, however, neither process stability nor fiber morphology were affected. Additionally, continued stirring of the emulsions makes stable electrospinning possible for much longer dwell times (no visible phase separation after 48 h).

PCL degrades significantly in the AA/FA solvent system with prolonged dwell times in solution [5]. Therefore, the polymer blend degradation was investigated for longer dwell times in solution through viscosity measurements (Figure S2). Although the reduction in solution viscosity is not as large as for pure PCL (about 40% after 48 h in 70/30 AA/FA), 50/50 PCL/Gt blends are also characterized by a decrease in viscosity with increasing dwell time in the acid solution (about 20% after 48 h in 70/30 AA/FA). This is probably mainly due to PCL degradation [20], resulting in slightly smaller fiber diameters (276 ± 67 nm after 3 h in solution vs. 233 ± 47 nm after 48 h in solution). Despite the small change in fiber morphology, process stability was not affected by PCL degradation. Using a continuously stirred PCL/Gt emulsion in 70/30 AA/FA, uniform bead-free nanofibrous membranes can thus be produced over the course of 48 h, with only a slight influence on fiber diameter.

As the electrospinning remains stable for a prolonged time, PCL/Gt blend electrospinning using the AA/FA solvent system was readily scalable to produce large nanofibrous membranes. Homogeneous nanofibrous membranes with a nominal size of 300 × 300 mm^2 were produced on a multinozzle setup. Their uniformity was validated by using SEM and FTIR analysis at several points throughout the membrane. No significant differences in fiber morphology or diameter were observed. Similarly, for all PCL/Gt blend ratios, FTIR spectra were identical within a single membrane, even though they were electrospun using emulsions. Pure PCL and pure gelatin are characterized by a few non-overlapping peaks in their FTIR spectra (Figure S3 and Table S1) [22,26,27]. Therefore, a qualitative indication of the PCL/Gt composition of the nanofibrous membranes is given by the ration of the carbonyl stretching peak of PCL and the amide I peak of gelatin. The 70/30 AA/FA solvent system thus allows for stable, reproducible, and scalable production of PCL/Gt blend nanofibers with high flexibility in fiber composition and diameter.

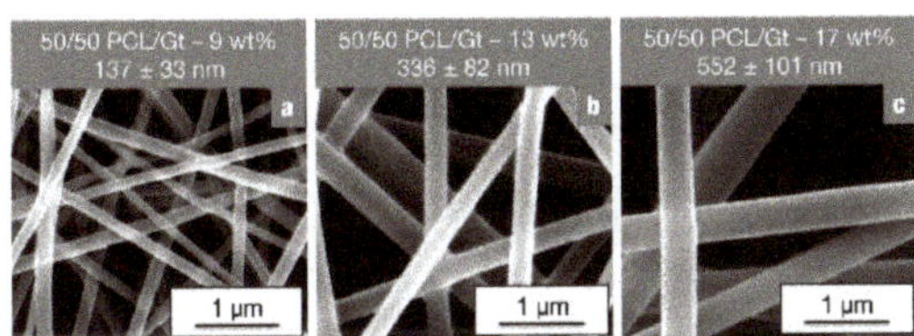

Figure 2. SEM images of 50/50 PCL/Gt blend nanofibers electrospun using 70/30 AA/FA (TCD of 12.5 cm, FR of 1 mL·h^{-1} and E adjusted for stable electrospinning) with a total polymer concentration of (**a**) 9 wt%, (**b**) 13 wt% and (**c**) 17 wt%. Varying the total polymer concentration significantly affects fiber diameters.

3.2. Stabilizing the Electrospinning Emulsions by Tuning the Solvent System

Water could be added to the electrospinning solutions up to 5 vol% without affecting solubility of the polymers. It stabilizes the solutions and hinders the phase separation of the PCL/Gt blends into two completely separated phase (the gelatin-rich phase settles at the bottom of the container without continued stirring). This makes electrospinning of bead-free PCL/Gt blend nanofibers possible for low gelatin concentrations without continued stirring (Figure 3b). Moreover, using a 70/25/5 AA/FA/water solvent system, the settling phenomenon can be prevented for gelatin concentrations up to 20 wt% of the total polymer mass. A similar effect has been reported by Feng et al., who illustrated that addition of a small amount of AA to a trifluoroethanol solution increases PCL-gelatin miscibility [22].

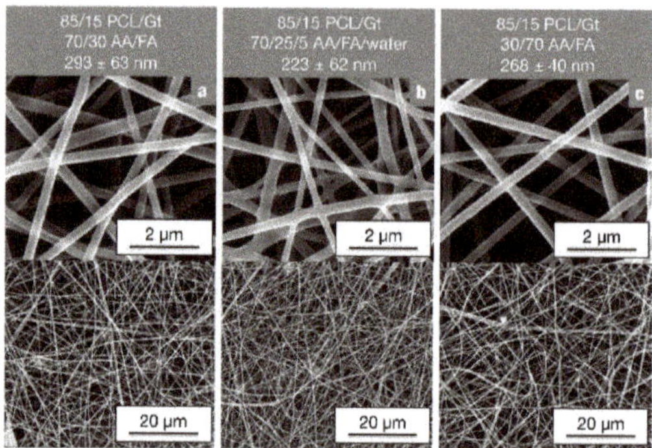

Figure 3. SEM images of 85/15 PCL/Gt nanofibers produced using a polymer concentration of 13 wt% but with differing solvent systems: (**a**) An emulsion unstable in time; (**b**) An emulsion stable in time, and; (**c**) A clear solution stable in time.

Alternatively, the effect of a changing the AA/FA solvent ratio was investigated. With an increasing FA concentration, miscibility is slightly improved. Although the gelatin-rich phase still settles on the bottom of the container when stirring is stopped, the PCL-rich upper phase contains more gelatin with decreasing AA/FA ratio of the solution. This is clearly illustrated by FTIR analysis of solution cast films of the upper and lower phase of an emulsion kept at rest (Figure 4). The gelatin peaks of the PCL-rich upper phase are much more pronounced when dissolved in 30/70 AA/FA (A_{PCL}/A_{Gt} = 1.5) than in 70/30 AA/FA (A_{PCL}/A_{Gt} = 21.3) (Figure 4 red curves, b vs. a), pointing to a higher gelatin concentration. Due to this higher miscibility in the PCL-rich phase, clear solutions stable in time were obtained for gelatin concentrations up to 30 wt% of the total polymer mass when using the 30/70 AA/FA solvent system. These clear and stable solutions were well electrospinnable, giving rise to bead-free nanofibers with reproducible diameters (Figure 3c). FTIR analysis of the obtained nanofibrous membranes, however, showed no differences in composition compared to the membranes electrospun using emulsions in 70/30 AA/FA (Figure S4). Consequently, the increased miscibility in 30/70 AA/FA does not affect nanofiber composition on the mm-scale.

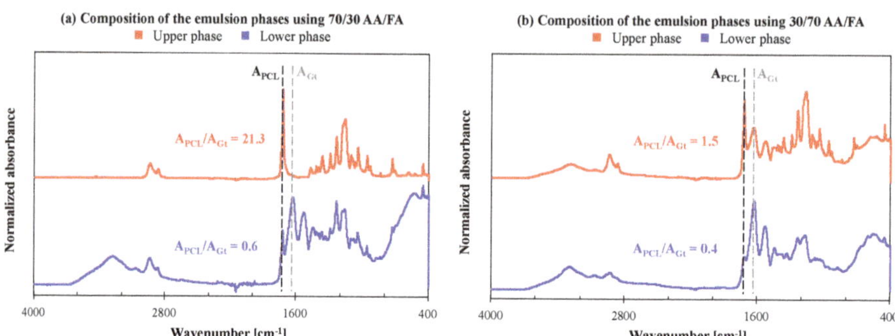

Figure 4. Normalized Attenuated Total Reflectance-Fourier-transform infrared spectroscopy (ATR-FTIR) spectra of films obtained by solution casting the upper (red) or lower (blue) phase, representing the continuous and dispersed phase respectively, of a PCL/Gt emulsion in (**a**) 70/30 AA/FA or (**b**) 30/70 AA/FA left at rest.

3.3. Interactions between the PCL and Gt Components by Thermal Analysis

The interactions and miscibility of polymer components in a blend often affect the thermal properties of the blend compared to their pure counterparts. Indeed, in a polymer blend containing a crystallizable component, a decrease in melting temperature and/or crystallinity is often observed; interaction between chains and chain segments of the blend components can cause a decrease in lamellar thickness of the crystals and/or a decrease in the amount of crystallizable material [28–31]. Additional to changes in melting behavior, also the glass transition is possibly affected by blending [28,32], with completely miscible polymer blends showing only one intermediate T_g. Therefore, the PCL/Gt blend nanofibers were analyzed using DSC and DMA and compared to the melting behavior and glass transition to pure PCL nanofibers.

While studying PCL/Gt blend nanofibers using DSC, care has to be taken when analyzing the PCL melting endotherm, since dissociation of gelatin triple helices occurs within the same temperature range [20]. However, the heat effect of this triple helix dissociation is far less than the melting enthalpy measured for pure PCL nanofibers, namely about $4\,J\cdot g^{-1}$ and about $70\,J\cdot g^{-1}$ respectively. The endothermic transition can thus mainly be attributed to the PCL component. As expected, DSC analysis of PCL/Gt blend nanofibers shows a clear decrease in overall melting enthalpy (Figure 5a). This decrease is in line with the decreasing PCL concentration. Indeed, recalculation of the overall melting enthalpy as a function of PCL mass results in values of about $70\,J\cdot g^{-1}$ for all samples (Table 1). The value for 15/85 PCL/Gt nanofibers is slightly higher, probably due to a more significant overlapping heat effect of the triple helix dissociation within the gelatin component and a larger error. Overall, however, the PCL melting enthalpy does not seem to be significantly decreased due to blending. Additionally, the melting trace showed a peak value (T_p) of $55 \pm 1\,°C$ for all samples. These results indicate that blending of PCL with gelatin does not significantly affect the melting behavior of the PCL component and interaction between the components is thus low. This also confirms that the analyzed nanofibers contain the expected amount of PCL and gelatin, pointing to a homogeneous blend composition throughout the membranes. Similar results were obtained for all investigated solvent systems (70/30 AA/FA, 30/70 AA/FA and 70/25/5 AA/FA/water). Nanofiber composition, uniformity and PCL-gelatin interactions are thus not affected when electrospinning a clear solution compared to an emulsion. DMA analysis on 85/15 PCL/Gt blend nanofibers illustrated that the glass transition of PCL is hardly affected compared to pure PCL (Figure 5b). This again points to weak interactions and low miscibility between PCL and gelatin.

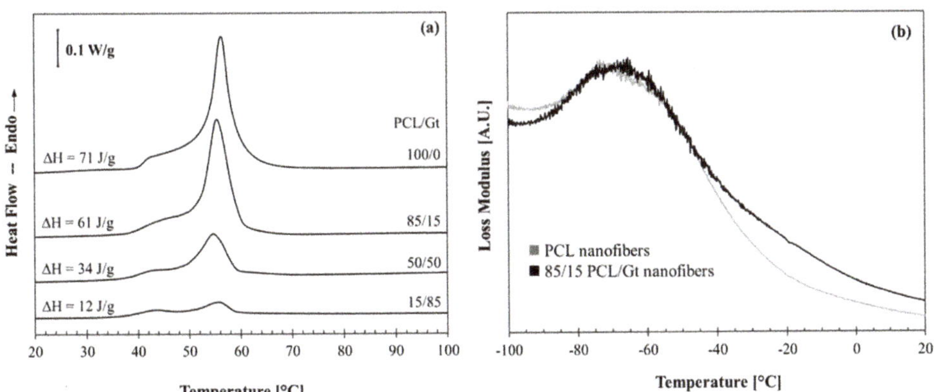

Figure 5. The effect of blending with gelatin, using a 70/30 AA/FA solvent system, on (**a**) the melting behavior and (**b**) the glass transition of PCL-based nanofibers measured by DSC at $2.5\,K\cdot min^{-1}$ and by DMA respectively.

Table 1. Melting enthalpy measured using DSC (Figure 5a), as a function of total polymer mass and recalculated as a function of PCL mass.

Nanofibrous Membranes	ΔH_m (Joules per Gram Fiber)	ΔH_m (Joules per Gram PCL)
PCL	71	71
85/15 PCL/Gt	61	72
50/50 PCL/Gt	29	68
15/85 PCL/Gt	12	80

3.4. Cold-Water Solubility of the Gelatin Component

As discussed in our previous paper, gelatin nanofibers are cold-water-soluble and dissolve to form a hydrogel in water at room temperature when they are not cross-linked [20]. Blending with PCL could affect this property. The 85/15 PCL/Gt nanofibers do not show significant changes in nanofiber morphology when submerged in cold water, suggesting the gelatin component to be not dissolved. With increasing gelatin concentration, however, a fiber-reinforced hydrogel is obtained, where gelatin partly gels but a nanofibrous network mainly consisting of PCL remains. Although water stability through cross-linking can easily be obtained [33], nanofiber-reinforced gelatin hydrogels could prove to be a promising biomedical material [34–36], combining a soft and elastic consistency while allowing for cell support on a nanofibrous membrane with increased porosity. To characterize the nanofibrous reinforcing structure and gain insight into the phase morphology of PCL/Gt nanofibers, the water-soluble gelatin fraction was removed using water of 35 °C.

All nanofibrous membranes having a PCL concentration of $\geq$50% still showed structural integrity after water treatment. It was, therefore, possible to characterize them using SEM and FTIR (Figure 6 and Table 2). Additionally, the mass loss of the membranes was determined. Since the water treatment procedure does not affect pure PCL nanofibers (Figure 6a,e), this mass reduction can be attributed to a decrease in gelatin content, resulting in a new PCL/Gt ratio.

SEM analysis, the measured mass loss and FTIR analysis all clearly indicate that the rinsing procedure only has a minor influence on 85/15 PCL/Gt nanofibers. Indeed, fiber morphology before and after water treatment is comparable (Figure 6b,f), and the FTIR spectra were identical, resulting in similar A_{PCL}/A_{Gt} ratios (Table 2). A similar result was obtained for 85/15 PCL/Gt nanofibers electrospun using 30/70 AA/FA (clear solutions). The phase separation in the electrospinning solution thus only has a minor effect on the final PCL/Gt fiber morphology. Although PCL-gelatin interaction within the nanofibers is low according to our thermal analysis, the gelatin component no longer dissolves in water. This indicates that the gelatin-rich phase is finely dispersed throughout the nanofibers, so that the hydrophobicity of PCL prevents gelatin dissolution.

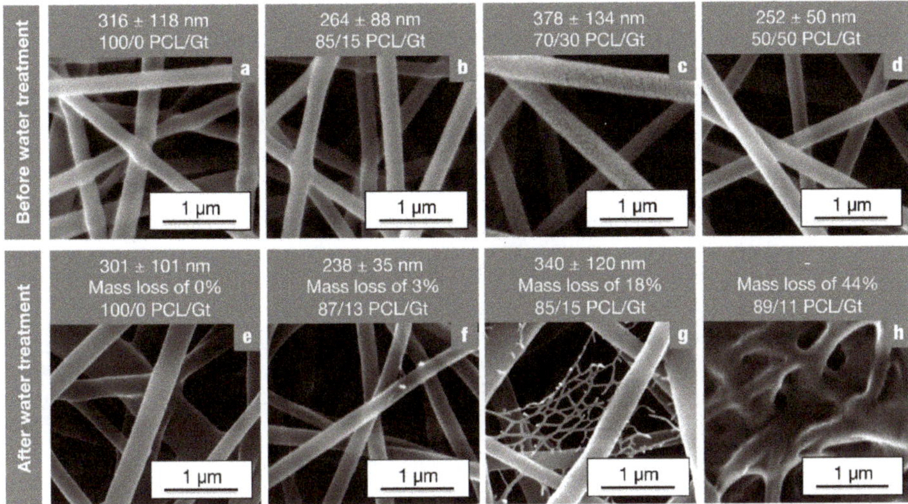

Figure 6. SEM images of PCL/Gt blend nanofibers (13 wt%, 70/30 AA/FA) with different PCL/Gt ratios (**a–d**) before and (**e–h**) after water treatment, i.e., rinsing with demineralized water at 35 °C. Mass loss is calculated by weighing the dried weight. Based on this mass loss of gelatin, a new PCL/Gt ratio was calculated for the rinsed samples.

Table 2. Composition of the nanofibrous membranes before and after water treatment, analyzed using the ratio of the carbonyl stretching peak of PCL and the amide I peak of gelatin measured using FTIR. After water treatment, the composition is comparable and reflecting a gelatin concentration of about 15 wt%.

Nanofibrous Membranes	A_{PCL}/A_{Gt} before Water Treatment	A_{PCL}/A_{Gt} after Water Treatment
85/15 PCL/Gt	4.9	5
70/30 PCL/Gt	2.3	4.7
50/50 PCL/Gt	0.9	3.1 *

* A slightly higher gelatin concentration is measured, probably due to the gelatin film covering the membrane, as demonstrated in Figure 6h.

With increasing gelatin content, there is a significant amount of gelatin dissolved by rinsing in demineralized water at 35 °C, as evidenced by the mass loss and the decreasing amide I peak of gelatin in the FTIR spectra (Figure 6 and Table 2 respectively). The influence on the fiber diameter; however, is only small, especially for 70/30 PCL/Gt nanofibers, and no porosity of individual nanofibers is observed (Figure 6g). Additionally, the new PCL/Gt ratio after extraction is about 85/15. It is, therefore, hypothesized that a 70/30 PCL/Gt nanofibrous membrane is built up by a mixture of PCL-rich and gelatin-rich nanofibers. The PCL-rich nanofibers within the membrane contain about 15 wt% of gelatin and are not affected by water treatment, similar to 85/15 PCL/Gt nanofibers. The gelatin-rich nanofibers are dissolved, possibly leaving some PCL residue upon rinsing (Figure 6g). Although there are clearly still intact nanofibers present in the 50/50 PCL/Gt nanofibrous membranes after water treatment, a gelatin film is observed in SEM analysis (Figure 6h). This indicates that the gelatin polypeptide chains are sufficiently immobilized by entanglements with the PCL component to prevent removal by dissolution. Moreover, the gelatin phase must be present more toward the shell of the nanofibers, causing film formation and a slightly higher A_{PCL}/A_{Gt} ratio in FTIR (Table 2).

These results indicate that PCL/Gt blend nanofibrous membranes have a high potential for advanced biomedical material design. On one hand, water-stable PCL/Gt blend nanofibrous membranes can be produced for gelatin concentrations up to 15 wt%. On the other hand, for higher amounts of gelatin, cold-gelling when in contact with water is possible, giving rise to a nanostructured hydrogel that is physically reinforced by PCL nanofibers. The nanofibers in these nanostructured reinforced hydrogels consist of PCL-rich nanofibers that still contain a significant amount of gelatin (about 10 to 15 wt%).

4. Conclusions

In conclusion, the AA/FA-based solvent system allows for stable, reproducible, and scalable PCL/Gt blend electrospinning with high flexibility, this while the toxicity of the solvent system is significantly reduced compared to traditionally utilized solvents. Indeed, dissolution of PCL/Gt blends in 70/30 AA/FA yields well electrospinnable emulsions for the whole PCL/Gt composition range. Although these emulsions are subject to settling of the gelatin-rich phase, uniform bead-free nanofibers can easily be produced within a time frame of a few hours or by continued stirring of the electrospinning solution. Additionally, the resulting fiber diameter can be tuned by altering the total polymer concentration. Using 30/70 AA/FA, stable and clear electrospinning solutions are obtained for gelatin concentrations up to 30 wt% due to increased miscibility. Further augmentation of gelatin concentration results in phase separation with the formation of an unstable emulsion. For all PCL/Gt ratios, process stability is best, and toxicity is minimized using 70/30 AA/FA. For gelatin concentrations up to 20 wt%, substitution of 5 vol% acid with water stabilizes the emulsion for over 48 h, making electrospinning possible without continued stirring or compromising the lower toxicity of the solvent system.

The resulting PCL/Gt nanofibrous membranes are water-stable up to gelatin concentrations of 15 wt%, irrespective of the possible phase separation in the electrospinning solution used for production (emulsion vs. clear miscible solution). It is hypothesized that membranes containing a higher amount of gelatin are built up of a mixture of nanofibers mainly consisting of PCL (~85/15 PCL/Gt) and nanofibers mainly consisting of cold-water-soluble and cold-gelling gelatin. The PCL-gelatin interactions within the nanofibers were low, as blending did not seem to affect the glass transition or melting of the PCL component. If the original nanofibrous membrane contains a high amount of gelatin (>15 wt%), cold-gelling when in contact with water is possible, giving rise to a nanofiber-reinforced physical gelatin hydrogel. This novel combination of material properties could prove to be a promising material for biomedical applications.

Supplementary Materials: The following are available online at http://www.mdpi.com/2079-4991/8/7/551/s1. Figure S1. Viscosity and conductivity measurements of PCL/Gt blend electrospinning solutions, using the 70/30 AA/FA solvent system (a) as a function of the PCL/Gt ratio and (b) as a function of the total polymer concentration. Figure S2. Viscosity measurements of the electrospinning solutions (PCL and/or Gt dissolved in 70/30 AA/FA) with increasing dwell time show that PCL degrades substantially whereas the Gt component remains quite stable. Figure S3. ATR-FTIR spectra of pure PCL and pure Gt, showing their characteristic peaks. Figure S4. Normalized ATR-FTIR spectra of 85/15 PCL/Gt blend nanofibers electrospun using an emulsion (dissolution in 70/30 AA/FA) or a clear solution (dissolution in 30/70 AA/FA). Table S1. Characteristic peaks of a PCL pellet and Gt powder in ATR-FTIR, as indicated in Figure S3.

Author Contributions: Conceptualization, L.D., I.S., H.R. and K.D.C.; Data Curation, L.D. and I.S.; Funding acquisition, K.D.C.; Investigation, L.D., I.S., E.S. and C.G.; Methodology, L.D. and I.S.; Project administration, K.D.C.; Resources, H.R. and K.D.C.; Supervision, H.R. and K.D.C.; Visualization, L.D. and I.S.; Writing—original draft, I.S.; Writing—review and editing, L.D., H.R. and K.D.C.

Funding: This research was funded by the Agency for Innovation by Science and Technology of Flanders (IWT) grant number 111158 and 141344.

Acknowledgments: The authors thank Rousselot for providing the free gelatin samples.

Conflicts of Interest: The authors declare no conflict of interest.

References

1. Lim, E.-H.; Sardinha, J.P.; Myers, S. Nanotechnology Biomimetic Cartilage Regenerative Scaffolds. *Arch. Plast. Surg.* **2014**, *41*, 231–240. [CrossRef] [PubMed]
2. Kai, D.; Jin, G.; Prabhakaran, M.P.; Ramakrishna, S. Electrospun synthetic and natural nanofibers for regenerative medicine and stem cells. *Biotechnol. J.* **2013**, *8*, 59–72. [CrossRef] [PubMed]
3. Ng, K.W.; Achuth, H.N.; Moochhala, S.; Lim, T.C.; Hutmacher, D.W. In vivo evaluation of an ultra-thin polycaprolactone film as a wound dressing. *J. Biomater. Sci. Polym. Ed.* **2007**, *18*, 925–938. [CrossRef] [PubMed]
4. Bölgen, N.; Menceloğlu, Y.Z.; Acatay, K.; Vargel, I.; Pişkin, E. In vitro and in vivo degradation of non-woven materials made of poly(epsilon-caprolactone) nanofibers prepared by electrospinning under different conditions. *J. Biomater. Sci. Polym. Ed.* **2005**, *16*, 1537–1555. [CrossRef] [PubMed]
5. Van der Schueren, L.; De Schoenmaker, B.; Kalaoglu, Ö.I.; De Clerck, K. An alternative solvent system for the steady state electrospinning of polycaprolactone. *Eur. Polym. J.* **2011**, *47*, 1256–1263. [CrossRef]
6. Herrero-Herrero, M.; Gómez-Tejedor, J.A.; Vallés-Lluch, A. PLA/PCL electrospun membranes of tailored fibres diameter as drug delivery systems. *Eur. Polym. J.* **2018**, *99*, 445–455. [CrossRef]
7. Alves da Silva, M.L.; Martins, A.; Costa-Pinto, A.R.; Costa, P.; Faria, S.; Gomes, M.; Reis, R.L.; Neves, N.M. Cartilage tissue engineering using electrospun PCL Nanofiber Meshes and MSCs. *Biomacromolecules* **2010**, *11*, 3228–3236. [CrossRef] [PubMed]
8. Ren, K.; Wang, Y.; Sun, T.; Yue, W.; Zhang, H. Electrospun PCL/gelatin composite nanofiber structures for effective guided bone regeneration membranes. *Mater. Sci. Eng. C* **2017**, *78*, 324–332. [CrossRef] [PubMed]
9. Venugopal, J.; Zhang, Y.Z.; Ramakrishna, S. Fabrication of modified and functionalized polycaprolactone nanofibre scaffolds for vascular tissue engineering. *Nanotechnology* **2005**, *16*, 2138–2142. [CrossRef] [PubMed]
10. Stafiej, P.; Küng, F.; Thieme, D.; Czugala, M.; Kruse, F.E.; Schubert, D.W.; Fuchsluger, T.A. Adhesion and metabolic activity of human corneal cells on PCL based nanofiber matrices. *Mater. Sci. Eng. C* **2017**, *71*, 764–770. [CrossRef] [PubMed]
11. Bhattarai, N.; Li, Z.; Gunn, J.; Leung, M.; Cooper, A.; Edmondson, D.; Veiseh, O.; Chen, M.H.; Zhang, Y.; Ellenbogen, R.G.; et al. Natural-synthetic polyblend nanofibers for biomedical applications. *Adv. Mater.* **2009**, *21*, 2792–2797. [CrossRef]
12. Beachley, V.; Wen, X. Polymer nanofibrous structures: Fabrication, biofunctionalization, and cell interactions. *Prog. Polym. Sci.* **2010**, *35*, 868–892. [CrossRef] [PubMed]
13. Cipitria, A.; Skelton, A.; Dargaville, T.R.; Dalton, P.D.; Hutmacher, D.W. Design, fabrication and characterization of PCL electrospun scaffolds—A review. *J. Mater. Chem.* **2011**, *21*, 9419. [CrossRef]
14. Gunn, J.; Zhang, M. Polyblend nanofibers for biomedical applications: Perspectives and challenges. *Trends Biotechnol.* **2010**, *28*, 189–197. [CrossRef] [PubMed]
15. Schiffman, J.D.; Schauer, C.L. A review: Electrospinning of biopolymer nanofibers and their applications. *Polym. Rev.* **2008**, *48*, 317–352. [CrossRef]
16. Metcalfe, A.D.; Ferguson, M.W.J. Tissue engineering of replacement skin: The crossroads of biomaterials, wound healing, embryonic development, stem cells and regeneration. *J. R. Soc. Interface* **2007**, *4*, 413–437. [CrossRef] [PubMed]
17. Zhang, Y.Z.; Venugopal, J.; Huang, Z.M.; Lim, C.T.; Ramakrishna, S. Crosslinking of the electrospun gelatin nanofibers. *Polymer* **2006**, *47*, 2911–2917. [CrossRef]
18. Wu, S.-C.; Chang, W.-H.; Dong, G.-C.; Chen, K.-Y.; Chen, Y.-S.; Yao, C.-H. Cell adhesion and proliferation enhancement by gelatin nanofiber scaffolds. *J. Bioact. Compat. Polym.* **2011**, *26*, 565–577. [CrossRef]
19. Panzavolta, S.; Gioffrè, M.; Focarete, M.L.; Gualandi, C.; Foroni, L.; Bigi, A. Electrospun gelatin nanofibers: Optimization of genipin cross-linking to preserve fiber morphology after exposure to water. *Acta Biomater.* **2011**, *7*, 1702–1709. [CrossRef] [PubMed]
20. Steyaert, I.; Rahier, H.; Van Vlierberghe, S.; Olijve, J.; De Clerck, K. Gelatin nanofibers: Analysis of triple helix dissociation temperature and cold-water-solubility. *Food Hydrocoll.* **2016**, *57*, 200–208. [CrossRef]
21. Zhang, Y.; Ouyang, H.; Chwee, T.L.; Ramakrishna, S.; Huang, Z.M. Electrospinning of gelatin fibers and gelatin/PCL composite fibrous scaffolds. *J. Biomed. Mater. Res. B* **2005**, *72*, 156–165. [CrossRef] [PubMed]
22. Feng, B.; Tu, H.; Yuan, H.; Peng, H.; Zhang, Y. Acetic-acid-mediated miscibility toward electrospinning homogeneous composite nanofibers of GT/PCL. *Biomacromolecules* **2012**, *13*, 3917–3925. [CrossRef] [PubMed]

23. Prabhakaran, M.P.; Venugopal, J.R.; Chyan, T.T.; Hai, L.B.; Chan, C.K.; Lim, A.Y.; Ramakrishna, S. Electrospun biocomposite nanofibrous scaffolds for neural tissue engineering. *Tissue Eng. A* **2008**, *14*, 1787–1797. [CrossRef] [PubMed]

24. Daelemans, L.; van der Heijden, S.; De Baere, I.; Rahier, H.; Van Paepegem, W.; De Clerck, K. Nanofibre bridging as a toughening mechanism in carbon/epoxy composite laminates interleaved with electrospun polyamide nanofibrous veils. *Compos. Sci. Technol.* **2015**, *117*, 244–256. [CrossRef]

25. Daelemans, L.; van der Heijden, S.; De Baere, I.; Rahier, H.; Van Paepegem, W.; De Clerck, K. Using aligned nanofibres for identifying the toughening micromechanisms in nanofibre interleaved laminates. *Compos. Sci. Technol.* **2016**, *124*, 17–26. [CrossRef]

26. Zha, Z.; Teng, W.; Markle, V.; Dai, Z.; Wu, X. Fabrication of gelatin nanofibrous scaffolds using ethanol/phosphate buffer saline as a benign solvent. *Biopolymers* **2012**, *97*, 1026–1036. [CrossRef] [PubMed]

27. Ghasemi-Mobarakeh, L.; Prabhakaran, M.P.; Morshed, M.; Nasr-Esfahani, M.H.; Ramakrishna, S. Electrospun poly(ε-caprolactone)/gelatin nanofibrous scaffolds for nerve tissue engineering. *Biomaterials* **2008**, *29*, 4532–4539. [CrossRef] [PubMed]

28. Mathot, V.B.F. *Calorimetry and Thermal Analysis of Polymers*; Carl Hanser Verlag: München, Germany, 1994.

29. Senda, T.; He, Y.; Inoue, Y. Biodegradable blends of poly(epsilon-caprolactone) with alpha-chitin and chitosan: Specific interactions, thermal properties and crystallization behavior. *Polym. Int.* **2002**, *51*, 33–39. [CrossRef]

30. Mano, F.; Alves, M.; Azevedo, J.V. Development of new poly(ε-caprolactone)/chitosan films. *Polym. Int.* **2013**, *62*, 1425–1432. [CrossRef]

31. Sarasam, A.; Madihally, S.V. Characterization of chitosan-polycaprolactone blends for tissue engineering applications. *Biomaterials* **2005**, *26*, 5500–5508. [CrossRef] [PubMed]

32. Young, R.J.; Lovell, P.A. *Introduction to Polymers*, 3rd ed.; CRC Press: Boca Raton, FL, USA, 2011.

33. Pereira, I.H.L.; Ayres, E.; Averous, L.; Schlatter, G.; Hebraud, A.; De Paula, A.C.C.; Viana, P.H.L.; Goes, A.M.; Oréfice, R.L. Differentiation of human adipose-derived stem cells seeded on mineralized electrospun co-axial poly(ε-caprolactone) (PCL)/gelatin nanofibers. *J. Mater. Sci. Mater. Med.* **2014**, *25*, 1137–1148. [CrossRef] [PubMed]

34. Tonsomboon, K.; Oyen, M.L. Composite electrospun gelatin fiber-alginate gel scaffolds for mechanically robust tissue engineered cornea. *J. Mech. Behav. Biomed. Mater.* **2013**, *21*, 185–194. [CrossRef] [PubMed]

35. Jaiswal, M.; Gupta, A.; Dinda, A.K.; Koul, V. An investigation study of gelatin release from semi-interpenetrating polymeric network hydrogel patch for excision wound healing on Wistar rat model. *J. Appl. Polym. Sci.* **2015**, *132*, 25. [CrossRef]

36. Shi, J.; Wang, L.; Zhang, F.; Li, H.; Lei, L.; Liu, L.; Chen, Y. Incorporating protein gradient into electrospun nanofibers as scaffolds for tissue engineering. *ACS Appl. Mater. Interfaces* **2010**, *2*, 1025–1030. [CrossRef] [PubMed]

Article

Fabrication of Sericin/Agrose Gel Loaded Lysozyme and Its Potential in Wound Dressing Application

Meirong Yang [1], Yejing Wang [1,2,*], Gang Tao [1], Rui Cai [2], Peng Wang [2], Liying Liu [1], Lisha Ai [1], Hua Zuo [3], Ping Zhao [1,4], Ahmad Umar [5], Chuanbin Mao [6,7] and Huawei He [1,4,*]

[1]   State Key Laboratory of Silkworm Genome Biology, Southwest University, Beibei, Chongqing 400715, China; yangmeirong@email.swu.edu.cn (M.Y.); taogang@email.swu.edu.cn (G.T.); l3341345@email.swu.edu.cn (L.L.); als123@email.swu.edu.cn (L.A.); zhaop@swu.edu.cn (P.Z.)

[2]   College of Biotechnology, Southwest University, Beibei, Chongqing 400715, China; cairui0330@email.swu.edu.cn (R.C.); modelsums@email.swu.edu.cn (P.W.)

[3]   College of Pharmaceutical Sciences, Southwest University, Beibei, Chongqing 400715, China; zuohua@swu.edu.cn

[4]   Chongqing Engineering and Technology Research Center for Novel Silk Materials, Southwest University, Beibei, Chongqing 400715, China

[5]   Department of Chemistry, College of Science and Arts and Promising Centre for Sensors and Electronics Devices (PCSED), Najran University, P.O. Box 1988, Najran 11001, Saudi Arabia; umahmad@nu.edu.sa

[6]   Department of Chemistry & Biochemistry, Stephenson Life Science Research Center, University of Oklahoma, 101 Stephenson Parkway, Norman, OK 73019, USA; cbmao@ou.edu

[7]   School of Materials Science and Engineering, Zhejiang University, Hangzhou 310027, China

*   Correspondence: yjwang@swu.edu.cn (Y.W.); hehuawei@swu.edu.cn (H.H.); Tel.: +86-23-6825-1575 (H.H.)

Received: 6 March 2018; Accepted: 4 April 2018; Published: 13 April 2018

Abstract: Sericin is a biomaterial resource for its significant biodegradability, biocompatibility, hydrophilicity, and reactivity. Designing a material with superabsorbent, antiseptic, and non-cytotoxic wound dressing properties is advantageous to reduce wound infection and promote wound healing. Herein, we propose an environment-friendly strategy to obtain an interpenetrating polymer network gel through blending sericin and agarose and freeze-drying. The physicochemical characterizations of the sericin/agarose gel including morphology, porosity, swelling behavior, crystallinity, secondary structure, and thermal property were well characterized. Subsequently, the lysozyme loaded sericin/agarose composite gel was successfully prepared by the solution impregnation method. To evaluate the potential of the lysozyme loaded sericin/agarose gel in wound dressing application, we analyzed the lysozyme loading and release, antimicrobial activity, and cytocompatibility of the resulting gel. The results showed the lysozyme loaded composite gel had high porosity, excellent water absorption property, and good antimicrobial activities against *Escherichia coli* and *Staphylococcus aureus*. Also, the lysozyme loaded gel showed excellent cytocompatibility on NIH3T3 and HEK293 cells. So, the lysozyme loaded sericin/agarose gel is a potential alternative biomaterial for wound dressing.

Keywords: silk sericin; agarose; lysozyme; composite gel; wound dressing

---

1. Introduction

Non-healing skin wounds exposed to bacterial infections are biologically characterized by lengthening inflammation, interfering re-epithelialization, disturbing collagen production, and finally delaying wound healing [1]. In wound care, wound dressing is an important biomedical material used to protect the wound from infection and facilitate wound healing [2]. Accompanied by the growing number of chronic diseases, the wound dressing market is evolving rapidly in the present healthcare system worldwide [3]. The ideal wound dressing should absorb wound exudate in a manner, allow gas exchange and maintain necessary moisture at wound interface without cytotoxicity and allergenic response [4]. Besides, it can promote wound healing by creating a suitable microenvironment to prevent bacterial infection and promote cell adhesion and proper proliferation [5]. Among all wound dressing materials, hydrogel is an attractive alternative in traditional therapeutic approaches for its multifunctional abilities such as hydrophilicity, swelling, drug delivery, and in situ gelling capacity [6].

Sericin (SS) is one of the major protein components of silk, which is discarded as a waste during the degumming process in the textile industry [7]. Sericin is a natural protein, exhibiting immense potential in the field of biomaterial owing to its biodegradability, easy availability, and hydrophilicity [8]. Sericin has numerous biological activities such as anti-oxidation, anti-bacterium, and anti-coagulation, promoting cell growth and differentiation [9]. However, sericin is physically fragile and highly soluble due to its amorphous nature [10], which is unsuitable for biomedical applications. Hence, in order to obtain desired material with improved properties for regenerative medicine application, sericin is mostly designed to copolymerize, crosslink, or blend with other polymers as it has polar side chains with diverse functional groups, such as amine, hydroxyl, and carboxyl groups [11–15]. Agarose (AR) is a transparent, neutrally charged, and thermo-reversible natural polysaccharide [16]. Agarose is used extensively in vitro cartilage tissue engineering as it provides a superior foundation for chondrogenesis and higher glycosaminoglycan deposition to produce constructs with functional properties approaching those of native articular cartilage [17,18]. Additionally, agarose gel is considered as a biological scaffold material for the central nervous system repair and regeneration due to its excellent mechanical properties which can well match the growth and control of nerve axis, the porous structure which is conducive to nutrient delivery, and the implant which does not cause adverse reactions [19]. However, agarose shows low cell adhesiveness and cell proliferation activity in vivo [20]. Therefore, blending agarose with other polymers such as chitosan and gelatin to overcome the valid drawbacks has escalated in recent years [21,22]. Consequently, the present study brought together the innate advantages of sericin and agarose to fabricate a blended hydrogel for prospective application in a wound dressing.

Nevertheless, the hydrogel is limited as a wound dressing material because it may paradoxically provide a preferred environment for infectious bacteria. To prevent bacterial infection on both skin wound and dressing material, antibiotics such as penicillin and methicillin have been widely used. However, the widespread and indiscriminate use of antibiotics now constitutes a major health concern worldwide due to the emergence of numerous resistant pathogens [23]. Therefore, tremendous attention has been paid to the discovery and development of alternative novel antibiotics, particularly with new modes of action to overcome the resistance. Lysozyme is a natural antibacterial agent that has been isolated from the cells and secretions of virtually all the living organisms [24]. Lysozyme plays the role of the anti-microbial agent through catalyzing the hydrolysis of β-1,4 glycosidic bonds between N-acetylmuramic acid and N-acetylglucosamine in peptidoglycans of the bacterial cell wall [25]. It is commercially available at low cost, and classified as GRAS grade by the Food and Drug Administration (FDA, US) and as a food additive by the European Union (E 1105) [26]. Lysozyme has been extensively applied as antibacterial agents in wound dressing and protein separation [27]. Accordingly, the development of lysozyme-based antimicrobial material is significantly important toward an environmentally benign antimicrobial field.

We herein developed an improved lysozyme (LZM) loaded sericin/agarose (SS/AR) gel (SS/AR/LZM). Scanning Electronic Microscopy (SEM), Attenuated Total Reflection Fourier Transform Infrared Spectroscopy (ATR-FTIR), X-ray Diffraction (XRD), Thermogravimetric Analysis (TGA), and swelling behavior test were performed to characterize the physicochemical properties of SS/AR gel. We successfully fabricated lysozyme loaded SS/AR composite biomaterials by solution impregnation method. We investigated the lysozyme loading and release, the antimicrobial activity of SS/AR/LZM gel against typical Gram negative/positive bacteria *Escherichia coli* (*E. coli*) and *Staphyloccus aureus* (*S. aureus*). In addition, the cytotoxicity of the lysozyme loaded SS/AR gel was evaluated on NIH3T3 and HEK293 cells. The results suggested that the SS/AR/LZM gel with antimicrobial activity and cytocompatibility has a great potential in wound dressing application.

2. Materials and Methods

2.1. Materials

Bombyx mori cocoons were provided by the State Key Laboratory of Silkworm Genome Biology, Southwest University (China, 400716). Lysozyme (20,000 U/mg) was obtained from Sangon Biotech Co. Ltd. (Shanghai, China). Agarose G-10 was purchased from Biowest (Nuaillé, France). Cell counting kit-8 (CCK-8) was from Beyotime (Beijing, China). LIVE/DEAD cell viability kit was from Thermo Fisher Scientific (Waltham, MA, USA). NIH3T3 (mouse embryonic fibroblast) and HEK293 (human embryonic kidney) cell lines were obtained from China Infrastructure of Cell Line Resources. Chemicals for cell culture such as Dulbecco's modified Eagle's medium (DMEM), Fetal Bovine Serum (FBS), Trypsin-EDTA and Penicillin/Streptomycin were from Gibco BRL (Gaithersburg, MD, USA). Ultrapure water was the product of Milli-Q Plus system from Millipore (Billerica, MA, USA). All other chemicals utilized were of analytical grade.

2.2. Fabrication of SS/AR Gel

Sericin was extracted from *Bombyx mori* cocoons as previous reports [28,29]. Briefly, silkworm cocoons were cut into pieces and autoclaved at 121 °C for 30 min to obtain sericin solution. Subsequently, sericin solution was freeze-drying to become sericin powder, and then dissolved in hot water. Sericin and agarose solution (2%, *w/v*) were mixed gently, and allowed to gel after casting into 24-well cell culture plates. These stable hydrogels were then frozen at −80 °C for 12 h followed by lyophilization for 24 h to become gels. According to the volume ratios of sericin and agarose solution, the corresponding SS/AR gels were termed as S100A0, S75A25, S50A50, S25A75, and S0A100, respectively.

2.3. Characterization of SS/AR Gel

SEM observation was performed on JSM-5610LV (Tokyo, Japan) with working voltage of 25 kV to examine the surface morphologies of the gel. The porosities of SS/AR gels were calculated according to the liquid displacement method [30]. Briefly, SS/AR gel was immersed into water (V1) in a graduated cylinder, the total volume including water and SS/AR gel was recorded as V2. The SS/AR gel was then removed from the cylinder and the residual water volume was recorded as V3. The porosity (*p*) of SS/AR gel was calculated using the following equation:

$$p = (V_1 - V_3)/(V_2 - V_3) \times 100\%. \tag{1}$$

ATR-FTIR spectra of sericin and SS/AR gel were determined in the wavenumber range of 650–4000 cm^{-1} at a resolution of 4 cm^{-1} on a Nicolet iz10 Infrared spectrophotometer from Thermo Fisher Scientific (Waltham, MA, USA). XRD of sericin and SS/AR gel were carried out by X'Pert powder X-ray diffraction system (PANalytical, Almelo, OV, Netherland) within a 2θ range of 10°–70°. The thermal behaviors of sericin and SS/AR gel were analyzed by a thermogravimetric analyzer TGA-Q50 (TA instruments, New Castle, DE, USA) under a nitrogen flow of 20 mL/min [31]. The specimens were heated from room temperature to 600 °C, at a heating rate of 10 °C/min.

2.4. Swelling Behavior

The swelling ability of the SS/AR gel was analyzed using a conventional gravimetric method [32]. Briefly, a pre-weighed dry gel (Wd) was immersed into water at 37 °C for 30 min to achieve equilibrium. The swollen weight of gel was recorded as Ws at specific time intervals. The experiment was repeated for three times under the same conditions. Swelling ratios (S) were determined as the following equation:

$$S = (Ws - Wd)/Wd \times 100\%. \tag{2}$$

2.5. Preparation of SS/AR/LZM Gel

To prepare the lysozyme loaded SS/AR gel, S50A50 was cut into a circular piece with a diameter of 1.5 cm and then immersed into lysozyme solution (20–75 mg/mL) for 16 h. Subsequently, the lysozyme loaded SS/AR gel was removed from lysozyme solution and freeze-dried. According to the lysozyme concentration, the resulting SS/AR/LZM gels were termed as S50A50L20, S50A50L50 and S50A50L75, respectively.

2.6. The Loading and Release of Lysozyme

Lysozyme has a specific absorption peak at 280 nm [33], which could be used to measure the loaded and released lysozyme concentration. Ultraviolet visible spectrophotometer was employed to analyze the loading and release of lysozyme. The loaded lysozyme content was determined by the difference of lysozyme concentration before and after the treatment. The circular SS/AR/LZM gel with a diameter of 1.5 cm was dispensed into a centrifuge tube containing 4 mL of 0.01 M PBS (pH 7.4) buffer at 37 °C. At special time points, an aliquot (1 mL) PBS buffer was collected to measure the absorbance at 280 nm to determine the released lysozyme contents. Then the gel was transferred into 4 mL fresh PBS solution for the next measurement. The cumulative release rate was determined according to the ratio of the released and loaded lysozyme. Various lysozyme concentrations (0.1–0.6 mg/mL) were prepared for the calibration curve. All experiments were performed in triplicate.

2.7. In Vitro Antibacterial Assay

The antibacterial activity was evaluated according to the previous procedures with a slight modification [34,35]. SS/AR and SS/AR/LZM gels were cut pieces with 1 mm in thick and 1.5 cm in diameter, subsequently sterilized with UV radiation for 30 min. *E. coli* and *S. aureus* were grown in Luria–Bertani (LB) media at 37 °C with continuous shaking until the optical density at 600 nm (OD_{600}) reached about 1.5. Bacteria (500 μL) were harvested by centrifugation at 1000 rpm for 5 min followed by washing with 0.01 M phosphate buffer saline (PBS, pH 7.4). Subsequently, bacteria were re-suspended and diluted with PBS buffer. Next, 50 μL of the diluted bacterial suspension was cultured at 37 °C for 2 h in the presence of SS/AR gel or SS/AR/LZM gel. Aliquots (1 μL) of the mixture were diluted (1:10,000) in PBS, and then manually spread on LB agar plates. After 16 h incubation at 37 °C, the units of colony formation in each agar plate were counted to check the antibacterial ability of the gels. Each independent experiment was performed in triplicate.

2.8. Cytocompatibility Assay

NIH3T3 and HEK293 cells were cultured in high glucose DMEM supplemented with 10% FBS and 1% penicillin/streptomycin in a humidified atmosphere of 95% and 5% CO_2 at 37 °C. To check the cell viability, NIH3T3 and HEK293 cells (100 μL) were loaded at the density of 1×10^4 cells/well in 96-well plates and incubated 12 h at 37 °C. SS/AR or SS/AR/LZM gel was sterilized by ultraviolet radiation overnight and then added to the cell plates. Non-treated cells were used as a control.

After treated with the gels for 12 h, 24 h and 36 h, CCK-8 assay was used to assess cell viability according to the manufacturer's instructions. CCK-8 solution (10 μL) was added into each well and then incubated at 37 °C for 1.5 h. The optical density (OD) of each well was measured at 450 nm on a microplate reader TECAN (Mannedorf, Switzerland). The cell viability is defined as the percentage of OD value of the treated and control wells. For each experiment, at least three samples were evaluated. The morphologies of NIH3T3 and HEK293 cells after incubation for 24 h in the absence and presence of SS/AR or SS/AR/LZM gel were observed on a fluorescence microscope EVOS FL Auto Cell Imaging System (Life, Bothell, WA, USA).

Additionally, LIVE/DEAD staining assay was carried out to further assess the effects of the gels on the cells viability. NIH3T3 and HEK293 cells were cultured and incubated at 37 °C as described above. The cells after treated with SS/AR or SS/AR/LZM gel for 24 h were incubated with staining solution at 37 °C for 15 min. Then the images were collected on EVOS FL Auto Cell Imaging System. For each sample, the experiment was done in triplicate.

3. Results and Discussion

3.1. Preparation of SS/AR/LZM Gel

In this study, we prepared the SS/AR/LZM gel with good antibacterial activity and cytocompatibility, as illustrated in Figure 1. Sericin and agarose solution were mixed and then freeze-dried to become a porous gel. Thereafter, lysozyme, a natural antimicrobial agent, was loaded into the SS/AR gel. As sericin is negatively charged, agarose is neutral, whereas lysozyme is positively charged, thus the adsorption of lysozyme into the SS/AR gel may be attributed to the electrostatic interactions between the opposite charges of sericin and lysozyme [36]. Also, the physical adsorption caused by free diffusion could promote the adsorption of lysozyme. In addition, lysozyme has carboxyl group, amino groups and four disulfide bonds [37], and sericin has hydroxyl, carboxyl, and amino groups [38]. The special hydrophilic/hydrophobic interactions between lysozyme and sericin are also able to enhance the adsorption of lysozyme. The resulting SS/AR/LZM gel with interconnected porous structures, high swelling ability, good antibacterial activity and cytocompatibility may be a prospective alternative for wound dressing.

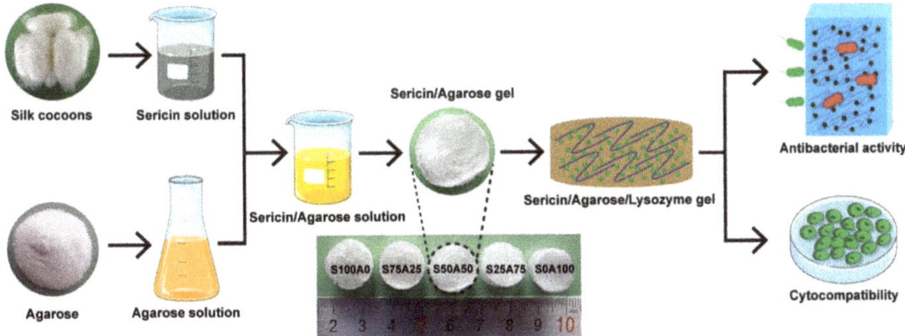

Figure 1. Schematic illustration of the fabrication of sericin (SS)/agarose (AR)/lysozyme (LZM) gel.

3.2. Morphology of SS/AR Gel

Porous materials provide space for cell growth and proliferation, and the microenvironment for the retention and release of bioactive molecules [39]. Furthermore, the porous structure affects the supply of nutrients and oxygen, and the removal of wastes [40], which is of utmost importance to wound dressing. In this study, the prepared gels had macro-porous "open-cell" structures (Figure 2A–D). The porosity of S75A25, S50A50, S25A75, and S0A100 gels were 53.17%, 49.54%, 46.17% and 59.75%, respectively (Figure 2E). Compared to other gels, S50A50 and S25A75 gels had significantly bigger pore sizes. This may be the fact that S50A50 and S25A75 gels could adsorb more water. After lyophilization, the space water occupied resulted in the formation of pores. Consequently, the pore size and porosity of gels were dependent on the ratio of sericin and agarose solution.

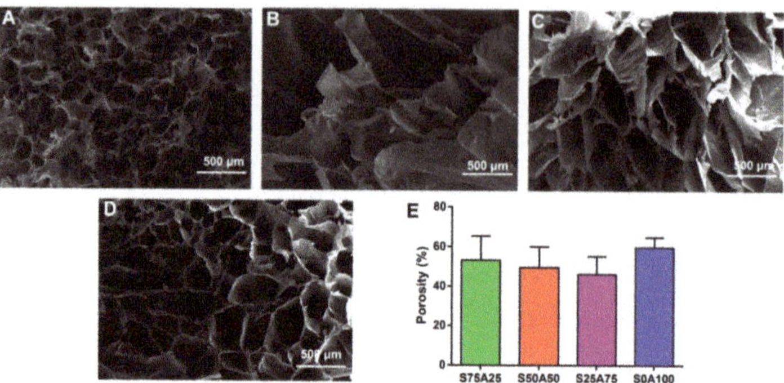

Figure 2. The porous microstructures of S75A25 (**A**), S50A50 (**B**), S25A75 (**C**), and S0A100 (**D**) gels. The porosity of gels with different ratios of sericin and agarose (**E**).

3.3. Characterization of SS/AR Gel

ATR-FTIR was employed to analyze the chemical interactions between sericin and agarose as ATR technique can probe to only a shallow depth and thus emphasize any surface coatings [41]. Sericin has a typical spectrum with distinctive peaks at 1600–1700 cm^{-1} (amide I, C=O stretching vibration), 1480–1575 cm^{-1} (amide II, N–H bending vibration), and 1229–1301 cm^{-1} (amide III) [42]. As shown in Figure 3A, sericin gel had characteristic peaks at 1621 cm^{-1}, 1521 cm^{-1}, and 1241 cm^{-1}, corresponding to amide I, amide II, and amide III, respectively. Pure agarose exhibited its characteristic peaks at 1068 cm^{-1} (C–O, axial deformation), 930 cm^{-1} (3, 6-anhydro-galactose), and 891 cm^{-1} (C–H, angular deformation of β anomeric carbon), the result was consistent with the previous study [43]. In the blended gels, the characteristic peaks of both agarose and sericin were recorded, which confirmed the presence of both components. Some slight shifts in amide I and amide II peaks, and the differences in the intensity of peaks were observed in case of the composite gels, which indicated that the backbone structures of sericin and agarose did not change. Lysozyme has characteristic peaks at 3295 cm^{-1} (NAH stretching of the free amino groups), and 2961 cm^{-1} (CAH stretching) as well7 as amide I (1600–1700 cm^{-1}), amide II (1500–1600 cm^{-1}) and amide III (1230–1320 cm^{-1}) [44]. Few peaks were found to overlap with the peaks of sericin.

The crystalline structure of the composite was analyzed by XRD. Silk protein has main diffraction peaks of Silk I (2θ = 12.2° and 28.2°), and Silk II (2θ = 18.9° and 20.7°) [45]. Similar XRD patterns of sericin with peaks at 2θ = 19.2° and 21.15° have been reported [46–49]. Sericin and SS/AR gel exhibited obvious diffraction peaks at 19.08° and 19.56°, respectively (Figure 3B), which indicated the existence of sericin in the SS/AR gel. The difference of 2θ between sericin and SS/AR gel reflect the conversion of the random coil to β-sheet structure for the presence of the intermolecular hydrogen bond in sericin [50].

The thermal stability of sericin and SS/AR gel were examined by TGA. Sericin and SS/AR gel underwent three stages of thermal degradation including dehydration, deploymerization, and decomposition (Figure 3C). The first stage was from room temperature to around 110 °C, where the mass loss revealed the removal of adsorbed water molecules in sericin and SS/AR gel. The second major decomposition was attributed to the degradation of sericin and agarose occurred in the temperature range of 120–410 °C. At this stage, the mass loss of SS/AR gel was faster than that of sericin, indicating that sericin could improve the thermal stability of the composite gels and delay the thermal degradation process. At the last stage, the mass loss occurred from 420 °C to 600 °C, which was associated with the breakdown of sericin and agarose.

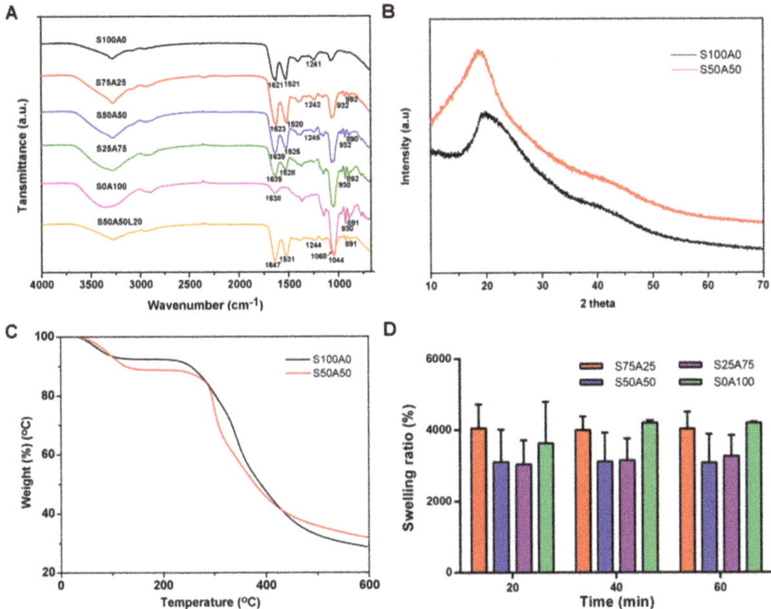

Figure 3. Characterizations of SS/AR gels. (**A**) Attenuated Total Reflection Fourier Transform Infrared Spectroscopy (ATR-FTIR); (**B**) X-Ray Diffraction (XRD); (**C**) Thermogravimetric Analysis (TGA); and (**D**) swelling ratio.

3.4. Swelling Behavior

Swelling ratio is a vital property to illustrate the uptake of liquids in wound dressing. The swelling ratio of various SS/AR gels were shown in Figure 3D. The results showed that all samples exhibited good swelling behavior. After 60 min, S75A25 and S0A100 gels had the swelling ratios of 3628–4196%, whereas S50A50 and S25A75 had the swelling ratios of 3040–3262%. The swelling ratios of S75A25 and S0A100 gels were higher than those of S50A50 and S25A75 as they had smaller pore size and higher porosities. The swelling of the gels had two stages, including the growing period and equilibrium period. In the initial 20 min, the swelling ratios of all samples increased quickly, indicating the excellent hydrophilicity of the composite gels. Thereafter, all gels quickly reached the swelling equilibrium.

The SS/AR gels with various ratios had a honeycomb structure, high porosity as well as excellent swelling capacity. In this study, we purposed to develop an alternative wound dressing through bringing together the innate advantages of sericin and agarose. High content of sericin will reduce the mechanical property of the gel, and high content of agarose may affect the cell adhesion and proliferation. Hence, we suggested the SS/AR gel with a ratio of 50:50 (S50A50) had moderate mechanical property and cytocompatibility, which may be suitable for wound dressing. The S50A50 gel was chosen for the further experiments.

3.5. Lysozyme Release

To avoid frequent replacement of the dressing and reduce the risk of overexposing wound to bacteria, a wound dressing should have a controllable drug release ability [51,52]. The loaded lysozyme contents of S50A50L20 and S50A50L50 gels were 24.94 mg and 50.78 mg, respectively (Figure 4A). And the loading efficiency of lysozyme was 62% and 51% for S50A50L20 and S50A50L50 gels, respectively (Figure 4B). Increasing lysozyme solution concentration could increase the loaded lysozyme contents, however, reduce its loading efficiency. To assess lysozyme release, we analyzed the correlation between lysozyme concentration and UV absorption by the standard curve. The results showed that UV absorption was tightly correlated with lysozyme concentration (Figure 4C). Obviously, lysozyme could be released from both S50A50L20 and S50A50L50 gels. Lysozyme release could be divided into burst stage and steady stage (Figure 4D). During the initial burst phase, the release rate was relatively high. This effect is associated with the diffusion of water molecules and desorption of lysozyme close to the surface of the gel [53]. At the steady phase, lysozyme was gradually released from SS/AR/LZM gels up to 60 h. The cumulative release of S50A50L20 and S50A50L50 gels reached 74% and 86% at 3 h, respectively. After 60 h, the cumulative release reached 98% and 99%, indicating lysozyme was almost entirely released from the composite gels. The results suggested that the SS/AR/LZM gel had a sustainable lysozyme releasing ability, which is required for a wound dressing.

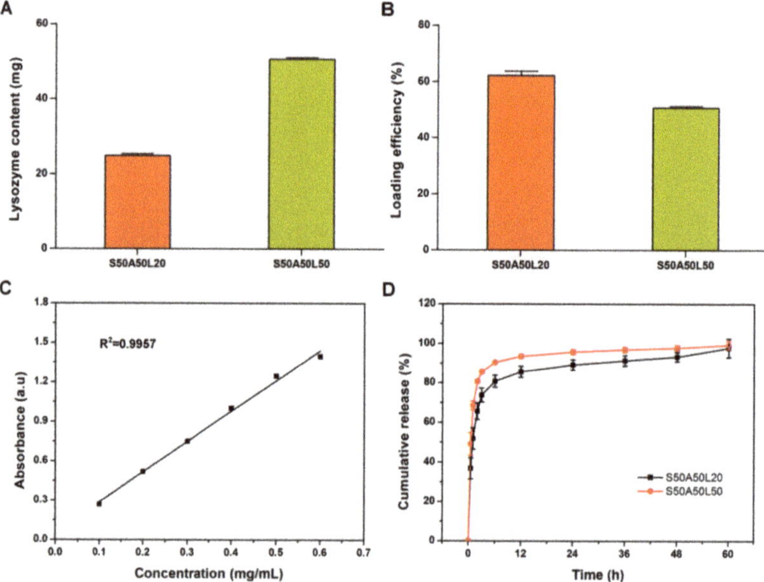

Figure 4. The loading and release of lysozyme. (**A**) Lysozyme contents loaded on SS/AR gel; (**B**) loading efficiency; (**C**) standard curves of UV intensity and lysozyme concentration; and (**D**) the cumulative release of lysozyme.

3.6. Antibacterial Activity

The antibacterial activity of the composite gels toward *E. coli* and *S. aureus* were shown in Figure 5A. Compared to the control, the total colonies number in the presence of SS/AR/LZM gels significantly decreased, indicating the SS/AR/LZM gel had good antibacterial activity. The S50A50L20 and S50A50L50 gels exhibited the bacteria reduction rates of 76%/87% and 84%/95% toward *E. coli*, *S. aureus*, respectively. The S50A50L75 gel completely inhibited the growth of the bacteria both for *E. coli* and *S. aureus*, suggesting increasing lysozyme solution concentration can improve the bactericidal ability of the composite gel as it increased the loaded and released lysozyme contents.

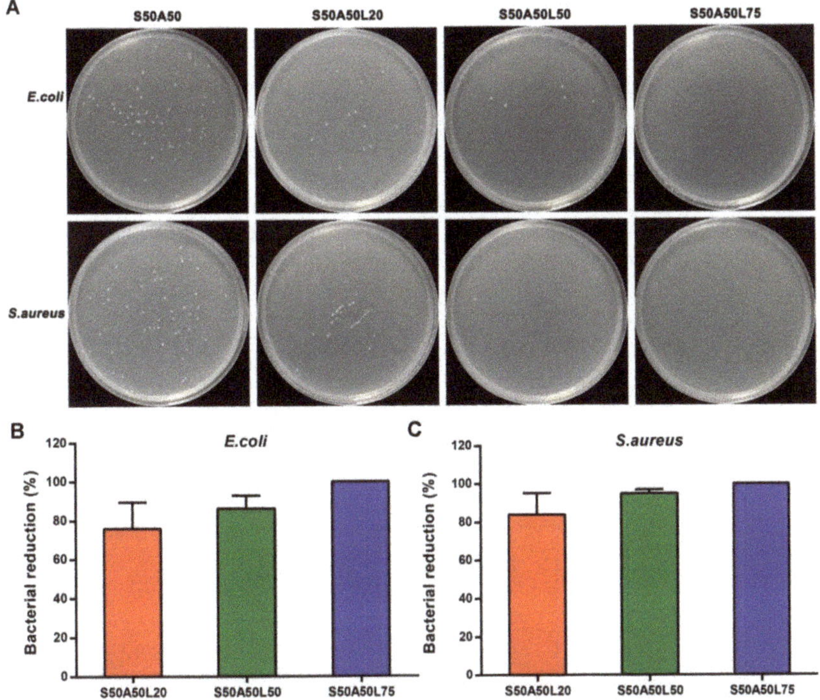

Figure 5. Antibacterial activities of SS/AR/LZM gels against *E. coli* and *S. aureus*. (**A**) Total bacterial colonies counting; (**B**, **C**) Bacterial colonies reduction rate.

Lysozyme is a natural antibacterial agent. Its antibacterial activity is mild compared with inorganic and organic antibacterial agents. To improve the antibacterial effect of lysozyme, some strategies such as physical and chemical modifications or synergistic action with other substances have been developed in recent years [26,54–56]. In this study, the SS/AR/LZM gels for antibacterial test had a dimension of 1 mm in thickness and 1.5 cm in diameter. The size and thickness determined the low loading content of lysozyme. In addition, the bacterial suspension volume affected the release efficiency of lysozyme from the gels. Therefore, SS/AR/LZM gels exhibited expected antibacterial effect; however, the loading content and the release efficiency of lysozyme determined its antibacterial activity.

Various antimicrobial materials have been developed for wound dressing application, such as metals/metal oxides [57], antibiotics [58], and peptides [2]. These materials are associated with a big concern about adverse effects on ecosystems. For instance, silver nanoparticle, one of the most classical and important materials, exert an adverse effect on environmental safety during its preparation and application. Also, the widespread and indiscriminate use of antibiotics response to the emergence of numerous resistant pathogens. The cost of the synthetic antibacterial peptide is very high. Here, the developed SS/AR/LZM material has good cytocompatibility and without toxic side effects and drug resistance. The main advantage of the lysozyme-based material is significantly safe to human and environment, which has exciting and expected potentials in biomedical materials such as wound dressing.

3.7. Cytocompatibility Assay

To evaluate the cytocompatibility of SS/AR and SS/AR/LZM gels, we tested the effects of SS/AR and SS/AR/LZM gels on the viability of NIH3T3 and HEK293 cells. In the CCK-8 test, a soluble formazan dye with the maximum absorbance at 450 nm produced as metabolically active cells react with a tetrazolium salt in the CCK-8 reagent. And, the higher optical density (OD) value indicates better cell viability and more live cells [59]. As shown in Figure 6A, B, after 12 h, there were no significant differences in the cells viabilities among SS/AR, SS/AR/LZM gels and the control. After 24 h, the cells viabilities for SS/AR, SS/AR/LZM, and control groups increased quickly. After 36 h, the cells in all groups exhibited significantly higher viabilities than those at 24 h. The results suggested the SS/AR/LZM gel had good cytocompatibility on NIH3T3 and HEK293 cells. This may be due to the fact that active sericin promotes cell growth for its cytoprotective and mitogenic abilities [9]. Furthermore, the morphologies of NIH3T3 and HEK293 cells were observed on an optical microscope after 24 h in the absence and presence of SS/AR and SS/AR/LZM gels. The results showed that morphologies of the cells were nearly identical to that of the control (Figure 6C), indicating SS/AR and SS/AR/LZM gels were not toxic on NIH3T3 and HEK293 cells.

After staining with the LIVE/DEAD fluorescent reagent, the alive cells were stained green while the apoptotic cells were stained red. The result showed most of the cells were green and only a few of cells were red (Figure 7), indicating that the SS/AR and SS/AR/LZM gels had good cytocompatibility on NIH3T3 and HEK293 cells. The fluorescence images of the LIVE/DEAD staining assay were in accordance with the results of CCK-8 assay.

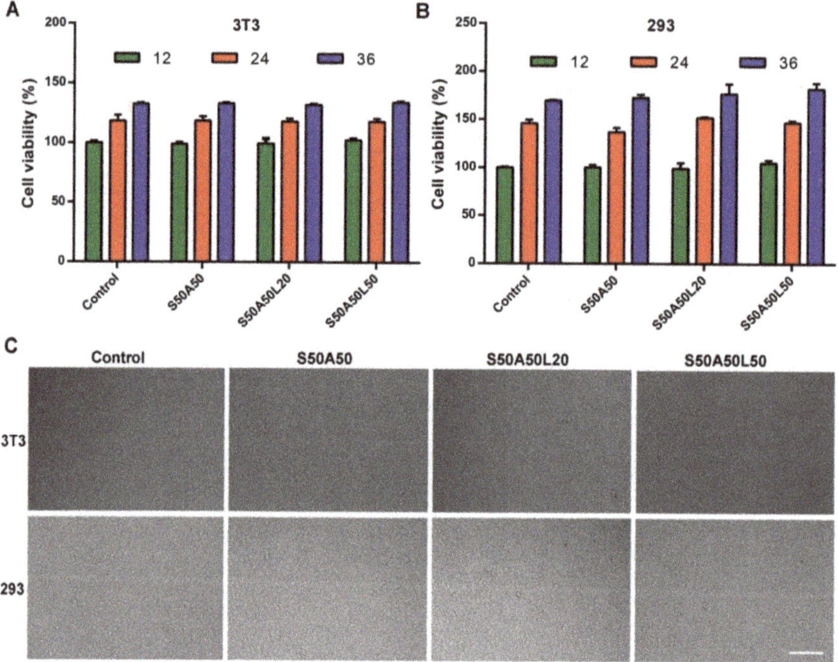

Figure 6. CCK-8 assay of the SS/AR/LZM gels. Cells viability of NIH3T3 (**A**) and HEK293 (**B**) in the presence of SS/AR gel or SS/AR/LZM gel, respectively. Microscopic analysis of NIH3T3 and HEK293 cells; ((**C**), scale bar, 200 μm).

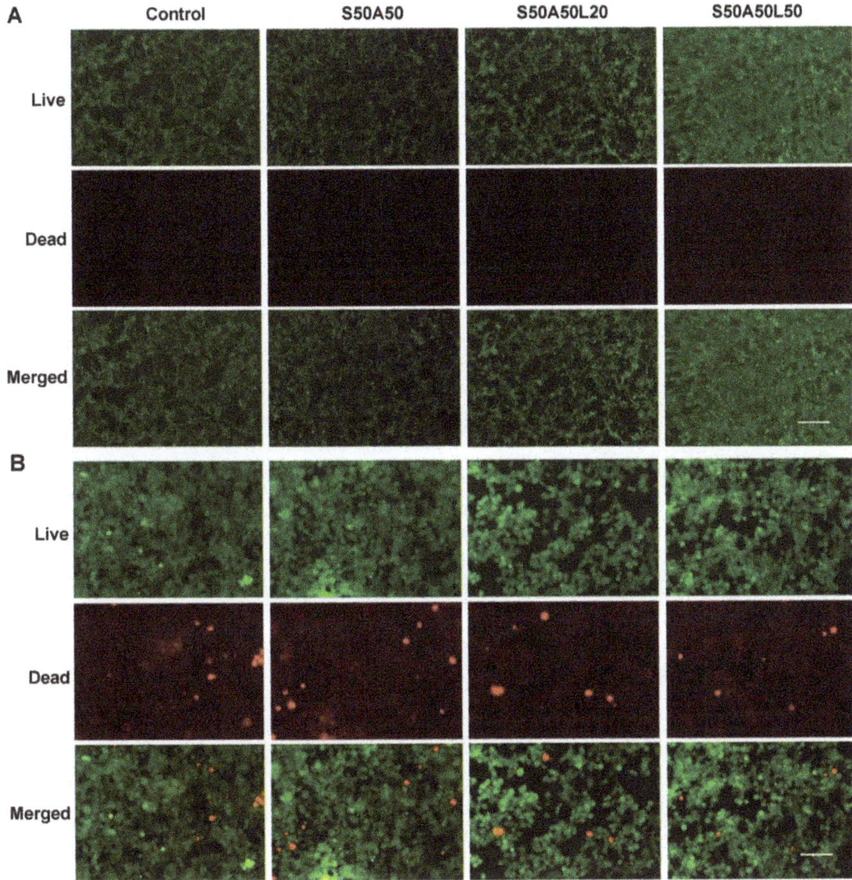

Figure 7. The LIVE/DEAD staining assay. NIH3T3 ((**A**), scale bar, 200 μm) and HEK293 ((**B**), scale bar, 100 μm) cells.

4. Conclusions

In summary, the SS/AR composite gel was successfully prepared by blending and freeze-drying. The SS/AR gel had highly porous and inter-connected structure, and good swelling behavior. The lysozyme loaded SS/AR gel had a sustainable lysozyme releasing ability and good antimicrobial activities against *E. coli* and *S. aureus* as well as excellent cytocompatibility on NIH3T3 and HEK293 cells. The SS/AR/LZM gel is expected to develop as an alternative for wound dressing.

Acknowledgments: This work was supported by the National Natural Science Foundation of China (31572465, 51673168), the State Key Program of National Natural Science of China (31530071), Chongqing Research Program of Basic Research and Frontier Technology (cstc2015jcyja00040, cstc2015jcyjbx0035), the Fundamental Research Funds for the Central Universities (XDJK2018B010, XDJK2018C063), the Graduate Research and Innovation Project of Chongqing (CYS17076, CYS17081, CYB17069), Open Project Program of Chongqing Engineering and Technology Research Center for Novel Silk Materials (silkgczx2016003), and the National Key Research and Development Program of China (2016YFA0100900).

Author Contributions: Meirong Yang, Yejing Wang and Huawei He conceived and designed the experiments; Meirong Yang, Peng Wang, Liying Liu, and Lisha Ai performed the experiments; Meirong Yang and Gang Tao analyzed the data; Gang Tao, Rui Cai, and Hua Zuo contributed reagents/materials/analysis tools; Meirong Yang,

Huawei He, Yejing Wang, Ping Zhao wrote the paper; Meirong Yang, Huawei He, Umar Ahmad and Chuanbin Mao revised the paper.

Conflicts of Interest: The authors declare no conflict of interest.

References

1. Edwards, R.; Harding, K.G. Bacteria and wound healing. *Curr. Opin. Infect. Dis.* **2004**, *17*, 91–96. [CrossRef] [PubMed]
2. Song, D.W.; Kim, S.H.; Kim, H.H.; Lee, K.H.; Ki, C.S.; Park, Y.H. Multi-biofunction of antimicrobial peptide-immobilized silk fibroin nanofiber membrane: Implications for wound healing. *Acta Biomater.* **2016**, *39*, 146–155. [CrossRef] [PubMed]
3. Sheikh, E.S.; Sheikh, E.S.; Fetterolf, D.E. Use of dehydrated human amniotic membrane allografts to promote healing in patients with refractory non healing wounds. *Int. Wound J.* **2014**, *11*, 711–717. [CrossRef] [PubMed]
4. Boateng, J.; Catanzano, O. Advanced therapeutic dressings for effective wound healing. *J. Pharm. Sci.* **2015**, *104*, 3653–3680. [CrossRef] [PubMed]
5. Stumpf, T.R.; Pertile, R.A.; Rambo, C.R.; Porto, L.M. Enriched glucose and dextrin mannitol-based media modulates fibroblast behavior on bacterial cellulose membranes. *Mater. Sci. Eng. C* **2013**, *33*, 4739–4745. [CrossRef] [PubMed]
6. Xie, Z.; Aphale, N.V.; Kadapure, T.D.; Wadajkar, A.S.; Orr, S.; Gyawali, D.; Qian, G.; Nguyen, K.T.; Yang, J. Design of antimicrobial peptides conjugated biodegradable citric acid derived hydrogels for wound healing. *J. Biomed. Mater. Res. A* **2015**, *103*, 3907–3918. [CrossRef] [PubMed]
7. Zhang, Y.-Q. Applications of natural silk protein sericin in biomaterials. *Biotechnol. Adv.* **2002**, *20*, 91–100. [CrossRef]
8. Kundu, S.C.; Dash, B.C.; Dash, R.; Kaplan, D.L. Natural protective glue protein, sericin bioengineered by silkworms: Potential for biomedical and biotechnological applications. *Prog. Polym. Sci.* **2008**, *33*, 998–1012. [CrossRef]
9. Lamboni, L.; Gauthier, M.; Yang, G.; Wang, Q. Silk sericin: A versatile material for tissue engineering and drug delivery. *Biotechnol. Adv.* **2015**, *33*, 1855–1867. [CrossRef] [PubMed]
10. Wang, Z.; Zhang, Y.; Zhang, J.; Huang, L.; Liu, J.; Li, Y.; Zhang, G.; Kundu, S.C.; Wang, L. Exploring natural silk protein sericin for regenerative medicine: An injectable, photoluminescent, cell-adhesive 3D hydrogel. *Sci. Rep.* **2014**, *4*, 7064. [CrossRef] [PubMed]
11. Kundu, B.; Kundu, S.C. Silk sericin/polyacrylamide in situ forming hydrogels for dermal reconstruction. *Biomaterials* **2012**, *33*, 7456–7467. [CrossRef] [PubMed]
12. Cho, K.Y.; Moon, J.Y.; Lee, Y.W.; Lee, K.G.; Yeo, J.H.; Kweon, H.Y.; Kim, K.H.; Cho, C.S. Preparation of self-assembled silk sericin nanoparticles. *Int. J. Biol. Macromol.* **2003**, *32*, 36–42. [CrossRef]
13. Nayak, S.; Talukdar, S.; Kundu, S.C. Potential of 2D crosslinked sericin membranes with improved biostability for skin tissue engineering. *Cell Tissue Res.* **2012**, *347*, 783–794. [CrossRef] [PubMed]
14. Aramwit, P.; Siritientong, T.; Kanokpanont, S.; Srichana, T. Formulation and characterization of silk sericin-PVA scaffold crosslinked with genipin. *Int. J. Biol. Macromol.* **2010**, *47*, 668–675. [CrossRef] [PubMed]
15. He, H.; Cai, R.; Wang, Y.; Tao, G.; Guo, P.; Zuo, H.; Chen, L.; Liu, X.; Zhao, P.; Xia, Q. Preparation and characterization of silk sericin/PVA blend film with silver nanoparticles for potential antimicrobial application. *Int. J. Biol. Macromol.* **2017**, *104*, 457–464. [CrossRef] [PubMed]
16. Bellamkonda, R.; Ranieri, J.P.; Bouche, N.; Aebischer, P. Hydrogel-based three-dimensional matrix for neural cells. *J. Biomed. Mater. Res.* **1995**, *29*, 663–671. [CrossRef] [PubMed]
17. Yodmuang, S.; McNamara, S.L.; Nover, A.B.; Mandal, B.B.; Agarwal, M.; Kelly, T.A.; Chao, P.H.; Hung, C.; Kaplan, D.L.; Vunjak-Novakovic, G. Silk microfiber-reinforced silk hydrogel composites for functional cartilage tissue repair. *Acta Biomater.* **2015**, *11*, 27–36. [CrossRef] [PubMed]
18. Lima, E.G.; Bian, L.; Ng, K.W.; Mauck, R.L.; Byers, B.A.; Tuan, R.S.; Ateshian, G.A.; Hung, C.T. The beneficial effect of delayed compressive loading on tissue-engineered cartilage constructs cultured with TGF-β3. *Osteoarthr. Cartil.* **2007**, *15*, 1025–1033. [CrossRef] [PubMed]
19. Stokols, S.; Tuszynski, M.H. Freeze-dried agarose scaffolds with uniaxial channels stimulate and guide linear axonal growth following spinal cord injury. *Biomaterials* **2006**, *27*, 443–451. [CrossRef] [PubMed]

20. Gruber, H.E.; Fisher, E.C., Jr.; Desai, B.; Stasky, A.A.; Hoelscher, G.; Hanley, E.N., Jr. Human intervertebral disc cells from the annulus: Three-dimensional culture in agarose or alginate and responsiveness to TGF-β1. *Exp. Cell Res.* **1997**, *235*, 13–21. [CrossRef] [PubMed]

21. Doulabi, A.; Mequanint, K.; Mohammadi, H. Blends and nanocomposite biomaterials for articular cartilage tissue engineering. *Materials* **2014**, *7*, 5327–5355. [CrossRef] [PubMed]

22. Bhat, S.; Tripathi, A.; Kumar, A. Supermacroprous chitosan-agarose-gelatin cryogels: In vitro characterization and in vivo assessment for cartilage tissue engineering. *J. R. Soc. Interfaces* **2011**, *8*, 540–554. [CrossRef] [PubMed]

23. Nikaido, H. Multidrug resistance in bacteria. *Annu. Rev. Biochem.* **2009**, *78*, 119–146. [CrossRef] [PubMed]

24. Benkerroum, N. Antimicrobial activity of lysozyme with special relevance to milk. *Afr. J. Biotechnol.* **2008**, *7*, 4856–4867.

25. Hamdani, A.M.; Wani, I.A.; Bhat, N.A.; Siddiqi, R.A. Effect of guar gum conjugation on functional, antioxidant and antimicrobial activity of egg white lysozyme. *Food Chem.* **2018**, *240*, 1201–1209. [CrossRef] [PubMed]

26. Gu, J.; Su, Y.; Liu, P.; Li, P.; Yang, P. An environmentally benign antimicrobial coating based on a protein supramolecular assembly. *ACS Appl. Mater. Interfaces* **2017**, *9*, 198–210. [CrossRef] [PubMed]

27. Zhao, J.; Wang, X.; Kuang, Y.; Zhang, Y.; Shi, X.; Liu, X.; Deng, H. Multilayer composite beads constructed via layer-by-layer self-assembly for lysozyme controlled release. *RSC Adv.* **2014**, *4*, 24369–24376. [CrossRef]

28. Yang, M.; Wang, Y.; Cai, R.; Tao, G.; Chang, H.; Ding, C.; Zuo, H.; Shen, H.; Zhao, P.; He, H. Preparation and characterization of silk sericin/glycerol films coated with silver nanoparticles for antibacterial application. *Sci. Adv. Mater.* **2018**, *10*, 1–8. [CrossRef]

29. He, H.; Tao, G.; Wang, Y.; Cai, R.; Guo, P.; Chen, L.; Zuo, H.; Zhao, P.; Xia, Q. In situ green synthesis and characterization of sericin-silver nanoparticle composite with effective antibacterial activity and good biocompatibility. *Mater. Sci. Eng. C* **2017**, *80*, 509–516. [CrossRef] [PubMed]

30. Nazarov, R.; Jin, H.J.; Kaplan, D.L. Porous 3-D scaffolds from regenerated silk fibroin. *Biomacromolecules* **2004**, *5*, 718–726. [CrossRef] [PubMed]

31. Tao, G.; Cai, R.; Wang, Y.; Song, K.; Guo, P.; Zhao, P.; Zuo, H.; He, H. Biosynthesis and characterization of agnps–silk/PVA film for potential packaging application. *Materials* **2017**, *10*, 667. [CrossRef] [PubMed]

32. Cai, R.; Tao, G.; He, H.; Guo, P.; Yang, M.; Ding, C.; Zuo, H.; Wang, L.; Zhao, P.; Wang, Y. In situ synthesis of silver nanoparticles on the polyelectrolyte-coated sericin/pva film for enhanced antibacterial application. *Materials* **2017**, *10*, 967. [CrossRef] [PubMed]

33. Jiang, S.; Qin, Y.; Yang, J.; Li, M.; Xiong, L.; Sun, Q. Enhanced antibacterial activity of lysozyme immobilized on chitin nanowhiskers. *Food Chem.* **2017**, *221*, 1507–1513. [CrossRef] [PubMed]

34. Chen, C.; Pan, F.; Zhang, S.; Hu, J.; Cao, M.; Wang, J.; Xu, H.; Zhao, X.; Lu, J.R. Antibacterial activities of short designer peptides: A link between propensity for nanostructuring and capacity for membrane destabilization. *Biomacromolecules* **2010**, *11*, 402–411. [CrossRef] [PubMed]

35. Wu, Y.-B.; Yu, S.-H.; Mi, F.-L.; Wu, C.-W.; Shyu, S.-S.; Peng, C.-K.; Chao, A.-C. Preparation and characterization on mechanical and antibacterial properties of chitsoan/cellulose blends. *Carbohydr. Polym.* **2004**, *57*, 435–440. [CrossRef]

36. Zhou, B.; Li, Y.; Deng, H.; Hu, Y.; Li, B. Antibacterial multilayer films fabricated by layer-by-layer immobilizing lysozyme and gold nanoparticles on nanofibers. *Colloids Surf. B* **2014**, *116*, 432–438. [CrossRef] [PubMed]

37. Zhou, T.; Huang, Y.; Li, W.; Cai, Z.; Luo, F.; Yang, C.J.; Chen, X. Facile synthesis of red-emitting lysozyme-stabilized ag nanoclusters. *Nanoscale* **2012**, *4*, 5312–5315. [CrossRef] [PubMed]

38. Chuang, C.-C.; Prasannan, A.; Huang, B.-R.; Hong, P.-D.; Chiang, M.-Y. Simple synthesis of eco-friendly multifunctional silk-sericin capped zinc oxide nanorods and their potential for fabrication of hydrogen sensors and UV photodetectors. *ACS Sustain. Chem. Eng.* **2017**, *5*, 4002–4010. [CrossRef]

39. Al-Abboodi, A.; Fu, J.; Doran, P.M.; Tan, T.T.Y.; Chan, P.P.Y. Injectable 3D hydrogel scaffold with tailorable porosity post-implantation. *Adv. Healthc. Mater.* **2014**, *3*, 725–736. [CrossRef] [PubMed]

40. Annabi, N.; Nichol, J.W.; Zhong, X.; Ji, C.; Koshy, S.; Khademhosseini, A.; Dehghani, F. Controlling the porosity and microarchitecture of hydrogels for tissue engineering. *Tissue Eng. Part B* **2010**, *16*, 371–383. [CrossRef] [PubMed]

41. Zhang, X.; Wyeth, P. Using FTIR spectroscopy to detect sericin on historic silk. *Sci. China Chem.* **2010**, *53*, 626–631. [CrossRef]

42. Verma, V.K.; Subbiah, S. Prospects of silk sericin as an adsorbent for removal of ibuprofen from aqueous solution. *Ind. Eng. Chem. Res.* **2017**, *56*, 10142–10154. [CrossRef]

43. Singh, Y.P.; Bhardwaj, N.; Mandal, B.B. Potential of agarose/silk fibroin blended hydrogel for in vitro cartilage tissue engineering. *ACS Appl. Mater. Inteerfaces* **2016**, *8*, 21236–21249. [CrossRef] [PubMed]

44. Wu, T.; Huang, J.; Jiang, Y.; Hu, Y.; Ye, X.; Liu, D.; Chen, J. Formation of hydrogels based on chitosan/alginate for the delivery of lysozyme and their antibacterial activity. *Food Chem.* **2018**, *240*, 361–369. [CrossRef] [PubMed]

45. Gupta, D.; Agrawal, A.; Rangi, A. Extraction and characterization of silk sericin. *Indian J. Fibre Text.* **2014**, *39*, 364–372.

46. Tao, G.; Liu, L.; Wang, Y.; Chang, H.; Zhao, P.; Zuo, H.; He, H. Characterization of silver nanoparticlein situsynthesis on porous sericin gel for antibacterial application. *J. Nanomater.* **2016**, *2016*, 1–8. [CrossRef]

47. Tao, G.; Wang, Y.; Liu, L.; Chang, H.; Zhao, P.; He, H. Preparation and characterization of silver nanoparticles composited on polyelectrolyte film coated sericin gel for enhanced antibacterial application. *Sci. Adv. Mater.* **2016**, *8*, 1547–1552. [CrossRef]

48. Cai, R.; Tao, G.; He, H.; Song, K.; Zuo, H.; Jiang, W.; Wang, Y. One-step synthesis of silver nanoparticles on polydopamine-coated sericin/polyvinyl alcohol composite films for potential antimicrobial applications. *Molecules* **2017**, *22*, 721. [CrossRef] [PubMed]

49. Cao, T.T.; Zhang, Y.Q. Processing and characterization of silk sericin from bombyx mori and its application in biomaterials and biomedicines. *Mater. Sci. Eng. C* **2016**, *61*, 940–952. [CrossRef] [PubMed]

50. Silva, V.R.; Hamerski, F.; Weschenfelder, T.A.; Ribani, M.; Gimenes, M.L.; Scheer, A.P. Equilibrium, kinetic, and thermodynamic studies on the biosorption of bordeaux s dye by sericin powder derived from cocoons of the silkwormbombyx mori. *Desalin. Water Treat.* **2015**, *57*, 5119–5129. [CrossRef]

51. Shemesh, M.; Zilberman, M. Structure-property effects of novel bioresorbable hybrid structures with controlled release of analgesic drugs for wound healing applications. *Acta Biomater.* **2014**, *10*, 1380–1391. [CrossRef] [PubMed]

52. Atar-Froyman, L.; Sharon, A.; Weiss, E.I.; Houri-Haddad, Y.; Kesler-Shvero, D.; Domb, A.J.; Pilo, R.; Beyth, N. Anti-biofilm properties of wound dressing incorporating nonrelease polycationic antimicrobials. *Biomaterials* **2015**, *46*, 141–148. [CrossRef] [PubMed]

53. Liu, Y.; Sun, Y.; Xu, Y.; Feng, H.; Fu, S.; Tang, J.; Liu, W.; Sun, D.; Jiang, H.; Xu, S. Preparation and evaluation of lysozyme-loaded nanoparticles coated with poly-gamma-glutamic acid and chitosan. *Int. J. Biol. Macromol.* **2013**, *59*, 201–207. [CrossRef] [PubMed]

54. Ibrahim, H.R.; Higashiguchi, S.; Juneja, L.R.; Kim, M.; Yamamoto, T. A structural phase of heat-denatured lysozyme with novel antimicrobial action. *J. Agric. Food Chem.* **1996**, *44*, 1416–1423. [CrossRef]

55. Bernkop-Schnürch, A.; Krist, S.; Vehabovic, M.; Valenta, C. Synthesis and evaluation of lysozyme derivatives exhibiting an enhanced antimicrobial action. *Eur. J. Pharm. Sci.* **1998**, *6*, 301–306. [CrossRef]

56. Wu, T.; Wu, C.; Fu, S.; Wang, L.; Yuan, C.; Chen, S.; Hu, Y. Integration of lysozyme into chitosan nanoparticles for improving antibacterial activity. *Carbohydr. Polym.* **2017**, *155*, 192–200. [CrossRef] [PubMed]

57. Yu, K.; Lu, F.; Li, Q.; Chen, H.; Lu, B.; Liu, J.; Li, Z.; Dai, F.; Wu, D.; Lan, G. In situ assembly of ag nanoparticles (agnps) on porous silkworm cocoon-based would film: Enhanced antimicrobial and wound healing activity. *Sci. Rep.* **2017**, *7*, 2107. [CrossRef] [PubMed]

58. Kamoun, E.A.; Kenawy, E.-R.S.; Tamer, T.M.; El-Meligy, M.A.; Mohy Eldin, M.S. Poly (vinyl alcohol)-alginate physically crosslinked hydrogel membranes for wound dressing applications: Characterization and bio-evaluation. *Arab. J. Chem.* **2015**, *8*, 38–47. [CrossRef]

59. Yao, Y.; Liu, H.; Ding, X.; Jing, X.; Gong, X.; Zhou, G.; Fan, Y. Preparation and characterization of silk fibroin/poly(L-lactide-co-ε-caprolactone) nanofibrous membranes for tissue engineering applications. *J. Bioact. Compat. Polym.* **2015**, *30*, 633–648. [CrossRef]

Article

Effect of Laminating Pressure on Polymeric Multilayer Nanofibrous Membranes for Liquid Filtration

Fatma Yalcinkaya [1,2,*] and Jakub Hruza [1,2]

[1] Department of Nanotechnology and Informatics, Institute of Nanomaterials, Advanced Technologies and Innovation, Technical University of Liberec, Studentska 1402/2, 46117 Liberec, Czech Republic; jakub.hruza@tul.cz

[2] Institute for New Technologies and Applied Informatics, Faculty of Mechatronics, Studentska 1402/2, 46117 Liberec, Czech Republic

* Correspondence: yenertex@hotmail.com; Tel.: +420-485-353-389

Received: 8 February 2018; Accepted: 23 April 2018; Published: 24 April 2018

Abstract: In the new century, electrospun nanofibrous webs are widely employed in various applications due to their specific surface area and porous structure with narrow pore size. The mechanical properties have a major influence on the applications of nanofiber webs. Lamination technology is an important method for improving the mechanical strength of nanofiber webs. In this study, the influence of laminating pressure on the properties of polyacrylonitrile (PAN) and polyvinylidene fluoride (PVDF) nanofibers/laminate was investigated. Heat-press lamination was carried out at three different pressures, and the surface morphologies of the multilayer nanofibrous membranes were observed under an optical microscope. In addition, air permeability, water filtration, and contact angle experiments were performed to examine the effect of laminating pressure on the breathability, water permeability and surface wettability of multilayer nanofibrous membranes. A bursting strength test was developed and applied to measure the maximum bursting pressure of the nanofibers from the laminated surface. A water filtration test was performed using a cross-flow unit. Based on the results of the tests, the optimum laminating pressure was determined for both PAN and PVDF multilayer nanofibrous membranes to prepare suitable microfilters for liquid filtration.

Keywords: nanofiber; lamination; water filtration

---

1. Introduction

Electrospun polymeric nanofiber web has gained increasing importance in the production of engineered surfaces with sub-micron to nano-scale fibers. The widely employed areas of electrospun nanofibers are tissue engineering [1,2], wound healing [3], drug delivery systems [4], composites [5], solar cells [6], protective clothing [7], lithium-ion batteries [8,9], sensors [10–12], gas sensors and separators [13,14], and air and water filtration [15–19], owing to their high surface-area-to-volume ratio, highly porous structure and extremely narrow pore size. The main factor influencing the application of nanofibers is their mechanical properties. An electrospun nanofiber has very poor mechanical strength due to low contact and adhesion between the fibers.

Several methods have been developed to provide suitable mechanical strength to electrospun nanofibers. One of the most common methods is to blend several polymers, the advantage of which is that it is easy and low-cost. However, it is necessary to use polymers which can dissolve in the same solvent system, of which there are only a few [20,21]. In another method, lamination was achieved using an epoxy composite. In this method, an electrospun layer was laid on the epoxy/curing agent in a mold and then the curing process was performed for a period of 16 h [22].

Nanofiber reinforced nanomaterials such as carbon nanotubes and graphene represent another route for improving the mechanical strength of nanofibers. However, this method is costly and in some cases has a low efficiency [23,24]. Charles et al. [25] used a dip coating method to improve the mechanical strength of nanofibers. They described the mechanical properties of a composite system comprising hydroxyapatite (HA)-coated poly (L-lactic acid) (PLLA) fibers in a poly (ε-caprolactone) (PCL) matrix. A biomimetic method was used to coat the fibers with HA, and a dip-coating procedure served for the application of PCL to the coated fibers. The composite was formed into a bar using compression molding at low temperatures. The disadvantage of this method is that it is a time- and chemical-consuming procedure. Xu et al. [26] developed a self-reinforcing method to enhance the strength of polycarbonate (PC) membranes. In this method, the PC nanofibers were immersed into a solvent (30%) and non-solvent (70%) mixture, which resulted in enhancement in the strength (128%) owing to the fusion of junction points. The entire porous structure on the PC nanofibers was destroyed, which greatly impaired the application of the membranes. The thermal lamination method is one of the most reliable, repeatable, time-saving, environmentally friendly and cost-effective methods used to adhere two surfaces. In this method, an adhesive polymer or web is usually applied between two surfaces. Using heat and pressure, the surfaces adhere to each other. There is a large amount of research related to improving the strength of nanofibers; however, the number of reports is still limited compared to those dealing with lamination technology. Jiricek [27,28] and Yalcinkaya et al. [29,30] used a bi-component polyethylene (PE)/polypropylene (PP) spunbond as a supporting layer for nanofiber layers. A fusing machine was used for the lamination process. A nanofiber layer was adhered on the outer surface of the bi-component due to the low melting point of PE. The resultant multilayer nanofibrous membranes were used for micro and nanofilters. The supporting material and the density of the nanofibers have an influence on the water and air permeability of the multilayer materials. Yoon et al. [7] laminated nanofibers with various densities of polyurethane (PUR) fiber onto different textile surfaces using an adhesive web. The results showed that the various multilayer nanofibrous membrane structure designs had a considerable influence on the degrees of breathability and waterproofness of the textile surfaces. In our previous work [31], it was observed that both the supporting material and the density of the nanofiber web have an influence on the water permeability of the multilayer nanofibrous membranes. The lower area weight with open structure supporting materials has higher water flux and permeability. Kanafchian et al. [32] used a heat-press technique to laminate the polyacrylonitrile (PAN) nanofiber on a polypropylene spunbond at various laminating temperatures. It was observed that, when the applied temperature is lower than the melting point of polypropylene spunbond, the nanofiber web remains unchanged. Moreover, the increase in temperature increased the adhesion between nanofiber while decreasing the air permeability.

Although the system and parameters of the electrospinning process have been well analyzed, there is still a lack of information about a proper lamination technique for nanofiber webs. So far, mainly the effect of temperature on the lamination of electrospun nanofiber webs has been investigated using heat-press methods [32–34]. Yao et al. [33] studied the effect of the heat-press temperature, pressure, and duration on the morphological and mechanical characteristics of an electrospun membrane for membrane distillation. The results showed that the temperature and duration of the heat-press play more important roles than the pressure in the heat-press treatment. However, the pressure varied between 0.7 and 9.8 kPa at 150 °C during a 2-h period, which is time- and energy-consuming and therefore a more comprehensive parametric study is required. The aim of this study is to consider the influence of laminating pressure on the properties of multilayer nanofibrous membranes. Polyvinylidene (PVDF) and polyacrylonitrile (PAN) are the most commonly used nanofiber layers owing to their chemical and thermal stability. Herein, both polymers were electrospun using a semi-industrial scale nanofiber production method. The nanofiber layers were laminated onto a nonwoven surface to improve their mechanical strength via the heat-press method, which is easy to scale and is an energy-saving method at various applied pressures. Other conditions, such as temperature and the duration of lamination, were kept stable. The effect of the laminating pressure on the nanofiber web has not yet been well reported. To investigate the effect of the pressure on the nanofiber/laminate process, surface morphology under an optical microscope, the minimum bursting pressure, air permeability, water permeability and contact angle tests were applied. Our objective was to optimize the lamination technology and produce multilayer nanofibrous membranes for suitable application in liquid filtration. Another novelty of this paper is that nanofiber layers were produced by using the industrial equipment, and that the layers strongly adhered on the supporting layer without any damage using various lamination pressures to improve their application in liquid filtration.

2. Materials and Methods

2.1. Preparation of Nanofibre Webs

13 wt. % PVDF (Solef 1015, Bruxelles, Belgium) was prepared in *N,N*-dimethylacetamide (DMAC) and 8 wt. % PAN (150 kDa H-polymer, Elmarco, Liberec, Czech Republic) was prepared in *N,N*-dimethylformamide. The solvents were purchased from Penta s.r.o. (Prague, Czech Republic). The solutions were stirred overnight using a magnetic stirrer. Nanofiber webs were prepared using the semi-industrial Nanospider electrospinning device (Elmarco, Liberec, Czech Republic) as shown in Figure 1.

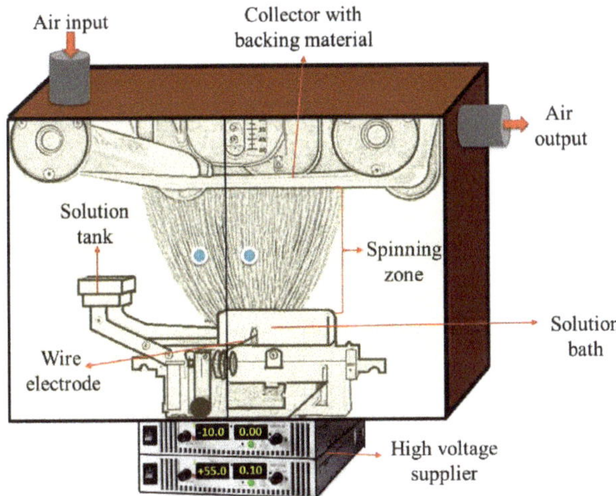

Figure 1. Schematic diagram of an electrospinning unit.

The solution is placed in a solution tank, which is a closed system and connected to a solution bath. The wire electrode passes along a metal orifice in the middle of the solution bath. The solution bath is moved back and forward to feed the surface of the wire electrode. The solution bath feeds the polymer solution on a moving stainless steel wire. The speed of the bath is 240 mm/s. A high voltage supplier is connected to a positively charged wire electrode (55 kV). A second wire electrode, which is connected to a negatively charged voltage supplier (−15 kV), is placed on the top of the spinning unit. A conveyor backing material is placed between the two electrodes. The spinning takes place between the two electrodes. The nanofiber web is collected on baking paper moving in front of the collector electrode. The distance between the electrodes is 188 mm. The distance between the second electrode and the supporting backing material is 2 mm. The speed of the backing material for PAN and PVDF is 15 mm/min and 20 mm/min, respectively. The amount of solution on the wire is controlled with the speed of the solution bath, the diameter of the wire (0.2 mm) and the diameter of the metal orifice (0.6 mm) in the middle of the solution bath. No solution dipping was observed. An air conditioning unit is used to control the humidity and temperature of the spinning in a closed chamber. The temperature and humidity of the input air are set to 23 °C and 20% by the air-conditioning system. The volumes of air input and output are 100 and 115 m^3/h, respectively. The area weight of the PVDF and PAN nanofibers was set at 3 g/m^2.

2.2. Lamination of Nanofibre Webs

The prepared nanofiber webs were cut into A4 size (210 × 297 mm^2). As a supporting layer, 120 g/m^2 of polyethylene terephthalate spunbond nonwoven and 12 g/m^2 of adhesive web were used (the supplier information is not given). Heat-press equipment (Pracovni Stroje, Teplice, Czech Republic) was used for the lamination process (Figure 2). In this equipment there are two metallic hot plates (upper and lower) used under pressure. The nanofibers, a copolyamide adhesive web and a polyethylene terephthalate spunbond supporting layer were placed between the two hot plates. Two silicon layers were used to block direct contact between the multilayer nanofibrous membranes and the hot plates. The heat was applied (130 °C) for a duration of 3 min. Pressures of 50, 75 and 100 kN were applied between the upper and lower plates. For each pressure, PVDF and PAN nanofiber webs were used. The resultant multilayer nanofibrous membranes were termed PAN50, PAN75, PAN100, PVDF50, PVDF75, and PVDF100 according to the pressure value.

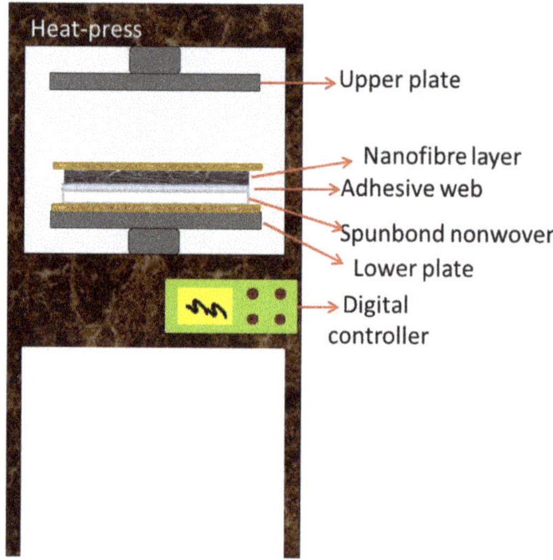

Figure 2. Schematic design of the heat-press equipment and replacement of the multilayer nanofibrous membranes.

2.3. Characterization of the Multilayer Nanofibrous Membranes

The surface morphology of the electrospun fibers and laminated multilayer nanofibrous membranes was observed using a scanning electron microscope (SEM, Vega 3SB, Brno, Czech Republic). From each sample, at least 50 fibers were measured. The fiber diameter was analyzed using the Image-J program (free online program). The Origin-Lab program was used to evaluate the diameter distribution. The surface contact angle of the samples was measured using a Krüss Drop Shape Analyzer DS4 (Krüss GmbH, Hamburg, Germany), at five different points, using distilled water (surface tension 72.0 mN m^{-1}) and ethylene glycol (surface tension 47.3 mN m^{-1}) on the clean and dry samples at room temperature. The air permeability of the multilayer nanofibrous membranes was tested using an SDL

ATLAS Air Permeability Tester (@200 Pa and 20 cm^2, Rock Hill, SC, USA). At least three measurements were taken for each sample.

The maximum, average and minimum pore sizes were determined by a bubble point measurement device, which was developed in our laboratory. The bubble point test allowed the size of the pores of the porous material to be measured. The pore flow means a set of continuous hole channels connecting the opposite sides of the porous material (see Figure 3). At least three measurements were taken.

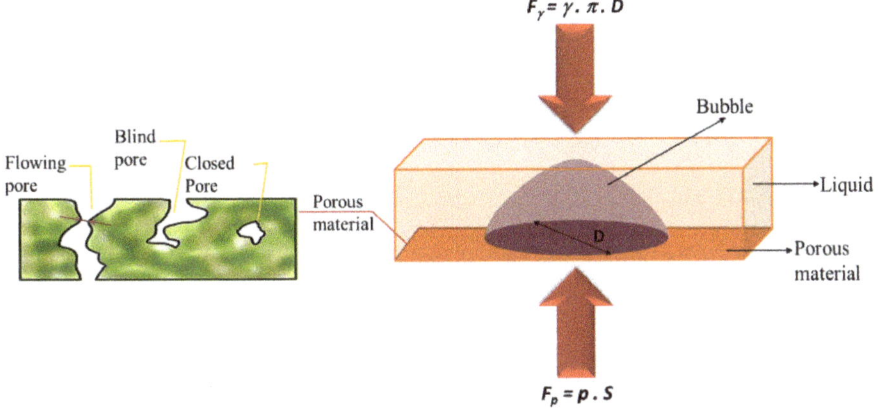

Figure 3. Schematic diagram of pore flow and forces acting on a pore.

The main part of the method was to control the pressure needed to pass a liquid through the tested porous material and for wetting the sample. This is because the wetting force (and hence the opposite force required to extrude the liquid) depends on the pore circumference. The principle for calculating the pore size is shown in Figure 3 and Equations (1) and (2).

$$F_\gamma = \gamma \pi D \tag{1}$$

$$Fp = pS \tag{2}$$

where F_γ is the force given by surface tension γ of the liquid around the perimeter of πD. The force Fp is given by external pressure p displacing the liquid from the pores and acting on the surface of pore S.

It is possible to calculate the magnitude of the force given by the surface tension and the force given by the pore pressurizing fluid. By increasing the air pressure and measuring its flow through the sample, the size of the average and minimum pores can also be determined. In this case, it is necessary to compare the pressure curve of the wet sample with the pressure curve of the dry sample (see Figure 4). The dry sample pressure curve required to determine the mean and minimum pores is also applicable for determining the air permeability coefficient (K) of the sample calculated according to the relation (Equation (3)):

$$K = Q/(\Delta p\, A) \tag{3}$$

where Q is the air flow rate (m^2/s), Δp is the pressure drop of the sample, and A is the area of air flow (m^2).

When the pressure increases in the dry sample, the flow rate also increases. Conversely, in the wet sample, at the beginning, there is no flow because all the pores are filled with the liquid. At a certain pressure, the gas empties the largest pore, which determines the minimum pore size, and gas begins to flow through the wet sample. The intersection between the calculated half-dry and the wet sample gives the mean flow pore size. When all the pores are emptied, an intersection between the wet and dry curve will be observed. This means the relation between the applied pressure and the detected flow becomes linear and the intersection of the wet and dry curve represents the detected minimum pore size.

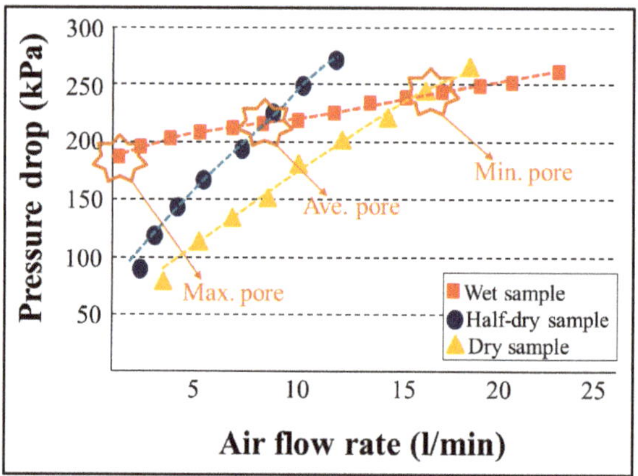

Figure 4. Example of a pressure drop determination to calculate the pore size.

The bursting strength of the multilayer nanofibrous membrane was tested, and the maximum delamination pressure was recorded. The testing device was developed in our laboratory as shown in Figure 5. In this test, the samples were placed between two rings, and the nanofiber side of the samples was placed on the upper side. The sample size was 47 mm in diameter. Pressurized water was sent to the membrane, and the hydrostatic pressure was measured using a pressure controller, which was placed in front of the membrane and connected to a computer. The hydrostatic pressure was increased gradually, and as soon as the nanofiber layer burst, the pressure value on the screen decreased sharply. The maximum pressure value was recorded as the bursting strength of the membrane. The testing samples are shown in Figure 5. After bursting, the nanofiber layer delaminated from the surface of the multilayer membrane. At least three measurements were taken for each membrane.

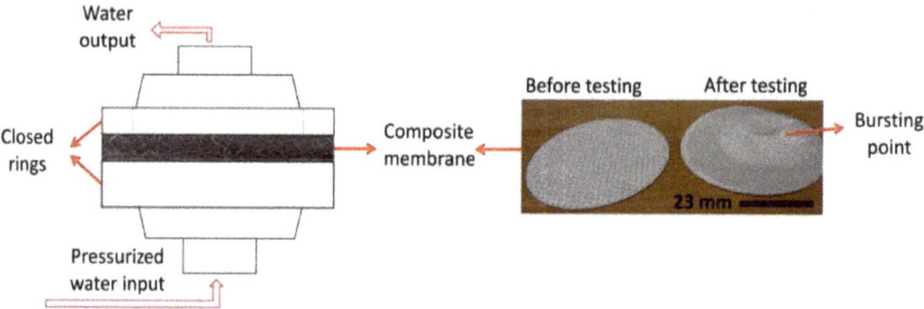

Figure 5. Bursting strength testing unit.

A lab-scale cross-flow filtration unit was developed as shown in Figure 6. Tap water was used as the feed solution. The maximum amount of feed solution was 1500 mL. The flux (F) and the permeability (k) of the membranes were calculated according to Equations (4) and (5) [31,35]:

$$F = \frac{1}{A}\frac{dV}{dt}$$

(4)

$$k = \frac{F}{p} \tag{5}$$

where A is the effective membrane area (m^2), V is the total volume of the permeate (L), p is the transmembrane pressure (bar), and t is the filtration time.

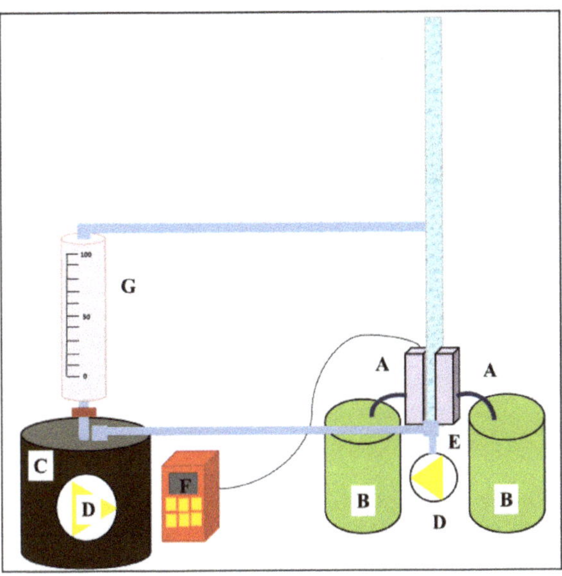

Figure 6. A cross-flow unit: (A) membrane cells; (B) permeate; (C) feed; (D) pump; (E) surface bubble cleaning; (F) pressure controller; (G) feed flow speed controller.

3. Results and Discussion

3.1. Characterization of Nanofibre Webs and Laminated Multilayer Nanofibrous Membranes

To characterize the nanofiber webs into the format of multilayer nanofibrous membranes, various aspects of their material properties were carefully considered. These properties include the fiber diameter, diameter distribution, mean pore size, wetting property, air permeability, and bursting strength.

The surface morphology of the nanofiber webs before and after lamination was imaged using a scanning electron microscope as shown in Figure 7. The average fiber diameter of the PAN and PVDF nanofibers before lamination was determined to be 171 nm and 221 nm, respectively. The diameter of the PVDF nanofibers was greater than that of the PAN nanofibers. The main reason was the difference in viscosity. In previous work [36], it was determined that a 14 wt. %. PVDF solution has a viscosity of 969 mPa.s, while 8 wt. %. PAN has 191 mPa.s. Based on the viscosity results, one can expect that a polymeric solution with a lower viscosity will have a lower fiber diameter. After the lamination process, neither the PVDF nor the PAN nanofiber diameters significantly changed at a pressure of 50 kN (Figure 8). However, significant changes were observed at laminating pressures of 75 kN and 100 kN. When the highest laminating pressure was applied (i.e., 100 kN), the diameter of the PAN and PVDF nanofibers increased by 14% and 25%, respectively. The fibers were flattened under heat and pressure, and the fiber diameter increased gradually. The highest fiber diameter changes were observed in the case of the PVDF nanofiber layer due to its lower glass transition and melting temperature compared to PAN [37,38].

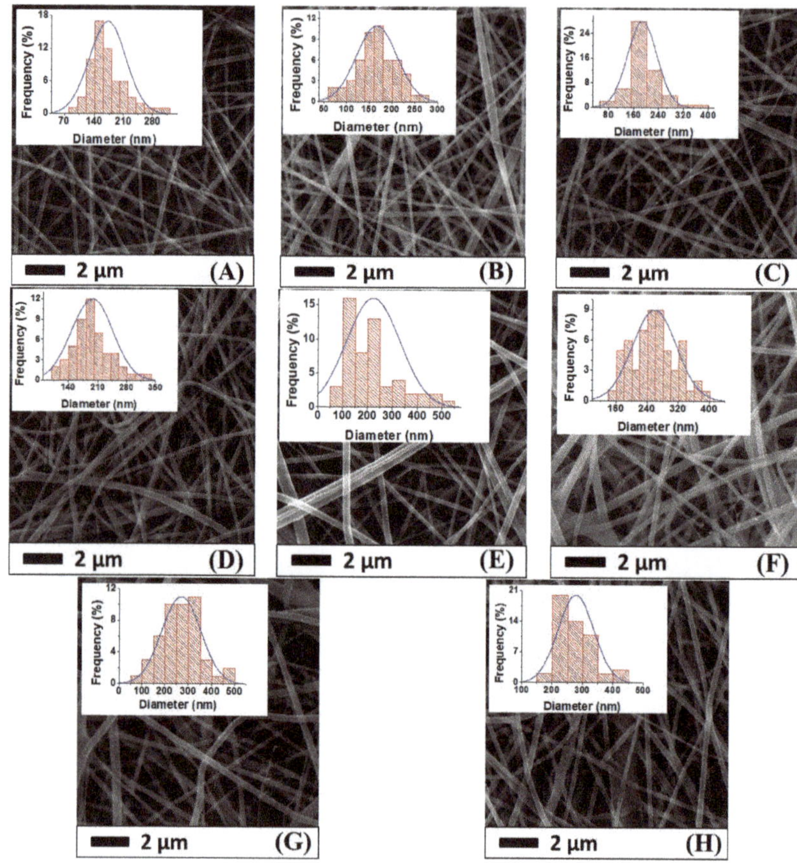

Figure 7. SEM images and fiber diameter distribution of (**A**) PAN nanofiber web before lamination; (**B**) PAN50; (**C**) PAN75; (**D**) PAN100; (**E**) PVDF nanofiber web before lamination; (**F**) PVDF50; (**G**) PVDF75; and (**H**) PVDF100.

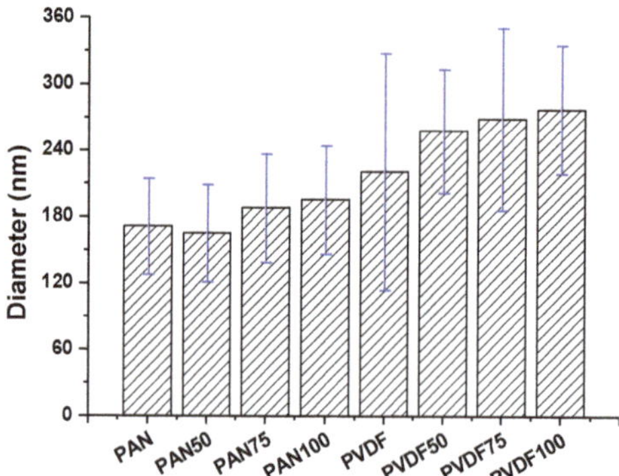

Figure 8. Fiber diameter of various multilayer nanofibrous membranes under different laminating pressures.

It was verified that there is a strong correlation between the electrospun fiber diameters and the polymer concentration, which has been well documented in the literature [39–41]. From the SEM images, the PAN and PVDF multilayer nanofibrous membranes exhibited bead-free surface morphology.

The average pore size of the membranes is given in Figure 9. Electrospun materials readily deform at low pressures. Since the tensile strength of PVDF and PAN nanofibers before lamination is quite low to withstand air pressure, their pore size was not measured.

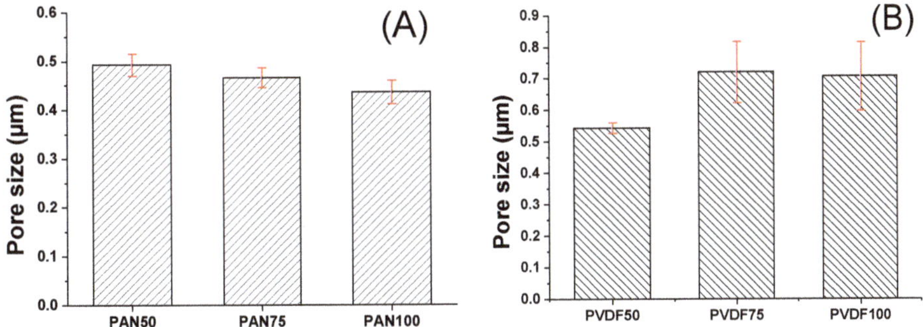

Figure 9. The relationship between the mean pore size and the laminating pressure of (**A**) PAN and (**B**) PVDF multilayer nanofibrous membranes.

In general, there is a correlation between the fiber diameter and the average pore size of the nanofibers. Reducing the fiber diameter increases the surface area and compact web structure, which results in a small pore size [42]. Bagherzadeh et al. [43] demonstrated a theoretical analysis to predict the pore size of electrospun nanofibers. According to their theory, at a given web porosity, increasing the fiber diameter and thickness of the web reduces the dimensions of the pores. This theory was validated experimentally, and the results were compared with the existing theory to predict the pore size distribution of nanofiber mats. Their results showed that the pore size significantly increased with an increase in fiber diameter, web porosity and density of the layers. In this work, the correlation between the diameter of the PVDF nanofibers and the mean pore size was compatible with the literature, while PAN showed an opposite correlation. The laminating pressure effect must be taken into consideration. The nanofiber layer did not change; only the fibers flattened after lamination due to the pressure. It was expected that a higher pressure would cause a lower pore size since the fibers flattened and melting adhesive filled more of the pores and covered the surface of the nanofibers as shown in Figure 10. The PAN multilayer nanofibrous membranes fulfilled this expectation while the PVDF did not. Figure 9B shows that the average pore size and the standard deviation of the pore size measurements increased with pressure, which could be due to possible damage of the PVDF nanofibers under high pressure. Gockeln et al. [44] investigated the influence of laminating pressures on the microstructure and electrochemical performance of the lithium-ion battery electrodes. The results indicated that all the laminated samples showed highly porous and homogeneous networks, while the pore size slightly decreased with an increase in laminating pressure. At higher pressures, the intrinsic electrical conductivity was improved due to more compression.

The water and ethylene glycol wettability of the PAN and PVDF multilayer nanofibrous membranes were examined by a contact angle measurement as shown in Figure 11. The surface energy and surface roughness are the dominant factors for the wettability. As can be seen from Figure 11, an increase in laminating pressure decreased the water and ethylene glycol contact angle of both PAN and PVDF multilayer nanofibrous membranes. Hence, ethylene glycol has a lower surface energy compared to water, with the differences in contact angle value being 20° for PAN and 30° for PVDF. Similar behavior was observed in the literature [45,46]. It is well known that when the surface energy is lowered, and surface roughness is raised, the hydrophobicity is enhanced [47–49]. With the help of heat, the higher laminating pressure on the surface may cause changes to the surface shape and make the surface flatter, which results in an increase in the surface wettability (Figure 11). The PVDF membranes showed hydrophobic characteristics at the lowest laminating pressure (i.e., 50 kN), while at higher laminating pressures they showed hydrophilic properties. By setting the lamination process parameters, one can prepare hydrophilic PVDF multilayer nanofibrous membranes without any surface modification.

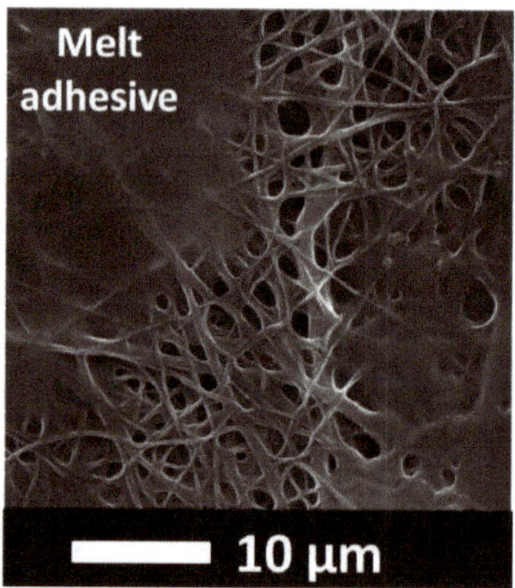

Figure 10. An illustration of adhesive melting over the surface of a nanofiber web, forming a non-porous film.

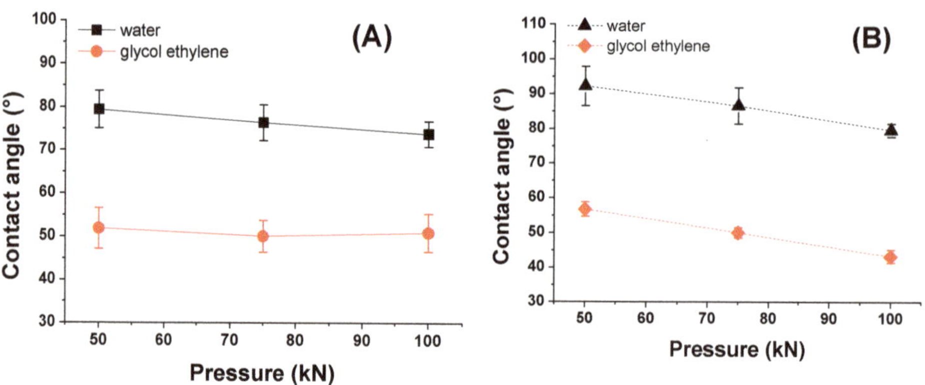

Figure 11. Contact angle vs. laminating pressure of (**A**) PAN; (**B**) PVDF multilayer nanofibrous membranes.

The morphology of the nanofiber webs, including their pore size, shape, size distribution and porosity, has a significant influence on the air permeability of the multilayer membrane. To investigate the effect of laminating pressure on the air permeability of the multilayer nanofibrous membranes, the samples were placed on a circular sample holder, and the air flow rates through the samples were measured (Figure 12A). Like the air flows, the areas of the sample and pressure drop remained constant during the measurement. Due to the weakness of neat nanofiber layers, the air permeability test was not performed. In a previous study [31], the tensile strength of the nanofiber layers was found to be between 3 and 4.33 (N/25 mm), which is extremely low to withstand any external force.

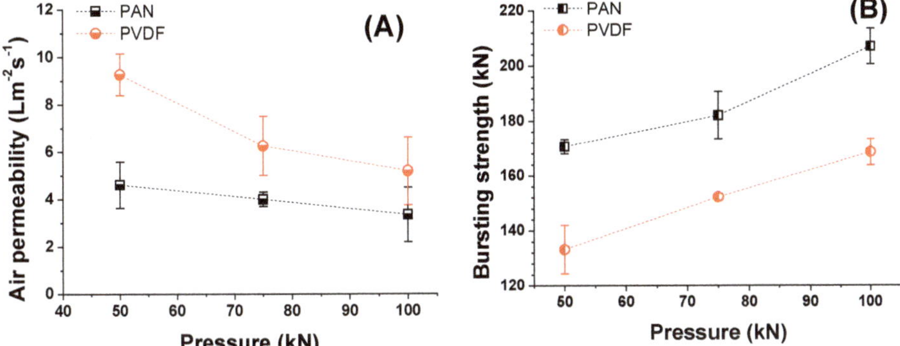

Figure 12. Influence of laminating pressure on (**A**) air permeability; and (**B**) bursting strength of multilayer nanofibrous membranes.

Abuzade et al. [50] studied the effects of the process parameters (e.g., concentration of solution, applied voltage) on the porosity and air permeability of an electrospun nanoweb. The results showed that the nanofiber diameter and size distribution are dominant parameters in controlling the pore sizes formed by the nanofiber intersections and air permeability of the electrospun web. Figure 12A showed that increasing the laminating pressure lowered the air permeability of the multilayer membranes. Compression of the melting adhesive, filling the pores of the nanofiber and nonwoven web, covered the surface of the thin nanofiber layer and created a non-porous film (Figure 10). As a result, the breathability of the membranes was decreased. Similarly, Kanafchian et al. [32] claimed that during the lamination, the melt adhesive penetrates through the nanofiber/fabric structure, which leads to filling of the pores of nanofibers and a decrease in air permeability. The PAN multilayer nanofibrous membrane has a lower air permeability than PVDF, mainly due to the lower fiber diameter of the PAN nanofibers compared to the PVDF nanofibers. Rajak et al. [51] prepared PAN nanofiber webs from various concentrations. The results indicated that changes in concentration affect the fiber diameter. At a higher concentration and fiber diameter, the air permeability has a higher value.

A bursting test was performed to determine the mechanical strength of the laminated layers, and the results are shown in Figure 12B. The test method has been developed in our laboratory. The maximum delamination point of the multilayer nanofibrous membranes was measured using hydrostatic pressure. The results showed that PAN nanofibers have a better adhesion to the supporting layer and a better bursting strength compared to PVDF. The adhesion between the layers is related to the material surface chemistry and its influence on adhesion, together with the properties of adhesive materials and interactions at the adhesive-substrate surface interface. Materials that can wet each other tend to have a better adhesion, and the wettability of the material is related to its surface energy. For instance, low surface energy materials such as poly(tetrafluoroethylene), ceramics, and silicon, are resistant to wetting and adhesive bonding [52]. Lee et al. [53] found that the surface energy of PAN is around 44 mJ/m^2, while this value was calculated as 54.1 mJ/m^2 by Pritykin et al. [54]. On the other hand, PVDF has a very low surface energy value of around 26 mJ/m^2 [55]. Due to the lower surface energy of PVDF compared to PAN, the adhesion between the layers is weaker, which results in low lamination strength. The results show that laminating pressure plays an important role in the bursting strength. By increasing the laminating pressure under heat, the melted adhesive fills the pores of the nanofibers and nonwovens and penetrates through the layers. A better mechanical strength is achieved due to the entanglement of the adhesive web and the layers. The results showed that the bursting strength of a material can be improved by adjusting the lamination conditions.

3.2. Evaluation of Liquid Filtration by Cross-Flow Filtration

Taking their practical applications into consideration, laminated multilayer nanofibrous membranes were used to further investigate their water permeability performance due to their hydrophilic, porous, small pore size and predominant mechanical properties for liquid filtration. A cross-flow filtration unit was prepared in our laboratory. Using Equation (5), the water permeability of the PAN and PVDF multilayer nanofibrous membranes was calculated (Figures 13 and 14).

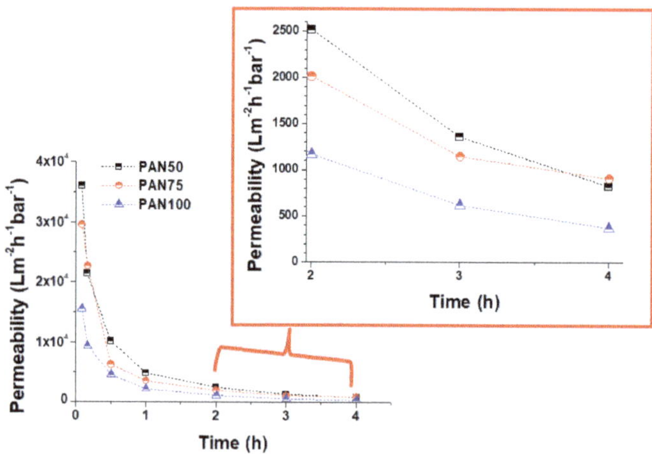

Figure 13. Permeability of PAN multilayer nanofibrous membranes at various laminating pressures over time.

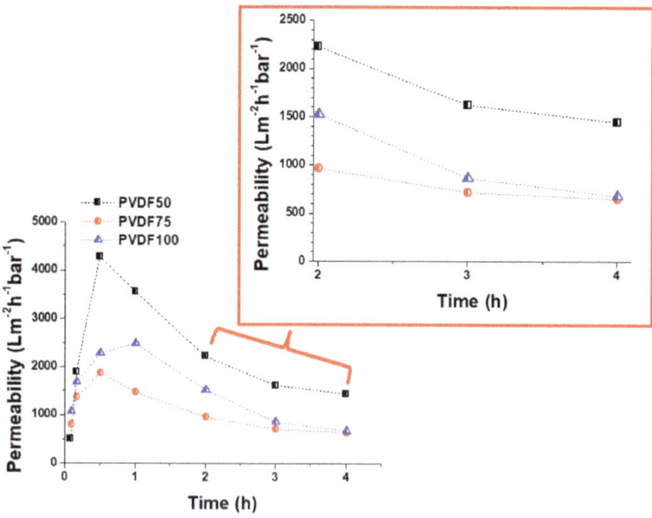

Figure 14. Permeability of PVDF multilayer nanofibrous membranes at various laminating pressures over time.

A decrease in permeability was observed for both the PAN and PVDF multilayer nanofibrous membranes depending on the operation time as shown in Figures 13 and 14. There are a few possible reasons for the decrease in permeability during liquid filtration. The first reason is concentration polarization, which is a consequence of the selectivity of the membrane. When the liquid passes through the membrane, the solute is retained by the membrane surface with a relatively high concentration. Moreover, the hydrophilicity of the membrane decreases over time during filtration due to membrane fouling and concentration polarization. Since tap water is not pure, dissolved molecules, suspended solids, and organics may be contained in the water, which can cause a decrease in the water flux due to fouling. The second reason is that close to the membrane surface, the effective transmembrane pressure (TMP) driving force reduces due to an osmotic pressure difference between the filtrate and the feed solution. TPM is generally observed in the case of ultra-filtration (UF) membranes. Another reason may be related to the compression/collapse of membrane pores, thereby causing a reduction in water permeability. The operating conditions (feed pressure, temperature, pH, flow rate, etc.) are also effective factors in membrane permeability. In general, the flux decline is caused by a decreased driving force and/or an increased resistance of the membrane, raw water characteristics, and particulate matter levels [56–58].

At the beginning, all the PAN membranes had the highest permeability (Figure 13). After a 4-h filtration test, the flux declined to 824, 909, and 375 $Lm^{-2}h^{-1}bar^{-1}$ in the case of PAN50, PAN75, and PAN100, respectively. The results indicated that laminating pressure has a huge impact on the water permeability of the multilayer membranes. The laminating pressure and the permeability of the membranes showed a non-linear relationship in the case of the PAN membranes. PAN50 and PAN75 multilayer nanofibrous membranes showed the best water permeability. On the other hand, all the PVDF membranes showed very low initial permeability at the beginning due to the hydrophobic nature of the PVDF nanofibers, and the melted adhesive web partially occupied the membrane pores, increasing the hydraulic resistance to filtration (Figure 14). The results of 4 h of filtration of PVDF membranes showed that the highest permeability (1444 $Lm^{-2}h^{-1}bar^{-1}$) was only achieved at the lowest laminating pressure (50 kN). PVDF75 and PVDF100 had almost the same permeability value (650 and 681 $Lm^{-2}h^{-1}bar^{-1}$, respectively) after the 4-h filtration test. Li et al. [59] reported a simple strategy to improve the waterproof/breathable performance and mechanical properties of electrospun PVDF fibrous membranes using a thermo-pressing system. It was found that the effect of temperature and pressure on PVDF has a synergistic effect on the fiber morphology and crystal structure. By properly adjusting the temperature and pressure, robust mechanical properties and excellent waterproof/breathable performance of PVDF membranes were achieved.

In terms of water permeability, PAN75 has the best results from the PAN membranes. PVDF50 showed the best permeability results after the 4-h filtration test from all the PAN and PVDF membranes. The results showed that after proper lamination multilayer nanofibrous membranes are suitable for future application in liquid filtration.

4. Conclusions

There is a huge demand for the filtration application of nanofiber layers due to their specific surface, low pore size and high porosity. In this study, the effect of laminating pressure on PAN and PVDF multilayer nanofibrous membranes was investigated to prepare suitable microfilters for liquid filtration. The surface morphology, average pore size, air permeability, water permeability, bursting strength, and the contact angle of the membranes were compared. Different performance levels were achieved by varying the laminating pressure of the multilayer nanofibrous membranes. The pressure effect had a considerable influence on air permeability, average pore size, contact angle, bursting strength, and water permeability. The surface morphology results showed that the fiber diameter slightly increased with an increase in laminating pressure, while the water and ethylene glycol contact angles decreased. The main effect of laminating pressure was observed on the average pore size, air permeability, bursting strength and water permeability of the membranes. PVDF50 showed the best water filtration of all the membranes. However, the bursting strength of PVDF50 is the lowest, which may cause possible damage and delamination of the layers under pressure over time. PAN nanofibers have a better adhesion to the surface of the multilayer. PAN75 was selected as the best candidate for liquid filtration due to its high water permeability and mechanical strength. PVDF multilayer nanofibrous membranes showed better air permeability than PAN, which may be better for the possible application of air filtration. These findings imply that to achieve the best permeable membrane results, the lamination process should be carefully optimized.

Author Contributions: Fatma Yalcinkaya conceived and designed the experiments; Fatma Yalcinkaya performed the experiments; Fatma Yalcinkaya analyzed the data; Jakub Hruza contributed reagents/materials/analysis tools and preparation of tools; Fatma Yalcinkaya wrote the paper.

Acknowledgments: This work was supported by project LO1201 obtained through the financial support of the Ministry of Education, Youth and Sports in the framework of the targeted support of the "National Program for Sustainability I" and the OPR&DI project CZ.1.05/2.1.00/19.0386.

Conflicts of Interest: The authors declare that there is no conflict of interest.

References

1. Gao, J.; Zhu, J.; Luo, J.; Xiong, J. Investigation of microporous composite scaffolds fabricated by embedding sacrificial polyethylene glycol microspheres in nanofibrous membrane. *Compos. Part A Appl. Sci. Manuf.* **2016**, *91*, 20–29. [CrossRef]

2. Tang, Y.; Chen, L.; Zhao, K.; Wu, Z.; Wang, Y.; Tan, Q. Fabrication of PLGA/HA (core)-collagen/amoxicillin (shell) nanofiber membranes through coaxial electrospinning for guided tissue regeneration. *Compos. Sci. Technol.* **2016**, *125*, 100–107. [CrossRef]

3. Zhang, W.; Ronca, S.; Mele, E. Electrospun nanofibres containing antimicrobial plant extracts. *Nanomaterials* **2017**, *7*, 42. [CrossRef] [PubMed]

4. Sill, T.J.; von Recum, H.A. Electrospinning: Applications in drug delivery and tissue engineering. *Biomaterials* **2008**, *29*, 1989–2006. [CrossRef] [PubMed]

5. Tan, L.; Gan, L.; Hu, J.; Zhu, Y.; Han, J. Functional shape memory composite nanofibers with graphene oxide filler. *Compos. Part A Appl. Sci. Manuf.* **2015**, *76*, 115–123. [CrossRef]

6. Zhao, X.G.; Jin, E.M.; Park, J.Y.; Gu, H.B. Hybrid polymer electrolyte composite with SiO_2 nanofiber filler for solid-state dye-sensitized solar cells. *Compos. Sci. Technol.* **2014**, *103*, 100–105. [CrossRef]

7. Yoon, B.; Lee, S. Designing waterproof breathable materials based on electrospun nanofibers and assessing the performance characteristics. *Fibers Polym.* **2011**, *12*, 57–64. [CrossRef]

8. Fehse, M.; Cavaliere, S.; Lippens, P.E.; Savych, I.; Iadecola, A.; Monconduit, L.; Jones, D.J.; Rozière, J.; Fischer, F.; Tessier, C.; et al. Nb-doped TiO_2 nanofibers for lithium ion batteries. *J. Phys. Chem. C* **2013**, *117*, 13827–13835. [CrossRef]

9. Aydın, H.; Çelik, S.Ü.; Bozkurt, A. Electrolyte loaded hexagonal boron nitride/polyacrylonitrile nanofibers for lithium ion battery application. *Solid State Ionics* **2017**, *309*, 71–76. [CrossRef]

10. Wang, X.; Drew, C.; Lee, S.H.; Senecal, K.J.; Kumar, J.; Samuelson, L.A. Electrospun nanofibrous membranes for highly sensitive optical sensors. *Nano Lett.* **2002**, *2*, 1273–1275. [CrossRef]

11. Liu, P.; Wu, S.; Zhang, Y.; Zhang, H.; Qin, X. A fast response ammonia sensor based on coaxial PPy–PAN nanofiber yarn. *Nanomaterials* **2016**, *6*, 121. [CrossRef] [PubMed]

12. Macagnano, A.; Perri, V.; Zampetti, E.; Bearzotti, A.; De Cesare, F.; Sprovieri, F.; Pirrone, N. A smart nanofibrous material for adsorbing and detecting elemental mercury in air. *Atmos. Chem. Phys.* **2017**, *17*, 6883–6893. [CrossRef]

13. Zhang, X.F.; Feng, Y.; Huang, C.; Pan, Y.; Yao, J. Temperature-induced formation of cellulose nanofiber film with remarkably high gas separation performance. *Cellulose* **2017**, *24*, 5649–5656. [CrossRef]

14. Zampetti, E.; Pantalei, S.; Bearzotti, A.; Bongiorno, C.; De Cesare, F.; Spinella, C.; Macagnano, A. TiO_2 nanofibrous chemoresistors coated with PEDOT and PANi blends for high performance gas sensors. In *Procedia Engineering*; Elsevier: Amsterdam, The Netherlands, 2012; Volume 47, pp. 937–940.

15. Ge, J.; Choi, N. Fabrication of functional polyurethane/rare earth nanocomposite membranes by electrospinning and its VOCs absorption capacity from air. *Nanomaterials* **2017**, *7*, 60. [CrossRef] [PubMed]

16. Desai, K.; Kit, K.; Li, J.; Michael Davidson, P.; Zivanovic, S.; Meyer, H. Nanofibrous chitosan non-wovens for filtration applications. *Polymer* **2009**, *50*, 3661–3669. [CrossRef]

17. Liao, Y.; Tian, M.; Wang, R. A high-performance and robust membrane with switchable super-wettability for oil/water separation under ultralow pressure. *J. Memb. Sci.* **2017**, *543*, 123–132. [CrossRef]

18. Li, Z.; Kang, W.; Zhao, H.; Hu, M.; Wei, N.; Qiu, J.; Cheng, B. A novel polyvinylidene fluoride tree-like nanofiber membrane for microfiltration. *Nanomaterials* **2016**, *6*, 152. [CrossRef] [PubMed]

19. Sood, R.; Cavaliere, S.; Jones, D.J.; Rozière, J. Electrospun nanofibre composite polymer electrolyte fuel cell and electrolysis membranes. *Nano Energy* **2016**, *26*, 729–745. [CrossRef]

20. Yalcinkaya, F. Preparation of various nanofiber layers using wire electrospinning system. *Arab. J. Chem.* **2016**. [CrossRef]

21. Ding, Y.; Zhang, P.; Long, Z.; Jiang, Y.; Xu, F.; Di, W. The ionic conductivity and mechanical property of electrospun P(VdF-HFP)/PMMA membranes for lithium ion batteries. *J. Memb. Sci.* **2009**, *329*, 56–59. [CrossRef]

22. Jahanbaani, A.R.; Behzad, T.; Borhani, S.; Darvanjooghi, M.H.K. Electrospinning of cellulose nanofibers mat for laminated epoxy composite production. *Fibers Polym.* **2016**, *17*, 1438–1448. [CrossRef]

23. Charles, L.E.; Kramer, E.R.; Shaw, M.T.; Olson, J.R.; Wei, M. Self-reinforced composites of hydroxyapatite-coated PLLA fibers: Fabrication and mechanical characterization. *J. Mech. Behav. Biomed. Mater.* **2012**, *17*, 269–277. [CrossRef] [PubMed]

24. Iqbal, Q.; Bernstein, P.; Zhu, Y.; Rahamim, J.; Cebe, P.; Staii, C. Quantitative analysis of mechanical and electrostatic properties of poly(lactic) acid fibers and poly(lactic) acid-carbon nanotube composites using atomic force microscopy. *Nanotechnology* **2015**, *26*, 105702. [CrossRef] [PubMed]

25. Charles, L.F.; Shaw, M.T.; Olson, J.R.; Wei, M. Fabrication and mechanical properties of PLLA/PCL/HA composites via a biomimetic, dip coating, and hot compression procedure. *J. Mater. Sci. Mater. Med.* **2010**, *21*, 1845–1854. [CrossRef] [PubMed]

26. Xu, Y.; Zhang, X.; Wang, X.; Wang, X.; Li, X.; Shen, C.; Li, Q. Simultaneous enhancements in the strength, modulus and toughness of electrospun polymeric membranes. *RSC Adv.* **2017**, *7*. [CrossRef]

27. Jiříček, T.; Komárek, M.; Chaloupek, J.; Lederer, T. Maximising flux in direct contact membrane distillation using nanofibre membranes. *Desalin. Water Treat.* **2017**, *73*, 249–255. [CrossRef]

28. Jiříček, T.; Komárek, M.; Chaloupek, J.; Lederer, T. Flux enhancement in membrane distillation using nanofiber membranes. *J. Nanomater.* **2016**, *2016*, 1–7. [CrossRef]

29. Yalcinkaya, B.; Yalcinkaya, F.; Chaloupek, J. Thin film nanofibrous composite membrane for dead-end seawater desalination. *J. Nanomater.* **2016**, *2016*, 1–12. [CrossRef]

30. Yalcinkaya, B.; Yalcinkaya, F.; Chaloupek, J. Optimisation of thin film composite nanofiltration membranes based on laminated nanofibrous and nonwoven supporting material. *Desalin. Water Treat.* **2017**, *59*, 19–30. [CrossRef]

31. Yalcinkaya, F.; Yalcinkaya, B.; Hruza, J.; Hrabak, P. Effect of nanofibrous membrane structures on the treatment of wastewater microfiltration. *Sci. Adv. Mater.* **2016**, *9*, 747–757. [CrossRef]

32. Kanafchian, M.; Valizadeh, M.; Haghi, A.K. A study on the effects of laminating temperature on the polymeric nanofiber web. *Korean J. Chem. Eng.* **2011**, *28*, 445–448. [CrossRef]

33. Yao, M.; Woo, Y.C.; Tijing, L.D.; Shim, W.-G.; Choi, J.-S.; Kim, S.-H.; Shon, H.K. Effect of heat-press conditions on electrospun membranes for desalination by direct contact membrane distillation. *Desalination* **2016**, *378*, 80–91. [CrossRef]

34. Mohammadian, M.; Haghi, A.K. Study on the production of a new generation of electrospun nanofiber webs. *Bulg. Chem. Commun.* **2014**, *46*, 530–534.

35. Yalcinkaya, F.; Siekierka, A.; Bryjak, M. Preparation of fouling-resistant nanofibrous composite membranes for separation of oily wastewater. *Polymers* **2017**, *9*, 679. [CrossRef]

36. Yalcinkaya, F.; Yalcinkaya, B.; Pazourek, A.; Mullerova, J.; Stuchlik, M.; Maryska, J. Surface modification of electrospun PVDF/PAN nanofibrous layers by low vacuum plasma treatment. *Int. J. Polym. Sci.* **2016**, *2016*, 1–9. [CrossRef]

37. Gopalan, A.I.; Santhosh, P.; Manesh, K.M.; Nho, J.H.; Kim, S.H.; Hwang, C.G.; Lee, K.P. Development of electrospun PVdF-PAN membrane-based polymer electrolytes for lithium batteries. *J. Memb. Sci.* **2008**, *325*, 683–690. [CrossRef]

38. Liu, T.-Y.; Lin, W.-C.; Huang, L.-Y.; Chen, S.-Y.; Yang, M.-C. Surface characteristics and hemocompatibility of PAN/PVDF blend membranes. *Polym. Adv. Technol.* **2005**, *16*, 413–419. [CrossRef]

39. Hammami, M.A.; Krifa, M.; Harzallah, O. Centrifugal force spinning of PA6 nanofibers—Processability and morphology of solution-spun fibers. *J. Text. Inst.* **2014**, *105*, 637–647. [CrossRef]

40. Essalhi, M.; Khayet, M. Self-sustained webs of polyvinylidene fluoride electrospun nano-fibers: Effects of polymer concentration and desalination by direct contact membrane distillation. *J. Memb. Sci.* **2014**, *454*, 133–143. [CrossRef]

41. Beachley, V.; Wen, X. Effect of electrospinning parameters on the nanofiber diameter and length. *Mater. Sci. Eng. C Mater. Biol. Appl.* **2009**, *29*, 663–668. [CrossRef] [PubMed]

42. Lowery, J.L.; Datta, N.; Rutledge, G.C. Effect of fiber diameter, pore size and seeding method on growth of human dermal fibroblasts in electrospun poly(ε-caprolactone) fibrous mats. *Biomaterials* **2010**, *31*, 491–504. [CrossRef] [PubMed]

43. Bagherzadeh, R.; Najar, S.S.; Latifi, M.; Tehran, M.A.; Kong, L. A theoretical analysis and prediction of pore size and pore size distribution in electrospun multilayer nanofibrous materials. *J. Biomed. Mater. Res. Part A* **2013**, *101*, 2107–2117. [CrossRef] [PubMed]

44. Gockeln, M.; Pokhrel, S.; Meierhofer, F.; Glenneberg, J.; Schowalter, M.; Rosenauer, A.; Fritsching, U.; Busse, M.; Mädler, L.; Kun, R. Fabrication and performance of $Li_4Ti_5O_{12}$/C Li-ion battery electrodes using combined double flame spray pyrolysis and pressure-based lamination technique. *J. Power Sources* **2018**, *374*, 97–106. [CrossRef]

45. Stiubianu, G.; Nicolescu, A.; Nistor, A.; Cazacu, M.; Varganici, C.; Simionescu, B.C. Chemical modification of cellulose acetate by allylation and crosslinking with siloxane derivatives. *Polym. Int.* **2012**, *61*, 1115–1126. [CrossRef]

46. Law, K.-Y.; Zhao, H. Wetting on rough surfaces. In *Surface Wetting*; Springer International Publishing: Cham, Switwerland, 2016; pp. 55–98.

47. Youngblood, J.P.; McCarthy, T.J. Ultrahydrophobic polymer surfaces prepared by simultaneous ablation of polypropylene and sputtering of poly(tetrafluoroethylene) using radio frequency plasma. *Am. Chem. Soc. Polym. Prepr. Div. Polym. Chem.* **1999**, *40*, 563–564. [CrossRef]

48. Chen, W.; Fadeev, A.Y.; Hsieh, M.C.; Öner, D.; Youngblood, J.; McCarthy, T.J. Ultrahydrophobic and ultralyophobic surfaces: Some comments and examples. *Langmuir* **1999**, *15*, 3395–3399. [CrossRef]

49. Wenzel, R.N. Surface roughness and contact angle. *J. Phys. Colloid Chem.* **1949**, *53*, 1466–1467. [CrossRef]

50. Abuzade, R.A.; Zadhoush, A.; Gharehaghaji, A.A. Air permeability of electrospun polyacrylonitrile nanoweb. *J. Appl. Polym. Sci.* **2012**, *126*, 232–243. [CrossRef]

51. Rajak, A. Synthesis of electrospun nanofibers membrane and its optimization for aerosol filter application. *KnE Eng.* **2016**, *1*. [CrossRef]

52. Von Fraunhofer, J.A. Adhesion and cohesion. *Int. J. Dent.* **2012**, *2012*, 1–8. [CrossRef] [PubMed]

53. Lee, L.-H. Relationships between surface wettability and glass temperatures of high polymers. *J. Appl. Polym. Sci.* **1968**, *12*, 719–730. [CrossRef]

54. Pritykin, L.M. Calculation of the surface energy of homo- and copolymers from the cohesion parameters and refractometric characterisics of the respective monomers. *J. Colloid Interface Sci.* **1986**, *112*, 539–543. [CrossRef]

55. Chen, N.; Hong, L. Surface phase morphology and composition of the casting films of PVDF-PVP blend. *Polymer* **2001**, *43*, 1429–1436. [CrossRef]

56. Kim, K.-M.; Woo, S.; Lee, J.; Park, H.; Park, J.; Min, B. Improved permeate flux of PVDF ultrafiltration membrane containing PVDF-g-PHEA synthesized via ATRP. *Appl. Sci.* **2015**, *5*, 1992–2008. [CrossRef]

57. Van den Berg, G.B.; Smolders, C.A. Flux decline in ultrafiltration processes. *Desalination* **1990**, *77*, 101–133. [CrossRef]

58. Meier-Haack, J.; Booker, N.A.; Carroll, T. A permeability-controlled microfiltration membrane for reduced fouling in drinking water treatment. *Water Res.* **2003**, *37*, 585–588. [CrossRef]

59. Li, X.; Lin, J.; Bian, F.; Zeng, Y. Improving waterproof/breathable performance of electrospun poly(vinylidene fluoride) fibrous membranes by thermo-Pressing. *J. Polym. Sci. Part B Polym. Phys.* **2017**, *56*, 1–10. [CrossRef]

 *nanomaterials*

Article

Porous Aluminum Oxide and Magnesium Oxide Films Using Organic Hydrogels as Structure Matrices

Zimei Chen [1,2], Dirk Kuckling [1],* and Michael Tiemann [2],*

[1]   Department of Chemistry—Organic and Macromolecular Chemistry, Paderborn University,
33098 Paderborn, Germany; zimei.chen@uni-paderborn.de
[2]   Department of Chemistry—Inorganic Functional Materials, Paderborn University,
33098 Paderborn, Germany
*   Correspondence: dirk.kuckling@uni-paderborn.de (D.K.); michael.tiemann@uni-paderborn.de (M.T.);
Tel.: +49-525-160-2171 (D.K.); +49-525-160-2154 (M.T.)

Received: 25 February 2018; Accepted: 21 March 2018; Published: 22 March 2018

Abstract: We describe the synthesis of mesoporous Al_2O_3 and MgO layers on silicon wafer substrates by using poly(dimethylacrylamide) hydrogels as porogenic matrices. Hydrogel films are prepared by spreading the polymer through spin-coating, followed by photo-cross-linking and anchoring to the substrate surface. The metal oxides are obtained by swelling the hydrogels in the respective metal nitrate solutions and subsequent thermal conversion. Combustion of the hydrogel results in mesoporous metal oxide layers with thicknesses in the μm range and high specific surface areas up to 558 $m^2 \cdot g^{-1}$. Materials are characterized by SEM, FIB ablation, EDX, and Kr physisorption porosimetry.

Keywords: mesoporous; Al_2O_3; MgO; poly(dimethylacrylamide); hydrogel; thin film; spin coating; SEM; FIB; Kr physisorption

1. Introduction

The synthesis of metal oxides with uniform mesopores is often achieved by utilization of porogenic structure directors or matrices. For example, micellar aggregates of amphiphilic species—such as surfactants or block co-polymers—are frequently utilized as porogens. They form spontaneously by self-organization and serve as pore fillers or even as structure-directing species during the formation of the inorganic phase by a sol–gel-based synthesis ('soft templating') [1,2]. This synthesis method is applicable to a limited variety of inorganic products, such as silica and some other oxidic materials, including aluminum oxide (Al_2O_3) [3–5]. For uniform, continuous layers ('solid films') of mesoporous metal oxides at a substrate surface the soft-templating approach is usually the method of choice, because the spontaneous self-aggregation into micellar units can take place inside a liquid film that contains both the amphiphilic species and the inorganic precursor compounds. For this purpose, the micellization is induced by evaporation of the solvent (evaporation-induced self-assembly, EISA) [6,7]. It needs to be stressed, though, that several metal oxides cannot be obtained in this way as their formation may go along with phase-separation and segregation from the amphiphilic species. As an alternative, the concept of using solid, porous structure matrices ('hard templates') has been shown to be a more versatile option [8,9]. This method, often called 'nanocasting', comprises the synthesis of the desired products within the pores of a silica or carbon matrix, followed by selective removal of the matrix; the product is obtained as a 'replica' of the pore system in the matrix. Nanocasting can be used for the fabrication of a multitude of metal oxides, including Al_2O_3 [10–12], as well as those that have so far not been obtained by soft templating, e.g., magnesium oxide (MgO) [12–15]. However, the nanocasting concept still has its limitations when it comes to the synthesis of porous films, since the removal of the structure matrix may cause detachment of the replica film from the substrate.

We have recently described the synthesis of mesoporous metal oxides by using poly(dimethylacrylamide) hydrogels as matrices [16,17]. Hydrogels are three-dimensional structures composed of hydrophilic polymer chains, which can absorb and hold large quantities of water in the spaces between the chains [18]. They can be fabricated via physical or chemical cross-linking [19] and have been used as matrices for porous inorganic materials [16,17,20–22]. Their utilization as porogenic matrices may be regarded as halfway between 'soft' and 'hard templating'. The hydrogel forms a continuous network that can take up the inorganic precursor species (such as a metal salt) with no risk of phase-separation, similar to a hard structure matrix. At the same time, the swollen hydrogel is a highly flexible phase; the (cross-linked) polymer strands are more or less loosely arranged and displaceable, like a soft matrix. In fact, the porogenic impact may even occur when the water-soluble polymer strands are not even cross-linked, but only sterically entangled [23]. We have rationalized that thick bundles of polymer chains (rather than single, individual chains) in poly(dimethylacrylamide) hydrogels form the porogenic entities [16]. The products obtained so far were powders with somewhat uniform mesopores and high specific surface areas.

Here we report on utilizing the same kind of porogenic hydrogels for mesoporous layers (solid films) of aluminum oxide (Al_2O_3) and magnesium oxide (MgO) at the surface of silicon wafer substrates. Photo-cross-linked poly(dimethylacrylamide) hydrogels are attached to the substrate by chemical bonding and serve as matrices for the metal oxides (Scheme 1). Porous Al_2O_3 and MgO with a high surface-to-volume ratio play an important role in separation [24,25] and heterogeneous catalysis [26–29]. Especially for the latter application immobilized layers of the catalyst (MgO) or support (Al_2O_3) materials with large pores are considered advantageous to facilitate easy access of the reactants by diffusion.

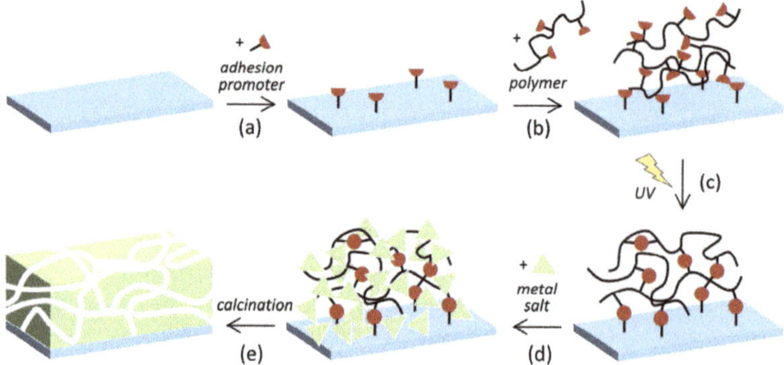

Scheme 1. Preparation of porous metal oxide (Al_2O_3, MgO) layers: (**a**) anchoring of the adhesion promoter on the Si wafer substrate; (**b**) spreading of the polymer by spin coating; (**c**) hydrogel formation and immobilization on the substrate by photo-induced cross-linking; (**d**) swelling in metal salt solution ($Al(NO_3)_3$, $Mg(NO_3)_2$); (**e**) formation of the porous metal oxide and combustion of the hydrogel by calcination.

2. Materials and Methods

Materials: Acryloyl chloride (Alfa Aesar, Karlsruhe, Germany, 96%), allylamine (Sigma-Aldrich, Taufkirchen, Germany, 98%), aluminum nitrate nonahydrate (Sigma-Aldrich, ≥98.0%), ammonia solution (Stockmeier, Bielefeld, Germany, 25%), bicyclohexyl (Acros, Geel, Belgium, 99%), chloroform (Stockmeier), chlorodimethylsilane (Alfa Aesar, 97%), 1,2-diaminoethane (Acros, >99%), 2,3-dimethylmaleic anhydride (Acros, 97%), di-*tert*-butyl dicarbonate (Boc₂O, Acros, 97%), ethanol, absolute (Sigma-Aldrich), hydrochloric acid, conc. (Stockmeier, 37%), hydrogen peroxide (Stockmeier, 35%), magnesium nitrate hexahydrate (Sigma-Aldrich, ≥97%), magnesium sulfate (Grüssing, Filsum, Germany, 99%), platinum(0)-1,3-divinyl-1,1,3,3-tetramethyldisiloxane complex solution in xylene (Sigma-Aldrich, Pt ~2%), 4″-silicon-wafer (Plano, Wetzlar, Germany), sulfuric acid, conc. (Stockmeier, ≥98%), thioxanthone (Sigma-Aldrich, 98%), and triethylamine (TEA, Grüssing, 99%) were used as received. Acetone (Stockmeier), diethyl ether (Hanke+Seidel, Steinfurt, Germany), ethyl acetate (Stockmeier), *n*-hexane (Stockmeier), methanol (Stockmeier), *n*-pentane (Stockmeier), silica gel (VWR), sodium bicarbonate (Stockmeier), and sodium chloride (Stockmeier) were of technical grade and used as received. 1,4-Dioxane (Carl Roth, Karlsruhe Germany, ≥99.5%), *N,N*-Dimethylacrylamide (DMAAm, TCI, Eschborn, Germany, 99%), tetrahydrofuran (THF, BASF, Ludwigshafen, Germany), and toluene (Grüssing, 99.5%) were distilled under low pressure. α,α'-Azobisisobutyronitrile (AIBN, Fluka, Seelze, Germany, >98%) was recrystallized from methanol. Cyclohexanone (Sigma-Aldrich, ≥99.0%) was distilled. Dichloromethane (Stockmeier) was dried over $CaCl_2$ and distilled.

Characterization: ^{1}H and ^{13}C NMR spectra were recorded on a Bruker AV 500 spectrometer at 500 MHz and 125 MHz, respectively. Reference solvent signals at 7.26 and 2.56 ppm were used for spectra in $CDCl_3$ (99.8 atom % Deuterium) and DMSO-d_6 ($O=S(CD_3)_2$, 99.9%), respectively. Gel permeation chromatography (GPC) was performed in chloroform for PDMAAm at 30 °C and at a flow rate of 0.75 mL·min^{-1} on a Jasco 880-PU Liquid Chromatograph connected to a Shodex RI-101 Detector. The instrument was equipped with four consecutive columns (PSS-SDV columns filled with 5 μm gel particles with a defined porosity of 10^6 Å, 10^5 Å, 10^3 Å and 10^2 Å, respectively) and were calibrated by poly(methyl methacrylate) standards. Krypton (Kr) physisorption analysis was performed at 77 K on a Quantachrome Autosorb 6B instrument. The masses of the films were determined by weighing the wafer substrates before and after film synthesis. Several samples of identical films (7 × 7 mm substrate dimensions) were combined for each sorption measurement to provide sufficient overall film masses (1–200 mg). Samples were degassed at 120 °C for 12 h prior to measurement. The specific surface areas were assessed by multi-point BET analysis [30] in the range $0.1 \leq p/p_0 \leq 0.3$. Scanning electron microscopy (SEM) and energy-dispersive X-ray (EDX) spectroscopy were performed on a Zeiss NEON® 40 microscope connected with an UltraDry detector from Thermo Fisher Scientific (Waltham, MA, USA).

Cross-Linker Synthesis: 2-(Dimethyl maleimido)-*N*-ethyl-acrylamide (DMIAAm) was synthesized through a four-step reaction as described in the literature [31] and can be found in detail in the Supplementary Materials.

Polymer Synthesis: Poly(DMAAm-*co*-DMIAAm) was synthesized with DMAAm monomer and DMIAAm cross-linker by free radical polymerization initiated with AIBN in an analogous fashion as described in the literature [31]. DMAAm (95 mol %) and DMIAAm (5 mol %) and about 0.002 mol % AIBN relative to the total amount of monomer were dissolved in 1,4-dioxane and purged with argon for 20 min. The total monomer concentration was 1 mol·L^{-1}. The polymerization was carried out at 70 °C for 7 h under argon atmosphere. Afterwards, the polymer was precipitated in diethyl ether and re-precipitated from tetrahydrofuran into diethyl ether for purification. Finally, the polymer was dried in high vacuum and characterized by NMR spectroscopy and GPC. ^{1}H NMR (500 MHz, $CDCl_3$): δ (ppm) = 1.51–1.83 (m, CH_2), 1, 94 (s, CH_3), 2.3–2.75 (m, CH), 2.77–3.19 (m, N-CH_2, NH-CH_2, N-CH_3), 3.6 (b, NH). Yield: 88%, M_n: 39,000 g·mol^{-1}, DMIAAm composition: 5 mol % (Feed)/4.8 mol % (NMR), PD: 5.1.

Synthesis and Immobilization of the Adhesion Promoter: 1-[3-(Chloro-dimethyl-silanyl)-propyl]-3,4-dimethyl-maleimide was synthesized as described in the literature [32] (see Supplementary Materials). A Si wafer (7 mm × 7 mm) was activated with a mixture (7:3 vol.) of concentrated sulfuric acid (H_2SO_4) and 30% hydrogen peroxide (H_2O_2) solution at 90 °C for 1 h. After repeated rinsing with water and ethanol and drying in argon stream the adhesion promoter was absorbed from 1 vol % solution in bicyclohexyl for 24 h. Finally the wafer was rinsed with chloroform and abs. ethanol and dried in an argon flow.

Preparation of Hydrogel Films: Solutions of the polymer in cyclohexanone (2 mL) with 2 wt % thioxanthone as a sensitizer were spin-coated on a pre-treated Si wafer, using variable spin velocities and polymer concentrations (see Results and Discussion section); polymer solutions were first spread at 250 rpm for 25 s, followed by 60 s of spinning at the final velocity. The polymer layer on the wafer was irradiated with UV light for one minute by using a 200 W mercury short arc lamp with an intensity of 266 mW·cm^{-2}.

Preparation of Porous Al_2O_3 and MgO layers: For the preparation of Al_2O_3, the hydrogel film was re-swelled in saturated aqueous aluminum nitrate solution (1.9 mol·L^{-1}) overnight and then treated with the vapor of an aqueous ammonia solution (12.5 wt %) for 3 h at 60 °C to convert $Al(NO_3)_3$ to $Al(OH)_3$/AlO(OH), followed by drying overnight at 60 °C. The material was calcined in a tube furnace for 4 h at 500 °C with a heating rate of 1 °C·min^{-1} to combust the polymer and to form a porous Al_2O_3 film on the Si wafer. For the preparation of MgO the hydrogel film was re-swelled in saturated aqueous magnesium nitrate solution (4.9 mol·L^{-1}) overnight and then dried at 120 °C. The material was calcined in a tube furnace for 2 h at 300 °C and 2 h at 500 °C with a heating rate of 1 °C·min^{-1} to combust the polymer and to form a porous MgO film on the Si wafer.

In an alternative approach, the above-described preparation of the hydrogel film was modified by dissolving the polymer in methanol (instead of cyclohexanone) and by adding aluminum nitrate or magnesium nitrate to this solution before (instead of after) spin-coating (2500 rpm final spin velocity) and subsequent photo-cross-linking. Otherwise, the same synthesis protocol was used.

3. Results and Discussion

Photo-cross-linked hydrogel films were used as structure matrices for the preparation of porous alumina (Al_2O_3) and magnesia (MgO) layers. The polymer for the hydrogels was synthesized by free radical polymerization of *N,N*-dimethylacrylamide (DMAAm) and 2-(dimethyl maleimido)-*N*-ethyl-acrylamide (DMIAAm) (Scheme 2a). The synthesized polymers have a molecular weight (M_n) of ca. 39,000 g·mol^{-1}. DMIAAm served as a photo-cross-linker to form a three-dimensional polymer network (Scheme 2b) by a reaction mechanism that can be primarily described as a [2+2] cycloaddition; however, other mechanisms are also possible [32]. According to NMR data the DMIAA fraction in the polymers is 4.8 mol %, slightly less than the feed composition (5 mol %), which is in accordance with previous findings [33]. To covalently attach the hydrogel to a silicon wafer substrate, 1-[3-(chloro-dimethyl-silanyl)-propyl]-3,4-dimethyl-maleimide was used as an adhesion promoter. The promoter was applied to the wafer prior to coating with the polymer. For this purpose, the wafer surface was chemically activated by oxidative treatment with piranha solution (H_2SO_4/H_2O_2). The promoter bonds to the surface via its reactive chloro-silane function (Scheme 2c); the maleimide function can react with the polymer during photo-induced cross-linking.

Porous Al_2O_3 or MgO layers were created by pre-fabricating hydrogel films on the substrate and then adding the inorganic precursor species in a second step (Scheme 1). The polymer was spin-coated on the pretreated Si wafer by using cyclohexanone as a solvent. The polymer concentration and spin velocity were varied in order to obtain variable film thicknesses. Photo-cross-linking of the polymer film was then achieved by UV irradiation as described in the Experimental Section. The cross-linked network forms a thin hydrogel film at the Si wafer surface. Figure 1 shows example scanning electron microscopic (SEM) images of dry films exhibiting high degrees of homogeneity. (further examples are shown in Figure S1 in the Supplementary Materials). The film thicknesses were analyzed by focused ion beam (FIB) ablation. Figure 1d shows a rectangular hole cut out of the film. The image was taken from a tilted angle (ca. 45° to the film surface), showing both the section through the film and the underlying substrate. This way, the average thickness of the film can be measured; depending on polymer concentration and spin velocity it ranges from 0.187 μm to 0.851 μm (Table 1).

The hydrogel film was then impregnated with $Al(NO_3)_3$ or $Mg(NO_3)_2$ by swelling in a saturated aqueous solution of the respective salt. The Al salt was transformed to $Al(OH)_3$/AlO(OH) by exposure to ammonia vapor and subsequently calcined to create Al_2O_3; this procedure is frequently applied for the structure-directed synthesis of Al_2O_3 [10,11]. The Mg salt was directly transformed to MgO by calcination. In both cases, the calcination procedure leads to the thermal combustion of the hydrogel matrix, leaving behind metal oxide layers that remain attached to the Si wafer substrates (presumably by Si-O-Al bonds in case of Al_2O_3 and by ionic interaction with the charged oxidized Si surface in case of MgO, respectively). Identification of the metal oxide phases by XRD was not feasible due to the very low thickness of the layers (see below), but previous studies [16,17,23] have shown that the applied synthesis conditions lead to formation of γ-Al_2O_3 (with low crystallinity) and MgO, respectively.

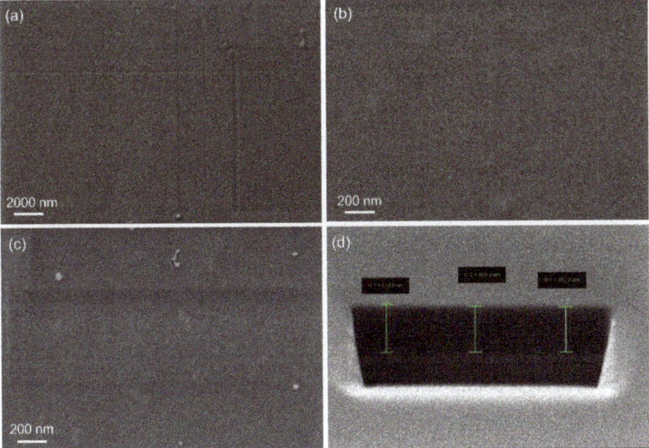

Scheme 2. (**a**) Synthesis of the polymer serving as the precursor for the hydrogel films; (**b**) photo-cross-linking of the polymer; (**c**) schematic [19] of the adhesion promoter attachment to the Si wafer surface.

Figure 1. SEM images of dry hydrogel films prepared by spin-coating with variable polymer concentration and spin velocity (average film thicknesses: (**a**) 0.187 μm; (**b**) 0.588 μm; (**c,d**) 0.851 μm; see Table 1). Image (**d**) shows an example of the FIB ablation analysis of a film (average thickness: 0.851 μm; green bars: 0.8796, 0.8439, 0.8320 μm).

Table 1. Characteristics of dry hydrogel films obtained by spin-coating of the polymer and subsequent photo-cross-linking.

Spin Velocity (rpm)	Polymer Conc. (wt %)	Film Thickness (μm)
2500	5	0.187
2500	7.5	0.306
2500	10	0.588
1000	5	0.607
1000	7.5	0.801
1000	10	0.851

Figure 2 shows SEM images with FIB analysis of two examples of porous Al_2O_3 and MgO layers. (further examples are shown in Figure S2 in the Supplementary Materials) EXD analysis confirms the approximate stoichiometry of Al/O = 1.5 and Mg/O = 1, respectively (Table S1 in the Supplementary Materials). The Al_2O_3 layer (Figure 2a,b) exhibits a fairly smooth and homogeneous texture and an average thickness of 1.77 μm, three times the thickness of the non-swollen (dry) hydrogel film that was used as the matrix (0.588 μm). This difference reflects the swelling of the hydrogel and also indicates a certain degree of porosity in the Al_2O_3 layer, as will be substantiated below.

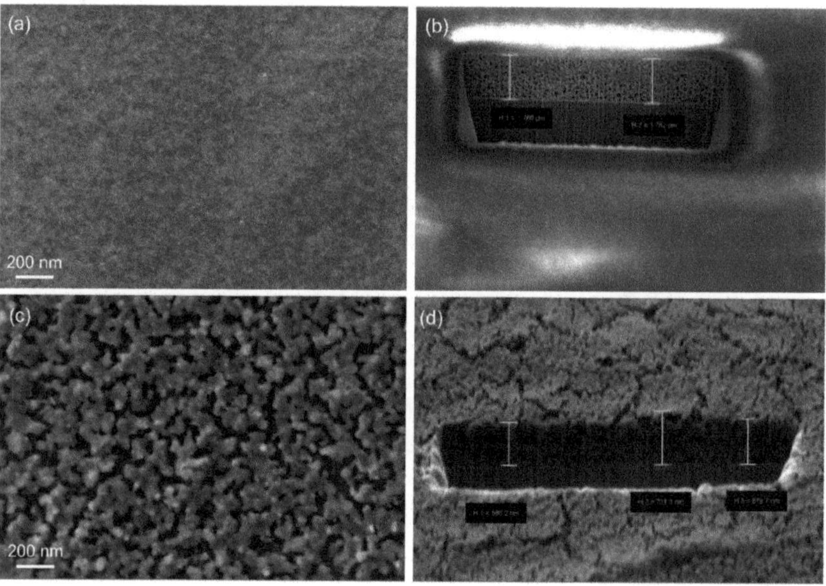

Figure 2. SEM images and FIB ablation analysis of two example layers of Al_2O_3 ((**a**,**b**); average thickness: 1.77 μm, prepared with a hydrogel film of 0.588 μm thickness; green bars: 1.766 and 1.782 μm) and MgO ((**c**,**d**); average thickness: 0.646 μm, prepared with a hydrogel film of 0.851 μm thickness; green bars: 0.5862, 0.7313, 0.6197 μm).

Assessment of the pore size distribution by nitrogen (N_2) or argon (Ar) physisorption analysis was not possible due to the low overall amount of material (as frequently encountered for thin layers of porous material), but krypton (Kr) physisorption allowed a five-point BET analysis, as shown in Figure 3a. The isotherm showing the adsorbed amount of Kr is shown in Figure S3 in the Supplementary Materials. The specific surface area of the Al_2O_3 layer is 370 $m^2 \cdot g^{-1}$, corresponding to 0.259 $m^2 \cdot cm^{-2}$ if normalized to the covered area of the substrate. The latter value incorporates the respective film thickness, while the former value is independent of the film dimensions. This large surface area confirms that the Al_2O_3 layer is indeed porous. As mentioned in the Introduction section, we have recently reported on the synthesis of γ-Al_2O_3 materials synthesized by the same procedure (using the same type of hydrogels), but in form of powders rather than as thin layers [16,17,23]. The powder samples exhibited similar BET surface areas (250–370 $m^2 \cdot g^{-1}$) with narrow pore size distributions around ca. 4 nm and mesopore volumes in the range of 0.4–0.5 $cm^3 \cdot g^{-1}$. Hence, it is fair to assume similar mesopores for the Al_2O_3 layer presented here. The origin of these mesopores is the porogenic impact of bundles of polymer strands in the hydrogel; the combustion of the hydrogel creates disordered, tubular mesopores, as previously described [16]. The MgO layer (Figure 2c,d), on the other hand, is significantly less homogeneous than the Al_2O_3 layer; it exhibits a rough surface with raptures and an almost granular texture. Its average thickness is 0.646 µm, which is actually less than the thickness if the respective non-swollen (dry) hydrogel film (0.851 µm). This indicates a lower degree of porosity which is confirmed by a low BET surface area of 112 $m^2 \cdot g^{-1}$ (0.025 $m^2 \cdot cm^{-2}$; Figure 3b). Obviously, the polymer network does not have a strong porogenic impact in this case. This may be due to the fact that the hydrogel matrix starts to decompose before a sufficiently stable network of MgO has formed.

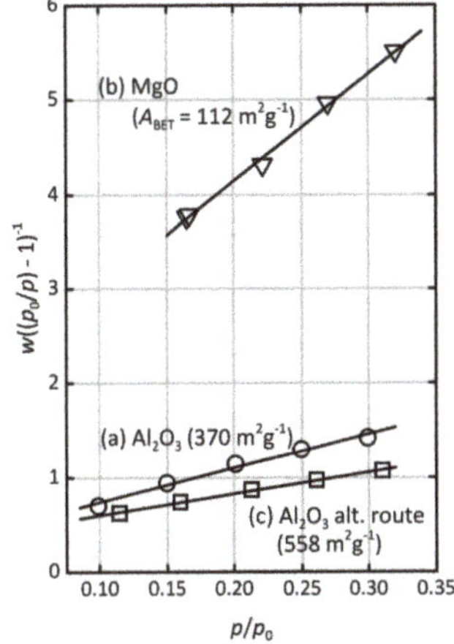

Figure 3. BET plots of Kr physisorption data of porous metal oxide layers: (**a**) Al_2O_3 layer (1.77 µm thickness); (**b**) MgO (0.646 µm); (**c**) Al_2O_3 (prepared by pre-mixing the polymer).

In an alternative synthesis approach, we simplified the process by dissolving the metal salts and the precursor polymer in methanol before applying them to the silicon wafer by spin-coating and subsequent photo-cross-linking of the polymer. Hence, no drying of the hydrogel films and subsequent re-swelling in metal salt solutions is necessary which facilitates the overall procedure. The rest of the synthesis was carried out in the same way as before. Two example SEM images of the resulting Al_2O_3 and MgO layers are shown in Figure 4. Further examples are listed in Table S2 and shown in Figure S4 in the Supplementary Materials. The Al_2O_3 layer (Figure 4a) exhibits an average thickness of 0.364 µm and shows a fairly homogeneous texture, although not as smooth as in case of the synthesis procedure described above (i.e., by re-swelling the pre-fabricated hydrogel films with metal salt solutions). However, it exhibits a higher BET surface area of 558 $m^2 \cdot g^{-1}$. Figure 3c; the surface area per substrate area, 0.080 $m^2 \cdot cm^{-2}$, is lower as a consequence of a lower layer thickness. The MgO layer (Figure 4b) is even less homogeneous; it appears seriously granular and rough, with a similar BET surface area as for the material prepared by the first route (112 $m^2 \cdot g^{-1}$). In summary, the alternative synthesis approach, despite being simpler and easier to carry out, cannot be regarded as equally successful as the first route in terms of the homogeneity and smoothness of the resulting metal oxide layers.

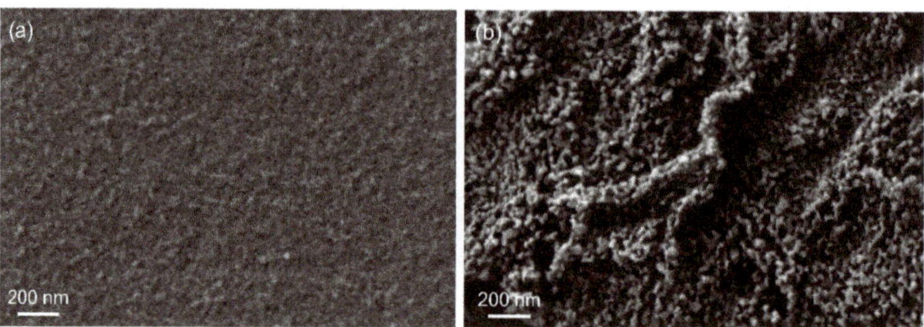

Figure 4. SEM images of two example layers of (**a**) Al_2O_3 and (**b**) MgO prepared by pre-mixing the polymer ((a): 200 mg; (b): 150 mg) with the metal salts ((a): 600 mg $Al(NO_3)_3 \cdot 9H_2O$; (b): 450 mg $Mg(NO_3)_2 \cdot 6H_2O$) prior to spin-coating and subsequent photo-cross-linking and calcination.

4. Conclusions

In summary, we have shown that the concept of using poly(dimethylacrylamide) hydrogels as porogenic matrices for the synthesis of mesoporous metal oxides can be applied to the preparation of porous layers on silicon substrates by anchoring the hydrogel to the substrate via chemical bonding. Homogeneous mesoporous layers of Al_2O_3 with high specific surface areas (up to 558 $m^2 \cdot g^{-1}$) are obtained. The MgO layers display lower homogeneity and porosity.

Supplementary Materials: The following are available online at http://www.mdpi.com/2079-4991/8/4/186/s1, Figure S1: SEM images of dry hydrogel films, Figure S2: SEM images of Al_2O_3 and MgO layers, Figure S3: Kr physisorption isotherms, Figure S4: SEM images of Al_2O_3 and MgO layers prepared by pre-mixing, Table S1: EDX analysis, Table S2: Amounts of polymer and metal salts for pre-mixing synthesis.

Acknowledgments: We thank Marc Hartmann for help with the Kr physisorption measurements and Manuel Traut for help with SEM/FIB.

Author Contributions: Z.C., M.T., and D.K. conceived and designed the experiments; Z.C. performed the experiments; Z.C., M.T., and D.K. wrote the paper.

Conflicts of Interest: The authors declare no conflict of interest.

References

1. Gu, D.; Schüth, F. Synthesis of non-siliceous mesoporous oxides. *Chem. Soc. Rev.* **2014**, *43*, 313–344. [CrossRef] [PubMed]
2. Smarsly, B.; Antonietti, M. Block copolymer assemblies as templates for the generation of mesoporous inorganic materials and crystalline films. *Eur. J. Inorg. Chem.* **2006**, *6*, 1111–1119. [CrossRef]
3. Bagshaw, S.A.; Pinnavaia, T.J. Mesoporous alumina molecular sieves. *Angew. Chem. Int. Ed.* **1996**, *35*, 1102–1105. [CrossRef]
4. Čejka, J. Organized mesoporous alumina: Synthesis, structure and potential in catalysis. *Appl. Catal. A* **2003**, *254*, 327–338. [CrossRef]
5. Morris, S.M.; Fulvio, P.F.; Jaroniec, M. Ordered mesoporous alumina-supported metal oxides. *J. Am. Chem. Soc.* **2008**, *130*, 15210–15216. [CrossRef] [PubMed]
6. Brinker, C.J.; Lu, Y.; Sellinger, A.; Fan, H. Evaporation-Induced Self-Assembly: Nanostructures Made Easy. *Adv. Mater.* **1999**, *11*, 579–585. [CrossRef]
7. Grosso, D.; Cagnol, F.; Soler-Illia, G.J.A.A.; Crepaldi, E.L.; Amenitsch, H.; Brunet-Bruneau, A.; Bourgeois, A.; Sanchez, C. Fundamentals of mesostructuring through evaporation-induced self-assembly. *Adv. Funct. Mater.* **2004**, *4*, 309–322. [CrossRef]

8. Tiemann, M. Repeated templating. *Chem. Mater.* **2007**, *20*, 961–971. [CrossRef]

9. Ren, Y.; Ma, Z.; Bruce, P.G. Ordered mesoporous metal oxides: Synthesis and applications. *Chem. Soc. Rev.* **2012**, *41*, 4909–4927. [CrossRef] [PubMed]

10. Liu, Q.; Wang, A.; Wang, X.; Zhang, T. Ordered crystalline alumina molecular sieves synthesized via a nanocasting route. *Chem. Mater.* **2006**, *18*, 5153–5155. [CrossRef]

11. Haffer, S.; Weinberger, C.; Tiemann, M. Mesoporous Al_2O_3 by nanocasting: Relationship between crystallinity and mesoscopic order. *Eur. J. Inorg. Chem.* **2012**, 3283–3288. [CrossRef]

12. Weinberger, C.; Roggenbuck, J.; Hanss, J.; Tiemann, M. Synthesis of Mesoporous Metal Oxides by Structure Replication: Thermal Analysis of Metal Nitrates in Porous Carbon Matrices. *Nanomaterials* **2015**, *5*, 1431–1441. [CrossRef] [PubMed]

13. Roggenbuck, J.; Tiemann, M. Ordered mesoporous magnesium oxide with high thermal stability synthesized by exotemplating using CMK-3 carbon. *J. Am. Chem. Soc.* **2005**, *127*, 1096–1097. [CrossRef] [PubMed]

14. Roggenbuck, J.; Koch, G.; Tiemann, M. Synthesis of mesoporous magnesium oxide by CMK-3 carbon structure replication. *Chem. Mater.* **2006**, *18*, 4151–4156. [CrossRef]

15. Roggenbuck, J.; Waitz, T.; Tiemann, M. Synthesis of Mesoporous Metal Oxides by Structure Replication: Strategies of Impregnating Porous Matrices with Metal Salts. *Microporous Mesoporous Mater.* **2008**, *113*, 575–582. [CrossRef]

16. Birnbaum, W.; Weinberger, C.; Schill, V.; Haffer, S.; Tiemann, M.; Kuckling, D. Synthesis of mesoporous alumina through photo cross-linked poly(dimethylacrylamide) hydrogels. *Colloid Polym. Sci.* **2014**, *292*, 3055–3060. [CrossRef]

17. Weinberger, C.; Chen, Z.; Birnbaum, W.; Kuckling, D.; Tiemann, M. Photo-cross-linked polydimethylacrylamide hydrogels as porogens for mesoporous alumina. *Eur. J. Inorg. Chem.* **2017**, *2017*, 1026–1031. [CrossRef]

18. Döring, A.; Birnbaum, W.; Kuckling, D. Responsive hydrogels—Structurally and dimensionally optimized smart frameworks for applications in catalysis, micro-system technology and material science. *Chem. Soc. Rev.* **2013**, *40*, 7391–7420. [CrossRef] [PubMed]

19. Kuckling, D.; Hoffmann, J.; Plötner, M.; Ferse, D.; Kretschmer, K.; Adler, H.-J.P.; Arndt, K.-F.; Reichelt, R. Photo cross-linkable poly(*N*-isopropylacrylamide) copolymers III: Micro-fabricated temperature responsive hydrogels. *Polymer* **2003**, *44*, 4455–4462. [CrossRef]

20. Kurumada, K.; Nakamura, T.; Suzuki, A.; Umeda, N.; Kishimoto, N.; Hiro, M. Nanoscopic replication of cross-linked hydrogel in high-porosity nanoporous silica. *J. Non-Cryst. Solids* **2007**, *353*, 4839–4844. [CrossRef]

21. Cui, X.; Tang, S.; Zhou, H. Mesoporous alumina materials synthesized in different gel templates. *Mater. Lett.* **2013**, *98*, 116–119. [CrossRef]

22. Jiang, R.; Zhu, H.-Y.; Chen, H.-H.; Yao, J.; Fu, Y.-Q.; Zhang, Z.-Y.; Xu, Y.-M. Effect of calcination temperature on physical parameters and photocatalytic activity of mesoporous titania spheres using chitosan/poly(vinyl alcohol) hydrogel beads as a template. *Appl. Surf. Sci.* **2014**, *319*, 189–196. [CrossRef]

23. Chen, Z.; Weinberger, C.; Tiemann, M.; Kuckling, D. Organic Polymers as Porogenic Structure Matrices for Mesoporous Alumina and Magnesia. *Processes* **2017**, *5*, 70. [CrossRef]

24. Li, L.; Wen, X.; Fu, X.; Wang, F.; Zhao, N.; Xiao, F.; Wie, W.; Sun, Y. MgO/Al_2O_3 Sorbent for CO_2 Capture. *Energy Fuels* **2010**, *24*, 5773–5780. [CrossRef]

25. Wie, J.; Ren, Y.; Luo, W.; Sun, Z.; Cheng, X.; Li, Y.; Deng, Y.; Elzatahry, A.A.; Al-Dahyan, D.; Zhao, D. Ordered mesoporous alumina with ultra-large pores as an efficient absorbent for selective bioenrichment. *Chem. Mater.* **2017**, *29*, 2211–2217.

26. Martín-Aranda, R.; Čejka, J. Recent Advances in Catalysis Over Mesoporous Molecular Sieves. *Top. Catal.* **2010**, *53*, 141–153. [CrossRef]

27. Choudary, B.M.; Mulukutla, R.S.; Klabunde, K.J. Benzylation of aromatic compounds with different crystallites of MgO. *J. Am. Chem. Soc.* **2003**, *125*, 2020–2021. [CrossRef] [PubMed]

28. Trueba, M.; Trasatti, S.P. γ-Alumina as a support for catalysts: A review of fundamental aspects. *Eur. J. Inorg. Chem.* **2005**, *17*, 3393–3403. [CrossRef]

29. Yu, X.; Yu, J.; Cheng, B.; Jaroniec, M. Synthesis of Hierarchical Flower-like AlOOH and TiO_2/AlOOH Superstructures and their Enhanced Photocatalytic Properties. *J. Phys. Chem. C* **2009**, *113*, 17527–17535. [CrossRef]

30. Brunauer, S.; Emmett, P.H.; Teller, E. Adsorption of gases in multimolecular layers. *J. Am. Chem. Soc.* **1938**, *60*, 309–319. [CrossRef]
31. Vo, C.D.; Kuckling, D.; Adler, H.-J.; Schönhoff, M. Preparation of thermosensitive nanogels by photo-cross-linking. *Colloid Polym. Sci.* **2002**, *280*, 400–409. [CrossRef]
32. Yu, X.; Corten, C.; Görner, H.; Wolff, T.; Kuckling, D. Photodimers of *N*-alkyl-3,4-dimethylmaleimides-product ratios and reaction mechanism. *J. Photochem. Photobiol. A* **2008**, *198*, 34–44. [CrossRef]
33. Kuckling, D.; Adler, H.-J.; Ling, L.; Habicher, W.D.; Arndt, K.-F. Temperature sensitive polymers based on 2-(dimethylmaleinimido)-*N*-ethyl-acrylamide: Copolymers with N-isopropylacrylamide. *Polym. Bull.* **2000**, *44*, 269–276. [CrossRef]

 *nanomaterials*

Article

Comparison of Surface-Bound and Free-Standing Variations of HKUST-1 MOFs: Effect of Activation and Ammonia Exposure on Morphology, Crystallinity, and Composition

Brandon H. Bowser [†], Landon J. Brower [†], Monica L. Ohnsorg, Lauren K. Gentry, Christopher K. Beaudoin and Mary E. Anderson *

Department of Chemistry, Hope College, 35 E. 12th Street, Holland, MI 49422, USA;
bhbowser@gmail.com (B.H.B.); landon.brower@hope.edu (L.J.B.); monica.ohnsorg@gmail.com (M.L.O.);
Lauren.gentry@memphistr.org (L.K.G.); chrbeaudoin@gmail.com (C.K.B.)
* Correspondence: meanderson@hope.edu; Tel.: +1-616-395-7425
† These authors contributed equally to this work.

Received: 28 June 2018; Accepted: 21 August 2018; Published: 23 August 2018

Abstract: Metal-organic frameworks (MOFs) are extremely porous, crystalline materials with high surface area for potential use in gas storage, sequestration, and separations. Toward incorporation into structures for these applications, this study compares three variations of surface-bound and free-standing HKUST-1 MOF structures: surface-anchored MOF (surMOF) thin film, drop-cast film, and bulk powder. Herein, effects of HKUST-1 ammonia interaction and framework activation, which is removal of guest molecules via heat, are investigated. Impact on morphology and crystal structure as a function of surface confinement and size variance are examined. Scanning probe microscopy, scanning electron microscopy, powder X-ray diffraction, Fourier-transform infrared spectroscopy, and energy dispersive X-ray spectroscopy monitor changes in morphology and crystal structure, track ammonia uptake, and examine elemental composition. After fabrication, ammonia uptake is observed for all MOF variations, but reveals dramatic morphological and crystal structure changes. However, activation of the framework was found to stabilize morphology. For activated surMOF films, findings demonstrate consistent morphology throughout uptake, removal, and recycling of ammonia over multiple exposures. To understand morphological effects, additional ammonia exposure experiments with controlled post-synthetic solvent adsorbates were conducted utilizing a HKUST-1 standard powder. These findings are foundational for determining the capabilities and limitation of MOF films and powders.

Keywords: metal-organic framework; microscopy; thin films; powders

1. Introduction

Highly porous crystalline materials known as metal-organic frameworks (MOFs) have been the focus of much attention over recent years due to their promising potential for a wide range of applications, including gas storage, separation, catalysis, and sensing [1–3]. Careful selection of the inorganic nodes and organic linkers that compose the chemical structure of the material can lead to a variety of highly tunable pore sizes and chemical functionality [4]. Varying synthetic conditions, such as processing solvent [5], temperature [6], concentration [7], etc., also allows one to tailor MOF properties. This careful control makes it possible to embed MOF materials within an array of hierarchical architectures and composites [8]. Once synthesized, MOF crystallinity and morphology are typically evaluated to determine quality. However, when testing the materials in the presence of different gases, it is rare to find studies that monitor possible morphological changes, even though changes in crystal structure are often investigated [9]. This lack of structural understanding needs to be addressed for a variety of MOF systems and host-guest interactions if they are to become a viable option for integration into more complex designs. This study herein characterizes the morphological impact of one such interaction for the HKUST-1 MOF, providing previously neglected insights into how the MOF may perform in modern applications.

The most widely studied MOF is HKUST-1, also known as Cu_3BTC_2 or MOF-199, which was first discovered by Chui et al. [10] and uses copper(II) ions and benzene-1,3,5-tricarboxylate (BTC) as the inorganic and organic building blocks, respectively. One of the reasons HKUST-1 has garnered so much attention is because of its ability to capture and store a wide variety of gases, including carbon monoxide, carbon dioxide, nitric oxide, nitrogen, hydrogen, and sulfur dioxide [11–13]. Furthermore, HKUST-1 has been shown to be a suitable candidate for the sequestration of harmful gases like hydrogen sulfide, arsine, and ammonia [14,15]. One of the key features of HKUST-1 is that it allows for the adsorption of these different gases in the copper paddlewheel unit, depicted in Figure 1. In this paddlewheel unit, Cu^{2+} ions are connected to form Cu^{2+} dimers through a weak bond and are bridged by four carboxylate units. The as-synthesized framework typically contains solvent molecules weakly bound to the axial coordination sites of the Cu^{2+} ions. These solvent molecules can be removed through a simple activation procedure (i.e., heating under vacuum), thus enabling the Cu^{2+} ions to act as Lewis acidic sites for the binding of molecules possessing basic character, such as ammonia. Herein, this research explores the effect of residual solvent molecules on the adsorption of ammonia in HKUST-1, highlighting for the first time how the presence of solvent (or lack thereof) affects morphology.

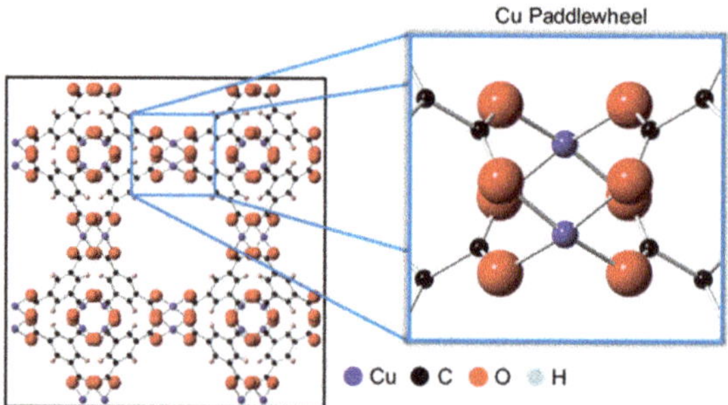

Figure 1. Unit cell (left) of the HKUST-1 metal-organic framework system ((100) crystal face) [10]. HKUST-1 is composed of four benzene-1,3,5-tricarboxylate organic ligands coordinated to Cu^{2+} dimer. Highlighted is the copper paddlewheel structure (right) characteristic of the HKUST-1 system. Each Cu^{2+} dimer completes its octahedral coordination sphere with two axial positions (vacant here) opposite of the Cu-Cu vector.

The synthesis of HKUST-1 can be varied so as to achieve a wide range of geometries, morphologies, and composites, allowing the MOF to be tailored toward specific applications [16]. For example, HKUST-1 can be tethered to a substrate as a surface-anchored MOF (surMOF) or a drop-cast thin film; and it can be synthesized to be a free-standing material (bulk powder). However, most studies for HKUST-1 only characterize and analyze one of its possible variations, typically its powder form. This is especially true in the investigation of the HKUST-1 interaction with ammonia gas. There have been studies that compare pure HKUST-1 powder to composite materials that contain it, but there is a deficiency of ammonia studies that directly compare different MOF variations [17].

The research herein compares the behavior of surface-bound and free-standing HKUST-1 materials. A study by Nijem et al. [18] suggested that aspects of the HKUST-1 interaction with ammonia may vary depending on whether the MOF is confined to a substrate. Also, the effect of size on MOF behavior is studied; the three MOF variations investigated herein all had the same chemical composition but were on different size scales. The aim of the study is to investigate how these different variations respond to ammonia in order to gain insights into the effects of size and substrate confinement on morphology, composition, and crystallinity. This is important because films of different size regimes are required depending on the application [19,20]. For example, films prepared on the nanometer scale may be useful as dielectric layers in transistors [21], whereas gas filtration technologies might require micron-sized films or larger [22]. For the creation of successful MOF-based devices it should not be assumed that material properties and reactivity will remain consistent across size regimes; therefore, more comprehensive studies such as the one conducted herein are needed in order to better understand the scope and broad applicability of various host-guest interactions.

Specifically, this study compares the three different variations of HKUST-1 mentioned above: the surMOF, drop-cast thin film, and bulk powder. Metal-organic frameworks anchored to a surface are able to integrate both the versatility and tunability of MOF systems into hierarchical architectures for potential applications in electronic or sensing devices [23,24]. Bulk powders have shown promise in their capabilities for incorporation into MOF-graphene hybrid materials and real-world gas mask filtrations [15,25]. The bulk powder used in this study was synthesized, yielding large, microscale, free-standing MOF crystals [26]. The surMOF was prepared via a well-established layer-by-layer approach where nanoscale MOF crystallites were deposited in a controlled fashion onto a carboxylic acid-functionalized gold substrate [27,28]. The formation of MOF crystals on the substrate using this method was recently found to occur via a Volmer–Weber growth mechanism, giving rise to a discontinuous surface of nucleating MOF crystallites [29,30]. Lastly, the drop-cast thin film was also formed on a functionalized gold substrate, yielding a uniform film of highly-oriented crystals [26]. This is the first time that a parallel study has investigated these three MOF variations to determine whether or not they exhibit uniform behavior independent of substrate confinement and size variance.

In this study, these three variations were used to probe the interaction of HKUST-1 with ammonia gas. Metal-organic framework interactions with ammonia have been the subject of much attention over recent years because they have been show to outperform other materials in their ability to adsorb ammonia [14,31,32]. The high-performance capabilities for the HKUST-1 system have been attributed to the copper paddlewheel, a preferential binding site for the chemisorption of Lewis bases such as ammonia [33]. In addition, ammonia is also able to interact with various other parts of the MOF as well, such as the organic linkers [17]. The HKUST-1 is known to react with ammonia in different ways depending on the presence of water, making it a useful test case for examining the effect of activation and gas exposure [34]. The HKUST-1 has been shown to adsorb more ammonia under humid conditions because the ammonia can bind to the Lewis acidic Cu ions and is also able to dissolve into films of water that are contained within the pores [35]. Peterson et al. [36] showed that under dry conditions, ammonia reacted with the open-metal binding sites of the copper paddlewheel to form a copper(II)-diamine species. That same study also showed that under humid conditions, the MOF underwent a more severe degradation owing to the reaction of ammonia with the BTC linkers to form $Cu(OH)_2$ and $(NH_4)_3BTC$ species. After exposure under both wet and dry conditions, the ammonia was removed via heating under vacuum to regenerate the material, which could then successfully re-uptake ammonia upon subsequent exposures. The research herein examines morphology throughout multiple cycles of regeneration and re-exposure for the activated surMOF material, which adds new and practical information about the viability of using surface-constrained MOF nanomaterials.

The interaction of ammonia with HKUST-1 has been modeled by simulations [17,37] and has been monitored via breakthrough studies [14]; IR spectroscopy [38]; X-ray crystallography [39]; nuclear magnetic resonance spectroscopy [36]; X-ray photoelectron, extended X-ray absorption fine structure, X-ray absorption near edge structure, UV−vis, electron paramagnetic resonance spectroscopies [34]; and microcalorimetry [40]. More recent work has shifted focus toward preventing or reducing MOF degradation upon ammonia exposure [41,42] and creating systems in the bulk that could function as filters in real-world applications [15]. Although positive approaches in these areas are underway, there are still fundamental aspects of the HKUST-1 ammonia interaction that need to be explored. A major shortcoming in those studies is the lack of details about the morphological changes that occur under different conditions. Therefore, for the first time ever, this study presents microscopic images for three HKUST-1 variations that show the morphological transformations upon exposure to ammonia for both the as-synthesized and activated material. This type of knowledge pertaining to changes in morphological structure has thus far been overlooked in the literature, but is essential if HKUST-1 is to be used in combination with another material, such as graphene-MOF composites [43], or integrated into device structures.

In addition, this study examines the effect of residual solvent adsorbates on the HKUST-1 ammonia interaction. For most synthetic routes towards HKUST-1, the solvent used for processing is still bound to the as-synthesized MOF either via chemisorption or physisorption [44]. It is well known that the solvent used for the processing of MOFs has an influence on the properties of the resulting framework [11]. For example, the size and porosity of HKUST-1 was shown to be altered by varying the ratio of solvents used during synthesis [45]. The presence of solvent can also hinder the ability for many MOFs to perform their desired function due to competition and blockage of binding sites [46]. The presence or absence of solvent plays a major factor in how the MOF interacts with ammonia, as evidenced by the effect of water on the reaction discussed prior. Research described herein will further explain the impact that solvent has on MOF degradation by exploring how different crystallographic and morphological changes occur depending on whether solvent molecules are within pores prior to ammonia exposure. Solvent is easily removed from HKUST-1 via activation by heat at reduced pressures, resulting in a visual color change of turquoise (as-synthesized) to dark blue (activated MOF) [10,11]. Multiple studies have confirmed that this activation process does not disrupt crystallinity [47], but potential effects on substrate morphology are not reported. Research described herein examines the morphological stability upon activation of the HKUST-1 surMOF.

Three variations of the HKUST-1 system and its interaction with ammonia gas were interrogated for both substrate-bound and free-standing materials. Ammonia gas exposure was investigated for as-synthesized and activated samples to explore the effect of residual solvent molecules. Further studies were done to gain insights into ammonia removal and re-exposure to the surMOF material. Additionally, ammonia exposure experiments, utilizing a purchased standard of HKUST-1 powder (Basolite® C 300), were completed to explore the effect of additional processing solvents bound to the framework. Fourier-transform infrared spectroscopy (FT-IR), scanning electron microscopy (SEM), energy dispersive X-ray spectroscopy (EDS), scanning probe microscopy (SPM), and powder X-ray diffraction (XRD) characterized samples before and after ammonia exposure. These techniques were used to track uptake and release of ammonia, monitor changes in surface morphology, examine elemental composition, and observe effects on overall crystal structure. These fundamental studies of HKUST-1 ammonia gas interactions elucidate which effects are inherent to the MOF system, as opposed to effects that are dependent on substrate confinement or size variance, and are critical for the successful incorporation of HKUST-1 into hierarchical architectures.

2. Materials and Methods

2.1. Materials

Copper (II) acetate monohydrate and copper (II) nitrate hemi(pentahydrate) were purchased from Fisher Scientific (Fair Lawn, NJ, USA). Absolute, anhydrous ethanol (200 proof, ACS/USP Grade) was obtained from Pharmco–Aaper (Shelbyville, KY, USA). Trimesic acid (TMA, H_3BTC) (95%), dimethyl sulfoxide (DMSO) (spectrophotometric grade), and 16-mercaptohexadecanoic acid (MHDA) (90%) were purchased from Aldrich (St. Louis, MO, USA). The DMSO was purged with nitrogen and passed through columns of molecular sieves. Gold substrates composed of silicon wafers with 5 nm Ti adhesion layer and 100 nm Au were purchased from Platypus Technologies (New Orleans, LA, USA). Anhydrous ammonia gas was obtained from Alexander Chemical Company (Kingsbury, IN, USA).

2.2. Sample Preparation

Three variations of MOFs were fabricated: surface-anchored MOFs (surMOF) via layer-by-layer deposition, MOF thin films by drop-cast method, and MOF powder synthesized to produce microcrystals.

2.2.1. surMOF

The surMOFs were fabricated according to literature precedent on a gold substrate functionalized by a self-assembled monolayer (SAM) using alternating, solution-phase deposition [27–29]. The gold substrate was first immersed in a 1 mM ethanol solution of MHDA for 1 h to form the foundational SAM that anchors the framework to the substrate. The substrate was then rinsed with ethanol and dried with nitrogen. Next, the sample was submerged in a 1 mM ethanol solution of copper acetate monohydrate. The substrate was removed after 30 min, rinsed, and dried as before. The sample was then submerged in a 0.1 mM ethanol solution of TMA for 1 h. Again, the substrate was rinsed, dried, and returned to the 1 mM copper acetate monohydrate solution. Four deposition cycles of the copper and TMA solutions yielded a film with an average thickness, roughness, and surface coverage of 10 nm, 20 nm, and 24%, respectively [29]. For all experiments herein, solutions were held at room temperature. After film formation, the substrate was characterized by SPM and IR.

2.2.2. Thin Film

The MOF thin film was prepared according to a modified drop-cast procedure [26]. The starting reagents, 2.8 mmol (0.59 g) TMA and 5.3 mmol (1.2 g) copper (II) nitrate hemi (pentahydrate), were combined in 5 mL of DMSO. This solution underwent sonication and stirring until powders were completely dissolved. A 1 mL portion of the resulting blue solution was then diluted to 5 mL by the addition of DMSO. A gold substrate that had been functionalized with a SAM of MHDA (according to the aforementioned method) was placed on a hotplate. A pipet transferred the diluted solution onto the gold to form a liquid layer that completely covered the substrate. The sample was covered with a glass beaker, the hotplate was ramped up to 100 °C over a 10-min period, and the sample was held at that temperature for an additional 10 min. As the solvent evaporated, a thin blue-green film was formed. Once the solvent was fully evaporated, the substrate was removed from the hotplate and placed onto an aluminum heat sink to cool the film to room temperature for characterization by XRD, SEM, and IR. This procedure yielded surface-confined crystals ranging in size from 1 to 10 microns. A quantitative study of crystal size and homogeneity was not undertaken.

2.2.3. Powder

The MOF bulk powder was prepared according to an optimized procedure [26]. First, 2.8 mmol (0.59 g) TMA and 5.1 mmol (1.2 g) copper (II) nitrate hemi(pentahydrate) were sonicated in 5 mL DMSO solution until completely dissolved. The solution was then heated in a beaker to 100 °C on a hotplate for 2 h, which resulted in the formation of blue-green crystals. After cooling, the precipitate was sonicated in ethanol, dried by vacuum filtration, and washed with additional ethanol solvent. The powder was transferred to a test tube, suspended in dichloromethane, and centrifuged for 6 min at 3300 rpm. This process was repeated three times, after which the resulting blue powder was dried overnight under high vacuum and then stored in a vacuum desiccator to be characterized by XRD, SEM, and IR. This procedure yielded free-standing crystals ranging in size from 10 to 100 microns. A quantitative study of crystal size and homogeneity was not undertaken.

2.3. Ammonia Exposure

Ammonia exposure was investigated for as-synthesized as well as activated samples. The activation process was undertaken by heating the sample under vacuum to evacuate the pores and remove any solvent (residual or coordinated). All samples that were regenerated and re-exposed underwent the same regeneration procedure.

2.3.1. Activation Process

Samples were placed under high vacuum at 180 °C for 2 h. The samples were then cooled under vacuum for 30 min prior to exposure.

2.3.2. Ammonia Exposure

For ammonia exposure, a sample under high vacuum at 25 °C was exposed to 360 torr of ammonia gas for 1 h. Samples exposed as-synthesized were held under high vacuum for 5 min at 25 °C prior to exposure. All samples (with and without prior activation) were exposed without breaking vacuum.

2.3.3. Regeneration and Re-Exposure

To undergo regeneration, samples were held under high vacuum at 180 °C for 1 h. The samples were then cooled under vacuum for 30 min prior to characterization. Prior to re-exposure, samples were left open to the atmosphere for at least 1 h. Each sample that was re-exposed underwent the same exposure method to which it was originally exposed.

2.4. Standard Powder Investigation of Solvent Effects

Basolite® C 300 powder was obtained from Aldrich (St. Louis, MO, USA) and was used as a standard for comparison to the HKUST-1 MOF powders fabricated as described above. The powder underwent ammonia exposures with and without activation in the manner outlined previously. Additional experimental exposures were undertaken to expose the Basolite® C 300 powder to solvents prior to ammonia exposures.

2.4.1. H_2O Exposure

Standard samples underwent H_2O exposure before subsequent ammonia exposure. The powder was exposed for 5 min to high vacuum conditions, and then for 2 h to H_2O vapor that had been heating to 85 °C under vacuum. Next, the sample was exposed for 5 min to high vacuum. The powder was then characterized and subsequent ammonia exposure was undertaken as described in the above section "ammonia exposure".

2.4.2. DMSO Exposure

Standard samples underwent exposure to DMSO prior to subsequent ammonia exposure. A 100–250 mg portion of powder was suspended in 10–15 mL DMSO. The mixture was stirred for 2 h, after which the powder was dried via vacuum filtration and overnight house vacuum. Once dry, the powder was characterized and exposed to ammonia gas without activation.

2.5. Characterization

All samples underwent characterization by microscopy and infrared spectroscopy. The surMOF samples were characterized by scanning probe microscopy. The drop-cast thin films and the powder samples were characterized by scanning electron microscopy with energy dispersive spectroscopy, as well as by powder X-ray diffraction.

2.5.1. Scanning Probe Microscopy (SPM)

A Dimension Icon Scanning Probe Microscope (Bruker, Santa Barbara, CA, USA) that operated in peak force tapping mode was used to obtain several images (512 × 512 pixels) for each sample, both before and after exposures, at 5 μm × 5 μm and 500 nm × 500 nm. Etched silicon tips, SCANASYST-AIR (Bruker, Santa Barbara, CA, USA), with a spring constant range of 0.2–0.8 N/m and a resonant frequency range of 45–95 kHz were used. Scan parameters were as follows: 1 Hz scan rate, 12 μm z-range, 250–370 mV amplitude set point, and 100–450 mV drive amplitude. Image analysis was carried out by Nanoscope Analysis software (Bruker, Santa Barbara, CA, USA).

2.5.2. Infrared Spectroscopy (IR)

Infrared spectra were collected from 3800–600 cm^{-1} in ATR (Attenuated Total Reflectance) mode and used a Thermo Scientific Nicolet iS50 instrument. The spectra were collected at a resolution of 4 cm^{-1} and used a bare gold substrate as the background for the surMOF studies and ambient air as the background for the powder and thin-film studies.

2.5.3. Scanning Electron Microscopy (SEM) and Energy Dispersive X-ray Spectroscopy (EDS)

The SEM images were collected by a Hitachi TM-3000 tabletop microscope and used an accelerating voltage of 15 kV with the detection of back-scattered electrons. The EDS data were collected at this voltage by this microscope coupled with a Bruker XFlash MIN SVE detector and scan generator for EDS capability.

2.5.4. Powder X-ray Diffraction (XRD)

The XRD patterns were collected by a Rigaku Miniflex X-ray diffractometer and used Cu Kα radiation at 30 kV and 15 mA. Data were collected from 10.00° to 79.99° for the powder and 10.00° to 34.00° for the drop-cast thin film (range selected to avoid the intense gold peaks from the underlying substrate). All samples were analyzed at a sampling width of 0.03° and scan speed of 3°/min.

3. Results and Discussion

To investigate how ammonia interacts with the HKUST-1 metal-organic framework, three different variations of the material (surMOF, drop-cast thin film, and powder) were exposed to ammonia; and the effect was monitored via microscopy, IR, and powder XRD. For these MOF variations, both as-synthesized and activated (heated under vacuum to remove species within pores) samples were investigated. The SPM characterization was undertaken to monitor how ammonia interacts with the surMOF deposited by a layer-by-layer protocol onto a gold surface functionalized with a self-assembled monolayer. The SEM characterization was used to monitor both a thin-film framework fabricated via drop-casting onto a functionalized gold surface as well as the synthesized bulk powder of microcrystals. Infrared spectroscopy was utilized in all three cases to track the uptake and release of ammonia. For the thin film and powder, powder XRD patterns were collected to gain further insights into how ammonia affects the overall crystal structure; and EDS was used to examine elemental composition. Beyond monitoring the effect of the initial ammonia exposure, both SPM and IR were used to understand what happens when ammonia is removed and re-introduced to the surMOF material. To further investigate the findings from the aforementioned experiments and explore the effect of different processing solvents, experiments utilizing a purchased standard of HKUST-1 powder (Basolite® C 300) were conducted.

3.1. Morphological Characterization

Integral to understanding how the material is affected by its interaction with ammonia, SPM and SEM characterization reveal the morphological structure of the surMOF, thin film, and powder HKUST-1 variations. Images of these structures before ammonia exposure, after exposure without activation, and after exposure with prior activation are shown in Figure 2. In all three cases, when the framework was exposed to ammonia without activation, the material undergoes a dramatic morphological change. The surface of the surMOF material changed from smaller, evenly distributed crystallites (Figure 2a) to larger bundles of nanowire-like structures (Figure 2b). A similar change of surface morphology was observed in the thin film (Figure 2d,e). The powder material also underwent a complete morphological change from octahedral shaped crystals (Figure 2g) to the nanowire-like structures (Figure 2h).

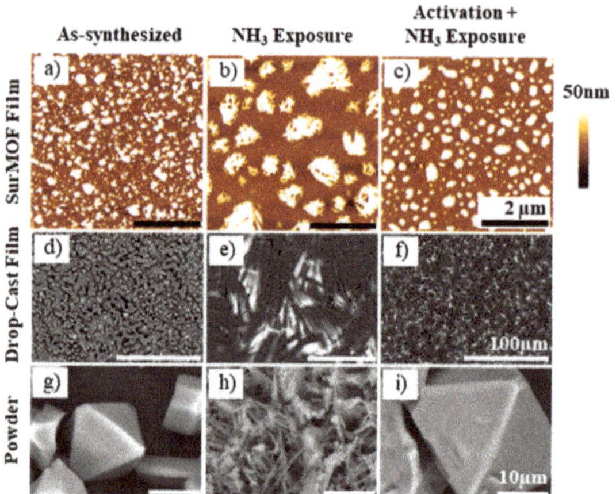

Figure 2. Top: Representative SPM images of HKUST-1 surMOFs (**a**) before exposure to NH_3; (**b**) after exposure without activation; and (**c**) after exposure with prior activation. Middle: SEM images of HKUST-1 drop-cast thin film (**d**) before exposure to NH_3; (**e**) after exposure without activation; and (**f**) after exposure with prior activation. Note: Due to high conductivity, underlying gold substrate appears bright in SEM images (especially prevalent in (**e**) and (**f**)). Bottom: SEM images of HKUST-1 powder (**g**) before exposure to NH_3; (**h**) after exposure without activation; and (**i**) after exposure with prior activation. All scale bars in (**a**–**c**) are 2 μm, in (**d**–**f**) are 100 μm, and in (**g**–**i**) are 10 μm.

A minor morphological change was observed when the framework was activated prior to ammonia exposure. The surMOF underwent a slight morphological change where the sharpness of the nanocrystallite features were reduced; and the size of the isolated structures on the surface increased (Figure 2c) relative to the film features before exposure (Figure 2a). The thin film and the powder materials did not undergo significant morphological changes after exposure to ammonia with activation (Figure 2f,i).

These images and observations provide new and important information as to how ammonia affects the morphology of the HKUST-1 crystals dependent on whether or not the material was activated (removing water or residual solvent) before the framework was exposed to the gas. When the sample was not activated, the ammonia-framework interaction resulted in a dramatic morphological change that was not observed when the framework was activated prior to exposure. This result is consistent with the findings of Peterson et al. [36] that the presence of water or residual solvent alters the reaction of the MOF with ammonia. Based on the morphological changes observed and the conclusions by Peterson, it can be hypothesized that the nanowires formed by exposure without prior activation could be $Cu(OH)_2$ crystallites and that the changes (or lack thereof) observed in the samples exposed with prior activation could be due to the formation of a copper(II)-diamine species. Regardless of the products formed, the images clearly show that different interactions are taking place dependent on whether the sample was activated.

Further, it was confirmed that the heating process involved did not cause a significant morphological change on the surMOF, which is an important observation because MOF materials are often activated prior to being used. This was shown by an experiment in which the as-synthesized sample was characterized following the activation process without subsequent ammonia exposure (shown in Section 3.5).

3.2. Compositional Characterization

Infrared spectra were obtained to investigate the chemical composition of the framework. This characterization technique was utilized to monitor uptake of ammonia and to examine how this exposure affected the framework. A comparison was conducted for when the sample underwent exposure to ammonia both without and with activation. Spectra of the framework before ammonia exposure, after exposure of an as-synthesized sample, and after exposure of an activated sample are shown in Figure 3 for the surMOF, thin film, and powder variations of the framework. The EDS was obtained for the powder and thin film samples to investigate the chemical identity of compounds captured within the framework, such as the DMSO solvent or the ammonia gas.

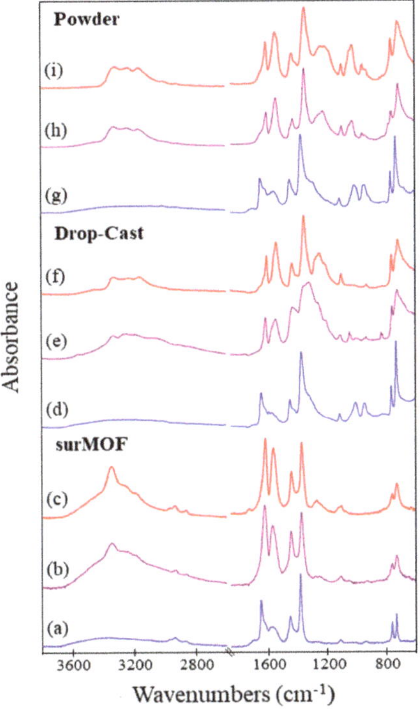

Figure 3. Infrared (IR) spectra of HKUST-1: (**a**) surMOF before exposure to NH_3; (**b**) surMOF after exposure without activation; (**c**) surMOF after exposure with prior activation; (**d**) drop-cast thin film before exposure to NH_3; (**e**) drop-cast thin film after exposure without activation; (**f**) drop-cast thin film after exposure with prior activation; (**g**) powder before exposure to NH_3; (**h**) powder after exposure without activation; and (**i**) powder after exposure with prior activation.

Characteristic IR peaks for HKUST-1 are present in all samples prior to exposure [27,34,38]; and a few small differences between the MOF variations are observed (Figure 3a,d,g). The asymmetric COO^- stretch (~1650 cm^{-1}) and the C–C aromatic vibration peak, corresponding to the linker molecule (~1440 cm^{-1}), are present in all spectra. All spectra for the as-synthesized materials also contain a very broad peak at ~3000–3400 cm^{-1}, indicative of water or deposition solvent within the framework (Figure 3a,d,g). Peaks at 2930 cm^{-1} and 2855 cm^{-1} are only present in the surMOF sample and correspond to the C–H stretch of the underlying self-assembled monolayer on the gold substrate (Figure 3a–c) [48]. Peaks at 950 cm^{-1} and 1000 cm^{-1} are present only in the thin film and powder spectra (Figure 3d,g) and correspond to the DMSO solvent, indicating that it is present within the framework. The DMSO also has broad peaks around 3000 cm^{-1}, which occur at the same vibrational frequency as water [49]. For both the as-synthesized thin film and powder sample, EDS data confirm the presence of the DMSO deposition solvent, showing a 1:1 ratio of the Cu:S (SI Figures 1 and 4). This suggests that each open axial copper paddlewheel position may be occupied by one DMSO ligand.

All of the data gleaned from the IR spectra indicate that ammonia was present after exposure regardless of prior activation and HKUST-1 variation. Changes occurred in the spectra when the framework was exposed to ammonia (Figure 3b,c,e,f,h,i) consistent with literature observations [34]. A new distinct peak appeared around 3300 cm^{-1} that corresponds to the N-H stretch of ammonia on the framework [50]. This was true regardless of whether the framework was activated prior to exposure, which indicated that the uptake of ammonia into the framework was not dependent on activating the framework. Other changes that occurred upon exposure to ammonia were the shifting of the peak at 1650 cm^{-1} to 1625 cm^{-1}, the increase in intensity of peak at 1560 cm^{-1}, the appearance of two peaks in the 1260–1210 cm^{-1} range, and the broadening/shifting of the 1370 cm^{-1} and 1450 cm^{-1} peaks. All of these observances are in accord with previous studies investigating the effect of ammonia gas on the HKUST-1 framework [34]. After ammonia exposure for both as-synthesized and activated samples, EDS for the thin film and powder variations confirmed the presence of nitrogen and the removal of sulfur (SI Figures 2, 3, 5 and 6). (Note that the low-energy -ray peak associated with nitrogen in the crowded region with carbon and oxygen did not permit quantitative analysis of the nitrogen composition, but did qualitatively confirm its presence.) The data show that the ammonia gas displaced any solvent molecules present within the as-synthesized framework.

3.3. Crystal Structure Characterization

Powder XRD data were obtained for the thin film and powder systems, but could not be obtained for the surMOF films investigated in this study due to the low density of material bound to the surface. The similarities observed via microscopy and IR analysis for all three variations suggest that crystal structure findings for the thin film and powder likely translate to the surMOF variation.

Powder XRD was important for confirming the HKUST-1 structure for the synthesized thin film and bulk samples, as well as for characterization of the crystal structure after ammonia exposure. Patterns for the material before ammonia exposure, after exposure without activation, and after exposure with prior activation are shown in Figure 4 for the thin film and powder variations of the framework. The peaks corresponding to the thin film and powder material pre-exposure (Figure 4b,e) match well with the reference pattern (Figure 4a), indicating that the HKUST-1 crystal structure was obtained. The MOF deposited as a thin film (Figure 4e) demonstrates a preferred (111) crystal orientation. Noteworthy is that the XRD for the powder has broader peaks in comparison to the film, revealing the highly crystalline nature of the film.

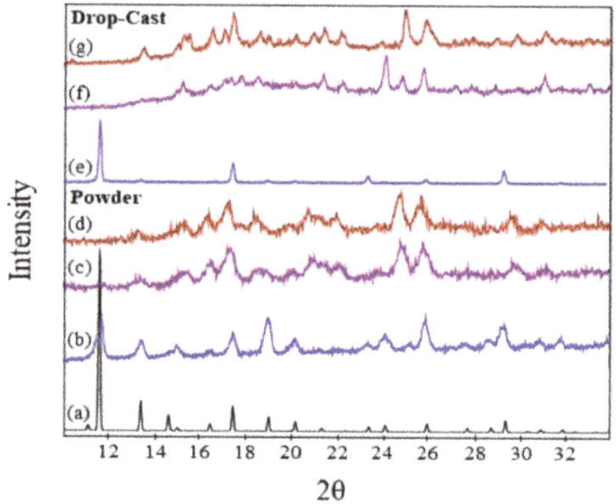

Figure 4. XRD patterns of (**a**) reference for HKUST-1 powder (ICDD pdf number 00-062-1183); (**b**) powder before exposure to NH$_3$; (**c**) powder after exposure without activation; (**d**) powder after exposure with prior activation; (**e**) drop-cast thin film before exposure to NH$_3$; (**f**) drop-cast thin film after exposure without activation; and (**g**) drop-cast thin film after exposure with prior activation.

Ammonia exposure with and without activation resulted in a change in crystal structure as observed by powder XRD (Figure 4c,d,f,g); and the resulting patterns are in general agreement with previous studies investigating the effect of ammonia on HKUST-1 [10,36,46]. For the HKUST-1 powder, the patterns for both as-synthesized and activated samples after ammonia exposures are consistent with one another (Figure 4c,d). This indicates that the activation step did not prevent the change in the crystal structure, despite what the SEM revealed with the crystal morphology remaining consistent. The same was observed for the as-synthesized and activated HKUST-1 thin films (Figure 4f,g). These observations highlight the importance of monitoring changes in both the morphology and the crystal structure, demonstrating that changes in crystallographic structure does not necessitate significant changes in substrate morphology. When comparing the ammonia exposures for the film and powder, the broadness of the peaks for the ammonia exposures to the film and powder are different, but the locations are consistent, revealing that the same crystal structure transformation is observed. The differences in the patterns are that the powder sample has broad peaks, while the film sample has sharper and therefore more distinguishable peaks. This is consistent with the samples before exposure with broader peaks observed for the synthesized powder (Figure 4b) and sharper peaks observed for the synthesized film (Figure 4e).

3.4. Standard Powder Investigation of Solvent Effects

Experiments with a standard HKUST-1 powder were undertaken in order to investigate a sample with and without residual solvent ligands within the framework. For the synthesized thin film and powder HKUST-1 samples, DMSO was found to be present in the as-synthesized materials and is likely coordinated to the open copper paddlewheel site. For the surMOF HKUST-1 sample, water is within the framework and likely bound at this same site. By conducting the experiments with a standard powder obtained commercially, this study investigated (1) the outcome of ammonia exposures on this standard powder void of residual solvent; (2) the effect of residual water and DMSO on this powder; and (3) the outcome of an ammonia exposure after the water or DMSO solvent were present within the framework.

3.4.1. Standard Powder as Received

The obtained standard powder was found to be consistent with literature precedent for HKUST-1 by SEM [11], IR [34], and XRD [10,36,46] characterization (Figures 5a, 6a and 7b). The SEM analysis of the crystal morphology for the standard powder was consistent with the powder synthesized by the procedure described herein. The EDS data showed no evidence of nitrogen present and minimal (2%–3%) sulfur (SI Figure 7). It is noteworthy that the standard powder matched the peak intensities of the HKUST-1 reference pattern (Figure 7a,b) more so than the synthesized powder shown in Figure 4b.

Exposures of ammonia gas for as-synthesized and activated samples were both successful, as IR confirmed the presence of ammonia (Figure 6b,c); and EDS data qualitatively showed similar uptake of nitrogen for both ammonia exposures (SI Figures 8 and 9). The SEM images reveal no morphological changes after either type of exposure (Figure 5b,c), suggesting that the standard powder was activated as received. (Note that the presence of silicon was observed via EDS (SI Figure 7), and this diatomaceous earth was observed in SEM images, which was likely an active desiccant to help maintain activation.) Despite the lack of morphological changes, the crystal structure observed by XRD did change upon exposure to ammonia (Figure 7b–d). The XRD patterns for the standard powder without and with activation (Figure 7c,d) are consistent with one another as well as with the XRD patterns for the ammonia exposures of the as-synthesized HKUST-1 powder (Figure 4c,d).

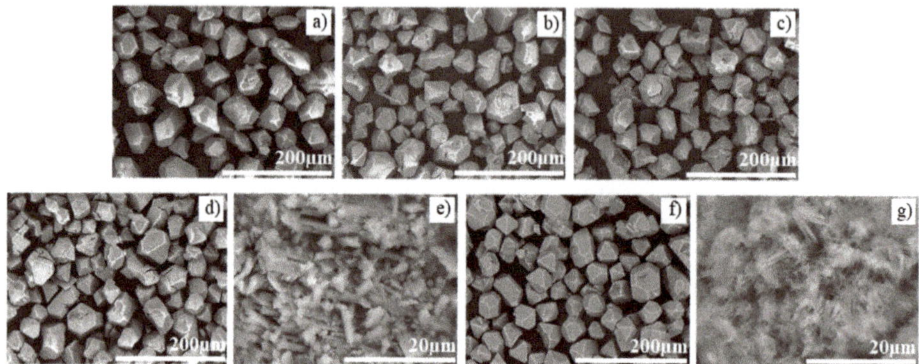

Figure 5. Representative SEM images of HKUST-1 powder (obtained commercially as Basolite® C 300) (**a**) before gas exposure; (**b**) after NH$_3$ exposure without activation; (**c**) after NH$_3$ exposure with prior activation; (**d**) after H$_2$O vapor exposure; (**e**) after H$_2$O vapor and subsequent NH$_3$ exposure without activation; (**f**) after DMSO exposure; (**g**) after DMSO and subsequent NH$_3$ exposure without activation.

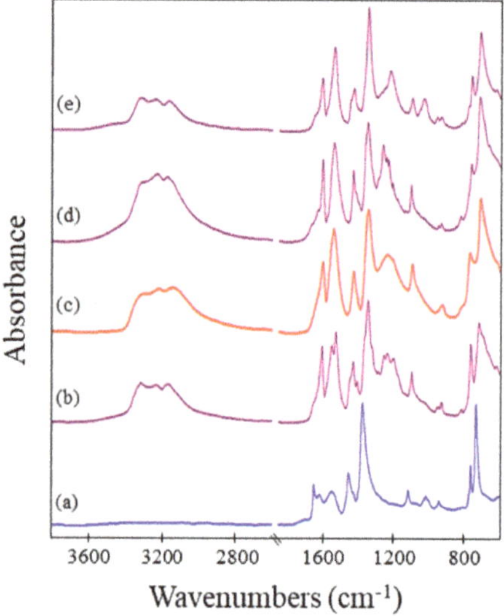

Figure 6. IR spectra of HKUST-1 powder (obtained commercially as Basolite® C 300) (**a**) before gas exposure; (**b**) after NH$_3$ exposure without activation; (**c**) after NH$_3$ exposure with prior activation; (**d**) after H$_2$O vapor and subsequent NH$_3$ exposure without activation; (**e**) after DMSO and subsequent NH$_3$ exposure without activation.

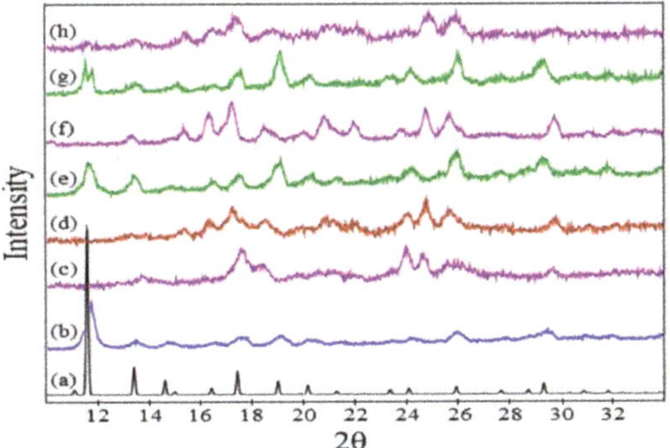

Figure 7. XRD patterns of (**a**) reference pattern for HKUST-1 powder (ICDD pdf number 00-062-1183); (**b**) commercially obtained HKUST-1 powder (Basolite® C 300) before gas exposure; (**c**) after NH_3 exposure without activation; (**d**) after NH_3 exposure with prior activation; (**e**) after H_2O vapor exposure; (**f**) after H_2O vapor and subsequent NH_3 exposure without activation; (**g**) after DMSO exposure; (**h**) after DMSO and subsequent NH_3 exposure without activation.

3.4.2. Standard Powder with Water Exposure

After exposing the standard powder to water vapor, the crystal morphology observed via SEM was unchanged (Figure 5d), and the crystal structure was maintained according to XRD (Figure 7e). Noteworthy, the peak intensities observed in the diffraction pattern were more similar to the synthesized powder (Figure 4b) rather than the standard powder as-received (Figure 5b). This is indicative that the water is bound to the framework of the standard powder after water exposure.

For this standard powder exposed to water vapor, an ammonia exposure resulted in a change in crystal morphology observed by SEM (Figure 5e), the uptake of ammonia detected by EDS (SI Figure 10) and IR (Figure 6d), and a change in crystal structure seen by XRD (Figure 7f). The morphology for this sample that had undergone an ammonia exposure after water vapor exposure (Figure 5e) was similar to that of the as-synthesized HKUST-1 powder upon exposure to ammonia (Figure 2h). The XRD data after ammonia exposure for this sample (Figure 7f) was consistent with the exposures on the standard powder (Figure 7c,d) and the synthesized powder (Figure 4c,d).

3.4.3. Standard Powder with DMSO Exposure

After exposing the standard powder to DMSO, the crystal morphology was unchanged as observed via SEM (Figure 5f). The crystal structure was maintained with the same small change in peak intensity (Figure 7g) as observed for the standard powder after water exposure (Figure 7e), which again was similar to that of the synthesized powder (Figure 4b). The IR shows peaks at approximately 950 cm^{-1} and 1000 cm^{-1} which are consistent with DMSO. The EDS shows a 1:1 ratio of Cu:S (SI Figure 8), which correlates with the same ratio found for the synthesized powder. This IR and EDS data suggests 1 DMSO may be coordinated to each axial copper paddlewheel site.

Upon subsequent ammonia exposure, IR and EDS data indicated that ammonia was captured (Figure 6e) (SI Figures 11 and 12). The SEM showed that the crystal morphology changed (Figure 5g) in a manner consistent with what had been observed for all ammonia exposures of as-synthesized samples (Figure 2b,e,h). The resulting XRD pattern (Figure 7h) is consistent with the patterns for other ammonia exposures (Figures 7c,d,f and 4c,d), especially those of the synthesized powder (Figure 4c,d). For the standard powder with water exposure before ammonia exposure, the XRD peaks (Figure 7f) were sharper, correlating with the size of the crystals observed by SEM (Figure 5f) relative to those for both the standard powder exposed to DMSO and the synthesized powder after ammonia exposure (Figures 5g and 2h, respectively).

3.4.4. Discussion of Solvent Effects on Standard Powder

This study shows that the presence of different solvent ligands results in the loss of crystal morphology after ammonia exposure, while the crystal structure is essentially the same independent of the solvent molecules presence. Why does the presence of the ligand result in loss of the crystal morphology? Based on this research, it is hypothesized that this may be that the ligand is bound to the copper paddlewheel, destabilizing the structure. This is specifically supported by the change in the XRD pattern of the standard powder after it is exposed to water and DMSO. The change in peak intensities may be indicative of this destabilization. In addition, if water or solvent is in fact bound to the copper paddlewheel as hypothesized, then this creates a competition for binding when ammonia is introduced [37]. This competition likely changes the way ammonia is able to interact with the various parts of the framework, explaining why the morphology change is different depending on the absence or presence of solvent. By activating the HKUST-1 before ammonia exposure, one is not able to prevent the change in crystal structure but is able to prevent the dramatic change in morphology, which remained unknown until now. This is advantageous because morphological changes destroy film continuity and would affect the packing and interaction of the powder within a composite material or device architecture. Understanding and preventing this degradation is fundamental toward the integration of MOF materials into hierarchical structures for the realization of their potential in applications as thin film sensors or for industrial gas sequestration.

3.5. Characterization of Regeneration and Re-exposure

This research has shown that the presence of residual solvent disrupts the morphology of three HKUST-1 variations upon exposure to ammonia and that this disruption can be prevented by activation prior to initial gas exposure. Will the original morphology be preserved upon the removal of the ammonia from within the framework and, if so, then what about after a second round of gas exposure? Toward future applications, it is necessary for the material to be morphologically stable not just upon initial exposure, but also for the ammonia uptake to be reversible for multiple cycles of exposure, removal, and re-exposure. This investigation was undertaken for the surMOF HKUST-1 variation with SPM and IR. This was done to determine if the ammonia can be removed after exposure and then whether the framework can re-uptake ammonia upon a subsequent exposure.

Scanning probe microscopy images and IR spectra in Figure 5 show that the activated surMOF (Figure 8b) remained morphologically stable after it was exposed to ammonia once (Figure 8c), regenerated by the removal of ammonia (Figure 8d), and exposed to ammonia a second and sixth time (Figure 8e,f). The SPM images reveal a minor loss in the sharpness of the features after the initial exposure, but the morphology did not change again even after multiple rounds of regeneration and re-exposure. The IR data support that the ammonia was removed by the regeneration process and re-incorporated upon re-exposure. The peak in the IR spectra (~3300 cm^{-1}) corresponding to ammonia was no longer present after regeneration (Figure 8d'), and the peak was again observed after the second (Figure 8e') and sixth (Figure 8f') exposures, indicating successful removal and re-addition of ammonia onto the framework. The observation that ammonia can re-bind to the substrate even after the material undergoes a morphological change is consistent with the results of Peterson et al., which showed that ammonia could still interact with the substrate after undergoing its initial changes in structure and composition [36].

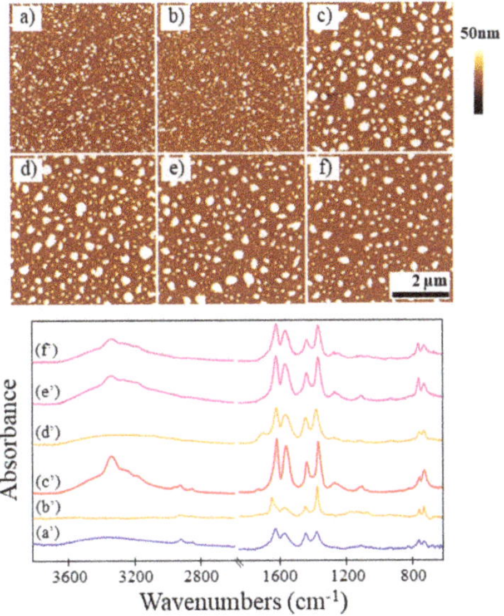

Figure 8. Top: Representative SPM images of HKUST-1 surMOFs (**a**) before exposure to NH$_3$; (**b**) after activation; (**c**) after exposure with prior activation; (**d**) after regeneration; (**e**) after second exposure with activation; and (**f**) after sixth exposure with activation. Bottom: Corresponding IR spectra (**a′–f′**).

4. Conclusions

For the successful incorporation of HKUST-1 into hierarchical architectures, fundamental research is necessary to investigate the effect of heat and gas on the material structure. This study explored how the morphology and crystal structure of the MOF respond to ammonia gas exposure, yielding insights into the effect of different residual solvent molecules within the HKUST-1 framework. In studying three variations of HKUST-1 (surMOF, drop-cast thin film, bulk powder), it was found that all responded similarly, independent of substrate confinement and size variance. While microscopy demonstrated a dramatic change in the morphological structure upon ammonia exposure for the as-synthesized material, activation of the framework by prior heating was demonstrated to rid the framework of solvent molecules and thus mitigates the disruption. An alteration in crystal structure, observed by XRD, was shown to occur in both the drop-cast film and powder, regardless of activation. Additionally, IR supported the successful uptake of ammonia, independent of activation prior to exposure. For activated surMOF films, ammonia uptake was shown to be reversible, permitting removal and recycling upon additional exposures, preserving framework morphology throughout. These findings provide new information about the HKUST-1 ammonia interaction specific to morphological stability and crystal structure change.

Building upon the research herein, additional avenues of study will be explored to expand the versatility and performance capability of HKUST-1 and other MOF systems for gas capture. Morphological stability studies involving exposure with additional gas varieties, such as arsine or hydrogen sulfide, are of interest to further understand the morphological impact induced by the host-guest interaction within both free-standing and surface-anchored HKUST-1 variations. Literature precedent has shown by XRD that HKUST-1 degrades upon arsine exposure without prior activation, whereas hydrogen sulfide is observed to disrupt the framework with or without prior activation [15]. Additional MOF systems, such as MOF-5, MOF-177, and UiO-66, shall be observed to determine morphological stability upon ammonia exposure with and without activation. Prior studies demonstrated that UiO-66 was stable according to XRD upon activated exposure [51], while both MOF-5 and MOF-177 systems were observed to degrade via XRD [52]. This area of research will yield further understanding regarding the potential capabilities and limitations of MOF materials, continuing toward the goal of incorporating MOF assemblies as smart interfaces for sensing applications or selective gas adsorption devices.

Supplementary Materials: The following are available online at http://www.mdpi.com/2079-4991/8/9/650/s1, SI Figures 1–12: Energy Dispersive X-ray Spectroscopy Data.

Author Contributions: Conceptualization, B.H.B., L.J.B., C.K.B. and M.E.A.; Data curation, L.J.B.; Formal analysis, B.H.B. and L.J.B.; Funding acquisition, M.E.A.; Investigation, B.H.B., L.J.B., M.L.O. and L.K.G.; Methodology, M.L.O., L.K.G. and C.K.B.; Project administration, M.E.A.; Resources, M.E.A.; Supervision, M.E.A.; Writing–original draft, B.H.B.; Writing–review & editing, L.J.B. and M.E.A.

Funding: This research was funded by NSF-CHE-158244-RUI, NSF-CHE-1263097-REU, ACS-PRF-54106-UNI5, and NSF-MRI-1126462. Additional funding was provided by the Arnold and Mabel Beckman Foundation Scholars Program (M.L.O.), the Towsley Foundation (M.E.A.), and Hope College.

Acknowledgments: We thank Jennifer Hampton for technical assistance.

Conflicts of Interest: The authors declare no conflict of interest.

References

1. Li, B.; Wen, H.-M.; Zhou, W.; Chen, B. Porous Metal–Organic Frameworks for Gas Storage and Separation: What, How, and Why? *J. Phys. Chem. Lett.* **2014**, *5*, 3468–3479. [CrossRef] [PubMed]

2. Liu, J.; Chen, L.; Cui, H.; Zhang, J.; Zhang, L.; Su, C.-Y. Applications of Metal–Organic Frameworks in Heterogeneous Supramolecular Catalysis. *Chem. Soc. Rev.* **2014**, *43*, 6011–6061. [CrossRef] [PubMed]

3. Yi, F.-Y.; Chen, D.; Wu, M.-K.; Han, L.; Jiang, H.-L. Chemical Sensors Based on Metal–Organic Frameworks. *Chempluschem* **2016**, *81*, 675–690. [CrossRef]

4. Lu, W.; Wei, Z.; Gu, Z.-Y.; Liu, T.-F.; Park, J.; Park, J.; Tian, J.; Zhang, M.; Zhang, Q.; Gentle, T.; et al. Tuning the Structure and Function of Metal–Organic frameworks via Linker Design. *Chem. Soc. Rev.* **2014**, *43*, 5561–5593. [CrossRef] [PubMed]

5. Li, P.Z.; Wang, X.J.; Li, Y.; Zhang, Q.; Tan, R.H.D.; Lim, W.Q.; Ganguly, R.; Zhao, Y. Co(II)-Tricarboxylate Metal-Organic Frameworks Constructed from Solvent-Directed Assembly for CO_2 Adsorption. *Microporous Mesoporous Mater.* **2013**, *176*, 194–198. [CrossRef]

6. Jiang, H.; Zhou, J.; Wang, C.; Li, Y.; Chen, Y.; Zhang, M. Effect of Cosolvent and Temperature on the Structures and Properties of Cu-MOF-74 in Low-Temperature NH_3-SCR. *Ind. Eng. Chem. Res.* **2017**, *56*, 3542–3550. [CrossRef]

7. Kim, D.; Song, X.; Yoon, J.H.; Lah, M.S. 3,6-Connected Metal–Organic Frameworks Based on Triscarboxylate as a 3-Connected Organic Node and a Linear Trinuclear $Co_3(COO)_6$ Secondary Building Unit as a 6-Connected Node. *Cryst. Growth Des.* **2012**, *12*, 4186–4193. [CrossRef]

8. Liu, X.-W.; Sun, T.-J.; Hu, J.-L.; Wang, S.-D. Composites of Metal–Organic Frameworks and Carbon-Based Materials: Preparations, Functionalities and Applications. *J. Mater. Chem. A* **2016**, *4*, 3584–3616. [CrossRef]

9. Mazaj, M.; Kaucic, V.; Logar, N.Z. Chemistry of Metal-Organic Frameworks Monitored by Advanced X-ray Diffraction and Scattering Techniques. *Acta Chim. Slov.* **2016**, *63*, 440–458. [CrossRef] [PubMed]

10. Chui, S.S.-Y.; Lo, S.M.-F.; Charmant, J.P.H.; Orpen, A.G.; Williams, I.D. A Chemically Functionalizable Nanoporous Material $[Cu_3(TMA)_2(H_2O)_3]_n$. *Science* **1999**, *283*, 1148–1150. [CrossRef] [PubMed]

11. Lin, K.-S.; Adhikari, A.K.; Ku, C.-N.; Chiang, C.-L.; Kuo, H. Synthesis and Characterization of Porous HKUST-1 Metal Organic Frameworks for Hydrogen Storage. *Int. J. Hydrogen Energy* **2012**, *37*, 13865–13871. [CrossRef]

12. Bordiga, S.; Regli, L.; Bonino, F.; Groppo, E.; Lamberti, C.; Xiao, B.; Wheatley, P.S.; Morris, R.E.; Zecchina, A. Adsorption Properties of HKUST-1 Toward Hydrogen and Other Small Molecules Monitored by IR. *Phys. Chem. Chem. Phys.* **2007**, *9*, 2676–2685. [CrossRef] [PubMed]

13. Dathe, H.; Peringer, E.; Roberts, V.; Jentys, A.; Lercher, J.A. Metal Organic Frameworks Based on Cu^{2+} and Benzene-1,3,5-Tricarboxylate as Host for SO_2 Trapping Agents. *Comptes Rendus Chim.* **2005**, *8*, 753–763. [CrossRef]

14. Britt, D.; Tranchemontagne, D.; Yaghi, O.M. Metal-organic Frameworks with High Capacity and Selectivity for Harmful Gases. *Proc. Natl. Acad. Sci. USA* **2008**, *105*, 11623–11627. [CrossRef] [PubMed]

15. Peterson, G.W.; Britt, D.K.; Sun, D.T.; Mahle, J.J.; Browe, M.; Demasky, T.; Smith, S.; Jenkins, A.; Rossin, J.A. Multifunctional Purification and Sensing of Toxic Hydride Gases by CuBTC Metal−Organic Framework. *Ind. Eng. Chem. Res.* **2015**, *54*, 3626–3633. [CrossRef]

16. Stock, N.; Biswas, S. Synthesis of Metal-Organic Frameworks (MOFs): Routes to Various MOF Topologies, Morphologies, and Composites. *Chem. Rev.* **2012**, *112*, 933–969. [CrossRef] [PubMed]

17. Petit, C.; Huang, L.; Jagiello, J.; Kenvin, J.; Gubbins, K.E.; Bandosz, T.J. Toward Understanding Reactive Adsorption of Ammonia on Cu-MOF/Graphite Oxide Nanocomposites. *Langmuir* **2011**, *27*, 13043–13051. [CrossRef] [PubMed]

18. Nijem, N.; Fursich, K.; Bluhm, H.; Leone, S.R.; Gilles, M.K. Ammonia Adsorption and Co-Adsorption with Water in HKUST-1: Spectroscopic Evidence for Cooperative Interactions. *J. Phys. Chem. C* **2015**, *119*, 24781–24788. [CrossRef]

19. Bétard, A.; Fischer, R.A. Metal-Organic Framework Thin Films: From Fundamentals to Applications. *Chem. Rev.* **2012**, *112*, 1055–1083. [CrossRef] [PubMed]

20. Shekhah, O.; Liu, J.; Fischer, R.A.; Wöll, C. MOF Thin Films: Existing and Future Applications. *Chem. Soc. Rev.* **2011**, *40*, 1081–1106. [CrossRef] [PubMed]

21. Gu, Z.-G.; Chen, S.-C.; Fu, W.-Q.; Zheng, Q.; Zhang, J. Epitaxial Growth of MOF Thin Film for Modifying the Dielectric Layer in Organic Field-Effect Transistors. *ACS Appl. Mater. Interfaces* **2017**, *9*, 7259–7264. [CrossRef] [PubMed]

22. Zhao, Z.; Ma, X.; Kasik, A.; Li, Z.; Lin, Y.S. Gas Separation Properties of Metal Organic Framework (MOF-5) Membranes. *Ind. Eng. Chem. Res.* **2013**, *52*, 1102–1108. [CrossRef]

23. Eslava, S.; Zhang, L.; Esconjauregui, S.; Yang, J.; Vanstreels, K.; Baklanov, M.R.; Saiz, E. Metal-Organic Framework ZIF-8 Films As Low-κ Dielectrics in Microelectronics. *Chem. Mater.* **2013**, *25*, 27–33. [CrossRef]

24. Kreno, L.E.; Hupp, J.T.; Van Duyne, R.P. Metal-Organic Framework Thin Film for Enhanced Localized Surface Plasmon Resonance Gas Sensing. *Anal. Chem.* **2010**, *82*, 8042–8046. [CrossRef] [PubMed]

25. Travlou, N.A.; Singh, K.; Rodríguez-Castellón, E.; Bandosz, T.J. Cu-BTC MOF/Graphene-Based Hybrid Materials as Low Concentration Ammonia Sensors. *J. Mater. Chem. A* **2015**, *3*, 11417–11429. [CrossRef]

26. Ameloot, R.; Cobechiya, E.; Uji-i, H.; Martens, J.A.; Hofkens, J.; Alaerts, L.; Sels, B.F.; DeVos, D.E. Direct Patterning of Oriented Metal-Organic Framework Crystals via Control Over Crystallization Kinetics in Clear Precursor Solutions. *Adv. Mater.* **2010**, *22*, 2685–2688. [CrossRef] [PubMed]

27. Shekhah, O.; Wang, H.; Kowarik, S.; Schreiber, F.; Paulus, M.; Tolan, M.; Sternemann, C.; Evers, F.; Zacher, D.; Fischer, R.A.; et al. Step-By-Step Route for the Synthesis of Metal-Organic Frameworks. *J. Am. Chem. Soc.* **2007**, *129*, 15118–15119. [CrossRef] [PubMed]

28. Shekhah, O.; Wang, H.; Zacher, D.; Fischer, R.; Wöll, C. Growth mechanism of metal-organic frameworks: Insights into the nucleation by employing a step-by-step route. *Angew. Chemie Int. Ed.* **2009**, *48*, 5038–5041. [CrossRef] [PubMed]

29. Ohnsorg, M.L.; Beaudoin, C.K.; Anderson, M.E. Fundamentals of MOF Thin Film Growth via Liquid-Phase Epitaxy: Investigating the Initiation of Deposition and the Influence of Temperature. *Langmuir* **2015**, *31*, 6114–6121. [CrossRef] [PubMed]

30. Summerfield, A.; Cebula, I.; Schröder, M.; Beton, P.H. Nucleation and Early Stages of Layer-by-Layer Growth of Metal Organic Frameworks on Surfaces. *J. Phys. Chem. C* **2015**, *119*, 23544–23551. [CrossRef] [PubMed]

31. Kajiwara, T.; Higuchi, M.; Watanabe, D.; Higashimura, H.; Yamada, T.; Kitagawa, H. A Systematic Study on the Stability of Porous Coordination Polymers Against Ammonia. *Chem. A Eur. J.* **2014**, *20*, 15611–15617. [CrossRef] [PubMed]

32. Rieth, A.J.; Dinca, M. Controlled Gas Uptake in Metal−Organic Frameworks with Record Ammonia Sorption. *J. Am. Chem. Soc.* **2018**, *140*, 3461–3466. [CrossRef] [PubMed]

33. DeCoste, J.B.; Peterson, G.W. Metal−Organic Frameworks for Air Purification of Toxic Chemicals. *Chem. Rev.* **2014**, *114*, 5695–5727. [CrossRef] [PubMed]

34. Borfecchia, E.; Maurelli, S.; Gianolio, D.; Groppo, E.; Chiesa, M.; Bonino, F.; Lamberti, C. Insights into Adsorption of NH_3 on HKUST-1 Metal−Organic Framework: A Multitechnique Approach. *J. Phys. Chem. C* **2012**, *116*, 19839–19850. [CrossRef]

35. Petit, C.; Mendoza, B.; Bandosz, T.J. Reactive Adsorption of Ammonia on Cu-Based MOF/Graphene Composites. *Langmuir* **2010**, *26*, 15302–15309. [CrossRef] [PubMed]
36. Peterson, G.W.; Wagner, G.W.; Balboa, A.; Mahle, J.; Sewell, T.; Karwacki, C.J. Ammonia Vapor Removal by Cu$_3$(BTC)$_2$ and Its Characterization by MAS NMR. *J. Phys. Chem. C* **2009**, *113*, 13906–13917. [CrossRef] [PubMed]
37. Huang, L.; Bandosz, T.; Joshi, K.L.; Van Duin, A.C.T.; Gubbins, K.E. Reactive Adsorption of Ammonia and Ammonia/Water on CuBTC Metal-Organic Framework: A ReaxFF Molecular Dynamics Simulation. *J. Chem. Phys.* **2013**, *138*, 034102. [CrossRef] [PubMed]
38. Shekhah, O.; Wang, H.; Strunskus, T.; Cyganik, P.; Zacher, D.; Fischer, R.; Wöll, C. Layer-by-Layer Growth of Oriented Metal Organic Polymers on a Functionalized Organic Surface. *Langmuir* **2007**, *23*, 7440–7442. [CrossRef] [PubMed]
39. Bashkova, S.; Bandosz, T.J. Effect of Surface Chemical and Structural Heterogeneity of Copper-Based MOF/Graphite Oxide Composites on the Adsorption of Ammonia. *J. Colloid Interface Sci.* **2014**, *417*, 109–114. [CrossRef] [PubMed]
40. Petit, C.; Wrabetz, S.; Bandosz, T.J. Microcalorimetric Insight into the Analysis of the Reactive Adsorption of Ammonia on Cu-MOF and its Composite with Graphite Oxide. *J. Mater. Chem.* **2012**, *22*, 21443–21447. [CrossRef]
41. DeCoste, J.B.; Denny, M.S.; Peterson, G.W.; Mahle, J.J.; Cohen, S.M. Enhanced Aging Properties of HKUST-1 in Hydrophobic Mixed-Matrix Membranes for Ammonia Adsorption. *Chem. Sci.* **2016**, *7*, 2711–2716. [CrossRef] [PubMed]
42. DeCoste, J.B.; Peterson, G.W.; Smith, M.W.; Stone, C.A.; Willis, C.R. Enhanced Stability of Cu-BTC MOF via Perfluorohexane Plasma-Enhanced Chemical Vapor Deposition. *J. Am. Chem. Soc.* **2012**, *134*, 1486–1489. [CrossRef] [PubMed]
43. Petit, C.; Bandosz, T.J. Synthesis, Characterization, and Ammonia Adsorption Properties of Mesoporous Metal–Organic Framework (MIL(Fe))–Graphite Oxide Composites: Exploring the Limits of Materials Fabrication. *Adv. Funct. Mater.* **2011**, *21*, 2108–2117. [CrossRef]
44. Bhunia, M.K.; Hughes, J.T.; Fettinger, J.C.; Navrotsky, A. Thermochemistry of Paddle Wheel MOFs: Cu-HKUST-1 and Zn-HKUST-1. *Langmuir* **2013**, *29*, 8140–8145. [CrossRef] [PubMed]
45. Zhang, B.; Zhang, J.; Liu, C.; Sang, X.; Peng, L.; Ma, X.; Wu, T.; Han, B.; Yang, G. Solvent Determines the Formation and Properties of Metal–Organic Frameworks. *RSC Adv.* **2015**, *5*, 37691–37696. [CrossRef]
46. Schlichte, K.; Kratzke, T.; Kaskel, S. Improved Synthesis, Thermal Stability and Catalytic Properties of the Metal-Organic Framework Compound Cu$_3$(BTC)$_2$. *Microporous Mesoporous Mater.* **2004**, *73*, 81–88. [CrossRef]
47. Prestipino, C.; Regli, L.; Vitillo, J.G.; Bonino, F.; Damin, A.; Lamberti, C.; Zecchina, A.; Solari, P.L.; Kongshaug, K.O.; Bordiga, S. Local Structure of Framework Cu(II) in HKUST-1 Metallorganic Framework: Spectroscopic Characterization upon Activation and Interaction with Adsorbates. *Chem. Mater.* **2006**, *18*, 1337–1346. [CrossRef]
48. Daniel, T.A.; Uppili, S.; McCarty, G.; Allara, D.L. Effects of Molecular Structure and Interfacial Ligation on the Precision of Cu-Bound a,w-Mercaptoalkanoic Acid "Molecular Ruler" Stacks. *Langmuir* **2007**, *23*, 638–648. [CrossRef] [PubMed]
49. NIST Chemistry WebBook Dimethyl Sulfoxide. Available online: http://webbook.nist.gov/cgi/cbook.cgi?ID=C67685&Type=IR-SPEC&Index=2#Top (accessed on 28 June 2018).
50. NIST Chemistry WebBook Ammonia. Available online: http://webbook.nist.gov/cgi/cbook.cgi?ID=C7664417&Type=IR-SPEC&Index=1 (accessed on 28 June 2018).
51. Morris, W.; Doonan, C.J.; Yaghi, O.M. Postsynthetic Modification of a Metal-Organic Framework for Stabilization of a Hemiaminal and Ammonia Uptake. *Inorg. Chem.* **2011**, *50*, 6853–6855. [CrossRef] [PubMed]
52. Saha, D.; Deng, S. Ammonia Adsorption and its Effects on Framework Stability of MOF-5 and MOF-177. *J. Colloid Interface Sci.* **2010**, *348*, 615–620. [CrossRef] [PubMed]

 *nanomaterials*

Article

In-Situ Growth of NiAl-Layered Double Hydroxide on AZ31 Mg Alloy towards Enhanced Corrosion Protection

Xin Ye [1], Zimin Jiang [2], Linxin Li [1] and Zhi-Hui Xie [1,*]

[1] Chemical Synthesis and Pollution Control Key Laboratory of Sichuan Province, China West Normal University, Nanchong 637002, China; xinchem4@163.com (X.Y.); Linxinyj@163.com (L.L.)

[2] College of Foreign Language Education, China West Normal University, Nanchong 637002, China; jenny8116@126.com

* Correspondence: zhxie@cwnu.edu.cn; Tel./Fax: +86-817-256-8081

Received: 20 May 2018; Accepted: 1 June 2018; Published: 7 June 2018

Abstract: NiAl-layered double hydroxide (NiAl-LDH) coatings grown in-situ on AZ31 Mg alloy were prepared for the first time utilizing a facile hydrothermal method. The surface morphologies, structures, and compositions of the NiAl-LDH coatings were characterized by scanning electron microscopy (SEM), three dimensional (3D) optical profilometer, X-ray diffractometer (XRD), Fourier transform infrared spectrometer (FT-IR), and X-ray photoelectron spectroscopy (XPS). The results show that NiAl-LDH coating could be successfully deposited on Mg alloy substrate using different nickel salts, i.e., carbonate, nitrate, and sulfate salts. Different coatings exhibit different surface morphologies, but all of which exhibit remarkable enhancement in corrosion protection in 3.5 wt % NaCl corrosive electrolyte. When nickel nitrate was employed especially, an extremely large impedance modulus at a low frequency of 0.1 Hz ($|Z|_{f = 0.1\,Hz}$), 11.6 MΩ cm^2, and a significant low corrosion current density (j_{corr}) down to 1.06 nA cm^{-2} are achieved, demonstrating NiAl-LDH coating's great potential application in harsh reaction conditions, particularly in a marine environment. The best corrosion inhibition of NiAl-LDH/CT coating deposited by carbonate may partially ascribed to the uniform and vertical orientation of the nanosheets in the coating.

Keywords: Mg alloy; LDH; corrosion; deposition; coating

1. Introduction

Mg alloy has excellent physical and mechanical properties but a high chemical reactivity and susceptibility to corrosion, which hinders its practical application and development in more fields [1]. Efforts by formation of protective coatings on Mg alloy surface to decrease the corrosion rate and extend the serving time have aroused an increasing interest in the area of surface engineering [1,2]. Many traditional coating methods are adopted to protect Mg alloy from corrosion, such as electroless Ni plating [3,4]. However, more and more attendant problems, especially environmental issues, emerge with increasing usage of these techniques, such as the high cost and complexity in dealing with the waste plating bath, and the detrimental effect of Cr to the ecosystem [3]. Thus, it is highly urgent to find state-of-the-art alternative coatings with low pollution emissions but with efficient corrosion inhibition.

Layered double hydroxides (LDHs) have received extensive attention for potential application in supercapacitor [5], catalysis [6], adsorbents [7], and corrosion protection [8] because of their diversity in compositions and structures. The charge of the metal layer in the LDHs is compensated by interlayer anions such as CO_3^{2-}. Once CO_3^{2-} anions are intercalated, it cannot be exchanged by most corrosive anions, such as SO_4^{2-} and Cl^-, due to the high ion-exchange equilibrium constant of CO_3^{2-} anions, thereby the corrosion of the substrate under the LDHs coating will be delayed. From this point of view, CO_3^{2-} anions intercalated LDHs are an ideal choice for obtaining a LDHs coating with high corrosion inhibition capacity. The most widely reported CO_3^{2-} intercalated LDHs on Mg alloy matrix is MgAl-LDHs coating, which is obtained, in most cases, by in-situ growth technology owing to the simplification of operation and strong adhesion of the coating [9]. Recently, MgFe-LDH and MgCr-LDH films were also obtained on anodized AZ31 Mg alloy by in-situ growth measurement [10]. However, the orders of magnitude of the j_{corr} of these in-situ grown LDH coatings are higher than -8, mostly range from -5 to -7 (Table 1) [8,11–14]. To break the bottleneck of j_{corr}, it is advisable to prepare carbonate-based LDHs film by attempting more different divalent or trivalent metallic cations. Though most recent research results proved that NiAl-LDH nanoparticles possess good photocatalytic performance [15], in-situ growth of NiAl-LDH film on Mg alloy for corrosion protection has not yet been reported to date.

Table 1. Corrosion inhibition of different layered double hydroxide (LDH) coatings on Mg alloy from most recent published literature.

Substrate	LDH Coatings	Corrosive Medium	E_{corr} V vs. SCE	j_{corr} (A cm^{-2})	Ref.
Mg alloy	MgAl-NO$_3^-$	Phosphate buffer saline	-1.53	3.63×10^{-7}	[14]
Anodized AZ31 Mg alloy	MgAl-NO$_3^-$	3.5 wt % NaCl solution	-0.47	9.48×10^{-7}	[8]
Anodized AZ31 Mg alloy	MgAl-VO$_3^-$	3.5 wt % NaCl solution	-0.40	2.48×10^{-7}	[8]
Anodized AZ31 Mg alloy	MgAl-NO$_3^-$	3.5 wt % NaCl solution	-1.34	1.18×10^{-7}	[10]
AZ31 Mg alloy	MgAl-5-fluorouracil	Phosphate buffer saline	-1.12	3.34×10^{-5}	[13]
Plasma electrolytic oxidation pretreated AZ31 Mg alloy	MgAl-5-fluorouracil	Phosphate buffer saline	-1.20	3.92×10^{-6}	[13]
AZ31 Mg alloy	MgAl-5-fluorouracil	Phosphate buffer saline	-1.42	3.27×10^{-5}	[16]
Anodized AZ31 Mg alloy	MgFe-NO$_3^-$	3.5 wt % NaCl solution	-1.44	1.09×10^{-6}	[10]
Anodized AZ31 Mg alloy	MgCr-NO$_3^-$	3.5 wt % NaCl solution	-1.47	2.16×10^{-6}	[10]
AZ91D Mg alloy	ZnAl-VO$_3^-$	3.5 wt % NaCl solution	-1.30	2.21×10^{-6}	[17]
AZ91D Mg alloy	ZnAl-Cl$^-$	3.5 wt % NaCl solution	-1.39	2.52×10^{-6}	[17]
AZ91D Mg alloy	ZnAl-NO$_3^-$	3.5 wt % NaCl solution	-1.42	1.33×10^{-5}	[17]
AZ31 Mg alloy	MgAl-CO$_3^{2-}$	3.5 wt % NaCl solution	-0.36	8.40×10^{-7}	[18]
Anodized AZ31	MgAl-CO$_3^{2-}$	3.5 wt % NaCl solution	-0.29	3.50×10^{-7}	[18]

Herein, we report a facile hydrothermal measurement to in-situ growth of NiAl-LDH nanocomposite on Mg alloy by use of alkaline solutions with three different nickel salts for preparation of a highly enhanced corrosion-resistant coating with an extraordinary low corrosion rate (Figure 1). The NiAl-LDH coating prepared by nickel carbonate exhibits uniformly and vertically aligned nanoarrays with an extremely large impedance modulus at a low frequency of 0.1 Hz ($|Z|_{f = 0.1\,Hz}$), 11.6 MΩ cm^2, and a significantly low j_{corr} down to 1.06 nA cm^{-2} in 3.5 wt % NaCl corrosive electrolyte, which outperforms the values of the foregoing achieved LDH coating on Mg alloy (Table 1) [8,10,13,14,16–18]. The NiAl-LDH coatings were characterized, and the enhanced corrosion inhibition mechanism was proposed and discussed.

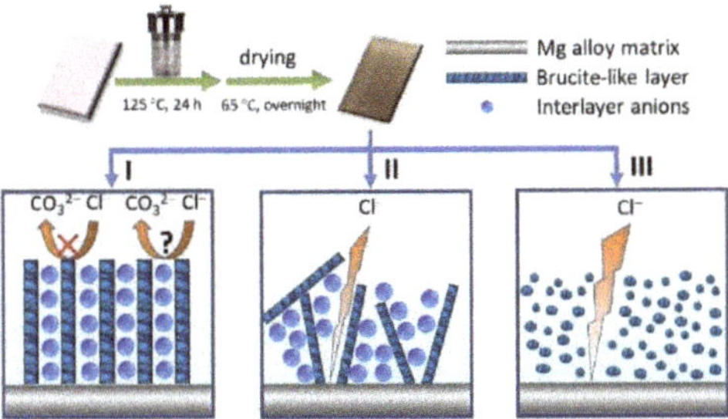

Figure 1. Schematic illustration of various microstructures of NiAl-LDH coatings prepared with different nickel salts. (I) nickel carbonate; (II) nickel nitrate; and (III) nickel sulfate. The interlayer anions are not given in case III.

2. Methods

2.1. Materials and Reagents

The matrix used is AZ31 Mg alloy with a chemical composition in wt %: 2.75 Al, 1.15 Zn, 0.16 Mn, and Mg balance. Primary chemicals include nickel carbonate (AR), nickel nitrate ($\geq$98%), nickel sulphate ($\geq$98.5%), sodium carbonate ($\geq$99.8%), and sodium hydroxide ($\geq$98%). Chemicals were acquired from Aladdin Industrial Inc. (Shanghai, China) and Sinopharm Chemical Reagent Co., Ltd. (Shanghai, China), and all chemicals were used without further purification. Ultrapure water used in the experiments was obtained using a water purification system (UPT-II-10T, Ulupure Corporation, Chengdu, China) with a resistivity of 18.2 MΩ cm at 25 °C.

2.2. Preparation of NiAl-LDH Coatings on Mg Alloy

All the NiAl-LDH coatings were grown on Mg alloy substrate by a simple one step hydrothermal method with the same steps and conditions but with three different nickel salts. The obtained coatings by use of nickel carbonate, nickel nitrate, and nickel sulphate are denoted as NiAl-LDH/CT, NiAl-LDH/NT, and NiAl-LDH/ST coatings, respectively. The mole ratio of Ni^{2+}, Al^{3+}, and CO_3^{2-} in the precursor solution is 6:2:1. Take the preparation of NiAl-LDH/CT coating, for example, which is described as follows:

Pre-treatment of substrate: The AZ31 Mg alloy was ground mechanically with SiC waterproof sand paper and degreased in an alkaline solution at 65 °C for 10 min [3].

Preparation of precursor solution: 40 mL aluminum nitrate solution (0.002 mol $Al(NO_3)_3 \cdot 9H_2O$) was first added to a nickel solution (~40 mL) containing 0.002 mol $NiCO_3.2Ni(OH)_2 \cdot 4H_2O$. Then, 0.001 mol anhydrous Na_2CO_3 was added, followed by pH adjustment of the solution to about 12 by use of a NaOH solution (~3.7 mL). Finally, the above solution was diluted to a volume of 100 mL by adding ultrapure water.

Growth of NiAl-LDH film: The solution above was then transferred into a 100 mL Teflon-lined autoclave where a pre-treated Mg alloy has been placed. Then, the autoclave was placed and kept in an oven with a temperature of 125 °C for 24 h for in-situ growth of NiAl-LDH films on the matrix.

After that, the samples were taken out of the autoclave, rinsed with water, dried overnight in an oven at 65 °C, and designed as NiAl-LDH/CT coating. The amount of nickel salts is 0.006 mol for deposition of NiAl-LDH/NT and NiAl-LDH/ST coatings, which is the sole difference in comparison with preparation of NiAl-LDH/CT coating.

2.3. Characterization and Electrochemical Tests

The surface morphologies and roughness of the different NiAl-LDH coatings were observed by scanning electron microscopy (SEM, Hitachi S-4800, Hitachi High-Technologies

Corporation, Tokyo, Japan), and three dimensional (3D) optical profilometer (Bruker Contour GT-K, Billerica, MA, USA), respectively. The arithmetical mean deviation of the profile (R_a) is used to estimate the roughness. The microstructures of the coatings were identified by X-ray diffractometer (XRD, D8 Advance). Fourier transform infrared spectroscopic acquisition of the specimens were obtained (FT-IR, Nicolet 6700, Thermo Scientific, Waltham, MA, USA) in the range of 4000~500 cm^{-1}. The elemental compositions of the sample surface were recorded by X-ray photoelectron spectroscopy (XPS, 250Xi, Thermo Scientific, Waltham, MA, USA) and energy dispersive X-ray spectroscopy (EDS) equipped in SEM. The water contact angles were measured with a water drop volume of 5 µL, utilizing optical contact angle meter (JC2000D, Shanghai Zhongchen Digital Technology Apparatus Co., Ltd., Shanghai, China) at 298 K. The electrochemical impedance spectroscopy (EIS) and Tafel curves were achieved using a classical three-electrode system with saturated calomel electrode (SCE) as reference electrode, Pt foil as counter electrode, and freshly ground bare Mg alloy and as-prepared NiAl-LDH coatings as working electrode (exposed area: 1×1 cm^2) on an electrochemical workstation (CHI660E, Chenhua, Shanghai, China). The EIS frequencies range from 10^5~0.1 Hz using an AC perturbation of 10 mV versus the open circuit potential (OCP). The Tafel measurement was performed at a scan rate of 5 mV/s in the potential region of $\pm$500 mV versus OCP. All the electrochemical tests were carried out in a 3.5 wt % NaCl solution at 298 K.

3. Results and Discussion

The digital photos of bare Mg alloy and three different NiAl-LDH coatings are shown in Figure 2. The substrate shows a shiny surface after polishing and pre-treatment processes, as shown in Figure 2a. After coating, all the samples show a light bronze-like surface in color, suggesting the successfully deposition of NiAl-LDH coating on the Mg alloy (Figure 2b–d). The surface brightness of NiAl-LDH/CT and NiAl-LDH/NT coatings are very close, and both of which are brighter than that of NiAl-LDH/ST coating.

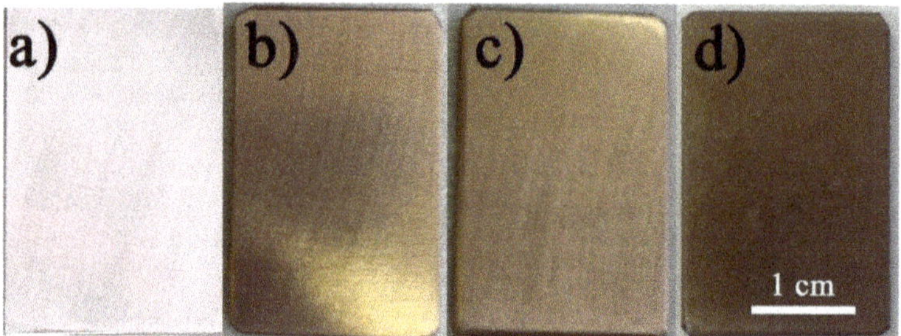

Figure 2. Digital photos of (**a**) bare Mg alloy; (**b**) NiAl-LDH/CT; (**c**) NiAl-LDH/NT; and (**d**) NiAl-LDH/ST coatings.

Figure 3a shows the XRD patterns of Mg alloy and NiAl-LDH coatings obtained with different nickel salts. For the substrate, three strong peaks at (2θ) 32.20°, 34.40°, and 36.62°, and several relatively weak peaks at 47.83°, 63.06°, and 72.50°, etc. are ascribed to the characteristic diffraction peaks of Mg (PDF 35-0821). These small and weak peaks at 18.59°, 38.02°, 58.64°, and 62.07° are the characteristic peaks of Mg(OH)$_2$ (PDF 07-0239) [19,20]. The XRD patterns of all coatings are almost identical and two new peaks appear at ca. 11.73° and 23.58°, which correspond to the (003) and (006) planes of NiAl-LDH (PDF 15-0087) [15]. Figure 3b shows that all samples present almost the same absorption peaks in the FT-IR spectra. The relatively strong peak at 3692 cm^{-1} is related to the O–H stretching vibration of Ni–O–H, and the broadening adsorption peaks at 3464 cm^{-1} and 1630 cm^{-1} are associated with the O–H stretching and bending vibrations from interlamellar water molecules [21] owing to formation of hydrogen-bond. The absorption bands located at 1368 and ca. 1070 cm^{-1} correspond to the asymmetrical and symmetrical stretching vibrations of C–O in CO_3^{2-}, respectively. The weak bands below 800 cm^{-1} are assigned to lattice vibrations of metal-oxygen (M–O) in the brucite-like layers [8]. XPS survey spectrum in Figure 3c shows that the primary elements of substrate are Mg and O, and C 1s peaks is hardly seen. After coating, two new peaks appeared at ca. 856.5 and 74.2 eV which are assigned to Ni 2p and Al 2s, respectively. In addition, significant intensification of C 1s peak at ca. 284 eV is observed. The Ni 2p high-resolution XPS spectra exhibit two major peaks along with two pairs of shake-up satellites (Figure 3d). The major peaks at 873.4 and 855.7 eV relating to Ni $2p_{1/2}$ and Ni $2p_{3/2}$ with a spin-energy separation of 17.7 eV are characteristics of Ni (II) in Ni(OH)$_2$. The Al 2p spectra showing peaks at ca. 74.4 eV (Figure 3e) are ascribed to Al^{3+} species (Al–O). The C 1s spectrum is deconvoluted into three separate peaks related to different types of carbon bonds including C–C at 284.8 eV owing to adventitious carbon, C–O (286.3 eV) and O–C=O (288.7 eV) from CO_3^{2-} (Figure 3f). These characterizations demonstrate the successful formation of NiAl-LDH crystal phase on the matrix.

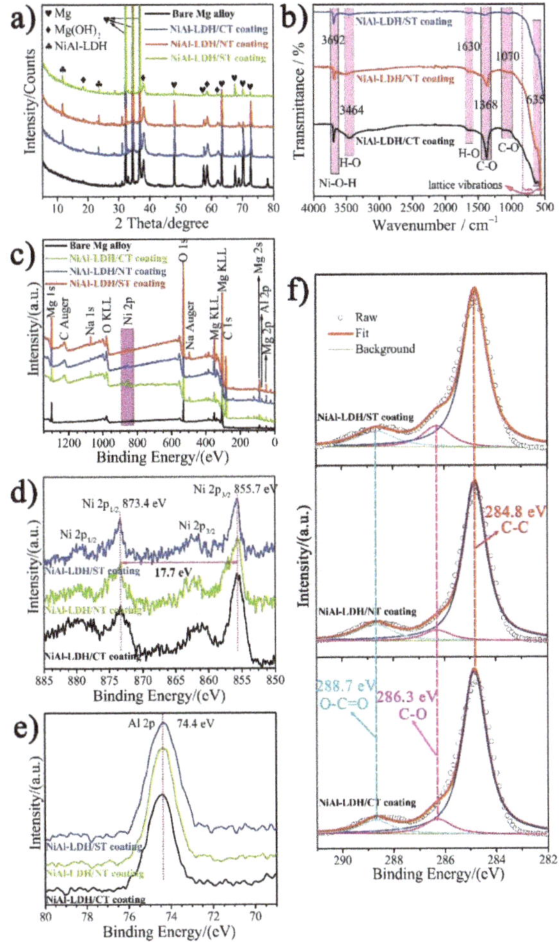

Figure 3. (a) XRD patterns, (b) FT-IR, and (c–f) XPS spectra of different NiAl-LDH coatings.

Figure 4a–c exhibits the matrix surface fully covered with vertically aligned nanoflake arrays for the NiAl-LDH/CT coating, both vertically and horizontally aligned, and randomly-inclined nanoflakes for the NiAl-LDH/NT coating, and nanonodules for the NiAl-LDH/ST coatings. These coatings are composed of Ni, Al, C, and O, and distribute uniformly all over the LDHs coating surfaces (Figure 4d–f). The thickness and side length of the nanoflakes are 63 and 425 nm, respectively, for NiAl-LDH/CT coating. The thickness of the nanoflakes for NiAl-LDH/NT coating is about 34 nm, and the length cannot be determined due to random orientation and irregular shape which result in a much higher R_a, 11.76 µm, than that of NiAl-LDH/CT coating (4.24 µm) (Figure 4g–i). The NiAl-LDH/ST coating exhibits a flat surface packed by uniform nanonodules with a size of ca. 17 nm, producing the lowest R_a, 2.01 µm, among the three types of NiAl-LDH coatings. The different shapes of the deposits result in different water contact angles (Figure 4j–l). The water contact angles of NiAl-LDH/CT and NiAl-LDH/NT coatings are 84.8° and 82.1°, respectively, while the value of NiAl-LDH/ST coating is only 67.9°. The variation in water contact angles led to a slight difference in corrosion inhibition for these coatings.

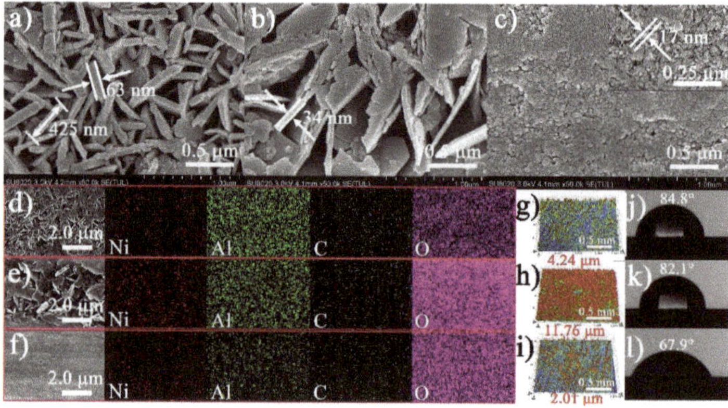

Figure 4. (**a–c**) SEM; (**d–f**) SEM-EDS mapping; and (**g–i**) 3D roughness images of the surface morphologies; and (**j–l**) water contact angles for (**a,d,g,j**) NiAl-LDH/CT, (**b,e,h,k**) NiAl-LDH/NT, and (**c,f,i,l**) NiAl-LDH/ST coatings.

Electrochemical impedance spectroscopy (EIS) was carried out in 3.5 wt % NaCl corrosive electrolyte to evaluate the corrosion resistance of the three NiAl-LDH coatings [22], which is depicted in Figure 5. EIS results show that Mg alloy has a two-time constant (Figure 5a), i.e., one capacitive loop at high frequency and one inductive loop with ranges from intermediate to low frequency region, which are ascribed to the electric double layer at the electrode/electrolyte interface and relaxation diffusion of corrosion products such as $Mg(OH)^+_{ads}$, respectively [20,23]. According to the fitting results by use of an equivalent circuit (EC) model as given in Figure 6a [20], the charge transfer resistance (R_{ct}) of the bare Mg alloy is only 369.20 Ω cm^2. After deposition of NiAl-LDH films, totally different Nyquist plots were observed. The fitting results based on an EC model consisting of R_f, R_{ct}, and Z_w (Figure 6b) are listed in Table 2. All the coatings show a very high R_f along with a remarkable increment of R_{ct}, manifesting the significant enhancement of corrosion protection by the coatings. In contrast, NiAl-LDH/CT coating exhibits the highest R_f (6.2 MΩ cm^2) and R_{ct} (3.5 MΩ cm^2), followed by NiAl-LDH/NT coating with medium R_f (3.7 MΩ cm^2) and R_{ct} (1.8 MΩ cm^2), and the NiAl-LDH/ST coating with the lowest R_f (1.4 MΩ cm^2) and R_{ct} (0.72 MΩ cm^2). The impedance modulus at a low frequency, such as $|Z|_{f=0.1\,Hz}$, which can be obtained directly from the Bode plots without fitting, also represent the corrosion-resistant capability of a coating [20,21]. It can be seen from Table 2 that the NiAl-LDH/CT coating also possesses the highest $|Z|_{f=0.1\,Hz}$, 11.6 MΩ cm^2, in comparison with that values of NiAl-LDH/NT (6.7 MΩ cm^2) and NiAl-LDH/ST (3.4 MΩ cm^2) coatings. The biggest radius of curvature in Figure 5b also confirms the best corrosion resistance of NiAl-LDH/CT coating.

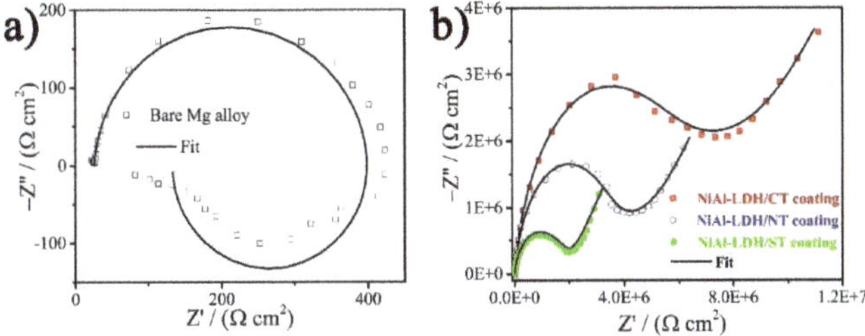

Figure 5. Nyquist plots of (**a**) bare Mg alloy and (**b**) different coatings.

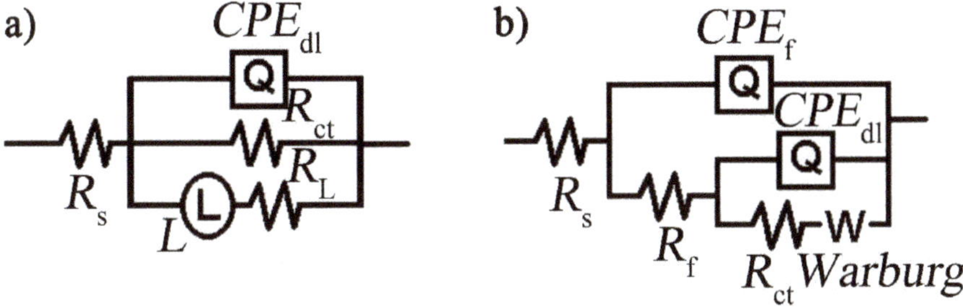

Figure 6. The equivalent circuit models used for fitting the EIS results of the (**a**) bare Mg alloy and (**b**) different NiAl-LDH coatings. The equivalent circuit (EC) model of the Mg alloy for fitting is $R_s(Q_{dl}R_{ct}(R_LL))$ where R_s is solution resistance, Q_{dl} is double layer capacitance, R_{ct} is charge transfer resistance, R_L is inductance resistance, and L is inductance. The EC model for coatings is $R_s(Q_f(R_f(Q_{dl}(R_{ct}Z_w))))$ where Q_f and R_f are capacitance and resistance of the coatings, respectively. Z_w is Warburg diffusion resistance.

Table 2. Parameters of EIS for three different NiAl-LDH coatings.

Samples	$Q_f/10^{-9}$ (S s^n cm^{-2})	R_f (MΩ cm^2)	$Q_{dl}/10^{-9}$ (S s^n cm^{-2})	R_{ct} (MΩ cm^2)	$W/10^{-7}$ (S s$^{0.5}$ cm^{-2})	$|Z|_{f=0.1\,Hz}$ (MΩ cm^2)	$\chi^2/10^{-3}$
NiAl-LDH/CT coating	2.1 ± 0.029	6.2 ± 0.17	62.8 ± 7.6	3.5 ± 0.45	3.5 ± 0.17	11.6	0.31
NiAl-LDH/NT coating	1.5 ± 0.021	3.7 ± 0.14	113.5 ± 9.8	1.8 ± 0.22	3.5 ± 0.35	6.7	0.22
NiAl-LDH/ST coating	0.84 ± 0.087	1.4 ± 0.15	24.5 ± 2.6	0.72 ± 0.076	6.9 ± 0.27	3.4	1.5

A further Tafel test was performed to ascertain the corrosion potential, E_{corr} vs. SCE, and j_{corr} of the samples [24,25], the results of which are shown in Figure 7 and listed in Table 3. The E_{corr} and j_{corr} of a bare Mg alloy are -1450 mV and 3242 nA cm^{-2}, respectively. After deposition of NiAl-LDH/CT film, the E_{corr} shifts positively by ca. 800 mV to -674 mV, and j_{corr} is decreased by a factor of 3058 in comparision with that value of the matrix. The j_{corr} of NiAl-LDH/NT and NiAl-LDH/ST coatings are 3.24 and 5.75 nA cm^{-2}, respectively, which are in positive agreement with the variation of $|Z|_{f=0.1\,Hz}$ because a coating with higher impedance always has lower j_{corr} and better corrosion resistance. However, it is worth noting that a material with a lower j_{corr} does not always possess a higher E_{corr}. For example, the NiAl-LDH/ST coating has the highest j_{corr} but the lowest E_{corr} (Table 3). That is because E_{corr} is a thermodynamic parameter, which stands for the corrosion tendency, while corrosion rate is proportional to the j_{corr}, i.e., there is no direct relationship between E_{corr} and j_{corr}.

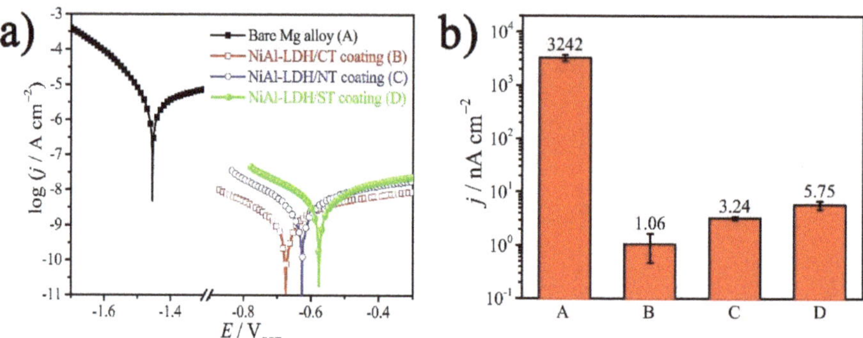

Figure 7. (**a**) Tafel curves of the bare Mg alloy and different NiAl-LDH coatings in NaCl corrosive medium. (**b**) The corresponding j_{corr} of substrate and coatings extracted from the Tafel curve.

Table 3. Parameters of Tafel curves for bare Mg alloy and different NiAl-LDH coatings.

Samples	E_{corr} (mV)	j_{corr} (nA cm^{-2})	β_a (mV dec^{-1})	$-\beta_c$ (mV dec^{-1})
Bare Mg alloy	-1450 ± 52	3242 ± 425	331	95
NiAl-LDH/CT coating	-674 ± 42	1.06 ± 0.59	190	171
NiAl-LDH/NT coating	-625 ± 96	3.24 ± 0.25	244	184
NiAl-LDH/ST coating	-577 ± 51	5.75 ± 1.03	270	188

The high capacity and slight difference in corrosion resistance of the three different NiAl-LDH coatings are ascribed to the high affinity of LDHs towards CO_3^{2-} in comparison with the corrosive Cl^-, the different orientations of the nanosheets, and the different shapes of the deposition, which is illustrated in Figure 1. The corrosive species cannot be exchanged with the brucite-like layer and have to exchange with the CO_3^{2-} anions to arrive at the substrate surface, but the ion-exchange is not easy due to the high affinity of LDHs to CO_3^{2-} anions, enabling high enhancement in corrosion inhibition in all the as-prepared NiAl-LDH coatings. Due to the vertically-aligned nanosheets in the NiAl-LDH/CT coating, the carbonate anions distribute evenly in the interlayer spaces, which further increase the difficulty of the ion-exchange between CO_3^{2-} anions and corrosive species (Case I in Figure 1). For the NiAl-LDH/NT coating, the basal spacing becomes larger at some locations and CO_3^{2-} anions distribute unevenly in the interlayer spaces owing to irregular orientation of the nanosheets, which increases the probability for corrosive species to arrive at the substrate surface by going through fewer CO_3^{2-} anions, resulting in reduction of corrosion inhibition (Case II in Figure 1). In general, the dense plate-like arrangement of LDHs is favorable for increasing corrosion resistance of the film due to lower exposure of CO_3^{2-} anions for ion-exchange. However, for the NiAl-LDH/ST coating in this work, nanomodules with many obvious voids rather than nanoflakes were formed, as demonstrated by the SEM image in Figure 4c, which decreases the corrosion-resistant capability of the LDHs film. In addition, the smaller water contact angle of NiAl-LDH/ST coating in comparison with that of NiAl-LDH/CT and NiAl-LDH/NT coatings should also be responsible for the decreasing corrosion inhibition of NiAl-LDH/ST coating. It is worth noting that although these explanations above account for the difference in corrosion resistance of the three different coatings, the influences of the different orientation and shapes of the deposition on the corrosion inhibition are not significant because all the j_{corr} of the coatings remain in the level of nA cm^{-2}.

4. Conclusions

A facile hydrothermal strategy was progressed to achieve in-situ NiAl-LDH coatings on Mg alloy to improve the corrosion protection. All the NiAl-LDH coatings obtained by different nickel salts show remarkable enhancement in corrosion inhibition in NaCl solution compared with Mg alloy substrate, which is attributed to the strongly affinity between charge-compensating CO_3^{2-} and brucite-like layers. The different orientation of the nanosheets and the different shapes of the deposition are mainly responsible for the slight difference in corrosion inhibition performance among the three different coatings. The NiAl-LDH/CT coating deposited by carbonate shows relatively the highest $|Z|_{f = 0.1\ Hz}$ and lowest j_{corr}, suggesting its best corrosion inhibition. It is believed that these findings may inspire the design and development of other LDH nanosheet arrays, such as NiCr-LDH arrays, as highly enhanced corrosion protection film for a susceptible lightweight metal matrix.

Author Contributions: The conceptualization of the work was conceived by Z.-H.X. The experiments were mainly carried out by X.Y. Z.J. revised and edited the article. Some electrochemical tests were carried out by L.L. Z.-H.X. supervised all the experiments, discussed the results with all the authors, and reviewed and revised the manuscript.

Funding: This research was funded by the National Natural Science Foundation of China (51501157, 21571148); the Fundamental Research Funds of China West Normal University (17B005); and the Chemical Synthesis and Pollution Control Key Laboratory of Sichuan Province (CSPC2014-4-2).

Conflicts of Interest: The authors declare no conflict of interest.

References

1. Xie, Z.-H.; Li, D.; Skeete, Z.; Sharma, A.; Zhong, C.-J. Nanocontainer-Enhanced Self-Healing for Corrosion-Resistant Ni Coating on Mg Alloy. *ACS Appl. Mater. Interfaces* **2017**, *9*, 36247–36260. [CrossRef] [PubMed]

2. Zhang, J.; Xie, Z.-H.; Chen, H.; Hu, C.; Li, L.; Hu, B.; Song, Z.; Yan, D.; Yu, G. Electroless Deposition and Characterization of a Double-Layered Ni-B/Ni-P Coating on AZ91D Mg Alloy from Eco-Friendly Fluoride-Free Baths. *Surf. Coat. Technol.* **2018**, *342*, 178–189. [CrossRef]

3. Xie, Z.H.; Chen, F.; Xiang, S.; Zhou, J.; Song, Z.; Yu, G. Studies of Several Pickling and Activation Processes for Electroless Ni-P Plating on AZ31 Magnesium Alloy. *J. Electrochem. Soc.* **2015**, *162*, D115–D123. [CrossRef]

4. Rajabalizadeh, Z.; Seifzadeh, D. Application of Electroless Ni-P Coating on Magnesium Alloy via CrO_3/HF Free Titanate Pretreatment. *Appl. Surf. Sci.* **2017**, *422*, 696–709. [CrossRef]

5. Duan, C.; Zhao, J.; Qin, L.; Yang, L.; Zhou, Y. Ternary Ni-Co-Mo Oxy-Hydroxide Nanoflakes Grown on Carbon Cloth for Excellent Supercapacitor Electrodes. *Mater. Lett.* **2017**, *208*, 65–68. [CrossRef]

6. Wang, L.; Wang, Y.; Feng, X.; Ye, X.; Fu, J. Small-Sized Mg-Al LDH Nanosheets Supported on Silica Aerogel with Large Pore Channels: Textural Properties and Basic Catalytic Performance after Activation. *Nanomaterials* **2018**, *8*, 113. [CrossRef] [PubMed]

7. Ding, Y.; Liu, L.; Fang, Y.; Zhang, X.; Lyu, M.; Wang, S. The Adsorption of Dextranase onto Mg/Fe-Layered Double Hydroxide: Insight into the Immobilization. *Nanomaterials* **2018**, *8*, 173. [CrossRef] [PubMed]

8. Zhang, G.; Wu, L.; Tang, A.; Zhang, S.; Yuan, B.; Zheng, Z.; Pan, F. A Novel Approach to Fabricate Protective Layered Double Hydroxide Films on the Surface of Anodized Mg-Al alloy. *Adv. Mater. Interfaces* **2017**, *4*, 1700163. [CrossRef]

9. Guo, L.; Wu, W.; Zhou, Y.; Zhang, F.; Zeng, R.; Zeng, J. Layered Double Hydroxide Coatings on Magnesium Alloys: A Review. *J. Mater. Sci. Technol.* **2018**. [CrossRef]

10. Wu, L.; Yang, D.; Zhang, G.; Zhang, Z.; Zhang, S.; Tang, A.; Pan, F. Fabrication and Characterization of Mg-M Layered Double Hydroxide Films on Anodized Magnesium Alloy AZ31. *Appl. Surf. Sci.* **2018**, *431*, 177–186. [CrossRef]

11. Zhang, F.; Liu, Z.-G.; Zeng, R.-C.; Li, S.-Q.; Cui, H.-Z.; Song, L.; Han, E.-H. Corrosion Resistance of Mg-Al-LDH Coating on Magnesium Alloy AZ31. *Surf. Coat. Technol.* **2014**, *258*, 1152–1158. [CrossRef]

12. Yan, L.; Zhou, M.; Zhang, X.; Huang, L.; Chen, W.; Roy, V.A.L.; Zhang, W.; Chen, X. A Novel Type of Aqueous Dispersible Ultrathin-Layered Double Hydroxide Nanosheets for in Vivo Bioimaging and Drug Delivery. *ACS Appl. Mater. Interfaces* **2017**, *9*, 34185–34193. [CrossRef] [PubMed]

13. Peng, F.; Wang, D.; Tian, Y.; Cao, H.; Qiao, Y.; Liu, X. Sealing the Pores of PEO Coating with Mg-Al Layered Double Hydroxide: Enhanced Corrosion Resistance, Cytocompatibility and Drug Delivery Ability. *Sci. Rep.* **2017**, *7*, 8167. [CrossRef] [PubMed]

14. Peng, F.; Li, H.; Wang, D.; Tian, P.; Tian, Y.; Yuan, G.; Xu, D.; Liu, X. Enhanced Corrosion Resistance and Biocompatibility of Magnesium Alloy by Mg-Al-Layered Double Hydroxide. *ACS Appl. Mater. Interfaces* **2016**, *8*, 35033–35044. [CrossRef] [PubMed]

15. Ni, J.; Xue, J.; Xie, L.; Shen, J.; He, G.; Chen, H. Construction of Magnetically Separable NiAl LDH/Fe$_3$O$_4$-RGO Nanocomposites with Enhanced Photocatalytic Performance under Visible Light. *Phys. Chem. Chem. Phys.* **2018**, *20*, 414–421. [CrossRef] [PubMed]

16. Peng, F.; Wang, D.; Cao, H.; Liu, X. Loading 5-Fluorouracil into Calcined Mg/Al Layered Double Hydroxide on AZ31 Via Memory Effect. *Mater. Lett.* **2018**, *213*, 383–386. [CrossRef]

17. Zhou, M.; Yan, L.; Ling, H.; Diao, Y.; Pang, X.; Wang, Y.; Gao, K. Design and Fabrication of Enhanced Corrosion Resistance Zn-Al Layered Double Hydroxides Films Based Anion-Exchange Mechanism on Magnesium Alloys. *Appl. Surf. Sci.* **2017**, *404*, 246–253. [CrossRef]

18. Zhang, G.; Wu, L.; Tang, A.; Weng, B.; Atrens, A.; Ma, S.; Liu, L.; Pan, F. Sealing of Anodized Magnesium Alloy AZ31 with MgAl Layered Double Hydroxides Layers. *RSC Adv.* **2018**, *8*, 2248–2259. [CrossRef]

19. Li, D.; Chen, F.; Xie, Z.-H.; Shan, S.; Zhong, C.-J. Enhancing Structure Integrity and Corrosion Resistance of Mg Alloy by a Two-Step Deposition to Avoid F Ions Etching to Nano-SiO$_2$ Reinforcement. *J. Alloy. Compd.* **2017**, *705*, 70–78. [CrossRef]

20. Xie, Z.-H.; Shan, S. Nanocontainers-Enhanced Self-Healing Ni Coating for Corrosion Protection of Mg Alloy. *J. Mater. Sci.* **2018**, *53*, 3744–3755. [CrossRef]

21. Zeng, R.-C.; Liu, Z.-G.; Zhang, F.; Li, S.-Q.; Cui, H.-Z.; Han, E.-H. Corrosion of Molybdate Intercalated Hydrotalcite Coating on AZ31 Mg Alloy. *J. Mater. Chem. A* **2014**, *2*, 13049–13057. [CrossRef]

22. Chen, Y.; Zhang, J.; Dai, N.; Qin, P.; Attar, H.; Zhang, L.-C. Corrosion Behaviour of Selective Laser Melted Ti-TiB Biocomposite in Simulated Body Fluid. *Electrochim. Acta* **2017**, *232*, 89–97. [CrossRef]

23. Gan, R.; Wang, D.; Xie, Z.-H.; He, L. Improving Surface Characteristic and Corrosion Inhibition of Coating on Mg Alloy by Trace Stannous (II) Chloride. *Corros. Sci.* **2017**, *123*, 147–157. [CrossRef]

24. Dai, N.; Zhang, J.; Chen, Y.; Zhang, L.-C. Heat Treatment Degrading the Corrosion Resistance of Selective Laser Melted Ti-6Al-4V Alloy. *J. Electrochem. Soc.* **2017**, *164*, C428–C434. [CrossRef]

25. Chen, Y.; Zhang, J.; Gu, X.; Dai, N.; Qin, P.; Zhang, L.-C. Distinction of Corrosion Resistance of Selective Laser Melted Al-12Si Alloy on Different Planes. *J. Alloy. Compd.* **2018**, *747*, 648–658. [CrossRef]

MDPI

St. Alban-Anlage 66

4052 Basel

Switzerland

Tel. +41 61 683 77 34

Fax +41 61 302 89 18

www.mdpi.com

Nanomaterials Editorial Office

E-mail: nanomaterials@mdpi.com

www.mdpi.com/journal/nanomaterials